MATHEMATICS CONNECTIONS

INTEGRATED AND APPLIED

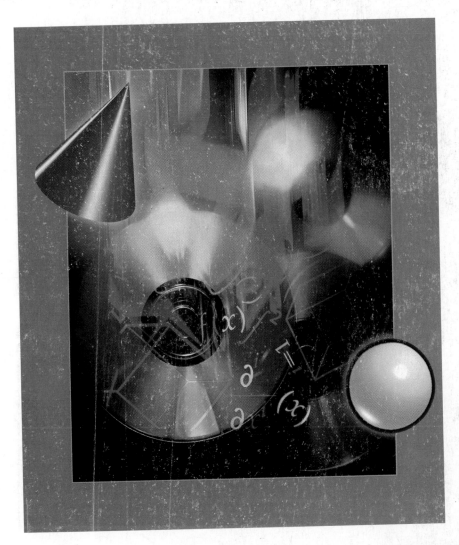

GLENCOE
McGraw-Hill

New York, New York   Columbus, Ohio   Mission Hills, California   Peoria, Illinois

CONSULTANT

Lee Yunker
Mathematics Department Head
West Chicago Community High School
West Chicago, Illinois

REVIEWERS

Wanda Sue Fowler
Mathematics Coordinator
Lee County Schools
Fort Myers, Florida

Sheryl G. Keith
Mathematics Teacher
Central High School
Evansville, Indiana

Patsy T. Malin
Secondary Mathematics Coordinator
El Dorado School District
El Dorado, Arkansas

Carl J. Minzenberger
Coordinator of Mathematics
Erie City School District
Erie, Pennsylvania

Dr. Charleen Mitchell DeRidder
Mathematics Supervisor
Knox County Schools
Knoxville, Tennessee

Roger L. O'Brien
Mathematics Supervisor
Polk County Schools
Bartow, Florida

Donna Marie Strickland
Mathematics Teacher
McGavock Comprehensive High School
Nashville, Tennessee

Susan C. Weaver
Mathematics Teacher
Hugo Junior High School
Hugo, Oklahoma

Printed in the United States of America.

Send all inquiries to:
Glencoe/McGraw-Hill
936 Eastwind Drive
Westerville, OH 43081

ISBN: 0-02-824795-7

3 4 5 6 7 8 9 10   RRW/LP   03 02 01 00 99 98 97

AUTHORS

Dr. Robert B. Ashlock is a Professor of Education at Covenant College in Lookout Mountain, Georgia. He authored the book *Error Patterns in Computation,* and is widely recognized as an authority on common mathematical errors. He also served as a consultant on *Merrill Mathematics.*

Dr. Mary M. Hatfield is an Associate Professor of Mathematics Education at Arizona State University, Tempe, Arizona. She is active in professional mathematics organizations. She speaks frequently at the National and Regional NCTM conferences. She recently served on the Board of Directors of the National Council of Teachers of Mathematics, as President of the Arizona Association of Teachers of Mathematics, and is developing a CD-ROM to enhance mathematics teaching.

Dr. Howard L. Hausher is chairperson of the Mathematics and Computer Science Department at the California University of Pennsylvania in California, Pennsylvania. He holds certification as math supervisor of grades K-12. He does inservice sessions for public and private secondary schools on implementing the *NCTM Standards.* He evaluates high school math departments.

Mr. John H. Stoeckinger has recently retired from Carmel High School in Carmel, Indiana, where he was a mathematics teacher and department head. During his career, he was actively involved in teaching all levels of mathematics for grades seven through twelve. He also served as an author for *Merrill Mathematics.*

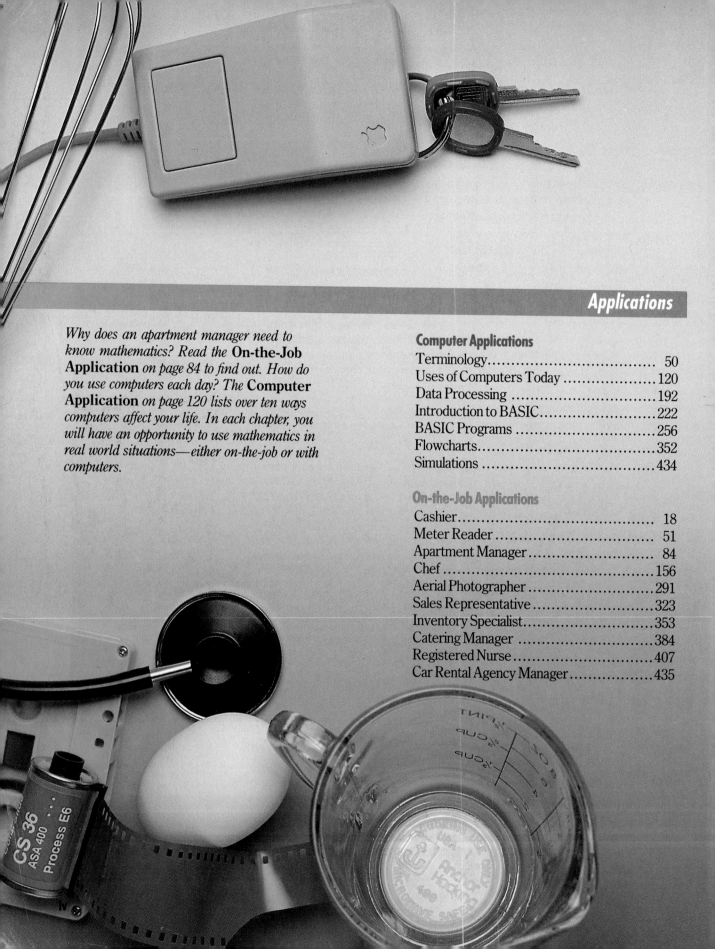

Why does an apartment manager need to know mathematics? Read the **On-the-Job Application** *on page 84 to find out. How do you use computers each day? The* **Computer Application** *on page 120 lists over ten ways computers affect your life. In each chapter, you will have an opportunity to use mathematics in real world situations—either on-the-job or with computers.*

TABLE OF CONTENTS

CONNECTIONS

TOPIC	PAGES
Algebra	24-29,31,43, 45,47,53,55, 59,61
Geometry	58-59
Statistics	5,9,49
Patterns	54-55
Technology	5,7,23,25,49, 50
Consumer Math	19,52-53,65
Business Math	18,51
Geography	35

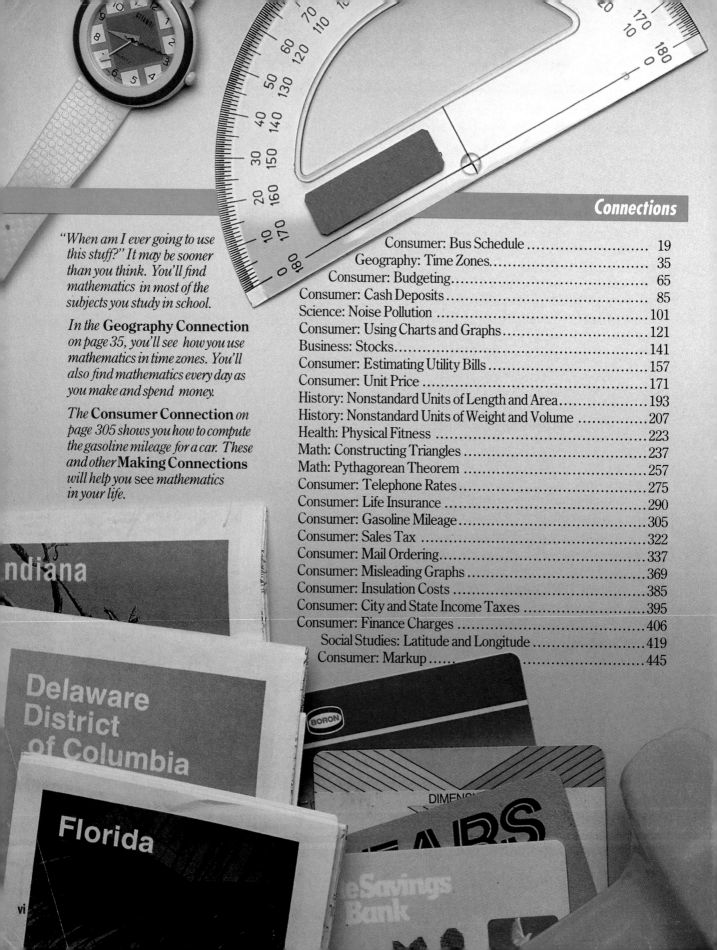

"When am I ever going to use this stuff?" It may be sooner than you think. You'll find mathematics in most of the subjects you study in school.

In the **Geography Connection** on page 35, you'll see how you use mathematics in time zones. You'll also find mathematics every day as you make and spend money.

The **Consumer Connection** on page 305 shows you how to compute the gasoline mileage for a car. These and other **Making Connections** will help you see mathematics in your life.

TECH PREP HANDBOOK

Have you thought about what career you want to pursue after you graduate from high school? Do you want to go on to a 2-year college, attend a vocational or technical school, or go to a university? There are many options open to you.

With technology on the move, more and more careers are being created that weren't here 10 years, or 5 years, or even 1 year ago. Most of these careers are in high demand for qualified people. The careers that are in demand require education beyond high school. A few of these careers are listed in the table of contents shown below.

Are you aware of what your skills and interests are? You will find questions in the **Tech Prep Handbook** *on pages 515-536 that will help you think about what you like to do and what you already know how to do. One goal of Tech Prep is to get you to think about your future, to make plans for it, and to prepare for it while you're in high school.*

Once you have read about the careers in the **Tech Prep Handbook***, do some research on your own at the public or school library or talk to your guidance counselor. Maybe your high school offers a Tech-Prep program, that you can consider enrolling in.*

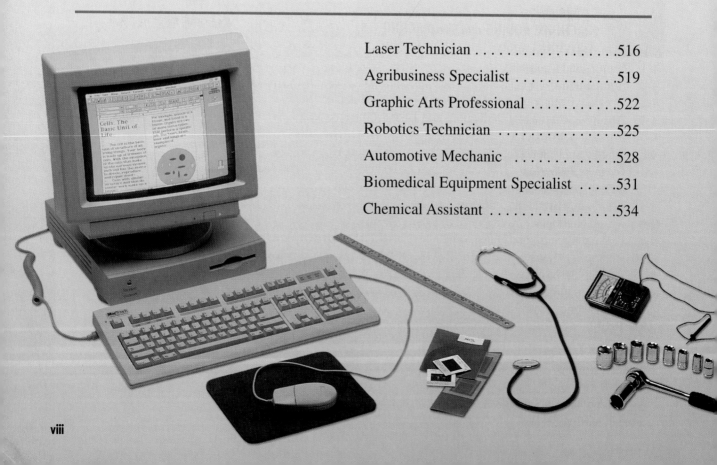

Have you ever wondered why there are customary measures such as feet and inches as well as metric measures such as meters and centimeters? In the **Activity** on page 177, you will have a chance to create your own measuring system. There are 14 such **Activities** in your textbook. You may investigate how to solve equations, make predictions, and discover many mathematical patterns. **Activities** let you do mathematics and let you find success in mathematics.

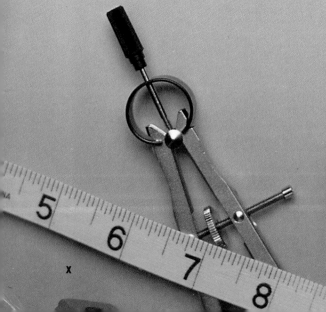

x

TOPIC	PAGES
Algebra	
	343, 347, 361,
Geometry	387, 389
	351, 355, 359,
Statistics	363, 383, 387,
	343-351,
	353-365, 369
	377, 383,
Patterns	391
Technology	364-365
	342, 345, 352,
Consumer Math	353, 381, 389
Business Math	369, 385, 395
	353, 384

BUYING A USED CAR

Doing Your Best at Decision-Making

Many people who say they "hate making decisions" don't realize how often they are already practicing this important skill. The fact is that all of us make decisions every single day. Some decisions, like deciding what to wear to school or eat for lunch, are easy. Others, like deciding what courses to take or whether to take a part-time job, are more difficult.

The best decisions are made after finding out as much as possible about a situation. Knowing the facts will give you a clearer picture of your alternatives. Then you can weigh your choices and arrive at a decision that's right for you.

An important decision nearly everyone must make at some point is deciding what kind of car to buy. Some people are fortunate enough to be able to afford a brand new automobile. Others—including most young people buying their first cars—must shop carefully to find the best possible used car for their money. However, a person buying a used car needs to be more cautious than a new car buyer. Why do you think this is true?

Besides purchase price, there are many other expenses related to owning and operating a car. Sales tax, the cost of transferring the car's title, and buying license plates all add to the initial expense. Auto insurance is also a major "upfront" expense—and it's one you must continue to pay as long as you own the car. Another ongoing expense for many people is repaying the money they borrowed to buy their cars. Maintenance, repairs, and operating expenses such as filling the gas tank and changing the motor oil are also regular expenses for car owners.

Clearly, deciding which car to buy is just one of many important decisions related to car ownership. In the next few pages, you will learn more about them all.

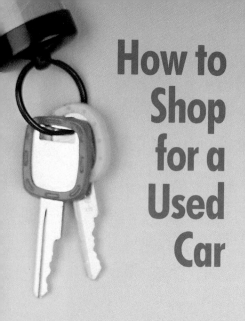

How to Shop for a Used Car

When you shop for new shoes, chances are you have something quite particular in mind. For example, if you need a pair of athletic shoes you arrive at the store knowing you prefer a certain style of shoe in a particular price range. You limit the types of shoes you choose from based on your preferences and what you can afford to pay.

This same common sense approach holds true for buying a used car. A new red sports car may be everyone's dream, but reality for most of us is usually something much less expensive and much more practical! Based on common sense, can you think of some important questions you need to ask yourself before proceeding to search for a used car?

Let's assume that before you begin shopping, you narrow your choices to buying a large, 4-door vehicle manufactured in 1985 or later that costs $5,500 or less, including tax, title, registration, and insurance. To save money, you decide to buy from a private owner rather than a dealer who's likely to have a higher markup. Carefully read the following ads to find cars that meet your requirements. Refer to the abbreviation chart if necessary.

Abbreviation Chart

AC—air conditioning	**int**—interior
auto—automatic transmission	**HT**—hardtop
EC—excellent condition	**PB**—power brakes
dr—door	**PS**—power steering
mpg—miles per gallon	**PW**—power windows
gd—good	**spd**—speed
ext—exterior	**cyl**—cylinder

'85 JEEP CHEROKEE
Red. Great shape! Sunroof, alum. mags, cloth int, A/C, cass./stereo, cruise control. $4,500. 241-0112 after 5 P.M.

'85 CHEVY CAVALIER
Liftback. 5 speed with air. AM/FM, 60,000 miles, 25-30 mpg, 1 owner, $2,250. 325-0012 after 6 P.M.

'86 OLDS CUTLASS CIERA WAGON
60,000 miles. Excellent condition. One owner. Loaded with extras. $4,300. 771-1030. Ask for Mark.

'87 CHEVROLET SPRINT
Great on gas! 2-door, 5 speed, air, AM/FM/cassette. 41,000 miles. $4,000 or best offer. 772-0860 anytime.

'87 HONDA ACCORD
2 dr. low mileage, 5 spd., cruise, A/C. Must sell. $7,400/best offer. 224-0967 eves.

---

1. What criteria does the Honda Accord fail to meet?

2. Which is the only ad that mentions specific gas mileage?

3. Why do you think the price of the Chevy Cavalier is so much lower than that of the other cars?

4. You regularly need to haul supplies for your job, as well as pick up four friends on the way to school. To which vehicles would you narrow your choices?

Decisions, Decisions!

Let's walk through the decision-making process you used to evaluate each vehicle and eliminate three of the cars from consideration.

• The '85 Jeep Cherokee meets the criteria: it's less than $4,500. It's large enough to carry large loads. Added plus: It's not a sports car, but it's red and many people find Jeeps sporty. Major question: How many miles has it been driven?

• The '85 Chevy Cavalier is certainly affordable but it's too small and it's got a lot of miles—which means expensive major parts may be wearing out.

• The '86 Cutlass Ciera wagon is within the price range and can carry a large load. It has as many miles as the Cavalier, but has had only one owner which implies the owner took good care of it. Major question: Does the owner have maintenance records to prove this?

• Despite the good mileage and price, the '87 Chevrolet Sprint won't do because it has only two doors—not practical for getting people and cargo in and out quickly.

• The Honda costs too much and has only two doors. Too bad! It's a nice car and a popular model.

Making Your Final Decision

Just as you try on new shoes before buying them, you must "try out" a car before making a purchase. Arrange with the seller to drive the car in the city and on the freeway. Ask yourself these questions: Does the car ride smoothly? Do the seat belts fasten properly in the front and rear seats? Do the doors and windows shut and lock properly? What other things should you check?

Even if everything looks and sounds right, never buy a used car without having it first checked out by a mechanic. The cost for this service varies, but count on paying at least $50. If the mechanic finds the car is basically sound but needs some minor repairs, use them as bargaining points to get the seller to lower the price. Unless the ad specifically states a firm price, most private sellers are quite willing—in fact they fully expect—to negotiate on their asking prices.

Let's Talk About Tax, Title, and Registration

The amount of paperwork involved in buying a car comes as a surprise to many first-time buyers. Some people are also surprised to learn that they must immediately pay a sales tax on their new or used cars, unless they live in a state where sales tax is not charged.

Here's how the process works: First, you must pay the seller "up front" for the full amount you, the buyer, and the seller have agreed upon. As previously mentioned, this may range from a couple of hundred to several hundred dollars below the seller's original asking price.

Used car dealers and new car dealers can help you arrange financing, a method for borrowing money to pay for the car. However, individual sellers—those private owners of cars who place ads in the newspapers—expect you to provide a check for the amount of the sale. Some buyers want a "certified" check, which means that your bank guarantees you have funds equal to the amount written on the check. Other sellers don't require a certified check, but will not release the car to you until your check "clears" the bank. When a check "clears," the amount of money specified on the check has been taken from the buyer's bank account and transferred to the seller's account.

Title Transfer

After receiving payment, the seller must have his or her signature notarized on the car title. The title is a paper that provides legal proof of ownership. A notarized signature is one witnessed by a "notary public," usually a Clerk of Courts in the county where the seller lives. The notary stamps the signature with an official seal. The seller gives or "transfers" the title to the buyer who then has his or her signature notarized to show that he or she is now owner of the car. The cost for transferring a title varies from county to county, but is usually around $5.

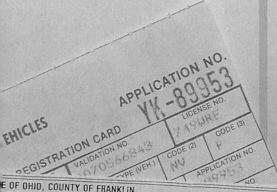

Sales Tax

When the title to the car is transferred to the new owner, the Clerk of Courts writes the price of the car on the back of the title. In some counties, the seller must produce a bill of sale showing what the buyer paid for the car. Sales tax, which usually ranges from 5 to 6 percent, is then figured on the cost of the transaction. The buyer must pay the sales tax to the Clerk of Courts at this time.

Let's assume that after having the '85 Jeep checked out by a mechanic, you and the seller agree to a sale price of $4,300—$200 lower than the asking price. Here is how you would figure a sales tax of 5 percent.

Multiple $4,300 by 0.05 because 5% = 0.05.

$$4300 \; \boxed{\times} \; 0.05 \; \boxed{=} \; 215$$

The sales tax on the Jeep would be $215.

Registration

Registration of cars in most states is handled through another county agency called the Office of the Deputy Registrar. Registration simply means getting license plates for the car. To do so, you show the title with your name on it to prove you are now the owner, and fill out a new registration card. The fee for this is usually about $5. If the county requires you as the new owner to buy license plates rather than transferring the plates used by the seller into your name, be prepared to pay another $45 or so.

1. Why do you think the seller insists on a certified check or waits until the buyer's check clears his or her bank before giving the buyer the car?

2. What is a certified check?

3. Besides selling price, list the other charges related to buying the '85 Jeep Cherokee.

4. What is the total you have spent so far on the '85 Jeep Cherokee?
   a. $4,420   b. $4,520   c. $4,620   d. $4,720

5. What is the purpose of a title for a car?

Find Out About Financing

Some people are able to save enough money to pay the full amount for the purchase of a used car. However, many others must borrow the difference between what they have saved and the total cost of the car.

Let's assume that over the past two years you have saved $2,000 from a part-time job toward the cost of a used car. You will need to borrow $3,200—the balance of the $5,500 you have set as the maximum price you can afford. The $5,500 must cover tax, title and registration fees, as well as the cost of insurance for one year.

Those under 18 will need to find a parent or other adult willing to take out a loan for them from a bank, savings and loan (S&L), or credit union. Interest rates and loan repayment terms vary, so it's wise for you and your parents to shop around. A quick way to get information is simply to phone the installment loan officer at various institutions and find out how much it will cost you, and how long you will have to repay the loan.

The borrower needs to fill out a loan application that provides information on income, credit history, and other debts. Most banks offer speedy turnout on installment loans—some are equipped to approve or deny the loan the very same day!

Because a loan agreement is a legal contract, no one under 18—the legal age of adulthood—is permitted to take out a bank loan, even if an adult co-signer agrees to be ultimately responsible. However, although the loan is in a parent's name, the bank will accept repayment directly from you!

Statement Savings

Banks base the repayment time period of the loan on the age of the car. Because it costs more than a used car, the length of time to repay money borrowed for a new car is often 60 months. Typically, a bank would allow up to 39 months to repay a loan on a 5-year-old car.

Repayment is made in monthly payments or "installments" for 39 months. To determine the monthly installment rate, the bank divides the amount of money to be borrowed—in this case, $3,200—by the number of months the borrower has to repay the loan. Into this monthly figure, the bank adds an interest charge. Suppose the interest rate is 16.45 percent APR (Annualized Percentage Rate). Your monthly repayment charge will be $100.67.

As you can see, it is expensive to borrow money! Banks distribute the interest over the "life" or length of the loan so that more interest is paid at the beginning than at the end. On a $3,200 loan, the interest would be about $3 a day at the beginning and drop to less than $1 a day toward the end of the repayment period. You can save a large amount of interest by paying off the loan early.

1. What's the most common repayment time period for a new car?

2. Why can't people under 18 take out bank loans?

3. If you paid off the $3,200 loan at the rate of $100.67 a month for 39 months, how much would you actually pay the bank?
   a. $3,765.24 b. $3,926.13 c. $4,006.24 d. $4,106.24

4. About how much money in interest will you pay the bank for the privilege of borrowing $3,200?
   a. $500   b. $700   c. $800   d. $900

5. How does a bank make money when they loan you money?

The Inside Story on Insurance

The high cost of insuring young drivers often comes as an unpleasant surprise to young people and their parents. However, insurance companies set higher rates for this age group for good reason: According to the Insurance Information Institute, more than 40 percent of the deaths among 16 to 22-year-olds result from auto accidents.

Rates for car insurance vary widely. Because they are involved in more accidents and get more traffic tickets, young unmarried males pay higher rates than any other group. Besides age and gender, these factors also affect the cost of car insurance:

• *How often you drive.* The more miles you drive each year, the higher the rate.

• *How well you drive.* Drivers with no accidents or citations have lower rates than others in their age group.

• *Where you live.* Because more auto accidents occur in cities, people who live in metropolitan areas generally pay more for insurance than those with similar driving records who live in the country.

• *What kind of car you drive.* Because they are more likely to be stolen—and to be driven faster and are therefore more likely to be involved in accidents—sports cars are much more expensive to insure than other types of automobiles.

How Young Drivers Can Save on Auto Insurance

Although young drivers as a group are charged higher auto insurance rates, individual drivers often qualify for discounts that lower the basic high cost for their age group. Here's how:

• *Qualify for a good student discount.* Offered by many insurance companies, good student rates provide a price break for students in the top 20% of their class, those with a "B" average, or those on the honor roll. Discounts vary and you must show proof that you qualify.

• *Drive the same car as your parents.* It's cheaper to "share the risk" with parents than to be the sole driver of a car in your own name.

• *Take a drivers' education course approved by the insurance company.* You will be asked to show proof you have passed the approved course. Most school-based courses qualify.

• *Maintain a good driving record.* Your rates will drop if you have no citations or violations three years or more after receiving your driver's license.

AUTO POLICY

What's Covered
by Auto Insurance?

In most states, all drivers are required to carry some type of insurance. It's important to talk with a qualified agent who will take time to explain all the types of coverage, from least to most expensive.

Briefly, here are some basic types of insurance coverage:

• **Collision** This pays for repairing your car if an accident was your fault. Usually you have to pay the first $100 or $200 and the insurance company picks up the rest.

• **Uninsured Motorist** If a person without insurance causes an accident that damages your car, this type of insurance will pay for the damages.

• **Bodily Injury** If the other driver or any passengers are injured in an accident that was your fault, this insurance will help pay their medical bills or lawsuits they file against you for damage to their bodies or minds resulting from the accident.

• **Comprehensive** This covers collision and accident-related claims as well as such noncollision situations, such as vandalism.

*Which type of coverage
is indicated in the following situations?*

1. Someone breaks into your car and steals a tape deck.

2. A passenger in your car is hurt when you run into a tree.

3. The driver whose car backs into yours has no insurance.

4. No one is hurt, but you crumple a fender when you back into a dumpster.

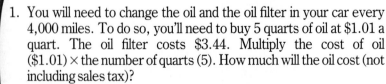

Taking Care of Your Car

Like a bicycle, lawnmower, or any other machine, a car needs regular maintenance to operate efficiently. Maintenance includes simple but important tasks you can do yourself, such as regularly checking the oil, wiper fluid, transmission fluid and antifreeze coolant levels, and replacing or filling them when necessary. You will probably need to pay a mechanic to do more complicated tasks such as changing the shock absorbers or fan belt and performing a general tune-up.

Car maintenance is expensive but necessary. Mechanics charge more than $30 an hour for labor, not including the cost of parts. Many first-time car owners are shocked at the prices they must pay to keep their cars in good condition.

The following is a list of car parts and their prices.

Car Parts/Fluids	Price
air filter	$4.44
alternator	47.53
antifreeze (gallon)	4.88
brake fluid (pint)	1.38
fan belt	2.99
fuel filter	1.73
headlight	4.89
oil (quart)	1.01
oil filter	3.44
PCV valve	1.68
shock absorber	10.88
spark plug	1.06
starter	31.24
transmission fluid (quart)	0.89
tune-up kit	6.29
water pump	17.44
wiper blade refills (pair)	3.08

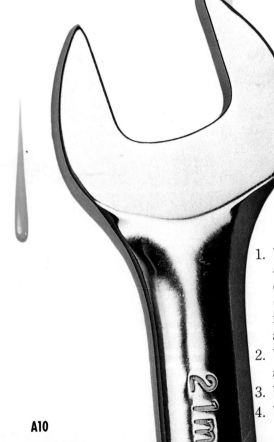

1. You will need to change the oil and the oil filter in your car every 4,000 miles. To do so, you'll need to buy 5 quarts of oil at $1.01 a quart. The oil filter costs $3.44. Multiply the cost of oil ($1.01) × the number of quarts (5). How much will the oil cost (not including sales tax)?
   a. $5.00   b. $5.05   c. $5.10   d. $5.15

2. What is the combined cost of the oil and the oil filter?
   a. $8.29   b. $8.39   c. $8.49   d. $8.59

3. What is the most expensive item on the chart?

4. What is the least expensive item on the chart?

Don't Forget
About Gas!

When budgeting for a car, be prepared not only for maintenance and repair, but for the cost of regularly buying gasoline. Clearly, buying a car is just the beginning! The costs associated with insuring, operating and maintaining a car are considerable.

It's wise to plan ahead and budget for these costs—a car won't be of any use to you if you can't afford to drive it! The majority of cars get gas mileage of between 13 and 18 miles per gallon of gas in the city and 21 to 28 miles per gallon on the highway. If you drive 60 miles a week to school and work and can drive 15 miles on each gallon of gas, you'll need to buy 4 gallons of gas to meet your needs.

60 miles divided by 15 miles for each gallon = 4 gallons

5. If gas costs $1.30 a gallon, what is the bare minimum you will need to spend each week on gas?
   a. $4.20   b. $5.20   c. $6.20   d. $7.20
6. About how much would your bare minimum monthly gas budget be? There are about 4½ weeks in a month.
   a. $10   b. $15   c. $20   d. $25

1

APPLYING NUMBERS AND VARIABLES

A bicycle ride!

What is the farthest distance you have ever bicycled? Some people enjoy bicycling just for the fun of it while others use the bicycle as their main mode of transportation. Why do you ride a bicycle? Have you ever been in a bicycle race?

Ted buys a racing bicycle for $324, a helmet for $32, and a pair of shoes for $25. About how much does he spend?

ACTIVITY: Place-Value Football

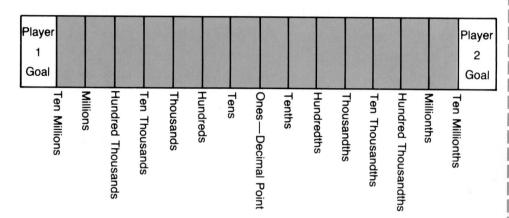

Materials: game markers, 15 index cards, pencil, paper or cardboard

Number of Players: 2

Making the Game Pieces

1. Write 5 as the first digit in each number from 50 million to 5 ten millionths on 15 index cards. For example: 50,000,000, 5,000,000, 500,000, . . . , 0.0000005.

2. On a piece of paper or cardboard, copy the gameboard shown above.

Playing the Game

3. Shuffle the index cards and place them face down.

4. Each player places a marker on the ones—decimal point yard line.

5. Player 1 turns over the first index card. Player 1 must identify the place-value position of the digit 5 on the card.

6. If Player 1 is correct, he or she moves one place-value yard line towards his or her goal. If Player 1 is incorrect, he or she moves one place-value yard line away from his or her goal.

7. Player 1 returns the index card to the bottom of the pile, and Player 2 takes a turn.

8. The first player to move into his or her goal area wins.

Variations of the Game

9. Draw a card and identify the place value that is two places to the right of the digit 5.

10. Draw two cards and name a place value that is between the two place values. If there is no place value between the two, name the larger place value.

Communicate Your Ideas

11. Create your own variation of the game. Make sure to adjust the rules to accommodate your changes.

1-1 PLACE VALUE

Objective

Use place value with whole numbers and decimals.

Mr. Robinson is a NASA engineer who plans communications for space travel. He needs to know how long it will take a radio signal to travel 6,943.857 kilometers. This number can be read as *six thousand, nine hundred forty-three and eight hundred fifty-seven thousandths.*

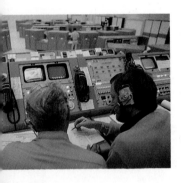

Notice that the decimal point is read as *and.*

A digit and its place-value position name a number. For example, in 6,943.857 the 9 in the hundreds place names the number 900. The 5 in the hundredths place names the number 0.05.

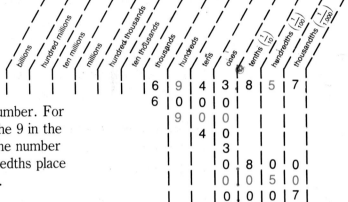

billions	hundred millions	ten millions	millions	hundred thousands	ten thousands	thousands	hundreds	tens	ones	tenths ($\frac{1}{10}$)	hundredths ($\frac{1}{100}$)	thousandths ($\frac{1}{1,000}$)
						6	9	4	3	8	5	7
						6	0	0	0			
							9	0	0			
								4	0			
									3			
									0	8	0	0
									0	0	5	0
									0	0	0	7

Example A

Name the place-value position for the digits 4 and 7 in 6,943.857.

4   tens   7   thousandths

Example B

Write each number in words.

Standard Form	Words
1,065	one thousand, sixty-five
10.430	ten and four hundred thirty thousandths

Example C

Zeros can be annexed to the right of the decimal point without changing the value.

$4 = 4.0$ $5.1 = 5.10$
$ = 4.00$ $ = 5.100$

Example D

Numbers in standard form can be renamed by using place value.

$200 = 2$ hundreds $0.030 = 30$ thousandths
$ = 20$ tens $ = 3$ hundredths
$ = 200$ ones

Guided Practice

Example A

Name the place-value position for each digit in 10,472.365.

1. 7   **2.** 0   **3.** 1   **4.** 3

5. 6   **6.** 5   **7.** 4   **8.** 2

| Example B | **Write each number in words.** |
| | **9.** 603   **10.** 164.05   **11.** 2,007.8 |

Example C

Annex zeros to write each number without changing their value.

12. 0.7   **13.** 0.42   **14.** 900   **15.** 0.04

Example D

Replace each ■ with a number to make a true sentence.

16. 0.050 = ■ thousandths   **17.** 5,000 = ■ hundreds

18. 600 = ■ tens   **19.** 0.080 = ■ hundredths

Exercises

Practice

Name the digit in each place-value position in 18,209.637.

20. tenths   **21.** tens   **22.** hundredths   **23.** hundreds

24. ones   **25.** thousandths   **26.** thousands   **27.** ten thousands

Write each number in words.

28. 6,010   **29.** 100.05   **30.** 0.112   **31.** 303.03

Write in standard form.

32. 4 thousand   **33.** 30 hundreds   **34.** 90 tens

35. 48 hundredths   **36.** 5 thousandths   **37.** 25 thousandths

Write Math

38. Write nine and six thousandths in standard form.

Applications

39. Rank the cities from 1 through 5, with 1 having the greatest population.

40. The chart shows the population of New York City as 7,300 thousand. What is the standard number for this number?

City	Population in Thousands
London	6,700
Moscow	8,800
New York City	7,300
Sydney	3,500
Tokyo	8,400

41. In Great Britain, 1,000,000,000 is read one thousand million. Read 73,000,000,000 as it is read in Great Britain and in the U.S.

Calculator

42. Enter 456,036 in your calculator. Change it to 456,836 by using addition. What number did you add?

43. Enter 24.539 in your calculator. Change it to 24.509 by using subtraction. What number did you subtract?

Make Up a Problem

44. Make up an addition problem using the digit 5 in two different place-value positions.

1-2 Exponents

Objective

Write numbers in expanded form using exponents.

Marie Alexander is a scientist. She often works with numbers in the billions and greater. To save time and space, she uses *exponents* to write numbers.

The product 10×10 can be written as 10^2.

$$10 \times 10 = 10^2 \longleftarrow \text{exponent}$$

2 factors base

The *exponent* is the number of times the *base* is used as a factor. Expressions written with exponents are called *powers*.

You read 10^2 as *10 to the second power* or *10 squared*.
You read 10^3 as *10 to the third power* or *10 cubed*.

THE MOON

Example A

Write 8^3 as a product and then find the number named.

$$8^3 = 8 \times 8 \times 8 \qquad 8 \times 8 \times 8 = 64 \times 8$$
$$= 512$$

8 is used as a factor 3 times

Example B

Write $3 \times 3 \times 3 \times 3$ using exponents.

$3 \times 3 \times 3 \times 3 = 3^4$ 3 is a factor 4 times.

The chart below shows how powers of 10 are related to place values. Each place value is 10 times the place value to its right.

1,000,000	100,000	10,000	1,000	100	10	1
10^6	10^5	10^4	10^3	10^2	10^1	10^0

Any nonzero number to the zero power is 1.

The number 26,097 is in **standard form**. The **expanded form** for a number can be written in several different ways.

Example C

Write 26,097 in expanded form using each of three ways.

● Use multiples of 1, 10, 100,. . . : $20,000 + 6,000 + 90 + 7$
● Use 1, 10, 100, . . . : $(2 \times 10,000) + (6 \times 1,000) + (9 \times 10) + (7 \times 1)$
● Use 10^0, 10^1, 10^2, . . . : $(2 \times 10^4) + (6 \times 10^3) + (9 \times 10^1) + (7 \times 10^0)$

Guided Practice

Example A

Write as a product and then find the number named.

1. 10^3 **2.** 5^2 **3.** 2^4 **4.** 12 squared   **5.** 20 cubed

Example B

Write using exponents.

6. $7 \times 7 \times 7$ **7.** 25×25 **8.** $5 \times 5 \times 5 \times 5$

Example C

Write in expanded form using 10^0, 10^1, 10^2, and so on.

9. 156   **10.** 2,934   **11.** 708   **12.** 4,076   **13.** 96,000

Write as a product and then find the number named.

14. 10^4 **15.** 3 cubed   **16.** 2 to the fifth power

Write using exponents.

17. 8×8 **18.** $7 \times 7 \times 7$ **19.** $3 \times 3 \times 3 \times 3$

20. 250×250 **21.** 11 squared   **22.** 10 to the sixth power

Write in standard form.

23. $400 + 70 + 5$ **24.** $(5 \times 10^3) + (6 \times 10^0)$

Write in expanded form using 10^0, 10^1, 10^2, and so on.

25. 345   **26.** 8,246   **27.** 3,705   **28.** 90,240

Calculator

You can use the square key, $\boxed{x^2}$, on a calculator to square a number. Simply enter the number to be squared and then press the square key.

f **UN with MATH**

Learn to send messages with your calculator.
See page 38.

Use a calculator to find the number named.

29. 56^2 **30.** 70^2 **31.** 814^2 **32.** 936^2

33. 426^2 **34.** $2,011^2$ **35.** $7,531^2$ **36.** $9,999^2$

Applications

37. One of the patterns of tiles in Josh's mosaic design has 5^2 tiles. How many tiles are in that pattern?

38. Sally is nearly 15 years old. She has been alive 362^2 hours. How many hours old is Sally?

Critical Thinking

39. Use the chart on page 6 to study how powers of ten are related to place values. Find the pattern for exponents. What power of ten would you write for 0.1?

Mixed Review

Lesson 1-1

Name the digit in each place-value position in 23,481.95.

40. thousands   **41.** ones   **42.** tenths   **43.** tens

Write in standard form.

44. 5 hundred   **45.** 16 hundreds   **46.** 5 tens

47. Maurice is offered a job with a computer company that will pay him 28 thousand dollars a year. What is the standard form of this number?

1-3 COMPARING WHOLE NUMBERS AND DECIMALS

Objective

Compare whole numbers and decimals.

George has $5.28 in his pocket. Jan has $5.92 in her purse. Who has more money?

The whole numbers 5 and 6 are shown on the number line. The interval between 5 and 6 is separated into tenths and hundredths.

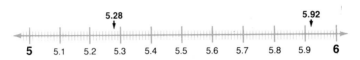

Numbers to the right are greater than numbers to the left.

5.28 *is less than* 5.92. 5.92 *is greater than* 5.28.
5.28 < 5.92   5.92 > 5.28

Decimals can also be compared by looking at the digits.

Method

1 ▶ Annex zeros, if necessary, so that each decimal has the same number of decimal places.

2 ▶ Starting at the left, compare digits in the same place-value position.

3 ▶ Use <, >, or =.

Examples

A *Compare 8.2 and 8.17. Use <, >, or =.*

1 ▶ Annex a zero so 8.2 has the same number of decimal places as 8.17.

2 ▶
```
8 . 2 0
8 . 1 7
```
same
  different: 2 > 1

3 ▶ So 8.2 > 8.17.

B *Compare 386 and 394. Use <, >, or =.*

1 ▶ Since there are no decimals, do not annex zeros.

2 ▶
```
3 8 6
3 9 4
```
same
  different: 8 < 9

3 ▶ So 386 < 394.

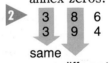

C *Order the following numbers from least to greatest.*

4.4, 2, 4.04, 6.775, 4.175, 4

1 ▶ Annex zeros so each number has three decimal places.

4.400   2 ▶   2.000
2.000   4.000
4.040   Order from   4.040
6.775   least to   4.175
4.175   greatest.   4.400
4.000   6.775

3 ▶ The order from least to greatest is 2, 4, 4.04, 4.175, 4.4, and 6.775.

Replace each ● with <, >, or = to make a true sentence.

1. 0.2 ● 0.8   **2.** 0.23 ● 0.230   **3.** 0.10 ● 0.01   **4.** 2.258 ● 20

5. 36 ● 89   **6.** 450 ● 405   **7.** 1,689 ● 1,689

Example C
Order from least to greatest.
8. 5.05, 5.51, 5.105, 5, 5.15   **9.** 372, 237, 723, 273, 327, 732

Replace each ● with <, >, or = to make a true sentence.

10. 0.201 ● 0.20   **11.** 0.03 ● 0.030   **12.** 4.923 ● 4.932
13. 7.017 ● 7.005   **14.** 13.7 ● 13.07   **15.** 0.612 ● 0.8

Order from least to greatest.
16. 0.02, 0.021, 0.002, 2.01, 2.1   **17.** 0.128, 1.28, 1.82, 0.821,
18. 52, 86, 34, 20, 63, 27, 38   **19.** 1,008, 1,035, 1,002, 1,156

20. The chart at the right lists winning times in the Olympic women's 100-meter hurdles. Rank the runners in order from 1 through 5. The fastest runner gets a rank of 1.

Runner	Time
Annelie Ehrhardt	12.59 s
Johanna Schaller	12.77 s
Vera Komisova	12.56 s
Patoulidou Paraskevi	12.64 s
Jordanka Donkova	12.38 s

21. Which runners had times faster than 12.6 seconds?

22. Which whole numbers are greater than 23.01 and less than 25.9?

23. Draw a number line that shows that 3.7 is greater than 3.2.

24. Karen and Bob are sister and brother. Karen has twice as many brothers as she has sisters. Bob has as many brothers as he has sisters. How many girls are in the family? How many are boys?

25. Find the heights of five tall buildings in your city or a city near you. Rank the height of the buildings in order from 1 through 5. The tallest building gets a rank of 1.

Name the place-value position for each digit in 41,687.25.
26. 8   **27.** 5   **28.** 1   **29.** 6

Write each expression using exponents.
30. $4 \times 4 \times 4 \times 4$ **31.** 15 cubed   **32.** $5 \times 5 \times 5$

33. Vince earns $3 an hour. In 4 years his hourly wage will be equal to the square of his hourly wage now. What will his hourly wage be in 4 years?

1-4 ROUNDING WHOLE NUMBERS AND DECIMALS

Objective

Round whole numbers and decimals.

Roger Kingdom set a new Olympic record in 1988 when he ran the 110-meter hurdles in 12.98 seconds. How fast was his time to the nearest second?

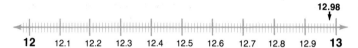

12.98

12   12.1   12.2   12.3   12.4   12.5   12.6   12.7   12.8   12.9   13

On a number line, numbers to the right are greater than numbers to the left. Note that 12.98 is closer to 13 than to 12. Therefore, 12.98 s rounded to the nearest second is 13 seconds.

Method

1. Look at the digit to the right of the place being rounded.

2. The digit remains the same if the digit to the right is 0, 1, 2, 3, or 4. Round up if the digit to the right is 5, 6, 7, 8, or 9.

Example A

Round 91,681 to the nearest thousand.

1. 91,681

   The digit to the right of the place being rounded is 6.

2. 91,681 → 92,000   91,681 rounded to the nearest thousand is 92,000.

Example B

Round 3.409 to the nearest whole number.

1. 3.409

   The digit to the right of the place being rounded is a 4.

2. 3.409 → 3   3.409 rounded to the nearest whole number is 3.

Example C

Round 28.395 to the nearest hundredth.

1. 28.395

   The digit to the right of the place being rounded is 5.

2. 28.395 → 28.40   To the nearest hundredth, 28.395 rounds to 28.40.

Example A

Round each number to the nearest thousand.

1. 3,840   **2.** 1,434   **3.** 587   **4.** 9,520   **5.** 44,623

Example B

Round each number to the nearest whole number.

6. 1.255   **7.** 100.5   **8.** 49.72   **9.** 0.2   **10.** 9.5

Example C

Round each number to the nearest hundredth.

11. 0.782   **12.** 12.861   **13.** 0.097   **14.** $8.552   **15.** $0.395

Exercises

Practice

Round each number to the underlined place-value position.

16. 16   **17.** 64   **18.** 736   **19.** 405   **20.** 230

21. 973   **22.** 5,848   **23.** 6,729   **24.** 9,915   **25.** 75,400

26. 83,912   **27.** 92,720,001   **28.** 69.5   **29.** $35.34   **30.** 0.752

31. 0.3   **32.** 399.6   **33.** 0.2749   **34.** 0.093   **35.** 0.995

Write Math

Write each number in words.

36. 820   **37.** 6,035   **38.** 1.37   **39.** 0.025   **40.** 10.401

Applications

41. Clay bought a jacket for $35.40. To the nearest dollar, how much did Clay pay for the jacket?

f **UN with MATH**

Where and when was the mini skirt created?
See page 39.

42. Pamela has only $10 bills. How many $10 bills must she give the sales clerk when she buys a skirt for $23.29?

43. A mutual fund reported a dividend of 6.954 shares to Mrs. McNair. To the nearest whole number, how many shares of stock did Mrs. McNair receive?

44. The Dow Jones stock index closed at 2,878.57. To the nearest tenth, what was the closing figure?

Round each distance to the nearest ten miles.

45. Baltimore to Memphis

46. Nashville to Baltimore

47. Atlanta to Charlotte

48. The least whole number that rounds to 40 is 35. What is the greatest whole number that rounds to 40?

Mileage Chart

	Atlanta	Baltimore	Charlotte	Memphis	Nashville
Atlanta		654	240	382	246
Baltimore	654		418	911	702
Charlotte	240	418		630	421
Memphis	382	911	630		209
Nashville	246	702	421	209	

Number Sense

Paper/ Pencil
Mental Math
Estimate
Calculator

Many times it is not necessary to use exact numbers. Rounded numbers provide an idea of how great a number is. Write whether each statement contains a rounded or an exact number.

49. The next flight for Orlando leaves at 9:49 A.M.

50. Nearly 76% of Americans live in urban areas.

51. Ty Cobb has a career record of 2,245 runs scored.

1-5 USING TAX TABLES

Objective

Use a tax table to determine amount of income tax.

Anyone who makes over a certain amount of money in a year must file a federal tax return. Instruction booklets provide tax tables to use in computing yearly federal income tax.

Gina Rivas earned $19,254 last year. She can find the tax she owes by locating her earnings on the table.

1. Look at the left side of the table. Locate the row that contains Gina's earnings.

   $19,254 is at least $19,250 but less than $19,300.

2. Look at the top of the table. Locate the column that indicates her filing status.

   Gina is single.

3. The amount of tax is at the intersection of the row and the column.

   Gina's federal income tax is $2,986.

If line 37 (taxable income) is—		And you are—			
At least	But less than	Single	Married filing jointly *	Married filing sepa-rately	Head of a house-hold
		Your tax is—			
19,000					
19,000	19,050	2,916	2,854	3,315	2,854
19,050	19,100	2,930	2,861	3,329	2,861
19,100	19,150	2,944	2,869	3,343	2,869
19,150	19,200	2,958	2,876	3,357	2,876
19,200	19,250	2,972	2,884	3,371	2,884
19,250	19,300	2,986	2,891	3,385	2,891
19,300	19,350	3,000	2,899	3,399	2,899
19,350	19,400	3,014	2,906	3,413	2,906
19,400	19,450	3,028	2,914	3,427	2,914
19,450	19,500	3,042	2,921	3,441	2,921
19,500	19,550	3,056	2,929	3,455	2,929
19,550	19,600	3,070	2,936	3,469	2,936
19,600	19,650	3,084	2,944	3,483	2,944
19,650	19,700	3,098	2,951	3,497	2,951
19,700	19,750	3,112	2,959	3,511	2,959
19,750	19,800	3,126	2,966	3,525	2,966
19,800	19,850	3,140	2,974	3,539	2,974
19,850	19,900	3,154	2,981	3,553	2,981
19,900	19,950	3,168	2,989	3,567	2,989
19,950	20,000	3,182	2,996	3,581	2,996

Understanding the Data

- Under what four categories do income tax filers fall?

- What are the least and greatest amounts in the tax table shown above?

- In which category did you file if you paid $2,884 in taxes?

Guided Practice

Find the tax for each income from the table. The filing status is single.

1. $19,525
2. $19,098
3. $19,147
4. $19,782
5. $19,240
6. $19,625
7. $19,987
8. $19,550

Practice

Find the tax for each income and filing status.

9. $19,418; married filing separately

10. $19,284; head of a household

11. $19,922; married filing jointly

12. Jim Sanders earned $19,827 as a drafter. He is single. What is his income tax?

13. Karen Johnson's income tax is $2,929. She files as head of a household. What is the most she could have earned?

14. For problem 13, explain how you determined the most Karen could earn.

15. Write an explanation to another student about how to use a tax table to determine income tax.

Cooperative Groups

16. Choose one person in your group to contact your local IRS office or your local library for a federal tax booklet 1040 EZ. Make up problems about yearly incomes. Then figure the amount of tax that would have to be paid.

Applications

17. Eric and Gladys Sneed file their taxes as married filing jointly. Eric earns $11,280 and Gladys earns $8,415. How much do they pay in taxes?

18. Lindsay Kulp earns $19,575 and files single. Ben and Marie Valdez earn $19,782 and file married filing jointly. Who pays more income taxes?

19. Max Turner earns $19,457 and files as head of household. Louise and Kevin Early earn $19,457 and file as married filing jointly. Who pays more income taxes?

Critical Thinking

20. Use these clues to find three gameboard scores. The scores are all perfect squares and the scores total 590. The first two scores have the same three digits.

Mixed Review

Lesson 1-3

Replace each ● with <, >, or = to make a true sentence.

21. 0.51 ● 0.42   22. 0.494 ● 0.497   23. 2.416 ● 2.410

Lesson 1-4

Round each number to the underlined place-value position.

24. 2̲3   25. 3̲42   26. 49̲,981   27. 0.87̲6

28. What are the least and greatest whole numbers that round to 120?

1-6 ESTIMATING SUMS AND DIFFERENCES

Objective

Estimate sums and differences.

Pat is collecting paper to sell to Recycling Central. During the first month she collected 578 pounds of paper, and during the second month she collected 412 pounds. *About* how many pounds of paper did she collect in the two months?

In this case, an exact answer is not needed. You can *estimate* to find the sum.

Method

1 Round each number to the same place-value position.

2 Add or subtract.

Examples

A *Estimate the sum of 578 and 412.*

1 Round to the nearest hundred.

$$
\begin{array}{r}
578 \rightarrow 600 \\
+\ 412 \rightarrow +\ 400 \\
\end{array}
$$

2 Add ⟶ 1,000

Pat collected *about* 1,000 pounds of paper.

B *Estimate the difference of 62,341 and 5,908.*

1 Round to the nearest thousand.

$$
\begin{array}{r}
62,341 \rightarrow 62,000 \\
-\ 5,908 \rightarrow -\ 6,000 \\
\end{array}
$$

2 Subtract ⟶ 56,000

The difference of 62,341 and 5,908 is *about* 56,000.

C *Estimate the sum of 5.23, 17.52, and 6.491.*

1 Round to the nearest whole number.

$$
\begin{array}{r}
5.23 \rightarrow 5 \\
17.52 \rightarrow 18 \\
+\ 6.491 \rightarrow +\ 6 \\
\end{array}
$$

2 Add. ⟶ 29

The sum of 5.23, 17.52, and 6.491 is *about* 29.

D *Estimate the difference of 0.504 and 0.17.*

1 Round to the nearest tenth.

$$
\begin{array}{r}
0.504 \rightarrow 0.5 \\
-0.17 \rightarrow -\ 0.2 \\
\end{array}
$$

2 Subtract. ⟶ 0.3

The difference of 0.504 and 0.17 is *about* 0.3.

Guided Practice

Examples A, C

Estimate each sum.

1.
$$
\begin{array}{r}
132 \\
+\ 361 \\
\end{array}
$$

2.
$$
\begin{array}{r}
9,136 \\
+\ 958 \\
\end{array}
$$

3.
$$
\begin{array}{r}
4,356 \\
+\ 237 \\
\end{array}
$$

4.
$$
\begin{array}{r}
0.035 \\
0.24 \\
+\ 0.471 \\
\end{array}
$$

5.
$$
\begin{array}{r}
7.09 \\
3.807 \\
+\ 0.48 \\
\end{array}
$$

6. $13.27 + $9.58

7. 86.12 + 57.9 + 4

Estimate each difference.

8.	697 − 406	**9.**	6,409 − 180	**10.**	10.651 − 5.524	**11.**	0.86 − 0.77	**12.**	$18.05 − 9.64

13. 65,495 − 46,485   **14.** $101.99 − $24

Exercises

Practice

Estimate.

15.	629 + 293	**16.**	126 − 90	**17.**	2,506 − 496	**18.**	6.8 + 9.2	**19.**	79.5 + 15.13

20.	71.31 − 24.528	**21.**	17.8 3.14 + 6.752	**22.**	0.5 0.84 + 0.174	**23.**	9.41 − 6.97	**24.**	58.3 − 7.95

25. 803 + 3,297   **26.** 421,172 − 90,900   **27.** $3.49 + $0.18

28. $20 − $5.77   **29.** 87.6 − 57   **30.** 3 − 1.63

31. Carolyn buys a cassette tape for $7.89. Estimate her change from a $10 bill.

Applications

32. Lee has $15. Estimate to determine if he can afford to buy tube socks for $2.10, tennis balls for $1.88, car wax for $3.94, and 35mm film for $4.88.

33. Jima has $40. Estimate to determine if she has enough money to buy a $27 sweater and $18 blouse.

Mental Math

34. Phyllis left for work at 7:35 A.M. If it takes her 27 minutes to drive to work, at what time will she arrive?

35. Buses stop at the corner of Fifth Street and Broad Street at 3:58 P.M., 4:10 P.M., 4:22 P.M. and so on until 8:00 P.M. Floyd leaves his office at 5:15 P.M. How much time does he have to get to the bus stop to catch the next bus?

Suppose

36. Suppose a number rounds to 9,500 to the nearest hundred and it rounds to 9,460 when rounded to the nearest ten. What is the number if 5 is in the ones place?

Determining Reasonable Answers

37. Peggy bought a jacket for $19.99 and a blouse for $34. The clerk charged her $67.56. Do you think the total is correct? Why?

1-7 ESTIMATING PRODUCTS AND QUOTIENTS

Objective

Estimate products and quotients.

Harry's Grocery sells a 15.7 pound turkey for $0.98 a pound. About how much does the turkey cost?

Sometimes an estimate, not an exact answer, is all that's needed when you multiply. You only need to know *about* how much the turkey costs.

Method

1 Round each factor to its greatest place-value position. Do *not* change 1-digit factors.

2 Multiply.

Examples

A *Estimate 15.7 × $0.98.*

1 15.7 → 20
   0.98 → 1

2 20 × 1 = 20
   The turkey costs *about* $20.

B *Estimate 472 × 5.*

1 472 → 500
   5 → 5

2 500 × 5 = 2,500
   The product of 472 and 5 is *about* 2,500.

Sometimes, you also need to estimate quotients.

Method

1 Round the divisor to its greatest place-value position. Do *not* change 1-digit divisors.

2 Round the dividend so it is a multiple of the divisor.

3 Divide.

Examples

C *Estimate 357 ÷ 62.*

1 62 → 60

2 Round 357 to 360 because 360 is a multiple of 60.

3 $\frac{6}{60)360}$ ← estimate

357 ÷ 62 is *about* 6.

D *Estimate 8,472.1 ÷ 26.5.*

1 26.5 → 30

2 Round 8,472.1 to 9,000 because 9,000 is a multiple of 30.

3 $\frac{300}{30)9,000}$ ← estimate

8,472.1 ÷ 26.5 is *about* 300.

Guided Practice

Estimate each product.

Examples A, B

1. 637 × 3

2. 56 × 3,179

3. 3,450 × 0.15

4. 1,292 × 567

5. $10.87 × 6.7

6. 15.7 × 0.98

Estimate each quotient.

7. $123 \div 58$ **8.** $735 \div 91$ **9.** $11.9 \div 3.2$ **10.** $369.7 \div 5.8$

11. $32\overline{)6,402}$ **12.** $3.2\overline{)9.07}$ **13.** $23.8\overline{)459.75}$ **14.** $1.17\overline{)248.9}$

Exercises

Practice

Estimate.

15. 59×2 **16.** 5×747 **17.** 738×29 **18.** 12×0.8

19. 34.15×1.78 **20.** $7,123 \times 4.6$ **21.** $602 \div 86$ **22.** $4,763 \div 51$

23. $55,114 \div 958$ **24.** $41.6 \div 3.9$ **25.** $\$31.85 \div 6$ **26.** $\$1.19 \div 24$

27. $\begin{array}{r} 528 \\ \times\ \ 36 \end{array}$ **28.** $\begin{array}{r} 8,275 \\ \times\ \ \ \ 66 \end{array}$ **29.** $\begin{array}{r} \$23.79 \\ \times\ \ \ \ \ 31 \end{array}$ **30.** $\begin{array}{r} 0.03 \\ \times\ 0.02 \end{array}$

31. $31\overline{)1,031}$ **32.** $691\overline{)3,532}$ **33.** $23.1\overline{)654.32}$ **34.** $1.23\overline{)457.8}$

35. Rachel estimates $4,254 \times 32$ to be about 1,200. What did Rachel do incorrectly?

Applications

36. Gregory uses 0.5 pounds of hamburger for each sandwich. About how many sandwiches can he make with 6.38 pounds of hamburger?

37. Marcia drives 231.3 miles on 11.4 gallons of gasoline. About how many miles per gallon is this?

Estimation

38. One factor of 4,763 is 51. Estimate the other factor.

Choose the most reasonable estimate for each situation.

39. number of seats on a school bus
   a. 60   **b.** 120   **c.** 180

40. population of the United States
   a. 225 billion   **b.** 225 million   **c.** 225 thousand

41. Make up a problem that involves an estimate of a product. Use at least one decimal.

CASHIER

Brian Hunter is a cashier for Paper Supply Outlet. In his job, he needs to know how to count change to the customer by counting from the amount of purchase to the amount given.

Cindy buys paper and supplies for $1.29. She gives Brian a $5 bill. Brian counts from $1.29 to $5.

The change is 1 penny, 2 dimes, 2 quarters, and 3 one-dollar bills, or $3.71. The art below shows how the cashier would count back the change.

The cashier says:

+ 1¢ $1.30   + 10¢ $1.40   + 10¢ $1.50   + 25¢ $1.75   + 25¢ $2.00   + $1 $3.00   + $1 $4.00   + $1 $5.00

Copy and complete the chart.

Amount of Purchase	Amount Given	Change							Total Change
		1¢	5¢	10¢	25¢	$1	$5	$10	
$1.29	$5.00	1		2	2	3			$3.71
1. $2.53	$5.00								
2. $1.74	$2.00								
3. $0.83	$10.00								
4. $7.34	$20.00								

5. Suppose your bill is $4.62 and you gave Brian $10.02. How much change should you receive?

6. In problem 5, why would you give Brian $10.02 instead of just giving him $10?

7. Suppose Susan's bill is $8.17. Why would she give Brian $10.17 rather than just giving him $10?

BUS SCHEDULE

The timetable below shows the departure (leave) and arrival (due) times for a bus route. The table gives times between 8:00 A.M. and 5:00 P.M.

| MON THROUGH FRI—WEST | | | | | | MON THROUGH FRI—EAST | | | | | |
| LEAVE | LEAVE | DUE | LEAVE | LEAVE | DUE | LEAVE | LEAVE | DUE | LEAVE | LEAVE | DUE |
COUNTRY CLUB AND MAIN	COLLEGE AND LIVINGSTON	BROAD AND HIGH	COUNTRY CLUB AND MAIN	COLLEGE AND LIVINGSTON	BROAD AND HIGH	BROAD AND HIGH	OHIO AND LIVINGSTON	COUNTRY CLUB AND MAIN	BROAD AND HIGH	OHIO AND LIVINGSTON	COUNTRY CLUB AND MAIN
8:22	8:39	9:01	12:22	12:39	1:01	8:12	8:23	8:52	12:12	12:23	12:52
8:42	8:59	9:21	12:42	12:59	1:21	8:32	8:43	9:12	12:32	12:43	1:12
9:02	9:19	9:41	1:02	1:19	1:41	8:52	9:03	9:32	12:52	1:03	1:32
9:22	9:39	10:01	1:22	1:39	2:01	9:12	9:23	9:52	1:12	1:23	1:52
9:42	9:59	10:21	1:42	1:59	2:21	9:32	9:43	10:12	1:32	1:43	2:12
10:02	10:19	10:41	2:02	2:19	2:41	9:52	10:03	10:32	1:52	2:03	2:32
10:22	10:39	11:01	2:22	2:39	3:01	10:12	10:23	10:52	2:12	2:23	2:52
10:42	10:59	11:21	2:42	2:59	3:21	10:32	10:43	11:12	2:32	2:43	3:12
11:02	11:19	11:41	3:02	3:19	3:41	10:52	11:03	11:32	2:52	3:03	3:32
11:22	11:39	12:01	3:22	3:39	4:01	11:12	11:23	11:52	3:12	3:23	3:52
11:42	11:59	12:21	3:42	3:59	4:21	11:32	11:43	12:12	3:32	3:43	4:12
12:02	12:19	12:41	4:02	4:19	4:41	11:52	12:03	12:32	3:52	4:03	4:32

Ms. McKenzie boards the bus at Broad and High at 11:32 A.M. When will she arrive at Country Club and Main?

Find the 11:32 A.M. departure at Broad and High and read across to the arrival time at Country Club and Main.

Ms. McKenzie will arrive at 12:12 P.M.

1. Suppose you board at Country Club and Main at 10:42 A.M. When will you arrive at Broad and High?

2. Mr. Juarez has a business appointment at Ohio and Livingston at 3:00 P.M. What is the latest he can board at Broad and High?

3. Suppose a bus broke down at Broad and High at 8:32 A.M. and it took one-half hour to repair it. How long would the bus have to wait after it was repaired to leave at the next scheduled time?

4. Suppose you left Broad and High at 10:52 A.M. and arrived at Country Club and Main at 11:32 A.M. Then you realized you forgot your billfold and decided to return to Broad and High immediately. Can you be back in 40 minutes? Why or why not?

BUS STOP

1-8 MATRIX LOGIC

Objective

Solve problems using matrix logic.

Steven, Diana, Janice, and Michael attend the same school. Each participates in a different school sport—basketball, football, track, or tennis. Diana does not like track or basketball. Steven does not participate in football or tennis. Janice prefers an indoor winter sport. Michael scored four touchdowns in the final game of the season. Name the sport in which each student participates.

What is given?

- Diana does not like track or basketball.
- Steven does not participate in football or tennis.
- Janice prefers an indoor winter sport. (basketball)
- Michael scored four touchdowns. (football)

Find each student's sport by listing the names and sports in a table. Then read each clue and write yes or no in the appropriate boxes.

	Basketball	Football	Track	Tennis
Diana	No	No	No	Yes
Steven	No	No	Yes	No
Janice	Yes	No	No	No
Michael	No	Yes	No	No

Matrix logic is a logic or reasoning problem. It uses a rectangular array with the appropriate number of columns and rows in which you record what you have learned from clues. Once all the rows and columns are filled, you can deduce the answer.

Write no beside Diana's name for track and basketball. Write no beside Steven's name for football and tennis. Write yes beside Janice's name for basketball. Since only one student plays basketball, write no in all the remaining boxes in that row and column. Steven's sport has to be track. All remaining boxes in that row and column are no. Since Michael plays football, Diana's sport is tennis.

Guided Practice

Solve. Use matrix logic.

1. Six friends live in consecutive houses on the same side of the street (#1-6). Lyn's house is #3. Diane's house is just beyond Al's. Lyn's closest neighbors are Aaron and Sue. Al's house is not #1-4. Aaron's house is before Sue's. If John lives in one of the houses, who lives in which house?

2. Dan, William, and Laurie have chosen careers as a teacher, a pediatrician, and a lawyer. If William does not like to be around children and Dan cannot stand the sight of blood, who has chosen which career?

Solve. Use any strategy.

3. Paul, David, and Ben order different individual pizzas for dinner. The pizzas delivered are: pepperoni and sausage; pepperoni and anchovies; and sausage. If David does not like fish, Paul does not like sausage, and Ben likes pepperoni, who ordered which pizza?

4. Four college graduates decide to have a reunion at a large hotel. Each person has a room on a different floor. Susan must ride the elevator down four floors to visit Corey. Greg is one floor below Kim. Corey has a room on the tenth floor. Kim must ride the elevator up six floors to visit Susan. Who is staying on which floor?

5. Tracy had cake left over from her graduation party. She gave her grandparents half of the remaining pieces and half of that amount to her friend, Jerrod. If Jerrod took seven pieces home, how many pieces were left over before she gave any away?

6. An odd number is greater than 7×3 and less than 9×4. Find the number if the sum of its digits is 11.

7. Three friends like three different kinds of music. If Jack does not like jazz, Kevin does not like rock, and Jason does not like heavy metal or rock, who likes which kind of music?

8. Jay has coins in his pocket consisting of nickels, dimes, and quarters. He has three fewer nickels than dimes and two more dimes than quarters. How much money does he have if he has three quarters?

9. If you step into an elevator on the fifth floor and ride up thirteen floors, down six floors, and up nine floors, on which floor will you be?

10. How many squares can you find in this rectangle? (Hint: there are more than 18.)

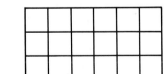

UN with MATH

How many grooves are on the edge of a dime?
See page 39.

Round each number to the nearest whole number.

11. 3.266   **12.** 46.019   **13.** 9.048   **14.** 63.51

Estimate.

15.   815   **16.**   9.68   **17.**   4,323   **18.**   634
     + 429   − 4.25   + 891   − 55

19. Ryan fills his tank with 12.62 gallons of gasoline. His car gets 21.2 miles per gallon. About how far can Ryan travel on one tank of gasoline?

1-9 ORDER OF OPERATIONS

Objective

Use the order of operations to find the value of an expression.

Cyd mows lawns during the summer. Monday, he mowed one lawn for $6 and four other lawns for $5 each. The expression $6 + 4 \times 5$ gives the total earned.

To be sure you find the correct value for an expression such as $6 + 4 \times 5$, use the following *order of operations*.

$$6 + 4 \times 5 \qquad\qquad 6 + 4 \times 5$$
$$10 \times 5 \qquad\qquad 6 + 20$$
$$50 \qquad\qquad\qquad 26$$

Incorrect   **Correct**

Method

1 ▶ Find the value of all powers.

2 ▶ Multiply and/or divide from left to right.

3 ▶ Add and/or subtract from left to right.

Cyd earned $26 mowing lawns.

Example A

Find the value of $18 - 4 \times 3^2 \div 6$.

1 ▶ $18 - 4 \times 3^2 \div 6 = 18 - 4 \times 9 \div 6.$ Find the value of 3^2.

2 ▶ $= 18 - 36 \div 6.$ Multiply 4 by 9.

$= 18 - 6$ Divide 36 by 6.

3 ▶ $= 12$ The value is 12.

When a different order of operations is needed, parentheses are used. First do operations inside the parentheses. Then do other operations.

Example B

Find the value of $(0.7 + 0.3) \times (13 - 5)$.

1 ▶ $(0.7 + 0.3) \times (13 - 5) = 1 \times (13 - 5)$ $0.7 + 0.3 = 1$

2 ▶ $= 1 \times 8$ $13 - 5 = 8$

3 ▶ $= 8$ The value is 8.

Guided Practice

Find the value of each expression.

Example A

1. $10 \times 4 + 5 \times 2$ **2.** $7 + 8 + 12 \div 4$ **3.** $54.3 - 42 \div 7$

4. $21 \div 7 + 4 \times 8$ **5.** $2^3 \div 4 \times 1.2$ **6.** $30 \div 6 - 8^2 \times 0$

Example B

7. $26 - (20 + 4)$ **8.** $4 \times (0.7 + 1.3)$ **9.** $(25 + 15) \div 4$

10. $6 \times (9.3 - 0.3)$ **11.** $(7 + 1) \times (3^2 - 1)$ **12.** $(7 - 3)^2$

Practice

Find the value of each expression.

13. $12 - 5 + 9 - 2$ **14.** $6 \times 6 + 3.6$ **15.** $12 + 20 \div 4 - 5$

16. $6 \times 3 \div 9 - 1$ **17.** $(4^2 + 2^3) \times 5$ **18.** $24 \div 8 - 2$

19. $3 \times (4 + 5) - 7$ **20.** $4.3 + (24 \div 6)$ **21.** $(5^2 + 2) \div 3$

22. $(40 \times 2) - (6 \times 10)$ **23.** $2 \times [5 \times (4 + 6) - 10]$

Mental Math

Find the value of each expression mentally.

24. $2 + 24 \div 6$ **25.** $2 \times (5^2 - 20)$ **26.** $24 \div (10 + 2)$

27. $10^2 - 2^2 \times 10$ **28.** $(20 + 80) \times (64 - 8^2)$

Applications

29. Bob buys six pairs of socks at $3.00 each. How much change does he receive from $20? Write an expression and solve.

30. Mrs. Anderson works 40 hours a week and makes $8.00 an hour. What is her take-home pay if $70.00 is withheld for taxes and insurance? Write an expression and solve.

Determining Reasonable Answers

31. Elliot worked this problem on his calculator. Without actually finding the correct answer, tell how you know that the answer he found is wrong.

A box of albums weighs 24 pounds. A box of tapes weighs 18 pounds. What would a shipment of 10 boxes of albums and 20 boxes of tapes weigh?

Elliot's answer: The shipment would weigh 360 pounds.

Critical Thinking

32. A clock strikes the number of hours each hour. How many times will the clock strike in a 24-hour day?

Calculator

33. Find the value of $2 + 5 \times 3$ on your calculator.

Is the result 17? If your result is not 17, can you find a way to get a result of 17?

Use a calculator to find the value of each expression.

34. $5 + 3 \times 7$ **35.** $12 \div 3 + 12$ **36.** $36 \div 4 \times 3$

Mixed Review

Lesson 1-6

Estimate.

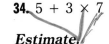

37. 42.35
       $- 16.48$

38. 0.6
       $+ 0.94$

39. 123.45
       $- 11.82$

40. $\$3.62$
       $+ 9.89$

Lesson 1-7

41. 16.4×2.8 **42.** $65.41 - 15.37$ **43.** 763×8

Lesson 1-8

44. Jenny, Betty, and Ashley are traveling to Daytona, Miami, and Ft. Myers for vacation. Jenny is not going to Miami, Betty is not going to Ft. Myers, and Ashley is traveling the farthest distance south. Who is traveling to which city?

1-10 EVALUATE EXPRESSIONS

Objective

Write and evaluate expressions using variables.

Just for Feet Shoe Store adds $5 to the wholesale cost of each pair of shoes to make a profit. What is the retail price of a pair of shoes?

You can let c represent the wholesale cost of any pair of shoes. Then $c + 5$ represents the retail price of the shoes.

A **mathematical expression** consists of numbers, operations like addition or multiplication, and sometimes variables. A **variable** is a symbol, usually a letter, used to represent a number. Any letter can be used.

Translate words into mathematical expressions.

Words	Mathematical Expression
five more than a number	$d + 5$
six less than a number	$m - 6$
three times a number	$3 \times y$ or $3y$
a number divided by three	$n \div 3$ or $\frac{n}{3}$

Example A
Example B
Example C
Example D

The value of an expression depends on the number that replaces each variable.

Method

1 ▶ Replace each variable with the given number.

2 ▶ Perform the indicated operations using the order of operations.

Example E

Find the value of $3n - 4$ if $n = 5$.

1 ▶ $3n - 4 = 3 \times 5 - 4$ Replace n with 5.

2 ▶ $= 15 - 4$ Multiply first.

      $= 11$ Subtract.   The value is 11.

Example F

Find the value of $a^2 + \frac{b}{3}$ if $a = 4$ and $b = 2$.

1 ▶ $a^2 + \frac{b}{3} = 4^2 + \frac{2}{3}$ Replace a with 4 and b with 2.

2 ▶ $= 16 + \frac{2}{3}$ Find 4^2.

      $= 16\frac{2}{3}$ Add.   The value is $16\frac{2}{3}$.

Write an expression for each phrase. Use n to represent the number.

Examples A-D

1. the sum of 7 and a number

2. 5 less than a number

3. 16 divided by a number

4. the product of 8 and a number

Examples E, F

Find the value of each expression if n = 3 and p = 5.

5. $42 - n$

6. $p^2 + 3$

7. $2np$

8. $\dfrac{2 + n}{p}$

Exercises

Practice

Write an expression for each phrase. Use n for the number.

9. 8 more than a number

10. 17 less than a number

11. n less than 3

12. a number divided by 2

13. the product of 7 and a number

14. a number increased by 3

15. a number decreased by 3

16. 4 times a number

Find the value of each expression if n = 9.

17. $8 + n$

18. $15.7 - n$

19. $n^2 + 7$

20. $3n - 4$

21. $\dfrac{2n}{3}$

Find the value of each expression if a = 2, b = 3, and c = 5.

22. $ab + c$

23. $\dfrac{a + b}{c}$

24. $a^2 + b$

25. a^2b

26. $(b + c)^2$

Write Math

Write a verbal phrase for each expression.

27. $a + 7$

28. $26 - y$

29. $n \div 9$

30. $18m$

31. $y - 26$

Suppose

32. Suppose six is squared, then the product of four and nine is subtracted. What number results?

Applications

Copy and complete each table.

33.

x	x + 4
3	7
0	▪
7	▪

34.

m	m + 11
2	13
22	▪
33	▪

35.

b	b − 3
8	5
▪	1
▪	14

Calculator

Most calculators have a memory key such as [STO] , [M+] , or [SUM] . Once a number has been put into memory, it can be recalled using a key such as [RCL] , [MR] , or [EXC] .

The memory is useful when the same variable is used more than once in an expression. For example, find the value of $p^2 + 4p$ if $p = 3.1$.

Enter: ∃.I [STO] This puts 3.1 into memory and clears the display.

Enter: [x²] [+] ५[×] [RCL] [=]  The value is 22.01.

Find the value of $p^2 + 4p$ for each value of p.

36. $p = 5$

37. $p = 1.1$

38. $p = 7.2$

39. $p = 3.14$

40. $p = 6.21$

1-11 SOLVING EQUATIONS USING ADDITION AND SUBTRACTION

Objective

Solve equations by using addition and subtraction.

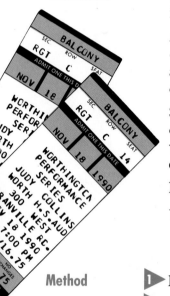

Lois knows she sold 80 tickets. She has 60 left. In order to find how many she had before she sold any, Lois writes an equation that contains a *variable*: $n - 80 = 60$. n represents how many tickets she had in the beginning.

A mathematical sentence with an equals sign is called an **equation.** An equation is like a scale in balance. If you add or subtract the same number on each side of an equation, the new equation is also true. It is *in balance.*

Some equations contain variables. **Solving the equation** means finding the correct replacement for the variable. To solve an equation, get the variable by itself on one side of the equals sign.

Lois had 140 tickets in the beginning.

$n - 80 = 60$

$n - 80 + 80 = 60 + 80$

$n = 140$

Use **inverse operations** to solve equations.

● If a number has been added to the variable, subtract.
● If a number has been subtracted from the variable, add.

Method

1 ▶ Identify the variable.

2 ▶ Add or subtract the same number on each side of the equation to get the variable by itself.

3 ▶ Check the solution by replacing the variable in the original equation.

Example A

Solve $n - 7 = 12$.

1 ▶ $n - 7 = 12$ 7 is subtracted from the variable n.

2 ▶ $n - 7 + 7 = 12 + 7$ Add 7 to each side.

$n = 19$

3 ▶ Check: $n - 7 = 12$ In the original equation, replace n with 19.

$19 - 7 \overset{?}{=} 12$

$12 = 12 \checkmark$ The solution is 19.

Example B

Solve $k + 0.5 = 0.9$.

1 ▶ $k + 0.5 = 0.9$ 0.5 is added to the variable k.

2 ▶ $k + 0.5 - 0.5 = 0.9 - 0.5$ Subtract 0.5 from each side.

$k = 0.4$

3 ▶ Check: $k + 0.5 = 0.9$ In the original equation replace k with 0.4.

$0.4 + 0.5 \overset{?}{=} 0.9$

$0.9 = 0.9 \checkmark$ The solution is 0.4.

Guided Practice

Example A

Name the number you would add to each side to solve each equation. Then solve and check each solution.

1. $m - 5 = 9$ **2.** $n - 20 = 30$ **3.** $k - 0.6 = 0.2$ **4.** $9 = x - 9$

Example B

Name the number you would subtract from each side to solve each equation. Then solve and check each solution.

5. $n + 3 = 5$ **6.** $4.7 + k = 7.9$ **7.** $90 = n + 40$ **8.** $4 + t = 9$

Exercises

Practice

Solve each equation. Check each solution.

9. $p - 8 = 8$ **10.** $h - 7 = 6$ **11.** $5 = u - 9$ **12.** $9 = t - 8$

13. $a + 6 = 8$ **14.** $8 + f = 11$ **15.** $18 = b + 7$ **16.** $6 + s = 15$

17. $16 = x - 2$ **18.** $r - 0.4 = 0.2$ **19.** $w - 0.5 = 0.5$

20. $15.3 = q - 0.2$ **21.** $z + 0.6 = 0.9$ **22.** $y - 500 = 800$

Using Equations

Write an equation. Then solve.

23. Seven increased by x is 11. What is the value of x?

24. b decreased by 0.6 is 0.5. What is the value of b?

25. Nine less than g is 7. What is the value of g?

Applications

Solve. Use an equation.

26. Ally is saving for a computer game that costs $60. She still needs to save $25. How much has Ally saved?

27. Greg owns 70 tapes and CDs. He owns 20 CDs. How many tapes does Greg own?

Talk Math

28. Ask this question in different ways: What number added to seven equals twelve?

Critical Thinking

29. Alex and Kevin organize a camping trip for ten boys. They have ten backpacks, but they intend to leave some backpacks empty for items the campers might bring. Five backpacks contain food, four contain camping supplies, and two contain both food and camping supplies. How many backpacks are empty?

1-12 SOLVING EQUATIONS USING MULTIPLICATION AND DIVISION

Objective

Solve equations by using multiplication and division.

The parking garage charges $3.00 per hour. How many hours can Oliver park if he has only $15? An equation can be written to solve this problem. The equation is in Example A.

Multiplication and division are *inverse operations*.

- If a variable is multiplied by a number, divide each side by that number.
- If a variable is divided by a number, multiply each side by that number.

Method

1. Identify the variable.
2. Multiply or divide each side of the equation by the same nonzero number to get the variable itself.
3. Check the solution by replacing the variable in the original equation with the solution.

Example A

Solve $3n = 15$. $3n$ means $3 \times n$.

1. $3n = 15$ 3 is multiplied by the variable n.

2. $\frac{3n}{3} = \frac{15}{3}$ Divide each side by 3.

   $n = 5$

3. Check: $3n = 15$

   $3 \times 5 \stackrel{?}{=} 15$ Replace n with 5.

   $15 = 15\checkmark$ Oliver can park for 5 hours.

Example B

Solve $\frac{t}{4} = 9$.

1. $\frac{t}{4} = 9$ The variable, t, is divided by 4.

2. $\frac{t}{4} \times 4 = 9 \times 4$ Multiply each side by 4.

   $t = 36$

3. Check: $\frac{t}{4} = 9$

   $\frac{36}{4} \stackrel{?}{=} 9$ Replace t with 36

   $9 = 9 \checkmark$ The solution is 36.

Example A

Name the number you would divide each side by to solve each equation. Then solve and check each solution.

1. $3 \times r = 12$ **2.** $7m = 21$ **3.** $80 = 20y$ **4.** $0.6 = 3n$

Example B

Name the number you would multiply each side by to solve each equation. Then solve and check each equation.

5. $\frac{n}{5} = 7$ **6.** $\frac{f}{10} = 30$ **7.** $0.3 = \frac{d}{2}$ **8.** $10 = \frac{r}{25}$

Practice

Solve each equation. Check each solution.

9. $9k = 72$ **10.** $10y = 1{,}000$ **11.** $72y = 0$ **12.** $25r = 25$

13. $2m = 0.08$ **14.** $180 = 3n$ **15.** $\frac{q}{8} = 6$ **16.** $\frac{m}{9} = 9$

17. $\frac{s}{6} = 7$ **18.** $9 = \frac{t}{6}$ **19.** $0.4 = \frac{r}{2}$ **20.** $30 = \frac{k}{3}$

21. $800 = 2b$ **22.** $\frac{s}{4} = 0.2$ **23.** $3n = 1{,}200$ **24.** $0.1 = \frac{p}{5}$

Using Algebra

Write an equation. Then solve.

25. The product of 6 and n is 72. What is the value of n?

26. Some number divided by 4 is 20. What is that number?

27. The sum of 200 and a number is 600. What is that number?

Applications

Solve. Use an equation.

28. Roger weighs 5 times what his brother weighs. Roger weighs 150 pounds. How much does his brother weigh?

29. Five friends share the cost of a lunch equally. Each person pays $4.00. What is the total cost of the lunch?

Research

30. Find out what the total operating budget for your school district is. Also find out how many students are enrolled in your school district. Determine the cost per student.

Number Sense

31. The graph shows that $400.9 billion was spent for construction in 1991. Write this amount of money in standard form.

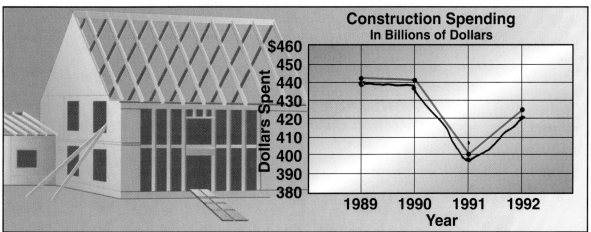

1-13   WRITE AN EQUATION

Objective
Solve verbal problems using an equation.

Marcy has delivered mail to 56 homes. She has to deliver mail to 32 more homes. How many homes does Marcy have on her route?

Method
You can use the four-step plan to solve any verbal problem.

 Explore

What facts are given?
● The part already delivered is 56.
● The part yet to be delivered is 32.

What fact do you need to find?
● To how many homes does Marcy deliver mail?

 Plan

Write an equation.

Let *n* be the total number of homes on Marcy's route.

$$\text{part} + \text{part} = \text{total}$$
$$56 + 32 = n$$

▶ **Solve**

Estimate. 60
         + 30
        ‾‾‾‾
          90

Add. 56
    + 32
   ‾‾‾‾
      88

Marcy has 88 homes on her route.

▶ **Examine**

The answer is reasonable because 88 is close to the estimate of 90.

Choose the correct equation to solve each problem. Then solve.

1. After it began raining, 346 people left the game. However, 632 people stayed. How many people attended the game?

   a. $632 + 346 = y$
   b. $632 - 346 = y$

2. Elin separates her audio tapes into five storage boxes. She puts 20 in each box. How many audio tapes does she have?

   a. $20 \times 5 = z$
   b. $20 \div 5 = z$

f UN with MATH
*How about making
a fun fruit salad?*
See page 39.

On pages 516–518, you can learn how laser technicians use mathematics in their jobs.

Applications

Critical Thinking

Mixed Review

Lesson 1-2

Lesson 1-10

Lesson 1-11

Write an equation to solve each problem. Then solve.

3. A television costs $530. The stand costs $60. What is the total cost of the television with the stand?

4. Carlos spent $50 today and has $30 left in his wallet. How much money did Carlos have to start?

5. Mr. David is allowed 1,600 calories a day in his diet. If he has already eaten 900 calories, how many more can he eat and stay on his diet?

6. Michele looks at three cars. The first costs $10,200, the second costs $10,900, and the third $10,400. What is the difference in cost between the most and least expensive car?

7. A stadium has 70 rows of seats with 30 seats in each row. How many seats are in the stadium?

8. The profit from a sale was divided equally among four people. Each received $200. What was the total profit?

9. Jill puts $10 into her savings account each week. How many weeks will it take her to save $260?

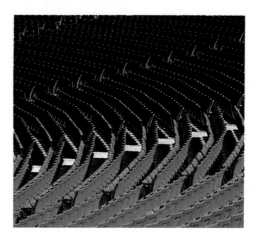

10. Mr. Sampsel bought 100 shares of Ajax stock and paid $3,400. What was the cost per share?

11. Mrs. Stigers paid $3,000 down when she ordered the car. She will pay the remaining $8,125 when the car is delivered. What is the cost of the car?

12. A portfolio represents samples of your work, collected over a period of time. Begin your portfolio by choosing an assignment that you feel shows your best work. Explain why you selected it. Date each item as you place it in your portfolio.

13. Draw the figure at the right without lifting up your pencil from the paper or crossing a line.

Write as a product and then find the number named.

14. 6^3 **15.** 4 cubed   **16.** 12 squared   **17.** 9^4

Write an expression for each phrase. Use n for the number.

18. 6 plus a number   **19.** 11 decreased by a number

20. 12 times n **21.** 23 increased by n

Solve. Use an equation.

22. In 14 years Frank will be 30 years old. How old is Frank now?

REVIEW

Vocabulary/ Concepts

Choose the word or number from the list at the right to complete each sentence.

1. In expanded form the number 604 is ___?___ .
2. Both 5^2 and 5^3 are ___?___ of five.
3. When we say a sum is *about* 600, we are ___?___ the sum.
4. Three less than a number can be written ___?___ .
5. Eight and thirty thousandths written in standard form is ___?___ .
6. When estimating a quotient, we round the ___?___ so it is a multiple of the divisor.
7. A mathematical sentence with an equals sign is called a(n) ___?___ .
8. Multiplication and division are ___?___ operations.
9. To solve an equation, we find the correct replacement for a(n) ___?___ .

dividend
estimating
equation
exponents
inverse
powers
variable
0.830
$3 - n$
$6 \times 100 + 4 \times 10$
$(6 \times 10^2) + (4 \times 10^0)$
8.030
$n - 3$

Exercises/ Applications

Lesson 1-1

Write in standard form.

10. 15 thousands   11. 36 tens   12. 6 tenths   13. 50 ones

Lesson 1-2

Write in expanded form using 10^0, 10^1, 10^2, and so on.

14. 500   15. 24   16. 1,821   17. 2,650   18. 20,340   19. 7,005

Lesson 1-3

Order from least to greatest.

20. 6.2, 6.02, 6.021, 6   21. 10.87, 10.708, 10.871, 10.78

Lesson 1-4

Round each number to the underlined place-value position.

22. <u>8</u>29   23. <u>9</u>,703   24. 36.<u>0</u>98   25. 10.6<u>5</u>4   26. 19.0<u>1</u>48

Lesson 1-5

Use the tax table on page 14 to find the tax for each income and filing status.

27. $19,505; single   28. $19,650; head of household

29. $19,167; married filing jointly   30. $19,400; married filing separately

Lessons 1-6 1-7

Estimate.

31. $409 + 2,803$ 32. $313,250 - 81,695$ 33. $2.39 + 3.64

34. $2,643 \times 3$ 35. 629×38 36. $408 \div 52$ **8**   37. $386 \div 3.704$

Lesson 1-8

Solve. Use matrix logic.

38. Spot, Lady, Fido, and Buffy each wear a collar. Spot's collar is not red. Lady's collar is not black or tan. Fido's fur is black and he wears a collar to match. Buffy's owner likes red but dislikes brown. What is the color of each dog's collar?

Lesson 1-9

Find the value of each expression.

39. $20 \times 4 + 2 \times 5$ **40.** $42 \div 6 + 6 \times 5$ **41.** $30.6 - 48 \div 8$

42. $(3 \times 7) - 1$ **43.** $15 - (64 \div 8)$ **44.** $(5 \times 8) + (60 \div 2)$

Lesson 1-10

Find the value of each expression if $a = 4$, $b = 5$, and $c = 10$.

45. $a^2 - b$ **46.** $c - (a + b)$ **47.** $\frac{ab}{c}$ **48.** $\frac{5a + c}{6}$ **49.** $(a + b)^2$

Lessons 1-11 1-12

Solve each equation. Check each solution.

50. $m - 7 = 7$ **51.** $y + 8 = 13$ **52.** $0.6 = n - 0.4$

53. $8k = 56$ **54.** $6 = \frac{a}{9}$ **55.** $420 = 6p$ **56.** $\frac{1.8}{t} = 0.6$

Lessons 1-11 1-12

Write an equation. Then solve.

57. Eight increased by m is 15. What is the value of m?

58. Some number decreased by 5 is 15. What is that number?

59. Six multiplied by y is 48. What is the value of y?

60. Some number divided by 3 is 600. What is that number?

Lesson 1-13

Write an equation to solve each problem. Then solve.

61. Beth spent $40 for a pair of shoes. She has $25 remaining in her purse. How much money did she have before buying the shoes?

62. Kevin places nine baseball cards in each page of his album. All 30 pages of his album are full. How many baseball cards are in the album?

63. Becky has $55. Does she have enough money to buy a $28.50 skirt and a $22.95 blouse?

64. The product of 8 and y is 7,200. What is the value of y?

65. Mr. Painter works 40 hours a week and makes $7 an hour. What is his take-home pay if $60 is withheld for taxes?

66. Sixty students moved into the Logan school district. Now there are 420 students attending the school. How many students were attending before the students moved in?

TEST

Write in standard form.

1. 70 tens

2. 30 hundredths

3. $(3 \times 10^2) + (7 \times 10^1)$

Write as a product and then find the number named.

4. 6^4

5. 3^3

6. 5^2

7. 2^6

Replace each ● with $<$, $>$, or $=$ to make a true sentence.

8. 36 ● 48

9. 3.8 ● 3.80

10. 0.6 ● 0.06

11. 78 ● 79

Round each number to the underlined place-value position.

12. 8̲76

13. 87̲.5

14. 1.09̲8

15. 1̲2,394

Solve.

16. John is single and earns $19,800. The tax table shows at least $19,750 but less than $19,800 is $3,126 or at least $19,800 but less than $19,850 is $3,140. How much tax does John owe?

Estimate.

17. $75 − $2.80

18. 3,168 + 425

19. 33 × 780

20. 245 ÷ 48

Solve. Use matrix logic.

21. Joan is baby-sitting Nancy, Bobby, Susie, and Timmy. For a snack, she gives them fruit. Each child gets a different fruit. Nancy only likes red fruit. Timmy does not like bananas or pears. Susie can choose either a pear or a banana. Bobby likes bananas but does not like apples or oranges. Who gets which fruit?

Find the value of each expression.

22. $8 + 8 \times 4$

23. $(40 + 60) \div 2$

24. $(9 \times 4) - 6^2$

Find the value of each expression if $r = 2$, $s = 5$, and $t = 10$.

25. $t \times (s - r)$

26. $2t \div 4$

27. $\frac{st}{r}$

28. rst

Solve each equation. Check each solution.

29. $m + 9 = 15$

30. $400 = n - 800$

31. $5,600 = 8q$

Write an equation to solve each problem. Then solve.

32. At Vick's Pizza, you can get a free pizza with 26 coupons. Sue has 20 coupons. How many more must she save to have enough for a free pizza?

33. Nick walks five times as many blocks to school as George walks. George walks three blocks. How many blocks does Nick walk?

▶ **BONUS:** **Describe the solution of each equation.**

a. $y = y$

b. $x + 4 = 2$

c. $x + 2 = x + 3$

TIME ZONES

Mr. Gunther owns a construction company. He bids for jobs all over the country. On Monday, Mr. Gunther makes a call from Denver at 10:30 A.M. to New York City. What time is it in New York? Use the map showing time zones.

Mr. Gunther's call is made from the west to the east. The time is one hour later for each time zone crossed.

10:30 A.M. in Denver ↔ 12:30 P.M. in New York City

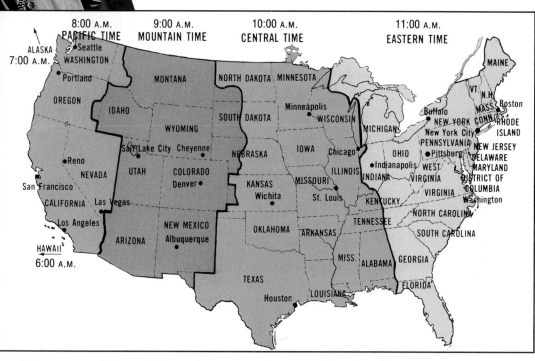

Suppose a call is made from the east to the west. The time is one hour earlier for each time zone crossed.

2:00 P.M. in East Florida ↔ 11:00 A.M. in California

For each city and time, give the time in the second city.

1. Boston   1:30 P.M. ↔ Wichita

2. Portland   11:00 P.M. ↔ Chicago

3. Cheyenne   8:35 P.M. ↔ Honolulu

4. Jessica makes a call from Indianapolis at 4:30 P.M. to Stockholm, Sweden. Stockholm, Sweden, is six time zones from Indianapolis. What time is her call received in Stockholm?

Free Response

Lesson 1-1

Replace each ■ with a number to make a true sentence.

1. $40 = $ ■ ones   **2.** $300 = $ ■ hundreds $2,000 = $ ■ tens

Name the digit in each place-value position in 9,105,832.467.

4. tens   **5.** thousandths   **6.** ten thousands

7. ones   **8.** tenths   **9.** hundredths

Write in standard form.

10. 6 hundred   **11.** 20 tens   **12.** 40 hundreds

Lesson 1-2

Write as a product and then find the number named.

13. 3^4 **14.** 15 squared   **15.** 7 cubed   **16.** 2^5 **17.** 10^6

Write in expanded form using 10^0, 10^1 10^2, and so on.

18. 562   **19.** 4,381   **20.** 10,601   **21.** 33,332

Lesson 1-3

22. Terri would like to purchase a compact disc that costs $12.95 or a cassette that costs $8.95. She has $10.52. Which can she afford to buy?

23. Which whole numbers are greater than 42.4 and less than 45.8?

24. Which cyclist had a time greater than 10.273 s?

Cyclist	Time
Bill Huck	10.153 s
Jens Fiedler	10.278 s
Ken Carpenter	10.283 s
Curtis Harrett	10.271 s

25. Which cyclist had the fastest time?

Lesson 1-4

Round each number to the underlined place-value position.

26. 3.021   **27.** 10.463   **28.** 0.642   **29.** 25.122

30. 87.5   **31.** 0.081   **32.** 28.195   **33.** 7.695

34. Janet's final grade average is 0.8273. To the nearest hundredth, what is her average?

Lesson 1-6
1-7

Estimate.

35. $6,189 + 523$ **36.** $906 - 32$ **37.** $0.715 + 0.328$

38. $43.17 - 6.55$ **39.** $0.85 - 0.2$ **40.** 32×19

41. 126×5 **42.** $360 \div 24$ **43.** $144 \div 18$

44. Denise is having a party with 86 invited guests. She would like to have enough cans of soda for each guest to have 2 cans. *About* how many cans of soda should she buy?

Multiple Choice

Choose the letter of the correct answer for each item.

1. What is the place-value position of the 8 in 184,972?
- **a.** tens
- **b.** hundreds
- **c.** thousands
- **d.** ten thousands

Lesson 1-1

6. What is the standard form of (6×10^4) + (5×10^3) + (7×10^0)?
- **e.** 657
- **f.** 65,007
- **g.** 65,700
- **h.** 650,007

Lesson 1-2

2. What is the numeral for 91 hundreds?
- **e.** 91
- **f.** 910
- **g.** 9,100
- **h.** 91,000

Lesson 1-1

7. Find the value of $30 - 9 \times (2 + 1)$.
- **a.** 3   **b.** 13
- **c.** 21   **d.** 63

Lesson 1-9

3. What is 93,740 rounded to the nearest hundred?
- **a.** 90,000
- **b.** 93,700
- **c.** 93,800
- **d.** 94,000

Lesson 1-4

8. Bobby gives 18 baseball cards away and he has 32 left. How many baseball cards did he have before he gave any away?
- **e.** 14
- **f.** 40
- **g.** 50
- **h.** 576

Lesson 1-11

4. Estimate the sum of 625 and 8,825.
- **e.** 1,400
- **f.** 9,400
- **g.** 14,000
- **h.** 15,000

Lesson 1-6

9. An odd number is greater than 6×5 and less than 6×7. What is the number if the sum of its digits is 8?
- **a.** 31
- **b.** 35
- **c.** 40
- **d.** 39

Lesson 1-8

5. Estimate the difference of 81,467 and 63,903.
- **a.** 2,000
- **b.** 8,000
- **c.** 20,000
- **d.** 140,000

Lesson 1-6

10. What facts are given in this problem?
   A concert hall has 100 rows of seats with 40 seats in each row. Each seat sells for $15.95. How much money is made if the concert is a sell-out?
- **e.** Who is giving the concert.
- **f.** The total number of seats.
- **g.** The number in attendance.
- **h.** The cost of parking.

Lesson 1-13

fun with MATH

First math problems found recorded on papyrus	880 BC	First use of plus (+) and minus (−) signs	1703
1650 BC	First inflatable swimming aid	1489 AD	Binary arithmetic (ones and zeros)

What does it mean to have 20/30 vision?

The figures 20/20 mean that at a distance of 20 feet, you can read letters of a size that is normal for that distance. If you have 20/30 vision, it means that you can read letters from a chart at 20 feet that a person with normal vision can read at 30 feet.

MATH M·E·N·U

FUN FRUIT SALAD

Mix 1 can (16 oz) fruit cocktail (drained), 1 can of mandarin oranges (drained), 1 cup miniature marshmallows, and ½ cup sour cream in a bowl. Place in refrigerator for 3 hours. Serve as a salad on lettuce, or in dessert dishes, each portion topped with a Maraschino cherry.

Did you know that a quarter has 119 grooves on its circumference? A dime has one less.

COMICS

CALVIN & HOBBES

Fun with Math

JOKE!

Q: If two's company and three's a crowd, what are four and five?

A: Nine.

How does a CD player work?

During recording, music is converted to a binary code, made up of ones and zeros, by a computer. A laser beam, carrying the code of ones and zeros, burns a pit along each track of a CD for each zero in the code. It leaves a non-pit, or 'land' for each one in the code. A laser beam in your CD player reads the pits and lands of a CD. The built-in microcomputer converts the binary number code back to music.

RIDDLE

Q. When can ten students stand under an umbrella without getting wet?

A. When it's not raining

TEASER

The Mayas used a number system of only two symbols—a dot and a line— to build numbers from 1 to 19. Complete the following.

1 ·	2 ··	(?)		
— 6	— 7	3 ···	4 ····	— 5
÷ 11	(?) 12	— 8	(?) 9	— 10
÷ 16	(?) 17	(?) 13	(?) 14	— 15
		18	19	

An oval below a number multiplied it by 20:
⬭·· = 2 × 20 = 40.
Write Mayan numbers for 60, 80, 100, and so on to 380.

QUIZ TIME

Substitute words for the numbers in the following story by doing the math on your calculator, then turning it upside down to read each word. $(632 \times 497) + 3433$ and $(2 \times 5 \times 7 \times 110) + 18$ went to the $(677 + 271 - 946) \div 100$. They walked up a $(59,982 \div 13)$ $(69,426 \div 9)$, $(11,283 + 17,584) \times 11$ broke the (193×38) of her $(1367 - 758) \times 5$ when $(69 \times 15) \div 3$ stepped in a $(49 \times 56) + (12 \times 80)$. (69×5) had to $(66,666 + 77,777 + 88,888 + 99,999 + 45,474)$ around and her (91×7) hurt. (227×34) almost stepped on a $(315,054 \div 9)$ $(3 \times 13 \times 17)$.

Chapter

2

PATTERNS: ADDING AND SUBTRACTING

▶ **Getting Around!**

Do you live in a large city, in a small town, or in the country? What forms of transportation are available in your community? Does it have a bus service, taxi service, subway, or elevated train? Did you know that the first regular bus service began in New York City in 1905? How do you get to school or work?

Trischa rides a city bus to school, and after school, she rides the bus to work. She rides 2.4 miles from home to school and 1.7 miles from school to work. How many miles does Trischa ride from home to work via school?

ACTIVITY: Exploring Decimals with Models

You can use decimal models to find the sum of decimals. Suppose you want to find the sum of 0.7 and 0.6.

Materials: decimal models (tenths and hundredths), pencil

1. Use two decimal models that are separated into tenths.

2. On the first decimal model, shade 7 tenths as shown.

3. On the second decimal model, shade 6 tenths as shown.

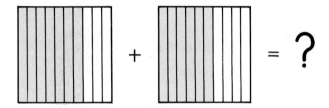

Cooperative Groups

Work together in groups of two or three.

4. Determine a way to use decimal models and shading to show the sum of 0.7 and 0.6. What is the sum?

5. Use decimal models to show the sum of 0.63 and 0.45. Discuss with your group how to represent the problem and the sum on decimal models. Explain how you arrived at the sum using decimal models.

6. Using decimal models, how would you find the sum of 0.3 and 0.09?

7. How could you show subtraction of decimals using decimal models? Use 0.7 − 0.4 as an example.

Communicate Your Ideas

8. Work with your group to draw a conclusion about how to add decimals without decimal models. Describe what happens to the decimal point.

9. What determines the type of decimal model, tenths, hundredths, and so on, you need to model a problem?

You will study more about adding and subtracting decimals in this chapter.

2-1 ADDING WHOLE NUMBERS

Objective
Add whole numbers.

At Rushmore High School, 33 students have signed up in the first week for volleyball intramurals. If there is room for 35 more students, how many will be able to play volleyball intramurals? To find the total number of students, add.

To add two or more whole numbers, you should begin at the right and add the numbers in each place-value position.

Method

1. Align the numbers by place-value position.
2. Add in each place-value position from right to left.

Example A

$33 + 35$ Estimate: $30 + 40 = 70$

1. $\begin{array}{r} 33 \\ + 35 \\ \hline \end{array}$ Align the numbers.

2. $\begin{array}{r} 33 \\ + 35 \\ \hline 8 \end{array}$ Add the ones. ⟶ $\begin{array}{r} 33 \\ + 35 \\ \hline 68 \end{array}$ Add the tens.

Compared to the estimate, is the answer reasonable?

There can be 68 students in Rushmore intramurals.

Example B

$249 + 28$ Estimate: $250 + 30 = 280$

1. $\begin{array}{r} 249 \\ + 28 \\ \hline \end{array}$

2. $\begin{array}{r} \overset{1}{2}49 \\ + 28 \\ \hline 7 \end{array}$ Write 17 ones as 1 ten and 7 ones. ⟶ $\begin{array}{r} \overset{1}{2}49 \\ + 28 \\ \hline 77 \end{array}$ ⟶ $\begin{array}{r} \overset{1}{2}49 \\ + 28 \\ \hline 277 \end{array}$

Is the answer reasonable?

Example C

$7,453 + 2,865$ Estimate: $7,000 + 3,000 = 10,000$

1. $\begin{array}{r} 7,453 \\ + 2,865 \\ \hline \end{array}$

2. $\begin{array}{r} \overset{1\ 1}{7,}453 \\ + 2,865 \\ \hline 10,318 \end{array}$ Based on the estimate, explain why the answer is reasonable.

Guided Practice
Add.

Example A

1. $\begin{array}{r} 43 \\ + 36 \\ \hline \end{array}$
2. $\begin{array}{r} 20 \\ + 18 \\ \hline \end{array}$
3. $\begin{array}{r} 81 \\ + 17 \\ \hline \end{array}$
4. $\begin{array}{r} 45 \\ + 83 \\ \hline \end{array}$
5. $\begin{array}{r} 62 \\ + 75 \\ \hline \end{array}$

Example B

6. $56 + 863$ 7. $94 + 786$ 8. $598 + 75$ 9. $545 + 57$

Example C

10. $4,726 + 28,538$ 11. $270 + 46,090$ 12. $76,118 + 768$

Practice

Add.

13.	25	14.	32	15.	53	16.	792	17.	847
	+ 71		+ 45		+ 30		+ 25		+ 990

18.	2,914	19.	4,803	20.	3,408	21.	62,924	22.	2,816
	+ 3,087		+ 297		+ 5,697		+ 6,085		+ 13,847

23. 408 + 694   **24.** 23,088 + 257   **25.** 94,670 + 69,033

26. Add 23 and 257.

27. Find the sum of 4,715 and 14,369.

Applications

28. The annual budget for the city of Circleville is $985,000. The voters approved a $53,000 increase in the budget. What is the amount of the increased budget?

29. John drove 238 kilometers from London, England, to Liverpool, England. Then he drove 52 kilometers from Liverpool to Manchester. How many kilometers did he drive in all?

30. Simsbury has three communities. Tariffville has 3,509 people, Weatogue has 2,878 people, and West Simsbury has 23,644 people. What is the total population of Simsbury?

31. Lin Ye has $420 in her savings account. She deposits $35. What is her new balance?

Critical Thinking

32. Fill in the appropriate whole numbers in the magic square so that each row, column, and diagonal add up to 39.

Cooperative Groups

33. In groups of two or three, make up five addition problems. Each problem must have a sum of 1,250. At least one of the problems must be a word problem. Exchange your problems with other groups to check that the problems are correct.

Mixed Review

Lesson 1-9

Find the value of each expression.

34. $5 + 6 - 3 \times 2$ **35.** $16 \div 4 + 8 \times 3$

36. $6 \times (4 \div 2) + 8$ **37.** $3 + 7 \times 2 - 9$

Lesson 1-11

Solve each equation. Check your solution.

38. $x + 7 = 41$ **39.** $13 + b = 22$ **40.** $52 = m - 30$

Lesson 1-13

41. Angelo found the price of pizzas at three competing pizza shops. Shop #1 has a one-item pizza for $7.95. Shop #2 has the same pizza for $8.50, and shop #3 has the pizza for $10.25. What is the difference in price between the most expensive and least expensive?

2-2 ADDING DECIMALS

Objective
Add decimals.

George Thompkin monitors transmitting equipment at Station 3 for Satellite Systems, Inc. Data beamed from Station 3 to a satellite takes 0.27 seconds. Data beamed back to Station 4 from the satellite takes 0.41 seconds. What is the total time it takes for data to travel from Station 3 to Station 4 through the satellite?

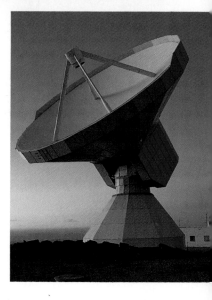

To add two or more decimals, begin at the right and add the numbers in each place-value position.

Method

1 Align the decimal points.

2 Add in each place-value position from right to left.

Example A

0.41 + 0.27   Estimate: 0.4 + 0.3 = 0.7

1
```
  0.41
+ 0.27
```
Align the decimal points.

2
```
  0.41
+ 0.27
─────
     8
```
Add the hundredths.

```
  0.41
+ 0.27
─────
  0.68
```
Add the tenths.

Based on the estimate, 0.7, explain why the answer is reasonable.

The data takes 0.68 seconds to travel from Station 3 to Station 4.

Example B

9.38 + 15 + 4.701   Estimate: 9 + 15 + 5 = 29

1
```
   9.380
  15.000
+  4.701
```
Align the decimal points. Annex zeros if you need to so each addend has the same number of decimal places.

2
```
  1 1
   9.380
  15.000
+  4.701
────────
  29.081
```

Is the answer reasonable?

Guided Practice

Add.

Example A

1.
```
  0.85
+ 0.12
```

2.
```
  7.946
+ 0.032
```

3.
```
  26.2
+  3.9
```

4.
```
  15.24
   9.18
+  6.37
```

Example B

5.
```
  48.6
+  9.31
```

6.
```
  76.7
+ 15.678
```

7.
```
   0.33
+ 21.7
```

8.
```
  126.2
+  71.08
```

9. 1.56 + 7.2 + 3.73

10. 5.98 + 12.4 + 3

11. 16.78 + 924 + 7.01

12. 21,195 + 18.2 + 46

Practice

Add.

13.   0.23
     + 0.18

14.   0.94
     + 1.23

15.   0.81
     + 1.9

16.   65.3
     + 4.91

17. $0.43
       0.97
     + 0.60

18. $0.38
       1.42
     + 0.66

19. $45.63
     + 5.96

20.   0.07
       0.19
     + 1.26

21.   0.275
       1.38
     + 6.91

22.   2.05
       0.156
     + 0.071

23. $355.92
       105.17
     + 481.37

24. 2.05 + 0.156 + 3

25. $3.51 + $14.62

26. 0.23 + 0.1 + 8

27. Find the sum of 0.036, 0.15, and 1.7.

28. What is the total of 490, 7.3, and 26.946?

Applications

29. The four members of a 400-meter relay team had times of 11.26 seconds, 11.07 seconds, 11.03 seconds, and 10.43 seconds. What was the total time for the relay team?

30. Pat bought a tennis racket for $84.98, a pair of shorts for $12.67, a pair of socks for $2.49, and a visor for $5.39. What was Pat's total bill not including sales tax?

31. Sarah's bank pays 5.1% interest on savings accounts. Since Sarah's balance is greater than $1,500, she receives an extra 0.3% interest. What is the total percent of interest Sarah receives?

Using Variables

Write an expression for each phrase.

32. the sum of a number and 43

33. 0.12 more than a number

Make Up a Problem

34. Make up a problem using decimals for which the sum is 23.46. Use three addends.

35. Make up a problem using four decimal addends that has a sum of 50.

Estimation

Deciding whether a computation will be over or under a certain value is called *reference point* estimation. For example, can you buy a 77¢ drink and a 26¢ pack of gum for $1.00? You know $0.75 plus $0.25 is $1.00. Since the items are more than $0.75 and $0.25, their sum will be more than $1.00. Therefore, you cannot buy both for $1.00. The reference point is $1.00.

36. 25 + 25 = 50, so 26 + 27 is ___?___ than 50. (greater, less)

37. 100 − 40 = 60, so 100 − 45 is ___?___ than 60. (greater, less)

38. 3 × 50 = 150, so 3 × 55 is ___?___ than 150. (greater, less)

2-3 SUBTRACTING WHOLE NUMBERS

Objective
Subtract whole numbers.

Wilson Golf Course has a fleet of 94 golf carts. By 10:30 A.M., Helen's list shows that 37 carts have been rented. She subtracts to find out how many carts she has left to rent.

To subtract whole numbers, you should begin at the right and subtract the numbers in each place-value position. Sometimes you must rename before subtracting.

Method

1 ▶ Align the numbers by place-value position.

2 ▶ Subtract in each place-value position from right to left. Rename first if necessary.

Example A

94 − 37   Estimate: 90 − 40 = 50

1 ▶   94   Align the
     − 37   numbers.

2 ▶   ⁸¹⁴
       9̶4̶   Rename 9 tens
     − 37   as 8 tens and
       ———   10 ones. Subtract
        7   the ones.

       ⁸¹⁴
       9̶4̶   Subtract
     − 37   the tens.
       ———
        57

Compared to the estimate, is the answer reasonable?

Helen has 57 carts left to rent.

Subtracting is the opposite of adding.

42 + 33 = 75 ⟋ 75 − 33 = 42
            ⟍ 75 − 42 = 33

Addition and subtraction are inverse operations.

So you can use addition to check subtraction.

Example B

601 − 378   Estimate: 600 − 400 = 200

1 ▶   601
     − 378

2 ▶   ⁵ ⁹¹¹
       6̶0̶1̶   Rename 60 tens
     − 378   as 59 tens
       ———   and 10 ones.
        3

       ⁵ ⁹¹¹
       6̶0̶1̶   Check by
     − 378   using addition.
       ———
        223

        ¹ ¹
        223
      + 378
      ———
        601 √

Based on the estimate, explain why the answer is reasonable.

Example C

6,014 − 878   Estimate: 6,000 − 900 = 5,100

1 ▶   6,014
     − 878

2 ▶   ⁵ ⁹¹⁰¹⁴
       6̶,0̶1̶4̶   Check by adding.
     −   878   5,136
       ———   + 878
       5,136   ———
               6,014 √   Is the answer reasonable?

Subtract.

Examples A, B

1.   72
        − 47

2.   76
        − 53

3.   67
        − 28

4.   985
        − 871

5.   563
        − 455

Example C

6. 3,723 $-\ \ \ 262$	**7.** 5,423 $-\ \ \ 718$	**8.** 4,540 $-\ \ \ 475$	**9.** 7,275 $-\ 3,866$	**10.** 3,838 $-\ 2,496$

Practice

Subtract.

11. 55 $-\ 23$	**12.** 95 $-\ 43$	**13.** 88 $-\ 65$	**14.** 800 $-\ 506$	**15.** 503 $-\ \ 85$

16. 29,006 $-\ 19,137$	**17.** 4,100 $-\ 3,308$	**18.** 6,209 $-\ \ \ 108$	**19.** 47,603 $-\ 37,501$	**20.** 2,506 $-\ \ \ 496$

21. $3,402 - 2,324$ **22.** $3,016 - 2,305$ **23.** $4,557 - 3,593$

24. Subtract 1,661 from 23,208.

25. Find the difference of 44,213 and 3,502.

Applications

26. Kathy needs $800 for the down payment on a car. She has saved $476. How much does she still need to save?

27. The elevation at the top of a ski slope is 2,123 feet. The elevation at the bottom of the slope is 941 feet. What is the vertical height of the slope?

28. Lake Huron occupies about 23,010 square miles, Lake Erie about 9,930 square miles. *About* how much larger is Lake Huron than Lake Erie?

f **UN with MATH**

Why worry about the rain forests?
See page 104.

29. When Kyle took Elise to the movies, he had a $20 bill in his wallet. He paid $5 each for two tickets, $1.00 each for two drinks, and $2.50 for the large bucket of popcorn. How much money did Kyle have left from the $20 bill?

Research

30. Find out how many unmanned lunar probes have been launched by the United States or by the Soviet Union. Estimate how many total space miles have been covered by the probes.

Critical Thinking

31. Robyn needs to thin 4 seedlings from a bed of 12 so that no more than 3 and no less than 1 remain in each row or column. Which seedlings does Robyn need to remove?

Mixed Review

Lesson 1-12

Solve each equation. Check your solution.

32. $14k = 42$ **33.** $60 = 5y$ **34.** $36 = t \div 3$

Lesson 2-1

Add.

35. 68 $+\ 19$	**36.** 163 $+\ \ 21$	**37.** 821 $+\ 601$	**38.** 32,181 $+\ 1,427$

Lesson 2-1

39. What is the least sum of two consecutive numbers that are greater than 499?

2-4 SUBTRACTING DECIMALS

Objective
Subtract decimals.

Cora is the costume designer for the Orleans Summer Playhouse. She has 3.7 yards of felt in storage and needs 2.1 yards for King Henry's hat and shoes. After cutting the hat and shoes, how much felt does Cora have left?

To subtract decimals, begin at the right and subtract the numbers in each place-value position. It may be necessary to annex zeros.

Method

1 Align the decimal points. If necessary, annex zeros so both numbers have the same number of decimal places.

2 Subtract in each place-value position from right to left.

Example A

3.7 − 2.1   Estimate: 4 − 2 = 2

1
$$\begin{array}{r} 3.7 \\ -\ 2.1 \end{array}$$
Align the decimal points.

2
$$\begin{array}{r} 3.7 \\ -\ 2.1 \\ \hline 1.6 \end{array}$$
Subtract tenths, then ones.

Compared to the estimate, is the answer reasonable?

Check:
$$\begin{array}{r} 2.1 \\ +\ 1.6 \\ \hline 3.7 \end{array}\ \checkmark$$

Cora has 1.6 yards of felt left.

Example B

6.391 − 4.62   Estimate: 6 − 5 = 1

1
$$\begin{array}{r} 6.391 \\ -\ 4.620 \end{array}$$
Align the decimal points. Annex a zero.

2
$$\begin{array}{r} \overset{5\ 13}{6.391} \\ -\ 4.620 \\ \hline 1.771 \end{array}$$
Subtract the digits in each place-value position.

Is the answer reasonable? Check by adding.

Example C

9.3 − 3.712   Estimate: 9 − 4 = 5

1
$$\begin{array}{r} 9.300 \\ -\ 3.712 \end{array}$$
Align the decimal points. Annex zeros so both numbers have three decimal places.

2
$$\begin{array}{r} \overset{8\ 12\ 9\ 10}{9.300} \\ -\ 3.712 \\ \hline 5.588 \end{array}$$
Subtract.

Is the answer reasonable? Check by adding.

Guided Practice — *Subtract.*

Examples A,B

1.
$$\begin{array}{r} 6.1 \\ -\ 4.3 \end{array}$$

2.
$$\begin{array}{r} 17.35 \\ -\ 9.17 \end{array}$$

3.
$$\begin{array}{r} 100.09 \\ -\ 22.67 \end{array}$$

4.
$$\begin{array}{r} 0.497 \\ -\ 0.16 \end{array}$$

5.
$$\begin{array}{r} 2.56 \\ -\ 1.4 \end{array}$$

Example C

6.
$$\begin{array}{r} 4.8 \\ -\ 1.85 \end{array}$$

7.
$$\begin{array}{r} 7 \\ -\ 1.63 \end{array}$$

8.
$$\begin{array}{r} 252.7 \\ -\ 38.95 \end{array}$$

9.
$$\begin{array}{r} 38.7 \\ -\ 9.82 \end{array}$$

10.
$$\begin{array}{r} 3,000 \\ -\ 789.5 \end{array}$$

Subtract.

11. 4.6
   – 2.2

12. 0.64
   – 0.21

13. 0.894
   – 0.172

14. 2.51
   – 0.8

15. 1.3
   – 0.48

16. 83.79
   – 51.16

17. 4.968
   – 2.3

18. 5.284
   – 3.197

19. 9.06
   – 7.752

20. 8.9
   – 1.45

21. 0.6 – 0.49   **22.** 21 – 4.09   **23.** 0.416 – 0.27

24. Subtract $23.25 from $68.93.

25. Find the difference of 568.335 and 34.008.

Applications

26. The Reisser Corporation posted a profit of $15.7 million for the second quarter. This was a $0.9 million gain over the first quarter. What was the profit for the first quarter? What was the total profit for the first two quarters.

27. The German mark closed at 1.645 on Wednesday. Six months ago the mark closed at 1.69. How much loss occurred over the six months?

Interpreting Data

28. How much is spent annually for the pet pest prevention and electronic bug killers?

f **UN with MATH**

How long ago were numbers first used?
See page 39.

29. How much more is spent annually on do-it-yourself products than on electronic bug killers?

30. If professional pest control companies make 10 million calls a year, what is the average cost of one professional call?

Dollars Spent Annually on Pest Control

Every year, ten million households use professional pest control companies.

Electronic bug killers $59 million
Pet pest prevention $250 million
Do-it-yourself household products $400 million
Professional pest control $3 billion

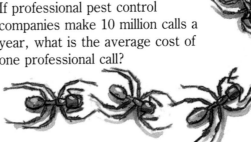

Number Sense

31. Write 1.111 as the sum of four different decimals.

Calculator

Place value and addition can be used to "build" a decimal number.

5000 + 300 + 20 + 8 + 0.1 + 0.02 + 0.004 = 5328.124

Add with a calculator to "build" each number.

32. 4,351.897   **33.** 946.12   **34.** 1,407.95   **35.** 9,720.822

Enter the first number in a calculator. Then add or subtract to get the remaining numbers in order.

36. 134.56 → 134.59 → 134.69 → 135.69

37. 8,345.22 → 8,345.28 → 8,345.88 → 8,345.888

TERMINOLOGY

A **computer** is an electronic device that can store, retrieve, and process data. The most widely used computer for educational purposes is the **microcomputer** or **personal computer (PC)**.

A computer must be given a set of instructions called a **program.** Computer programs are also referred to as **software.** The computer itself is called the **hardware.**

The heart of the computer is the **central processing unit (CPU).** The CPU contains an **arithmetic/logic unit** and a **control unit.** The arithmetic/logic unit performs standard arithmetic operations and comparisons. The control unit directs the overall actions of the computer.

Information that is entered into a computer is called **input.** A keyboard is the most common input device. Other input devices are light pens, disk drives, and cassette tapes.

The CPU handles a limited amount of information at one time. So programs and data are stored in computer **memory** and are called for by the CPU as they are needed.

The results of computer processing are called **output.** Two output devices are the cathode-ray tube (CRT) and the printer. The CRT, along with its controls, is called a **monitor.** A printer produces **hard copy** of the computer output.

1. What does PC stand for?
2. What is the set of instructions for a computer called?
3. Name and define the parts of a CPU.
4. Explain the difference between hardware and software.
5. Name the four basic parts of a computer system.
6. Name four input and two output devices.
7. What is a monitor? What does it display?
8. What is hard copy of computer output? How is it produced?

METER READER

Dawn Westrick is a meter reader for Southern Electric Company. On her route, Dawn reads electric meters and enters the readings in a hand-held computer. She also makes a quick check of the condition of the meters and connections.

Electric meters measure the number of kilowatt hours used. A **kilowatt hour (kWh)** is equal to 1 kilowatt of electricity used for 1 hour.

The electric meter shows 5 dials. Use the following steps to find the meter reading.

1. Choose the number that was just passed as the arrow rotates from 0 through 9. Note that the numbers on some dials run clockwise and some run counterclockwise.

2. Each dial is in its place-value position. This meter reads 26,453 kilowatt hours.

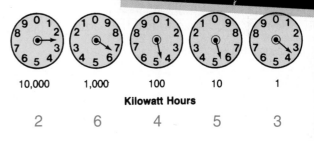

10,000	1,000	100	10	1
		Kilowatt Hours		
2	6	4	5	3

How many kilowatt hours are shown on each electric meter?

1.
2.
3.
4.
5.
6.

7. If electricity costs $0.028 per kWh, what is the monthly bill for a usage of 4,065 kilowatt hours? At this same rate of usage per month, what is the yearly cost for electricity?

2-5 CHECKING ACCOUNTS

Objective

Write a check and use a check register.

Rajiv Ramur opened a checking account after he began working part time. He uses checks to pay bills like car insurance with the money in his account. He uses deposit slips to add money to his account.

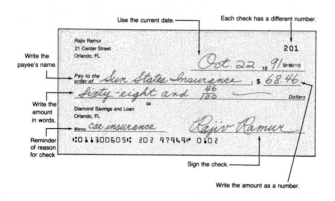

Use the current date.

Each check has a different number.

Write the payee's name.

Write the amount in words.

Reminder of reason for check

Sign the check.

Write the amount as a number.

Date of deposit

Amount of deposit

Use Other Side For Additional Listing

Total deposit

Rajiv keeps a check register to record checks and deposit amounts and to find the balance of his account.

Balance from previous page

CHECK NUMBER	DATE	DESCRIPTION OF TRANSACTION	PAYMENT/DEBIT (−)	√ TAX	DEPOSIT/CREDIT (+)	BALANCE FWD. $267 25
D	10/20	Deposit	$		$ 75 30	75 30
						342 55
201	10/22	Sun States Insurance	68 46			68 46
		Car insurance				274 09
202	10/22	Tommy's Electronics	82 10			82 10
		CD repair				191 99
203	10/25	Terri's birthday	21 98			21 98
		gift				170 01
D	10/27	Deposit			63 00	63 00
						233 01
204	10/29	Best, Inc. – frame	86 20			86 20
						146 81

Payee's name

Reminder of reason for check

Balance after each transaction

Understanding the Data

● What is the name of the payee on the check?

● What is the amount of the deposit made on October 20?

● What is Rajiv's balance from the check register on October 29?

To find the balance in a check register, add each deposit and subtract the amount of each check.

Check for Understanding

1. Why is the amount of the check in numbers written close to the dollar sign?

2. Why is the amount of the check in words written as far to the left as possible?

3. What word replaces the decimal point in the amount when it is written in words?

Practice

Write each amount in words as you would on a check.

4. $13.85　　**5.** $29.00　　**6.** $272.05　　**7.** $108.97

8. Copy and complete the balance column of the register.

9. What is the balance on June 21?

10. What is the balance on June 30?

Suppose

11. Suppose check number 329 was recorded incorrectly. The amount should have been $13.67. What should the corrected balance be on June 30?

CHECK NUMBER	DATE	DESCRIPTION OF TRANSACTION	PAYMENT/DEBIT (−)	√ TAX	DEPOSIT/CREDIT (+)	BALANCE FWD. $564 78
328	6/15	Golden Eagle groceries	$62 12	$		
329	6/17	Rosa's gift	13 76			
D	6/20	Deposit paycheck			720 35	
330	6/25	Asherton Corp. rent	515 00			
331	6/25	Diamond National Bank car payment	206 89			
332	6/25	Haven Fire + Casualty insurance	63 50			
D	6/30	Deposit dividend			27 31	

Research

12. Find out about checking account service charges and charges for using automated teller machines from literature available in banks or from the newspaper.

Using Variables

Write an equation and solve.

13. The service charge, *s*, at First National Bank is $0.10 for each check. Let *n* equal the number of checks. Write an equation to find the service charge for writing *n* checks.

14. If Karen writes 23 checks on her checking account at First National Bank, how much is the service charge? Use the equation in Exercise 13.

Critical Thinking

15. In two weeks, Katie savied $12.60. Each day she saved $0.10 more than the day before. How much did Katie save on the first day?

Mixed Review

Lesson 1-10

Write an expression for each phrase.

16. the product of 13 and *z*　　**17.** 5 less than *x*

18. 18 divided by a number　　**19.** a number increased by 14

Lesson 2-2

Add.

20.　　4.13
　　　　+ 6.85

21.　　18.627
　　　　+ 9.651

22.　　84.699
　　　　+ 24.276

23.　　0.392
　　　　+ 18.67

Lesson 2-3

24. At the beginning of Pat's trip his odometer read 25,481. When he had completed his trip the odometer read 26,715. How many total miles did Pat travel?

2-6 ARITHMETIC SEQUENCES

Objective

Identify and write arithmetic sequences.

Paula is mailing photographs to her pen pal in New Zealand. She notices that the costs for mailing are $0.45, $0.62, and $0.79 for the first, second, and third ounce respectively. If this pattern continues, how much will it cost Paula to mail a six-ounce package?

A list of numbers that follows a certain pattern is called a **sequence.** If the difference between consecutive numbers in a sequence is the same, the sequence is called an **arithmetic sequence.**

5, 10, 15, 20, ... Each number in the sequence is 5 *more than* the number before it.

60, 45, 30, 15, ... Each number in the sequence is 15 *less than* the number before it.

You can extend an arithmetic sequence by using the following steps.

Method

1 ▶ List the numbers of the sequence in order.

2 ▶ Find the difference between any two consecutive numbers.

3 ▶ Add or subtract the difference to find the next number.

Example A

Find the next three numbers in the sequence 0.45, 0.62, 0.79, ... to find the cost of a six-ounce package.

1 ▶ List the numbers in order. 0.45   0.62   0.79

2 ▶ Find the difference between consecutive numbers.   0.62 − 0.45 = 0.17   0.79 − 0.62 = 0.17
The difference is 0.17.

If the difference between consecutive numbers is *not* the same, it is *not* an arithmetic sequence.

3 ▶ Add the difference to the last number.   0.79 + 0.17 = 0.96   0.96 + 0.17 = 1.13
1.13 + 0.17 + 1.30
It will cost Paula $1.30 to mail a six-ounce package.

Example B

Find the next three numbers in the sequence 85, 75, 65, 55, ...

1 ▶ 85   75   65   55

2 ▶ 85 − 75 = 10   75 − 65 = 10   65 − 55 = 10
The difference is 10.

3 ▶ 55 − 10 = 45   45 − 10 = 35   35 − 10 = 25   Subtract since the sequence is decreasing.

The sequence is 85, 75, 65, 55, 45, 35, and 25.

1.

Guided Practice

Find the difference and the next three numbers for each sequence.

Example A

1. 6, 8, 10, 12, $\frac{?}{}$, $\frac{?}{}$, $\frac{?}{}$

2. 0.6, 3.6, 6.6, 9.6, $\frac{?}{}$, $\frac{?}{}$, $\frac{?}{}$

Example B

3. 9, 8, 7, 6, $\frac{?}{}$, $\frac{?}{}$, $\frac{?}{}$

4. 52, 49, 46, 43, $\frac{?}{}$, $\frac{?}{}$, $\frac{?}{}$

Write the next three numbers in each sequence.

5. 100, 92, 84, 76,

6. 10, 8.2, 6.4, ...

7. 4, 14, 24, 34, ...

8. 200, 300, 400, ...

9. 250, 225, 200, ...

10. 89, 78, 67, ...

11. 14.5, 16.2, 17.9, ...

12. 23.6, 49.1, 74.6, ...

State whether each sequence is arithmetic. If it is, write the next three terms. If it is not, write **no.**

13. 6, 30, 150, 750, ...

14. 200, 100, 50, 25, ...

15. 6.1, 11.3, 16.5, ...

16. 5.5, 5.2, 4.9, 4.6, ...

Copy and complete each arithmetic sequence.

17. 9, 18, _?_ , 36, _?_ , _?_

18. 81, _?_ , 91, 96, _?_ , _?_ , 111

19. 10.4, 26.2, _?_ , _?_ , 73.6, _?_

20. 26.9, _?_ , 40.5, _?_ , 54.1, _?_

21. _?_ , 6.3, _?_ , 4.9, _?_ , 3.5

22. 108.4, 83.4, _?_ , _?_ , 8.4

Applications

23. In the first three weeks of typing class Keith's speed increased from 28 words per minute to 32 words per minute to 36 words per minute. If Keith continues to improve at this rate, how many more weeks will it take him to type 48 words per minute?

24. Martin falls 16 feet in the first second, 48 feet in the second second, and 80 feet in the third second during a free-fall. At this rate, how many feet will he fall in the sixth second?

Number Sense

25. Copy and complete the table for the sequence.

First	Second	Third	Fourth	Fifth	Sixth
12	18	24	30		
$12 + 0 \times 6$	$12 + 1 \times 6$	$12 + 2 \times 6$	$12 + \underline{?} \times 6$	$12 + \underline{?} \times 6$	$12 + \underline{?} \times 6$

26. In Exercise 25 how would you write an expression for the seventh number? the eighth number?

Using Variables

27. How would you write an expression for the *n*th number of the sequence in Exercise 25?

Estimation

You can estimate differences by changing the numbers in the problem to numbers that are easy to subtract mentally. Any one of the problems given below could be used to estimate difference of 97,584 and 33,751.

$$
\begin{array}{r} 97{,}584 \\ -\ 33{,}751 \end{array}
\longrightarrow
\begin{array}{r} 95{,}000 \\ -\ 35{,}000 \\ \hline 60{,}000 \end{array}
\quad \text{or} \quad
\begin{array}{r} 95{,}000 \\ -\ 30{,}000 \\ \hline 65{,}000 \end{array}
\quad \text{or} \quad
\begin{array}{r} 100{,}000 \\ -\ 30{,}000 \\ \hline 70{,}000 \end{array}
$$

Estimate.

28. $\begin{array}{r} 697 \\ -\ 83 \end{array}$

29. $\begin{array}{r} 355 \\ -\ 103 \end{array}$

30. $\begin{array}{r} 727 \\ -\ 538 \end{array}$

31. $\begin{array}{r} 6{,}243 \\ -\ 4{,}564 \end{array}$

▶ Explore
▶ Plan
▶ Solve
▶ Examine

2-7 CHOOSING THE METHOD OF COMPUTATION

Objective

Choose an appropriate method of computation.

Sherri needs five pairs of socks for running camp. She saw an ad in the newspaper for a clearance sale at Sports Plus. If socks are on sale for $2.89 a pair, how much money should she take to the store?

For some problems, an estimate is enough to solve the problem.

Example A

Since Sherri doesn't need to know the exact amount, she estimates.

$2.89 is about $3 $3 × 5 = $15

Sherri decides to take $15 to the store.

Example B

The clerk in the store needs to know exactly how much five pairs of socks will cost. The cash register finds the cost of five pairs of socks.

$2.89 × 5 = $14.45

Five pairs of socks cost $14.45.

The following flowchart can be used to choose the method of computation.

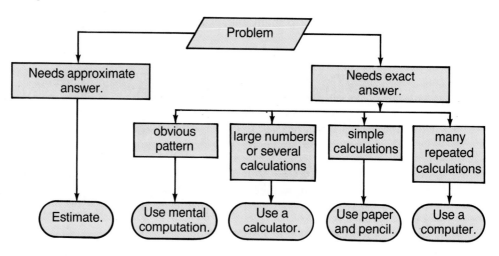

Guided Practice

Solve. State the method of computation used.

1. John is ordering parts from a catalog to fix his motorbike. The total bill must be prepaid with a check or money order. John needs a transmission for $97.95, a seat cover for $43.90, and two side baskets for $23.95 each. How much money should John send with his order?

2. Curtis is going to a movie with his friends. He wants to get popcorn and a drink before the movie starts. If a ticket costs $5.50, popcorn is $1.89, and a drink is $0.99, how much money should he take?

Paper/ Pencil
Mental Math
Estimate
Calculator

Solve. Use any strategy.

3. The senior class of Westfield High School wants to donate money to plant trees at the school entrance. There are 201 students donating $3.00 each. Determine how much money the class is donating and state the method of computation used.

4. Sean works part-time while he is in school. If he earns at least $2,000, he will be able to buy a car. He works four hours a day, three days a week, for $3.85 per hour. If he works 35 weeks this year, will he be able to buy the car?

5. A square, a circle, and a triangle are each a different color. The figure that is round is not green, one of the figures is red, and the blue figure has four equal sides. Find the color of each figure.

f**UN with MATH**

Try making Pretzel Cookies.
See page 104.

6. Darlene chooses a number between 1 and 10. If she adds 9, subtracts 8, then adds 6; the result is 12. What number did Darlene choose?

7. Susan's car uses approximately 9 gallons of gasoline each week. If she pays $1.29 a gallon, how much money should she plan to spend on gasoline each week? State the method of computation used.

8. Use each of the digits 1, 2, 3, and 4 only once. Find two 2-digit numbers that will give the greatest possible answer when multiplied.

Critical Thinking

9. In a farmyard containing chickens and pigs, there are 54 heads and 144 feet. How many chickens are in the enclosure? how many pigs?

Mixed Review

Lesson 2-1

Add.

10.   91
      + 18

11.   862
      + 714

12.   9,152
      + 3,601

13.   54,918
      + 16,875

Lesson 2-3

Subtract.

14.   67
      − 58

15.   876
      − 421

16.   953
      − 72

17.   7,658
      − 4,329

Lesson 2-4

18. On her last bank statement, Sarah's savings account had a total of $394.82. Since then, Sarah has made withdrawals of $48.00 and $27.50. What is the current balance of her savings account?

2-8 PERIMETER

Objective

Find the perimeter of a polygon.

Luke Adams is building a fence around his backyard to screen the swimming pool. He needs to know the perimeter of the yard to order fencing materials.

The **perimeter** (*P*) of any figure is the distance around the figure. It is found by adding the measures of the sides.

Example A

Find the perimeter of Luke's backyard shown at the right.

$$P = 100 + 50 + 80 + 20 + 20 + 30$$

$$P = 300$$

The perimeter is 300 feet.

Luke orders 300 feet of fencing.

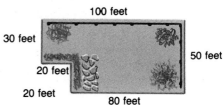

You can also find the perimeter of a rectangle by multiplying the length by 2 and the width by 2, and adding the products.

Perimeter = 2 × length + 2 × width or $P = 2\ell + 2w$

Example B

Find the perimeter of the rectangle at the right.

$$P = 2\ell + 2w \quad \ell = 43, w = 29$$

$$P = 2 \times 43 + 2 \times 29$$

$$P = 86 + 58$$

$$P = 144 \quad \text{You can check the answer by adding the lengths of the sides.}$$

The perimeter is 144 centimeters.

Marks on the sides of a figure represent sides of equal measure.

To find the perimeter of a square, multiply the length of one side by 4. Why?

Perimeter = 4 × length of one side or $P = 4s$

Example C

Find the perimeter of a square with a side that measures 1.3 meters.

$$P = 4s$$

$$P = 4 \times 1.3$$

$$P = 5.2 \text{ or } 5.2 \text{ meters.} \quad \text{Check by adding.}$$

Guided Practice

Examples A, B

Find the perimeter of each figure.

1.

2.

3.

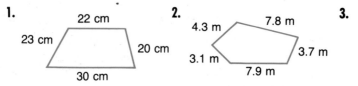

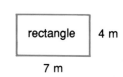

Example C

Find the perimeter of each square.

4. side, 35 mm   **5.** side, 12 cm   **6.** side, 22 ft   **7.** side, 14.1 cm

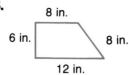
Exercises

Find the perimeter of each figure.

8.

8 in.

6 in. ⬜ 8 in.

12 in.

9.

⬜ 6 ft

square

10.

22 cm

24 cm / ⬜ \ 20 cm

28 cm

11. square; side, 14 ft

12. rectangle; ℓ, 12 m, w, 3 m

Applications

13. Pat is applying weatherstripping to a window on the north side of his house. If the window is 42 in. wide and 67 in. high, how many inches of weatherstripping does Pat need?

14. Miss Standish is outlining the bulletin board with ribbon using the school colors. The width is 8 feet and will be blue. The height is 4 feet and will be gold. How many feet of each color ribbon does she need?

Problem Solving

Paper/Pencil
Mental Math
Estimate
Calculator

15. Juan is planting flowers around 4 sides of his patio. He is planning for 3 plants in every foot. How many plants does Juan need to fill four sides?

8 ft 10 in.

10 ft 2 in.

16. Erica wants to fence a triangular area in the backyard for a garden. The three sides measure 45 ft, 57 ft, and 47 ft. How many 25-foot rolls of fence will she need?

17. A rectangular parking lot has a perimeter of 1,060 feet. Two of the sides measure 180 feet each. How long are the other two sides?

JOURNAL ENTRY

18. Draw a rectangle with a perimeter of 12 inches. Explain how you determined the length of each side.

Using Algebra

Write an equation and solve.

19. A triangular park has a perimeter of 936 m. If the length of the three sides are the same, what is the length of one side?

20. A rectangular stadium has a perimeter of 3,270 m. One side measures 545 m. What are the lengths of the other sides?

21. Explain how to find the perimeter of a six-sided figure if the lengths of each side are the same.

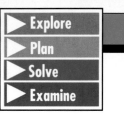

> ▶ Explore
> ▶ Plan
> ▶ Solve
> ▶ Examine

2-9 IDENTIFYING THE NECESSARY FACTS

Objective

Identify missing and extra facts.

At Rock Records Aaron sold 185 heavy metal tapes, 126 easy listening tapes, and 142 classical tapes on Thursday. How many heavy metal and easy listening tapes did he sell?

Always study a verbal problem carefully to determine what facts are given and what facts are needed. Be alert for extra facts you do not need to solve the problem.

▶ **Explore**

What is given?

● Aaron sold 185 heavy metal tapes, 126 easy listening tapes, and 142 classical tapes.

What is asked?

● How many heavy metal and easy listening tapes did Aaron sell?

The number of classical tapes sold is not needed.

▶ **Plan**

Add the sales for heavy metal and easy listening tapes to find the total sales.

▶ **Solve**

$$\begin{array}{r} 185 \\ +\ 126 \\ \hline 311 \end{array}$$ Estimate: 200 + 100 = 300

311 Aaron sold 311 tapes.

▶ **Examine**

The answer is close to the estimate, so 311 is a reasonable solution.

Sometimes facts that you need are missing from the problem.

Nick is saving money to buy a guitar. He has saved $120. He borrows $50. How much more money does Nick need?

▶ **Explore**

What is given?

● Nick has saved $120 and borrowed $50.
  Nick has a total of $170.

What is asked?

● How much more money does Nick need to buy the guitar?

▶ **Plan**

Find the difference between the cost of the guitar and the money Nick has saved.

▶ **Solve**

You do not know the cost of the guitar. The problem does not have all the facts you need to solve it.

If the problem has the necessary facts, solve. State any missing or extra facts.

1. Gasoline costs $1.28 a gallon and oil costs $1.69 a quart. What is the cost of two quarts of oil?

2. The Photography Club has bake sales to help pay for darkroom supplies. This week they made $34.80, and last week they made $29.40. How much more do they need to purchase darkroom supplies?

If the problem has the necessary facts, solve. State any missing or extra facts.

| Paper/ Pencil |
| Mental Math |
| Estimate |
| Calculator |

f **UN with MATH**
When and where were skis first used?
See page 104.

3. Frank wants to triple a recipe that calls for 2 cups of milk and 2 eggs. How many cups of flour should he use?

4. Claudia raises gerbils to sell to Petite Pets. If she sells 50 gerbils at $4 each, how much money will she earn?

5. It will cost the Ski Club $250 to hire a disc jockey, $125 to rent decorations, and $80 to buy refreshments for their dance. How much money does the club need to sponsor the dance?

6. The Choice Ticket agency sold 150 tickets for the better seats at $28 each. General admission tickets were sold out at $25 each. What was the income from the ticket sales?

Critical Thinking

7. You have 3 buckets. One holds 7 L and is filled with water. The other two buckets hold 5 L and 2 L and are empty. How can you pour water back and forth in order to end with 4 L in the largest bucket?

PORTFOLIO

8. Select an assignment from this chapter that shows your creativity and place it in your portfolio.

Mixed Review

Lesson 1-12

Solve.

9. $4m = 0.008$ 10. $500 = 2.5y$ 11. $40 = \frac{g}{5}$ 12. $0.9 = \frac{n}{2}$

Lesson 2-7

Solve. State the method of computation used.

13. Grant is going to order 3 magazine subscriptions for the next year. The magazines cost $38.95, $14.80, and $24.25. How much money does Grant need to pay for the magazines?

REVIEW

Vocabulary / Concepts

Choose the letter of the word at the right that best matches each description.

1. You can check a ___?___ problem by using addition.

2. When adding or subtracting decimal numbers, the ___?___ should be aligned.

3. When subtracting decimals, it is sometimes necessary to annex ___?___ .

4. A list of numbers that follows a certain pattern is called a ___?___ .

5. A check register is used to record ___?___ .

6. The perimeter of any figure is found by ___?___ the measures of the sides.

a. adding
b. decimal points
c. perimeter
d. sequence
e. subtraction
f. transactions
g. zeros
h. multiplying

Exercises / Applications

Lessons 2-1 2-2 2-3 2-4

Add or subtract.

7.	8.	9.	10.	11.
47 + 37	419 + 54	916 + 520	0.93 + 0.84	0.132 + 0.378

12.	13.	14.	15.	16.
0.132 − 0.017	60.7 − 42.3	375.3 − 190.4	5,343 − 602	5,047 − 2,085

17. $4.06 + 0.782$

18. $4.9 - 1.807$

19. $827 - 53$

20. $0.78 + 3.461$

21. $6.1 - 4.96$

22. $5,126 + 2,899$

23. $0.685 + 0.59 + 1.011$

24. $\$200 - \$133.62$

25. $38,876 + 3,603$

26. $9.84 + 0.27 + 3.6$

27. $43,040 - 5,503$

28. $58,316 - 29,957$

Lesson 2-5

Solve.

29. Edward has a balance of $86.09 in his checking account. If he deposits $356 and writes checks for $54.60 and $23.98, what is his new balance?

Copy and complete the balance column.

30. What is the balance on May 13?

31. What is the balance on May 16?

32. What is the balance on May 22?

CHECK NUMBER	DATE	DESCRIPTION OF TRANSACTION	PAYMENT/DEBIT (−)	√ TAX	DEPOSIT/CREDIT (+)	BALANCE FWD. $ 786.55
607	5/12	Supermart groceries	$ 57 31		$	
608	5/13	Gina's Boutique gift	29 70			
D	5/15	Deposit dividend			110 50	
609	5/16	Fraley's Hardware hardware	15 62			
D	5/20	Deposit paycheck			471 87	
610	5/22	Classic Rock CD's	26 63			

Lesson 2-6

State whether each sequence is arithmetic. If it is, write the next three numbers. If it is not, write no.

33. 11.3, 13.5, 15.7, ...

34. 110, 135, 160, ...

35. 85.4, 75.3, 65.2, ...

36. 54, 45, 36, ...

Lesson 2-7

| Paper/ Pencil |
| Mental Math |
| Estimate |
| Calculator |

Solve. State the method of computation used.

37. A sports car gets 32.5 miles per gallon of gasoline. *About* how far can it go on 12 gallons of gasoline?

38. Matt worked for 28 hours each week for the last two weeks and earned a total of $196 a week. How much was he paid per hour?

Lesson 2-8

Find the perimeter of each figure.

39.
12 in.
5 in.

40.
6 ft
10 ft
11 ft
7 ft
15 ft

41.
4 m
4 m   4 m
2 m   2 m
5 m   5 m
2 m

42. square; side, 4.6 m

43. rectangle; ℓ, 32 ft; w, 12 ft

Lesson 2-9

If the problem has the necessary information, solve. State any missing or extra facts.

44. In June, Troy could bench press 55.5 kg. By October, he could bench press an additional 5.75 kg. How much could Troy bench press in October?

45. Suzanne lost 4.6 pounds in February, 5.2 pounds in March, and 4.7 pounds in April. How much does she weigh now?

46. Benito's horse weighs 61 pounds at birth. The horse should weigh 1,100 pounds before he can carry Benito easily. How much weight does the horse need to gain?

47. The cheerleaders bought 350 pennants and 200 pompoms to sell. They received 282 orders for pennants and sold 35 pennants more during the football game. How many pennants do they have left to sell?

TEST

Add or subtract.

1.	7.6 − 3.9	**2.**	$21.71 + 3.42	**3.**	9.567 + 0.21	**4.**	180 − 19.7	**5.**	308 − 43.7

6.	7.74 − 3.48	**7.**	97.8 + 8.9	**8.**	85.3 + 71.52	**9.**	3,630 − 2,459	**10.**	356 + 2,522

11. 846 − 334   **12.** 256 + 16   **13.** 614 + 257   **14.** 630 − 225

15. 300 − 199   **16.** 359 + 2,330   **17.** 2,973 − 168   **18.** 680 + 3,945

19. 83.8 + 2.071   **20.** 78.4 − 9.47

State whether each sequence is arithmetic. If it is, write the next three numbers. If it is not, write no.

21. 3.1, 4.3, 5.5, …   **22.** 50, 42, 34, …

23. 3, 15, 75, …   **24.** 243, 81, 27, …

Find the perimeter of each figure.

25.

4 m

7 m

26.
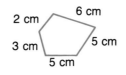
2 cm   6 cm   5 cm

3 cm   5 cm

27.
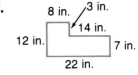
8 in.   3 in.

14 in.

12 in.   7 in.

22 in.

28. square; side, 6.8 cm   **29.** square; side, 8 ft

If the problem has the necessary facts, solve. State any missing or extra facts.

30. Randy deposited $25.67 to his checking account. He wrote checks for $12.35 and $36.11. His balance before these transactions was $632.75. What is his new balance?

31. If a city bus pass costs $2.10 a week, *about* how much will bus passes cost for one year? State the method of computation used.

32. Charles earns $6.50 an hour. How much does he earn in a week?

33. Ella finished the mile in 6.54 minutes. Rose finished in 6.32 minutes. Who won and by how many seconds?

▶ BONUS: How do you know that $P = 4 \times s$ will not always give the same answer as $P = 2\ell + 2w$, but $P = 2\ell + 2w$ will always give the same answer as $P = 4 \times s$?

BUDGETING

Marie Hiroto is a student at Central State College. She wants to buy a computer and has saved $900 for the down payment. The monthly payment on the computer is $68 for 24 months.

Marie makes $310 each month working in one of the dining halls on campus. She prepares and follows a **budget** so that she can afford the computer.

ITEM	MONTHLY AMOUNT
Food	$ 82
Computer	68
Recreation	40
Clothes	35
Savings	40
Miscellaneous	20
Transportation	25
TOTAL	$ 310

Find the average amount spent in one week for each item. Assume that there are 4 weeks in a month.

1. Food

2. Recreation

3. Clothing

4. Transportation

From which item of the budget should each expense be taken?

5. blouse

6. movie

7. birthday gift

8. dry cleaning

9. lunch

10. concert

11. If Marie takes $5 a month from her clothes allowance and $8 a month from her recreation allowance and adds both amounts to savings, how much will she save in 6 months?

12. At the end of the month, Marie finds she has $12 left over from her food allowance, $6 left over from her recreation allowance, and $5 left over from her miscellaneous allowance. Suppose Marie always has these extra amounts. How would you adjust her budget using these extra amounts?

13. Why is it a good practice to put money into savings on a regular basis?

14. Make up an income or allowance, and then plan a budget.

Free Response

Lesson 1-4

Round each number to the underlined place-value position.

1. 7<u>4</u>9   **2.** 3,<u>2</u>79   **3.** <u>6</u>33   **4.** 6<u>0</u>,280

5. 4,8<u>0</u>3   **6.** 90,<u>5</u>50   **7.** 3<u>9</u>3,488   **8.** <u>2</u>,683

Lesson 1-6

Estimate.

9. 697 + 77   **10.** 280 + 390   **11.** 655 + 4,848   **12.** 2,684 + 3,724   **13.** 96,278 + 3,843

Lesson 1-8

14. Mike, Steve, and Laura each have a baseball card collection. They have rookie cards of Willie Mays, Mickey Mantle, and Brooks Robinson. Mike does not have the Willie Mays card. Steve does not have the Mickey Mantle card and Laura does not have Mickey Mantle or Brooks Robinsons. Which person has which card?

Lesson 1-9

Find the value of each expression.

15. $(64 - 8) \div 4$ **16.** $144 \div 12 \times (4 - 1)$ **17.** $64 - 8 \div 4$

18. $36 \div (6 \times 3) - 2$ **19.** $9 \times (21 + 7)$ **20.** $9 \times 21 + 7$

Lesson 2-1

Add.

21. 12 + 67   **22.** 52 + 159   **23.** 342 + 4,181   **24.** 2,827 + 1,354   **25.** 6,184 + 56,424

26. $576 + 91$ **27.** $8,897 + 659$ **28.** $679 + 347$

29. During a half hour, 28 mini-vans, 97 sedans, and 36 sports cars passed Erika's house. How many passed in all?

Lesson 2-3

Subtract.

30. 53 − 19   **31.** 991 − 88   **32.** 1,471 − 927   **33.** 2,617 − 2,419   **34.** 10,451 − 8,367

35. $223 - 92$ **36.** $928 - 233$ **37.** $2,415 - 754$

38. Alice earned $45 last week. She bought new shoes that cost $27. How much money does she have left?

Lesson 2-6

39. Jude plans to collect CDs. Each month he plans to purchase 3 CDs. He begins his collection with 4 CDs. How many CDs will he have in 7 months?

Multiple Choice

Choose the letter of the correct answer for each item.

1. The first week he began his savings account, Caleb saved $7. The second week his total amount was $21. The third week he had a total of $35. How much will he have saved in 7 weeks?
- **a.** $77
- **b.** $105
- **c.** $91
- **d.** $133

Lesson 2-6

2. Find the value of $14 + 6 \times 3$.
- **e.** 60
- **f.** 32
- **g.** 18
- **h.** 46

Lesson 1-9

3. Estimate the product of 68 and 5,428.
- **a.** 35,000
- **b.** 350,000
- **c.** 3,500,000
- **d.** 35,000,000

Lesson 1-7

4. Margaret adds 150 to 874. What is her result?
- **e.** 724
- **f.** 824
- **g.** 924
- **h.** 1,024

Lesson 2-1

5. Which equation would you use to solve this problem?
Tim used $15 to purchase gas and oil for his car. If the oil cost $3.52, how much did he pay for gas?
- **a.** $15 + 3.52 = y$
- **b.** $3.52 + y = 15$
- **c.** $x - 3.52 = 15$
- **d.** none of these

Lesson 1-11

6. If Arnold spends $3,992 of $8,825 that he has in his savings account, how much does he have left?
- **e.** $4,833
- **f.** $4,933
- **g.** $5,173
- **h.** $5,833

Lesson 2-3

7. What is the difference of 37,153 and 1,279?
- **a.** 24,363
- **b.** 35,874
- **c.** 35,884
- **d.** 35,974

Lesson 2-3

8. What number, when multiplied by 3, equals 42?
- **e.** 14
- **f.** 39
- **g.** 45
- **h.** 126

Lesson 1-12

9. What is the sum of 1,436 and 93,806?
- **a.** 94,242
- **b.** 95,232
- **c.** 95,242
- **d.** 108,166

Lesson 2-1

10. Jonathan wants to purchase a computer for $899, a computer desk for $90, a printer for $350, and a box of computer paper for $32.50. Can Jonathon afford the computer system? What fact is missing?
- **e.** the amount of set-up time
- **f.** the cost of the computer with printer
- **g.** the amount Jonathan has to spend
- **h.** the size of the computer

Lesson 2-9

3

PATTERNS: MULTIPLYING AND DIVIDING

▶ Pick up the phone!

Did you know the cost of a long-distance telephone call depends on the time of day, the location of the caller, the city called, and the duration of the call? During business hours, rates are usually higher. The time of day for the *caller* is the time of day on which the rates are based.

During business hours, the cost of a call from Indianapolis to Columbus is $0.21 for the first minute and $0.22 for each additional minute. From 5:00 P.M. to 11:00 P.M. the cost of a call is $0.14 for the first minute and $0.15 for each additional minute. How much less is a 10-minute call from Indianapolis to Columbus after 5:00 P.M. than during business hours?

ACTIVITY: Multiplication with Decimal Models

In this chapter you will multiply whole numbers and decimals.

As with addition, you can use decimal models to multiply. Suppose you want to find the product of 0.3 and 0.5.

Materials: decimal models, pencil

1. Use a decimal model that is divided into hundredths.

2. Shade 3 tenths.   **3.** Shade 5 tenths.   This model represents the product.

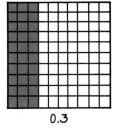 × 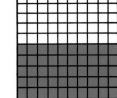 =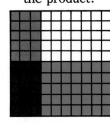

0.3   0.5

Cooperative Groups

Work together in groups of three or four.

4. Explain how the shaded models show the product of 0.3 and 0.5. What is the product? How did you find the product?

5. Discuss with your group how to show 0.7 × 0.4 with decimal models. Then find the product of 0.7 and 0.4 using models.

6. Discuss how to show the product of 3 and 0.2 using decimal models. Which model should you use? What is the product?

7. In Exercises 4, 5, and 6, how many decimal places were there in each factor? How many decimal places were there in each product?

Communicate Your Ideas

8. With your group, draw a conclusion about how the number of decimal places in a product is related to the number of decimal places in the factors.

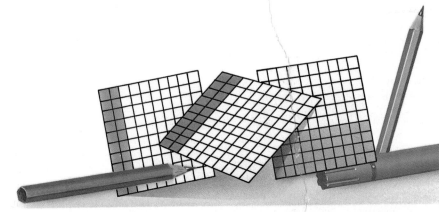

3-1 MULTIPLYING WHOLE NUMBERS

Objective

Multiply whole numbers.

Aaron likes to do three aerobic workouts each week. During one workout, he raises his heartbeat to 98 beats per minute. If he maintains this rate for 20 minutes, how many times will his heart beat? To find out, multiply 98 by 20.

Method

1 ▶ Align the numbers on the right.

2 ▶ Multiply by the number in each place-value position from right to left.

3 ▶ Add the products obtained in Step 2.

Example A

20 × 98   Estimate: 20 × 100 = 2,000

```
1 ▶    98       2 ▶    98                        1
     × 20            × 20                       98
                       0     0 × 98 = 0       × 20
                                              1,960
```

When multiplying by multiples of 10, there is no need to write another row.

Compared to the estimate, is the answer reasonable?

Aaron's heart will beat 1,960 times in twenty minutes.

Example B

27 × 53   Estimate: 30 × 50 = 1,500

```
1 ▶    53       2 ▶    53               3 ▶    53
     × 27            × 27                    × 27
                     371    7 × 53          371
                     106    2 × 53          1 06
                                           1,431    Is the answer reasonable?
```

Example C

65 × 238   Estimate: 70 × 200 = 14,000

```
1 ▶   238       2 ▶   238               3 ▶   238
    ×  65           ×  65                   ×  65
                    1190    5 × 238         1 190
                    1428    6 × 238        14 28
                                          15,470
```

Is the answer reasonable?

Guided Practice

Multiply.

Examples A, B

1. 63 × 20	**2.** 31 × 30	**3.** 75 × 19	**4.** 73 × 32	**5.** 26 × 58

Example C

6. 142 × 55	**7.** 427 × 14	**8.** 580 × 58	**9.** 232 × 10	**10.** 742 × 30

Practice

Multiply.

11.	76	**12.**	88	**13.**	12
	$\times\ 43$		$\times\ 35$		$\times\ 80$

14. 96
$\times\ 60$

15. 82
$\times\ 30$

16. 948
$\times\ 90$

17. 741
$\times\ 42$

18. 290
$\times\ 83$

19. 214
$\times\ 31$

20. 803
$\times\ 70$

21. 3,803
$\times\quad 52$

22. 2,341
$\times\quad 77$

23. 4,861
$\times\quad 30$

24. 3,219
$\times\quad 61$

25. $4{,}728 \times 83$　　　**26.** $70 \times 14{,}966$　　　**27.** $48 \times 26{,}091$

28. Find the product of 38 and 75.

Applications

29. Chuck is reading two books this month. One has 16 chapters. The other has 12 chapters. If the book with 16 chapters has 28 pages in each chapter, how many pages does the 16-chapter book have?

30. Pierre is spraying the yard for insects. He mixes 12 tablespoons of insect repellant with one gallon of water. He needs to make 15 gallons of spray. How many tablespoons of repellant will he need?

Using Equations

Solve each equation.

31. $4 \times y = 60$　　　**32.** $20 \times 12 = q$　　　**33.** $25c = 175$

34. $z \times 6 = 48$　　　**35.** $11 \times 12 = n$　　　**36.** $9 \times 8 = y$

Cooperative Groups

f **UN with MATH**

When was Rubik's cube invented?
See page 105.

37. Use four number cubes, with two cubes marked with the digits 0 through 5 and two cubes marked with the digits 4 through 9. Form groups of three or four. Each player takes a turn and rolls all four cubes. Arrange the digits rolled to form 2 two-digit numbers or a one-digit and a three-digit number that have the greatest product possible. The player with the greatest product is the winner.

Critical Thinking

38. Use the numbers 7, 1, 5, and 8 only once to write a whole number multiplication problem with a product as close as possible to 1,000.

Mental Math

Multiply $15 by 8.　　Think: $15 = $10 + $5

$8 \times \$15 \begin{cases} 8 \times \$10 = \$80 \\ 8 \times \$5\ \ = \$40 \end{cases} \$120$

Multiply. Write only the answers.

39. 7×18　　　**40.** 13×9　　　**41.** 12×15　　　**42.** 78×4

3-2 MULTIPLYING DECIMALS

Objective

Multiply using decimals.

Allison makes $4.25 an hour working at the snack bar in Fenway Park. She worked 6 hours on Tuesday and from 9:00 A.M. to 12:30 P.M. on Wednesday. To find out how much she earned on Tuesday, multiply $4.25 by 6.

When multiplying decimals, multiply as with whole numbers. The number of decimal places in the product is the same as the sum of the number of decimal places in the factors.

Method

1 ▶ Multiply as with whole numbers.

2 ▶ Find the sum of the number of decimal places in the factors.

3 ▶ To place the decimal point in the product, start at the right and count the number of decimal places needed. If more decimal places are needed, annex zeros on the left. Then insert the decimal point.

Example A

6 × $4.25   Estimate: 6 × 4 = 24

1 ▶ Align the numbers on the right and multiply.

$$
\begin{array}{r}
4.25 \\
\times \quad 6 \\
\hline
25.50
\end{array}
$$

2 ▶ 2 decimal places

3 ▶ Starting at the right, count 2 decimal places.

Check by adding.

$$
\begin{array}{r}
4.25 \\
4.25 \\
4.25 \\
4.25 \\
4.25 \\
+ \quad 4.25 \\
\hline
25.50
\end{array}
$$

Allison earned $25.50 on Tuesday.   Based on the estimate, explain why the answer is reasonable.

Example B

0.02 × 1.35   Estimate: 0 × 1 = 0

1 ▶
$$
\begin{array}{r}
1.35 \\
\times \; 0.02 \\
\hline
.0270
\end{array}
$$

2 ▶ 4 decimal places

3 ▶ To make 4 decimal places in the product, annex a zero on the left.

0.02 × 1.35 = 0.0270   How does this compare with the estimate?

Guided Practice — **Multiply.**

Example A

1.
$$
\begin{array}{r}
9.2 \\
\times \; 6 \\
\hline
\end{array}
$$

2.
$$
\begin{array}{r}
8.4 \\
\times \; 7 \\
\hline
\end{array}
$$

3.
$$
\begin{array}{r}
3.2 \\
\times \; 19 \\
\hline
\end{array}
$$

4.
$$
\begin{array}{r}
0.95 \\
\times \; 5 \\
\hline
\end{array}
$$

5.
$$
\begin{array}{r}
2.31 \\
\times \; 12 \\
\hline
\end{array}
$$

6. 0.3 × 0.2	**7.** 0.47 × 0.2	**8.** 0.03 × 1.5	**9.** 0.002 × 0.7	**10.** 0.012 × 0.1

Exercises

Practice

Multiply. **Encourage students to estimate the product first.**

11. 0.5 × 7	**12.** 0.4 × 0.9	**13.** 0.05 × 8	**14.** 4.5 × 0.06	**15.** 0.023 × 51

16. 2.6 × 4.7	**17.** 0.07 × 0.3	**18.** 0.881 × 0.5	**19.** 0.83 × 36	**20.** $8.45 × 27

21. 8.09 × 0.5	**22.** 332.5 × 7.3	**23.** 4.53 × 0.01	**24.** 239.75 × 0.1	**25.** 6.82 × 0.05

26. 147.5×0.3 **27.** 3.14×0.15 **28.** 64.2×0.021

29. Find the product of 0.3 and 52.   **30.** Multiply 0.368 and 0.6.

Application

31. Boy Scout Troop #672 is going to the baseball game. Each boy is allowed to buy a hot dog for $2.25, a drink for $1.25, and peanuts for $0.75. If 23 boys are going to the game, how much money is needed to buy all three items for the troop?

Use the chart for Exercises 32-33.

32. Suppose the owners of the White Sox decided to increase the seating capacity in Comisky Park by the number of seats in the Kingdome. If the expansion costs $257 per seat, how much will the total expansion cost?

Tickets, please!

About 50,000 fans can be seated in each Major League baseball stadium.

Seating Capacity

Kingdome 59,702 (Seattle Mariners)

Comisky Park 44,702 (Chicago White Sox)

f **UN with MATH**
 How do amoebas multiply?
 See page 104.

33. How many more seats are in the Kingdome than in Comisky Park? Suppose the owners of the Mariners claim that they have more than $1\frac{1}{3}$ the seating capacity of Comisky Park. Is this a correct statement? Explain your answer.

34. Use decimal models to show the product for Exercise 12.

Estimation

Estimate each product and decide if the decimal point is in the correct place. If not, write the correct product.

35. $2 \times 0.7 =$ ___14___   **36.** $2 \times 4.1 =$ ___8.2___

37. $\$1.23 \times 8 =$ ___$9.84___   **38.** $1.2 \times 71 =$ ___852___

3-3 BUYING ON CREDIT

Objective

Compute the finance charge for credit purchases.

Doug Halpern is buying the video camera described in the ad at the right on credit. After a down payment of $200, he will pay $50.50 a month for 24 months. What is the finance charge Doug will pay?

Many people buy expensive items on credit. They pay for the item over a period of time. Some purchases require a small portion of the credit price, called a **down payment**, to be paid at the time of purchase. The buyer makes a monthly payment for a given amount of time. The credit price of the item will be greater than the cash price. The difference between the credit price and the cash price is the **finance charge.**

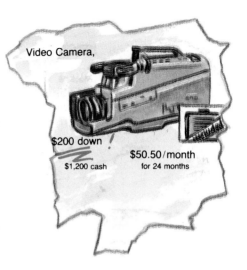

Video Camera,

$200 down
$1,200 cash

$50.50/month
for 24 months

Find the finance charge Doug will pay as follows. Multiply the amount of the monthly payment by the number of months.

$$50.50 \times 24 = 1212$$

Add the down payment, if there is one, to find the credit price.

$$1212 + 200 = 1412$$

To find the finance charge, subtract the cash price from the credit price.

$$1412 - 1200 = 212$$

Understanding the Data
- What is the cash price?
- What is the amount of the down payment?
- What is the amount of the monthly payment?
- What is the finance charge?

Guided Practice

Copy and complete the table.

	Cash Price	Down Payment	Monthly Payment	Number of Months	Credit Price	Finance Charge
1.	$384	$10	$36	12		
2.	$1,900	$250	$56.99	36		
3.	$5,996	$1,496	$156	36		
4.	$3,050	$600	$130.75	24		
5.	$1,200	$100	$83	18		

6. What is the credit price of this car?

7. How much can you save by paying cash for the camera?

8. How much is the finance charge if you purchase the stereo system on credit?

9. Jane Peters bought living room furniture for $400 down and $100 a month for 2 years. After making 6 payments, she wants to pay off the remaining cost so that she won't have monthly payments. What is the remaining cost of the furniture?

Car $10,790 cash

or $295 per month for 3 years
$1,000 down payment

Camera

$348 cash or $62 per month for 6 months
no down payment

Stereo System

$398 cash
or $35 per month for 12 months
$50 down payment

10. Refer to Exercise 9. If the store gives Jane $100 off the total price for paying off the remaining cost early, what is the total price Jane will pay for the furniture?

Collect Data

11. From newspaper ads, find the credit price of a car, a major appliance, or a stereo system. Find the finance charge if the cash price is not paid.

Make Up a Problem

12. Make up a problem in which the finance charge is $240 and there is no down payment.

Number Sense

13. The price of a car, in dollars, rounds to 9,460 when rounded to the nearest ten. What is the price of the car if 5 is in the ones place?

Mixed Review

Lesson 1-9

Find the value of each expression.

14. $14 \div 2 + 6 \times 3$

15. $16 \times 2 \div 4 + 20$

16. $18 \div 9 \times [3 \times (20 - 11)]$

Lesson 2-8

Find the perimeter.

17.

4 cm 5 cm
3 cm

18.

12 in.
4 in.
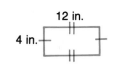

19.

6 mm

Lesson 3-1

20. Josiah began jogging and plans to jog 15 miles a week. How many miles will he jog in one year?

3-4 DIVIDING WHOLE NUMBERS

Objective

Divide whole numbers.

Marcella told her parents that she had 1,108 days left until she graduates from high school. How many weeks does she have left until graduation? To solve this problem, you must divide 1,108 by 7.

Method

1▶ Divide the greatest place-value position possible.

2▶ Multiply and subtract.

3▶ Repeat 1▶ and 2▶ as necessary.

Example A

$$1,108 \div 7 \rightarrow 7\overline{)1,108} \quad \text{Estimate: } \begin{array}{r} 200 \\ 7\overline{)1,400} \end{array}$$

$$1\blacktriangleright \quad 7\overline{)1,108}$$

$$2\blacktriangleright \quad \begin{array}{r} 1 \\ 7\overline{)1,108} \\ -7 \\ \hline 4 \end{array}$$

$$3\blacktriangleright \quad \begin{array}{r} 158 \text{ R2} \uparrow \\ 7\overline{)1,108} \quad \text{remainder} \\ -7 \\ \hline 40 \\ -35 \\ \hline 58 \\ -56 \\ \hline 2 \leftarrow \end{array}$$

Marcella has about 158 weeks left until graduation.

Compared to the estimate, is the answer reasonable?

Multiplication and division are inverse operations, so you can use multiplication to check division.

The answer is correct.

Check by multiplying. Add the remainder.

$$\begin{array}{r} 158 \text{ quotient} \\ \times \quad 7 \text{ divisor} \\ \hline 1,106 \\ + \quad 2 \text{ remainder} \\ \hline 1,108 \checkmark \text{ dividend} \end{array}$$

A *quotient* is the result of dividing one number, the *dividend,* by another, the *divisor.*

Use estimation to help determine the digits in the quotient.

Method

1▶ Round the divisor to the greatest place-value position.

2▶ Divide in each place-value position from greatest to least. Estimate each digit of the quotient using the rounded divisor. Revise the estimate if necessary. Then divide.

Example B

$$3,405 \div 36$$

1▶ $36 \rightarrow 40$

2▶ $$\begin{array}{r} 8 \\ 36\overline{)3,405} \\ -2\,88 \\ \hline 52 \end{array}$$

Note that 52 is greater than 36. So increase the estimate.

$$\begin{array}{r} 80 \\ 40\overline{)3,200} \end{array}$$

$$\begin{array}{r} 94 \text{ R21} \\ 36\overline{)3,405} \\ -3\,24 \\ \hline 165 \\ -144 \\ \hline 21 \end{array}$$

Is the answer reasonable?

$$\begin{array}{r} 4 \\ 40\overline{)160} \end{array}$$

Check:
$$\begin{array}{r} 36 \\ \times \quad 94 \\ \hline 144 \\ 3\,24 \\ \hline 3,384 \\ + \quad 21 \\ \hline 3,405 \end{array}$$

Divide. Check using multiplication.

1. $7\overline{)189}$ **2.** $8\overline{)768}$ **3.** $6\overline{)980}$ **4.** $2\overline{)457}$ **5.** $5\overline{)4,625}$

Use estimation to write the number of digits in each quotient. Then divide.

6. $29\overline{)89}$ **7.** $53\overline{)162}$ **8.** $47\overline{)310}$ **9.** $58\overline{)725}$ **10.** $23\overline{)165}$

Write the number of digits in the quotient.

11. $2,450 \div 540$ **12.** $645 \div 27$ **13.** $248 \div 19$ **14.** $820 \div 4$

Divide. Check using multiplication.

15. $5\overline{)270}$ **16.** $7\overline{)5,040}$ **17.** $8\overline{)512}$ **18.** $6\overline{)403}$

19. $39\overline{)176}$ **20.** $13\overline{)117}$ **21.** $19\overline{)115}$ **22.** $33\overline{)250}$

23. $810\overline{)3,258}$ **24.** $709\overline{)5,000}$ **25.** $450\overline{)3,750}$

26. $1,300 \div 28$ **27.** $2,700 \div 68$ **28.** $1,225 \div 13$

29. What is the quotient if 733 is divided by 62?

30. 137 adults and 79 players are guests at the sports award banquet. You decide to seat the guests in groups of 8 at each table. How many tables will be needed?

31. Jolanda wants to save $2,500 to buy a used car. She earns $96 a week from her paper route and babysitting. If she saves half of her earnings each week, how many weeks will it take her to save the money?

Complete. Use the graph at the right.

32. In the survey of 1,000 people, how many said they do not recycle their trash because it took too much time? (30% = 0.30)

33. Find the number of people in each category. The total number of people should be 1,000. (19% = 0.19; 12 % = 0.12; 8% = 0.08; 23% = 0.23)

34. Conduct a survey in your class, school, or neighborhood to find the reasons people do not recycle. Compare your results to this survey.

3-5 MULTIPLYING AND DIVIDING BY POWERS OF 10

Objective

Multiply and divide by powers of 10.

The automated packing system at Machine Builders Supply packs number 8 common nails in boxes weighing 3.7 pounds. If the machine packs 10 boxes in five minutes, how many pounds of nails does it pack in 10 boxes? How many pounds of nails does it pack in 100 boxes? in 1,000 boxes?

To multiply a number by 10, 100, or 1,000, move the decimal point to the right.

Method

▶ **1** Move the decimal point to the right one place for 10, two places for 100, or three places for 1,000.

▶ **2** Annex zeros on the right, if necessary, to move the correct number of decimal places.

Examples

A *Multiply.*

$3.7 \times 10^1 = 37.$
$3.7 \times 10^2 = 370.$
$3.7 \times 10^3 = 3,700.$

B *Multiply.*

$0.195 \times 10 = 1.95$
$0.195 \times 100 = 19.5$
$0.195 \times 1,000 = 195.$

The automated system packs 37 pounds of nails in 10 boxes, 370 pounds in 100 boxes, and 3,700 pounds in 1,000 boxes.

To divide a number by 10, 100, or 1,000, move the decimal point to the left.

Method

▶ **1** Move the decimal point to the left one place for 10, two places for 100, or three places for 1,000.

▶ **2** Insert zeros on left, if necessary, to move the correct number of decimal places.

Examples

C *Divide.*

$5,234.8 \div 10 = 523.48$
$5,234.8 \div 100 = 52.348$
$5,234.8 \div 1,000 = 5.2348$

D *Divide.*

$49 \div 10^1 = 4.9$
$49 \div 10^2 = 0.49$
$49 \div 10^3 = 0.049$

Guided Practice

Multiply.

Example A

1. 6.54×10^1 **2.** 8.204×10^3 **3.** 45×10^2 **4.** 340×10^3

Example B

5. 4.9×10 **6.** 5.12×100 **7.** $62.9 \times 1,000$ **8.** 3.7×100

Example C	Divide.			
	9. $4.48 \div 1,000$	**10.** $94.1 \div 10$	**11.** $14.5 \div 100$	**12.** $341.2 \div 100$
Example D	**13.** $63.7 \div 10^2$	**14.** $23.9 \div 10^3$	**15.** $39 \div 10^1$	**16.** $48 \div 10^3$

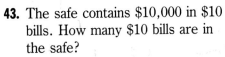

Exercises

Practice

Multiply.

17. 17.5×100 **18.** 4.59×10 **19.** 3.72×100 **20.** 9.1×10^1

21. $0.178 \times 1,000$ **22.** 5.15×10^3 **23.** $14.7 \times 1,000$ **24.** 50.38×10

25. 23.04×10^1 **26.** 72.9×10^2 **27.** 28.4×10^3 **28.** 0.58×10

Divide.

29. $0.93 \div 10$ **30.** $36.2 \div 10^2$ **31.** $5.29 \div 10$ **32.** $46.1 \div 100$

33. $281.7 \div 1,000$ **34.** $0.3 \div 10$ **35.** $11.8 \div 10^3$ **36.** $84.7 \div 100$

37. $6.05 \div 10^3$ **38.** $9.4 \div 10^2$ **39.** $50.7 \div 10^1$ **40.** $0.03 \div 10$

Write Math

41. Explain why the decimal point moves right when multiplying by powers of ten and why the decimal point moves left when dividing by powers of ten.

Applications

42. A safe contains \$1,000 in dimes. How many dimes are in the safe?

f **UN with MATH**

When and where was the decimal system first used? See page 104.

43. The safe contains \$10,000 in \$10 bills. How many \$10 bills are in the safe?

44. The diameter of Earth at its equator is 7,926 miles. The diameter of Jupiter at its equator is 88,700 miles. Jupiter's diameter is *about* how many times as large as Earth's?

Using Equations

Solve each equation.

45. $10x = 3.45$ **46.** $100y = 980.5$

47. $6.84q = 0.00684$ **48.** $42.23s = 4,223$

Mixed Review

Lesson 2-4

Subtract.

49. $\begin{array}{r} 4.83 \\ -\ 0.12 \end{array}$ **50.** $\begin{array}{r} 68.19 \\ -\ 53.02 \end{array}$ **51.** $\begin{array}{r} 98.008 \\ -17.012 \end{array}$ **52.** $\begin{array}{r} 16.019 \\ -\ 9.887 \end{array}$

Lesson 3-4

53. At the local library the librarian had 7,700 books in the children's department to arrange on 220 shelves. How many books did he put on each shelf if he divided them equally among the shelves?

3-6 LOOKING FOR A PATTERN

> ▶ Explore
> ▶ Plan
> ▶ Solve
> ▶ Examine

Objective

Solve problems by making a list and using the pattern.

Rita saves $20 during the first month at her new job. She plans to increase the amount she saves by an additional $4 a month for 10 months. What is the amount Rita will save in the tenth month?

You can make a list to find the pattern.

▶ **Explore**

What is given?
● Rita saves $20 the first month.
● Rita increases the amount an additional $4 each month.

What is asked?
● What is the amount Rita saves in the tenth month?

▶ **Plan**

Make a list and look for the pattern. Then apply it to month 10.

▶ **Solve**

Month	Amount Saved	Amount of Increase in Savings
1	$20	$ 0
2	$24	$ 4
3	$28	$ 8
4	$32	$12
5	$36	$16

Notice that the amount of increase from the first month is one *less* than the number of the month multiplied by 4. For each month add the increase to the first $20 to find the amount saved.

Applying the pattern to month 10, the expression $(10 - 1) \times 4 + 20$ is the amount saved.

$$(10 - 1) \times 4 + 20 = 9 \times 4 + 20$$
$$= 36 + 20 \text{ or } 56$$

▶ **Examine**

Rita saves $56 during month 10. You can check the solution by extending the list to month 10. The solution is correct.

Guided Practice

Solve. Make a list.

1. Jack sets patio blocks in a triangular area of his garden as shown. If he sets 10 rows, how many patio blocks will he use?

2. Find the sum of the first ten whole numbers that are multiples of three. Remember that zero is a whole number.

Row4 Row3 Row2 Row1

Solve. Use any strategy.

3. Cans of soup are stacked in a triangular arrangement in a supermarket display. There is one can in the top row, three in the next, and five in the third. How many cans are in a display that has eight rows?

4. Determine if the sequence 56, 48, 40 is arithmetic. If it is, find the next three numbers.

5. Sam plans to put one-half of his paycheck into his savings account and one-fourth of his paycheck into his checking account. Sam plans to spend one-eighth of his paycheck on cassette tapes and one-eighth of his paycheck on skateboard accessories. How much will Sam spend on cassette tapes if his paycheck is $372?

6. Glen, Doug, and Mike date three girls with eyes that are blue, green, and brown. Suppose Glen dates a girl with eyes the color of grass and Mike's date does not have brown eyes. Figure out the eye color of each boy's date.

On pages 519–521, you can learn how agribusiness specialists use mathematics in their jobs.

7. Andrea drove 4 hours at an average speed of 52 miles per hour. Then she drove 3 hours at an average speed of 43 miles per hour. How far did Andrea drive altogether?

8. Harry's Sporting Goods advertises socks at 2 pair for $5, running shorts for $9, and T-shirts for $14. If Berni needs 4 pairs of socks and 2 T-shirts, *about* how much money should she take to the store?

9. Gina says she ran a mile 0.5 seconds faster than Suzi and 0.7 seconds faster than Nicole. If Suzi ran the mile in 7 minutes 49.5 seconds, how fast did Nicole run the mile?

Critical Thinking

10. Find the 1-digit numbers # and * represent that will make the statement true. #*.# × #* = #*#.*

Mixed Review

Lesson 1-4

Round each number to the nearest hundredth.

11. 0.841   **12.** 11.976   **13.** 0.4901   **14.** 1.0097

Lesson 2-2

Add.

15. 68.45   **16.** 9.881   **17.** 0.381   **18.** 146.2
      + 1.79   + 6.728   + 4.2   + 91.093

Lesson 2-7

19. Darren chooses a number between 1 and 10. Starting with his number, he subtracts 3, adds 8, subtracts 12, and then adds 11. The result is 14. Which number did Darren choose?

3-7 DIVIDING DECIMALS

Objective
Divide decimals.

Jeff and three friends bought Darryl a video game for his birthday. How much will the gift cost each person if they share the total price of $46.28 equally? To solve this problem, divide $46.28 by 4.

Method

1▶ Write the decimal point in the quotient directly above the decimal point in the dividend.

2▶ Divide as with whole numbers.

Example A

$$4\overline{)46.28}$$ Estimate: $4\overline{)40}$ (10)

$$4\overline{)46.28}$$ 1▶ Write the decimal point in the quotient.

$$\begin{array}{r} 11.57 \\ 4\overline{)46.28} \\ -4 \\ \hline 6 \\ -4 \\ \hline 2\,2 \\ -2\,0 \\ \hline 28 \\ -\;28 \\ \hline 0 \end{array}$$ 2▶ Divide as with whole numbers.

The gift will cost each person $11.57.

Based on the estimate, explain why the answer is reasonable.

When the divisor is a decimal, multiply the divisor by 10, 100, 1,000, and so on, until the divisor is a whole number. Multiply the dividend by the same number.

Method

1▶ Move the decimal point in the divisor to the right until a whole number is obtained.

2▶ Move the decimal point in the dividend the same number of places. Then write the decimal point in the quotient directly above the adjusted decimal point.

3▶ Divide.

Example B

$$9.8 \div 0.002$$

$$0.002.\overline{)9.800.}$$
1▶ 2▶

Annex two zeros in the dividend so the decimal point can be moved.

3▶ $$\begin{array}{r} 4,900 \\ 2\overline{)9,800} \\ -8 \\ \hline 1\,8 \\ -1\,8 \\ \hline 0 \end{array}$$

Guided Practice — *Divide.*

Example A   **1.** $8\overline{)40.8}$ **2.** $2\overline{)8.6}$ **3.** $3\overline{)56.7}$ **4.** $13\overline{)39.39}$ **5.** $45\overline{)0.585}$

6. $0.4\overline{)7.6}$ **7.** $0.3\overline{)0.84}$ **8.** $4.8\overline{)1.2}$ **9.** $1.8\overline{)43.74}$ **10.** $0.1\overline{)6}$

Exercises

Divide.

Practice

11. $7\overline{)7.84}$ **12.** $4\overline{)9.2}$ **13.** $8\overline{)2.08}$ **14.** $8\overline{)0.024}$

15. $0.25\overline{)5}$ **16.** $0.4\overline{)5}$ **17.** $0.09\overline{)61.2}$ **18.** $1.6\overline{)44.8}$

19. $1.2\overline{)3.12}$ **20.** $1.25\overline{)200}$ **21.** $0.3\overline{)6.12}$ **22.** $0.5\overline{)32.15}$

23. $52 \div 1.3$ **24.** $3.755 \div 5$ **25.** $9.6 \div 12$ **26.** $16 \div 0.64$

27. $15 \div 0.5$ **28.** $0.312 \div 6$ **29.** $220.1 \div 3.1$ **30.** $\$0.72 \div 12$

Number Sense

Place the missing decimal point in the answer to each problem.

31. $5.6 \div 8 = 7$ **32.** $5.01 \times 6 = 3006$ **33.** $30.24 \div 27 = 112$

34. $73.2 \times 6 = 4392$ **35.** $0.4 \div 0.2 = 2$ **36.** $0.980 \div 4 = 245$

37. $2.9 \times 1.8 = 522$ **38.** $16 \div 0.64 = 25$ **39.** $6 \div 15 = 4$

Applications

40. A 3-ounce can of chili powder costs $1.29. How much does one ounce cost? At that price, how much do twenty ounces cost?

41. Shelley buys a piece of ribbon 21.6 yards long. She cuts it into 15 pieces. How long is each piece to the nearest tenth of a yard?

Research

42. Find the names of the five longest rivers in the world and list them in order from longest to shortest.

Mental Math

To add 0.9, you can add 1 and then subtract 0.1 because $0.9 = 1 - 0.1$.
$$2.57 + 0.9 = 2.57 + 1 - 0.1 = 3.57 - 0.1 = 3.47$$

Using a shortcut like the one shown above, find each sum mentally.

43. $1.76 + 0.8$ **44.** $7.2 + 1.9$ **45.** $0.85 + 0.9$

46. $4.2 + 1.9$ **47.** $3.4 + 0.9$ **48.** $14.5 + 0.8$

APARTMENT MANAGER

Carlen Jackson manages the Whispering Wood Apartments. He has an agreement with the landlord to live in an apartment, rent free, in return for his duties as a manager.

Mr. Jackson's duties include showing and renting apartments, performing routine maintenance, and collecting the rent each month.

During November, Mr. Jackson has 8 apartments rented for $540 each and 12 other apartments rented for $595 each. How much rent should Mr. Jackson collect for November?

Rent ×	Number of Rented Apartments =	Amount of Rent He Should Collect
$540	8	$ 4,320
$595	12	+ $ 7,140
	total rent	$11,460

Mr. Jackson should collect $11,460 for the month.

1. Ms. Evans manages a town house complex. In one month, she has six town houses rented for $600 each and nine rented for $655 each. How much rent should she collect for the month?

2. For a one-bedroom apartment, the rent is $475, the security deposit is the same as one month's rent, and the utility deposit is $150. What is the total deposit required? The total deposit and the first month's rent are collected at the same time. How much should the manager collect the first month?

3. Research the reasons that landlords require a security deposit and utility deposit. Find out if a security deposit is always equal to one month's rent. Find out if these are the only deposits that landlords require.

4. Managing Whispering Wood apartments is Mr. Jackson's parttime job. If the apartment Mr. Jackson lives in would normally rent for $595 a month, how much is he receiving annually as income in the form of free rent?

CASH DEPOSITS

People in many occupations or jobs, such as store managers, cashiers, or newspaper carriers, may handle large numbers of bills and coins.

Armen delivers newspapers in the morning to earn extra money. He collects from his customers on Thursday and pays his route manager on Friday. This week he collected the following bills and coins.

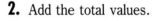

11–$5 bills
64–$1 bills
75–quarters
42–dimes
37–nickels

Find the total amount of money Armen collected.

1. Multiply the number of each type of bill or coin by its value.

11 × $5	=	$55.00
64 × $1	=	$64.00
75 × $0.25	=	$18.75
42 × $0.10	=	$ 4.20
37 × $0.05	=	$ 1.85

Armen collected $143.80.

2. Add the total values.

$55.00
64.00
18.75
4.20
+ 1.85
————
$143.80

---

1. The owner of Qwick Vending Machine Company deposits 1,643 quarters, 319 dimes and 740 nickels. What is the amount of money deposited?

2. A food store manager deposited 416 $20 bills, 329 $10 bills, 60 $5 bills, and 6,784 $1 bills. She also deposited 194 quarters, 137 dimes, 116 nickels, and 208 pennies. What was the amount of the deposit?

3. At noon, the manager counted $2,640.71. She deposited everything but 4 $10 bills, 20 $5 bills, 100 $1 bills, and 200 each of quarters, dimes, nickels, and pennies. What was the amount of the manager's deposit?

4. Name some types of businesses that have to handle large amounts of cash. Describe some of the problems involved with handling large amounts of cash.

3-8 SCIENTIFIC NOTATION

Objective

Change numbers from standard form to scientific notation and vice versa.

Mr. Andretti's computer holds 42 megabytes of data. One megabyte is one million bytes, so this computer holds $42 \times 1{,}000{,}000$ or 42,000,000 bytes.

Scientific notation is a way to write very large numbers, such as 42,000,000, or very small numbers.

To write a number using scientific notation, write the number as a product. One factor is a number that is at least 1 but less than 10. The other factor is a power of 10.

Method

▶1 Move the decimal point to find a number that is at least 1 but less than 10.

▶2 Count the number of places the decimal point was moved. This is the exponent for the power of ten.

Example A

Write 42,000,000 in scientific notation.

▶1 42,000,000   4.2 is at least 1 but less than 10.

▶2 The decimal point was moved 7 places.

$42{,}000{,}000 = 4.2 \times 10^7$ The computer holds 4.2×10^7 bytes.

Example B

Write 7,968 in scientific notation.

▶1 7,968   7.968 is at least 1 but less than 10.

▶2 The decimal point was moved 3 places.

$7{,}968 = 7.968 \times 10^3$

To write a number in standard form, remember the shortcut for multiplying by powers of ten.

Example C

Write 6.82×10^4 in standard form.

▶1 6.8200   ▶2 Move the decimal point 4 places to the right. Annex 2 zeros.

$6.82 \times 10^4 = 68{,}200$

Guided Practice

Write in scientific notation.

Examples A, B

1. 468   2. 14   3. 15,430   4. 266,000   5. 50,000

6. 140   7. 21,000,000   8. 3,500,000   9. 2,070   10. 540,060

Example C

Write in standard form.

11. 5.75×10^2 12. 3.1×10^3 13. 2.71×10^5 14. 7.68×10^4

15. 3.8×10^7 16. 1.09×10^6 17. 4.002×10^3 18. 1×10^5

Exercises

Practice

Write in scientific notation.

19. 350,000 **20.** 1,982 **21.** 92,372 **22.** 57,000,000

23. 6,700,000 **24.** 6,791 **25.** 15 **26.** 93,600

27. 6 million **28.** 149 million **29.** 417 billion

Write in standard form.

30. 6.9×10^2 **31.** 4.31×10^2 **32.** 9.07×10^3 **33.** 5.3×10^4

34. 8.27×10^3 **35.** 2.17×10^5 **36.** 3.43×10^6 **37.** 7.5×10^5

38. 3.694×10^5 **39.** 6.38×10^9 **40.** 4.6×10^8 **41.** 5.35×10^6

Applications

For Exercises 42-44, write your answer in standard form and in scientific notation.

42. It takes *about* 90,700 days for Pluto to complete one revolution around the Sun. There are 9.07×10^4 days in one Pluto year. How many Earth days are there in 10 Pluto years?

43. Earth's average distance from the sun is 9.296×10^7 miles. Venus is an average of 6.723×10^7 miles from the Sun. On the average, how much farther is Earth from the Sun than Venus?

44. How many seconds are there in a year (365 days)?

Critical Thinking

45. Have you lived a million seconds? Have you lived a billion seconds? Use a calculator to find out.

JOURNAL ENTRY

46. Complete this sentence. "The one thing I did not understand in this lesson was ____."

Calculator

Most calculators can display only eight digits at one time. So numbers with more than eight digits are usually shown in scientific notation.

Add 35,000,000 and 77,000,000 on a calculator.

Enter: 35000000 ⊞ 77000000 ⊟ $1.12 \ 08$

The readout 1.12 08 means 1.12×10^8.

The sum is 1.12×10^8 or 112,000,000.

Write each sum or product in scientific notation and then in standard form.

47. $68,000,000 + 74,000,000$ **48.** $99,999,999 + 1$

49. $80,000,000 + 80,000,000$ **50.** $99,999,000 \times 1,000$

51. $1,000,000 \times 100,000$ **52.** $18 \times 365 \times 24 \times 60 \times 60$

3-9 GEOMETRIC SEQUENCES

Objective

Identify and find geometric sequences.

Imagine that the Starlight Theater's ticket sales totaled $32,000 in 1992, $64,000 in 1993, and $128,000 in 1994. If ticket sales continue to grow at this rate, what amount will they total in 1997?

You know a list of numbers that follows a certain pattern is called a sequence. In a **geometric sequence,** consecutive numbers are found by multiplying the term before it by the same number. This number is called the **common ratio.**

1, 5, 25, 125, ... Each number in the sequence is 5 times the number before it.

96, 48, 24, 12, ... Each number in the sequence is 0.5 times the number before it.

The yearly ticket sales, $32,000, $64,000 and $128,000 form a geometric sequence.

To continue the pattern of numbers, find the common ratio. Then multiply the last number by the common ratio to find the next number.

Method

1 ▶ To find the common ratio, divide any number by the one before it.

2 ▶ Multiply the last number in the sequence by the common ratio to find the next number.

Example A

Find the common ratio and write the next three terms of the sequence 32,000; 64,000; 128,000; ...

1 ▶ Divide to find the common ratio.

$64,000 \div 32,000 = 2 \qquad 128,000 \div 64,000 = 2$
The common ratio is 2.

2 ▶ Multiply the last number in the sequence by the common ratio to find the next number.

$128,000 \times 2 = 256,000 \quad 256,000 \times 2 = 512,000$
$512,000 \times 2 = 1,024,000$

The sequence is 32,000; 64,000; 128,000; 256,000; 512,000; 1,024,000.

The ticket sales in 1997 will be $1,024,000.

Example B

Find the common ratio and write the next three numbers in the sequence 9,375; 1,875; 375; ...

1 ▶ Divide a number by the one before it to find the common ratio.

$1,875 \div 9,375 = 0.2 \qquad 375 \div 1,875 = 0.2$
The common ratio is 0.2

2 ▶ $375 \times 0.2 = 75 \quad 75 \times 0.2 = 15 \quad 15 \times 0.2 = 3$
The sequence is 9,375; 1,875; 375; 75; 15; 3.

Find the common ratio and write the next three terms in each sequence.

Example A
Example B

1. 4.3, 8.6, 17.2, ...
2. 56, 560, 5,600, ...
3. 7.5, 37.5, 187.5, ...
4. 91, 9.1, 0.91, ...
5. 200, 80, 32, ...
6. 20, 10, 5, ...

Exercises

Find the common ratio and write the next three numbers in each sequence.

Practice

7. 23, 92, 368, ...
8. 25, 75, 225, ...
9. 2, 10, 50, ...
10. 50, 10, 2, ...
11. 600, 240, 96, ...
12. 240, 144, 86.4, ...
13. 1.2, 1.44, 1.728, ...
14. 3, 12, 48, ...
15. 2, 6, 18, ...
16. 3,000, 900, 270, ...
17. 64, 32, 16, ...
18. 100, 90, 81, ...

Applications

19. The pitch of a musical note depends on the number of vibrations per second. A note one octave higher than a given note vibrates twice as many times as the given note. If the number of vibrations per second for middle C is 256, what is the number of vibrations per second for a note three octaves above middle C?

20. What is the number of vibrations per second for a note four octaves below middle C? Refer to Exercise 19.

21. Tom and Cheryl Jackson opened a savings account with $5 for their son Robert on his first birthday. On his second birthday, they put $15 in the account. Each year they deposit 3 times the amount of the previous year's deposit. On which of Robert's birthdays will they make the first four-digit deposit?

Number Sense

Is the answer about 1, about 10, or about 0.1?

22. 0.03 × 6
23. 3.1 × 0.3
24. 3.1 × 3.1
25. 6 ÷ 40
26. 5 ÷ 0.5
27. 1.1 ÷ 0.1

Critical Thinking

28. Julie has a collection of 12 math books. Seven of the books are about algebra and five of the books are about geometry. If she told you to choose any combination of books, one on each subject, how many possible choices would you have?

Mental Math

Find each product or quotient mentally.

29. 0.778 × 100
30. 0.778 × 1,000
31. 0.778 × 10,000
32. 12.309 × 10
33. 12.309 × 1,000
34. 12.309 × 100,000
35. 42.53 ÷ 10
36. 42.53 ÷ 100
37. 42.53 ÷ 1,000

3-10 FORMULAS

Objective

Use formulas to solve problems.

Mr. and Mrs. Perez travel 630 miles to the Grand Canyon. If they drive at an average rate of 60 miles per hour, how long will it take?

A **formula** shows how certain quantities are related. You can use a formula to find an unknown quantity if you know the other quantities.

Method

1 ▶ Write the appropriate formula.

2 ▶ Replace each variable with the appropriate number.

3 ▶ Solve.

A formula that relates distance, rate (speed), and time is $d = rt$. The variable d represents distance, r represents rate or speed, and t represents time.

Example A

Find the time it will take Mr. and Mrs. Perez to drive to the Grand Canyon using $d = rt$.

1 ▶ $d = rt$ ⠀⠀⠀⠀Write the formula.

2 ▶ $630 = 60t$ ⠀⠀Replace d with 630, r with 60.

3 ▶ $\frac{630}{60} = \frac{60}{60}t$ ⠀⠀Solve.

$10.5 = t$ ⠀⠀It will take 10.5 hours to drive to the Grand Canyon.

Determine the rushing average for Emmitt Smith of the Dallas Cowboys. If Smith carried the ball 373 times for 1,713 yards, what was his rushing average for the year?

The formula for a player's rushing average *(r)* with a total of yards rushed *(y)* in *n* carries of the ball is $r = y \div n$.

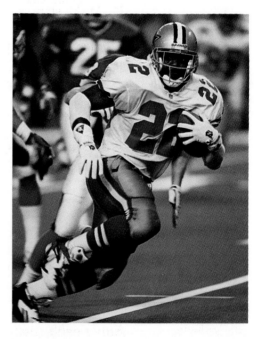

Example B

Find Smith's rushing average using $r = y \div n$.

1 ▶ $r = y \div n$ ⠀⠀$y = 1,713, n = 373$

2 ▶ $r = 1,713 \div 373$

3 ▶ $r = 4.592$

Smith's rushing average was 4.592 yards per carry.

Guided Practice

Example A

Solve for d, r, or t using the formula $d = rt$.

1. 264 miles, 55 mph ⠀⠀⠀⠀⠀⠀**2.** 2.5 mph, 2 hours

3. 30 miles, 12 mph ⠀⠀⠀⠀⠀⠀**4.** 650 mph, 2.2 hours

Example B

Solve for r, y, or n using the formula r = y ÷ n.

5. 165 yards, 15 carries

6. 10.5 yards per carry, 28 carries

7. 171 yards, 18 carries

8. 7.6 yards per carry, 190 yards

Exercises

Practice

Solve for d, r, or t using the formula d = rt.

9. 4 miles, 1 hour

10. 4.25 hours, 55 mph

11. 9 miles, 6 mph

12. 880 yards, 20 min

Solve for y, n, or r using the formula r = y ÷ n.

13. 216 yards, 7.2 yards per carry

14. 160 yards, 25 carries

15. 99 yards, 15 carries

16. 14 carries, 9 yards per carry

A formula for finding the gas mileage of a car is $s = \frac{m}{g}$, where s is gas mileage in miles per gallon (mpg), m is the number of miles driven, and g is the number of gallons of gas used. Find the gas mileage.

17. 220 miles, 10 gallons

18. 240 miles, 15 gallons

19. 120.5 miles, 5 gallons

20. 137.2 miles, 8 gallons

21. 374 miles, 8.5 gallons

22. 400 miles, 12.5 gallons

Applications

Write a formula. Then solve.

23. At the start of Guido's trip, the odometer read 5,123.7. At the end of the trip, the odometer read 5,339.7. If Guido's car used 8 gallons of gasoline, how many miles per gallon is this?

24. Bo Jackson ran for 168 yards in 24 carries during a football game in Oakland. What was his rushing average for the game?

Connections

25. An important formula in physics is Ohm's law. This formula relates voltage (volts, V), current (amperes, I), and resistance (ohms, R). Ohm's law states that voltage equals amperes of current multiplied by resistance in ohms, or $V = I \times R$. If a toaster draws 4 amperes of current when connected to a 120-volt circuit, what is the resistance?

26. If a 12-volt car battery has a resistance of 4 ohms with the engine running, what is the current? Use the formula in Exercise 25.

Mixed Review

Lesson 1-11

Solve each equation. Check your solutions.

27. $19 = y + 7$

28. $t - 60 = 11.2$

29. $m + 1.53 = 7$

Lesson 1-12

Solve each equation. Check your solutions.

30. $900 = 45b$

31. $\frac{n}{5} = 14$

32. $\frac{c}{6} = 1.5$

33. $18g = 180$

Lesson 3-7

34. The computer club purchased 450 floppy discs for a total of $648. What was the cost of each disc?

3-11 AREA OF RECTANGLES

Objective

Use formulas to find the area of rectangles and squares.

Mark fenced in a small rectangular pen for his puppy. He made the pen 5 feet long and 3 feet wide. How many square feet of area does the pen enclose?

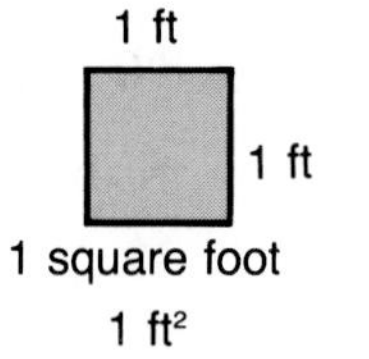

1 ft

1 ft

1 square foot
1 ft²

Area is the number of square units that covers a surface.

A drawing of the rectangular pen is shown below.

Example A

Find the area of the rectangle by counting the squares.

There are 15 square feet.

The area is 15 ft².

Mark's puppy's pen encloses an area of 15 square feet.

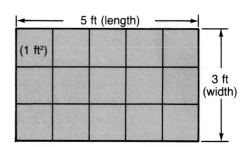

Example B

Find the area of the puppy pen. Use the formula A = ℓw.

A = area ℓ = length w = width

$A = \ell w$ length = 5, width = 3

$A = 5 \times 3$

$A = 15$ The area is 15 ft².

This is the same area found by counting squares.

A square is a rectangle in which all sides have the same length. Since the length and width are the same, use s to stand for both. So the formula for the area of a square is $A = s \times s$ or $A = s^2$.

Example C

Find the area of a square with sides 9 inches long.

$A = s^2$ $s = 9$

$A = 9^2$

$A = 81$ The area is 81 in².

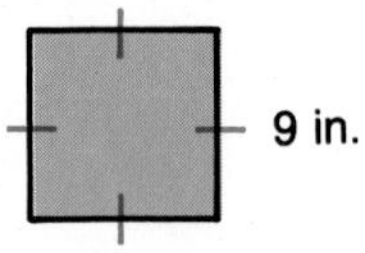

9 in.

Guided Practice

Example A

Find the area of each rectangle by counting the squares.

1.
1 in.²
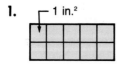

2.
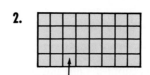
1 in.²

3.
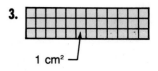
1 cm²

Examples B, C

Find the area of each rectangle.

4.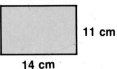
11 cm
14 cm

5.
16 m
16 m

6.
15 in.
6 in.

7. rectangle: $\ell = 20$ yd; $w = 12$ yd

8. square: $s = 40$ cm

Exercises

Practice

Find the area of each rectangle.

9.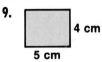
4 cm
5 cm

10.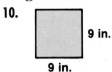
9 in.
9 in.

11.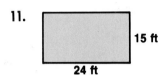
15 ft
24 ft

ƒUN with MATH
Challenge a friend to the game of Decimal Maze.
See page 105.

12.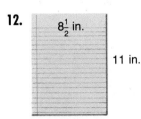
$8\frac{1}{2}$ in.
11 in.

13.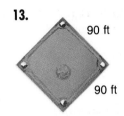
90 ft
90 ft

14.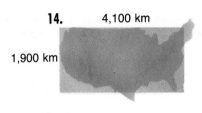
4,100 km
1,900 km

Applications

15. Kevin Lewis needs to replace sod for a section of lawn. The section is 10.5 feet long and 6 feet wide. If sod costs $2 a square foot, how much will it cost Kevin to replace the sod?

16. Eileen is making name tags out of poster board 22 inches by 28 inches. She would like each name tag to be 2 inches by 4 inches. How many name tags can she cut from one poster board?

Suppose

17. Suppose a square has a total area of 64 square feet. What is the length of each side?

18. Suppose a rectangle with an area of 884 square meters has a length of 34 meters. What is the width of the rectangle?

Show Math

19. Draw a square with an area of 16 square inches.

Critical Thinking

20. The area of the figure shown at the right is 384 square inches. The figure is made up of 6 squares. What is the perimeter of the figure?

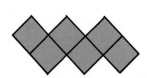

Calculator

You can use the ⌈EE⌉ key on a calculator to perform operations on numbers written in scientific notation.

Add 2×10^9 and 6×10^9. Enter:

The sum is $8 \times 10^9 = 8{,}000{,}000{,}000$ or 8 billion.

Write each answer in scientific notation and then in standard form.

21. $(3.51 \times 10^{15}) + (2.47 \times 10^{15})$

22. $(3.95 \times 10^{11}) - (2.31 \times 10^{11})$

3-12 AVERAGE (MEAN)

Objective

Find the mean of a set of data.

Vicki Riegle wants to buy a used car that gets high gasoline mileage. After looking at five cars, she makes a chart comparing their mileages to show her parents.

	Miles per Gallon
Sentra	33
Escort	33
Camaro	20
LeMans	28
Turcel	26

One average of a set of numbers is called the **mean.** Mean is the most common type of average.

Method

1. Find the sum of all the numbers.
2. Divide the sum by the number of addends.

Example A

Find the average for the mileages in the chart.

1. Find the sum of all the numbers.

$$33 + 33 + 20 + 28 + 26 = 140$$

2. There are 5 addends. Divide the sum by 5.

$$\frac{140}{5} = 28$$

The average of the mileages in the chart is 28 miles per gallon.

Example B

Find the average for these numbers: 3.4, 1.8, 2.6, 1.8, 2.3, and 3.2. Round to the nearest tenth.

1. $3.4 + 1.8 + 2.6 + 1.8 + 2.3 + 3.2 = 15.1$

2. $\frac{15.1}{6} \approx 2.51$ ≈ stands for "is approximately equal to."

The average, rounded to the nearest tenth, is 2.5.

Guided Practice

Find the average for each set of numbers.

Example A

1. 2, 3, 7, 8, 10   **2.** 1, 3, 5, 7   **3.** 1, 3, 4, 4, 7, 9, 14

4. 17, 18, 20, 13   **5.** 10, 40, 60, 30, 50   **6.** 23, 37, 20, 16

Example B

Find the average for each set of numbers. Round to the nearest tenth.

7. 58, 61, 60   **8.** 14, 13, 10, 9, 6   **9.** 129, 130, 100

10. 0.8, 0.5, 0.6   **11.** 4.8, 6.4, 7.2   **12.** 4.5, 2.3, 6.0, 3.5

Practice

Find the average for each set of numbers. Round to the nearest tenth.

13. 84, 111, 150   **14.** 63, 47, 130   **15.** 145, 230, 100, 225

16. 0.98, 0.85, 0.75   **17.** 1.1, 1.25, 1.34   **18.** 2.46, 3.7, 1.6, 2.55

19. 5, 8, 2, 7, 7, 9, 6, 4   **20.** 92, 87, 97, 80, 95

21. 476, 833, 729, 548   **22.** 801, 897, 843, 865, 854

23. 2.5, 4.73, 3.86, 3.39, 2.98, 4.24   **24.** 113.4, 6.6, 85, 16.16, 70.12

25. $50,000, $37,500, $43,900, $76,900, $46,000, $48,580

Write Math

26. Name a situation when it would be better to find the average as a rounded answer rather than as an exact answer.

27. Explain how you would use a calculator to find the average.

Applications

28. In Princeton, the total rainfall for each month in a recent year was as follows:

1.2 in., 1.8 in., 2.4 in., 3.9 in., 4.7 in., 5.5 in., 4.5 in., 4.7 in., 7.2 in., 6.8 in., 3.0 in., 1.1 in. What is the average?

29. Two basketballs are bought for $17.99 each. Another costs $32.99. What is the average cost of all three?

Critical Thinking

30. Emmy Explorer walked one mile north, then one mile west, and then one mile south. She arrived back where she started. Where did Emmy start?

Estimation

Sometimes the clustering strategy can be used to estimate the mean. For example, the numbers in the table cluster around 10,000,000. So 10,000,000 is a good estimate of the average number of vehicles sold each year in the United States between 1983 and 1988.

U.S. Motor Vehicle Retail Sales (rounded to the nearest thousand)	
1983	9,182,000
1984	10,390,000
1985	11,038,000
1986	11,460,000
1987	10,278,000
1988	11,225,000

Use the clustering strategy to estimate each average.

31. 82, 77, 82, 76, 79, 78, 81   **32.** 396, 391, 411, 407, 389

33. 6,637, 5,952, 5,848, 6,207   **34.** 3.6, 2.75, 3.8, 3, 1.6, 2.34

▶ Explore
▶ Plan
▶ Solve
▶ Examine

3-13 MULTI-STEP PROBLEMS

Objective

Solve multi-step problems.

Monica works at Monterey's Mexican restaurant as a waitress. During the first week of October, she worked 12 hours and made $52 in tips. If Monica earns $4.25 an hour, how much did she make that week?

You may need to use more than one operation to solve a problem.

▶ **Explore**

What is given?
● the hourly wage
● the number of hours worked
● the amount of tips earned
What is asked?
● Monica's total earnings during the first week of October

▶ **Plan**

Hourly wages are found by multiplying the number of hours worked by the hourly rate. Total earnings are found by adding hourly wages and tips.

▶ **Solve**

Estimate: $10 \times 4 = 40$ $40 + 50 = 90$

$12 \times 4.25 = 51$ Find the wages earned.

$51 + 52 = 103$ Add the tips.

Monica earned $103 during the first week of October.

▶ **Examine**

$103 is close to the estimate. The answer is reasonable.

Guided Practice

1. Brett saws seven 0.2-meter pieces, from a 2.5-meter board. How long is the remaining piece?

2. In his first three games, Larry scored 7, 15, and 21 points for Walnut Ridge. Geoff scored 22, 3, and 14 points in his first three games for DeSales. Which player has a better scoring average after three games?

3. Susan earns $28.70 working 7 hours on Saturdays at Laura's Gifts. In December, Susan worked an extra 6 hours a week. At this rate, what does Susan earn in one week during December?

4. Keith's test scores are 76, 89, 85, and 93. What is his average score?

5. Hal's lunch cost $4.85. He bought a hamburger for $2.50 and fries for $1.35. He also bought a large iced tea. What was the cost of the iced tea?

6. Refer to Exercise 5. If Hal leaves a tip of $0.75 and the tax on his lunch is $0.30, how much does he have left from $10?

7. Select one of the assignments from this chapter that you found particularly challenging. Place it in your portfolio.

Number Sense

8. Find the number of decimal places in each product. Use the answers to make a magic square. Each row, column, and diagonal should make the same sum.

0.06×0.05	5.5×3.2
2.4×5	0.29×1.4
7.420×8.0064	2.1×3.71272
0.124×0.65	3.30245×8.9632
0.00067×0.812	

9. When you buy a car, you have many additional expenses; one is automobile insurance. Make up a problem using the table. Exchange with a classmate to solve. Discuss your solution.

Average Auto Insurance Premiums
Premiums per Car

North Dakota	$343.85
Tennessee	$338.46
South Dakota	$324.90
Iowa	$292.51

National Average $517.71

Using Algebra

Write a mathematical expression for each verbal expression.

10. twice a number and 6 more

11. a number divided by four

12. the sum of a number and two, all times four

13. the sum of a number divided by three, and 10 more

Mixed Review

Lesson 3-1

Multiply.

14.
$$\begin{array}{r} 32 \\ \times\ 4 \\ \hline \end{array}$$

15.
$$\begin{array}{r} 481 \\ \times\ 12 \\ \hline \end{array}$$

16.
$$\begin{array}{r} 631 \\ \times\ 23 \\ \hline \end{array}$$

17.
$$\begin{array}{r} 7,801 \\ \times\ 18 \\ \hline \end{array}$$

Lesson 3-4

Divide. Check using multiplication.

18. $15\overline{)229}$ 19. $91\overline{)819}$ 20. $42\overline{)4,325}$ 21. $647\overline{)8,741}$

Lesson 3-7

22. Adrianne Hunter borrowed $10,000 from First Bank to buy a car. She will pay back a total of $13,728 that includes interest. Adrianne makes 48 equal payments. What is the amount of each payment?

REVIEW

Choose a word or number from the list at the right that best completes each sentence.

Vocabulary / Concepts

1. To divide by 100, move the decimal point __?__ places to the __?__ .

2. The area of a rectangle is found by multiplying the __?__ by the width.

3. If you move the decimal point two places in the divisor, you must move the decimal point two places in the __?__ .

4. To multiply by 1,000, move the decimal point __?__ places to the __?__ .

5. To find the common ratio in a geometric sequence, divide any number by the one __?__ it.

6. In standard form, 1.4×10^2 equals __?__ .

7. 9,520,000,000 in scientific notation is __?__ .

8. The result of multiplying two numbers is the __?__ .

9. One average of a set of data is the __?__ .

1.4
140
1,400
9.52×10^9
952×10^9
after
before
dividend
divisor
left
length
mean
product
quotient
right
three
two

Exercises / Applications

Lessons 3-1 3-2

Multiply. Estimate the products first.

10.	11.	12.	13.	14.
58	29	716	98	4,881
$\times\ 3$	$\times\ 18$	$\times\ 30$	$\times\ 27$	$\times\ 79$

15.	16.	17.	18.	19.
3.21	$79.52	$53.07	1.93	0.58
$\times\ 7$	$\times\ 8.5$	$\times\ 13$	$\times\ 7.8$	$\times\ 2.4$

Lesson 3-3

Solve.

20. A car is advertised at a cash price of $9,850. To purchase the car on credit, the down payment is $250 and the monthly payment is $282 for 36 months. What is the finance charge if the car is purchased on credit?

Lesson 3-4

Divide. Check using multiplication.

21. $7\overline{)494}$ 22. $7\overline{)642}$ 23. $5\overline{)7.285}$ 24. $16\overline{)3,841}$ 25. $98\overline{)588}$

Lesson 3-5

Multiply or divide.

26. $2.39 \div 10$ 27. 65.7×10^2 28. 4.382×10 29. $14.8 \div 1,000$

Lesson 3-6

Solve.

30. Suppose Mr. Gomez typed 1 page on Monday. Then on each successive day he types 2 additional pages. How many pages will he type on Friday?

Lesson 3-7

Divide.

31. $0.24\overline{)0.15}$ **32.** $1.3\overline{)0.91}$ **33.** $6.7\overline{)23.45}$ **34.** $0.5\overline{)4.8}$ **35.** $0.3\overline{)2.04}$

36. $0.6\overline{)0.507}$ **37.** $0.8\overline{)0.38}$ **38.** $0.9\overline{)3.078}$ **39.** $9.6\overline{)5.4}$ **40.** $2.9\overline{)9.28}$

Lesson 3-8

Write in scientific notation.

41. 79,900   **42.** 266,000   **43.** 86 million   **44.** 6,300,000,000

Write in standard form.

45. 4.9×10^2 **46.** 3.862×10^3 **47.** 9.7×10^3 **48.** 1.496×10^4

Lesson 3-9

Write the next three numbers in the sequence.

49. 800, 240, 72, . . .   **50.** 24, 12, 6, . . .   **51.** 26, 39, 58.5, . . .

Lesson 3-10

Solve for d, r, or t using d = rt.

52. $r = 45$ mph, $t = 3$ h   **53.** $r = 55$ mph, $d = 165$ mi

Lesson 3-11

Find the area of each rectangle.

54. 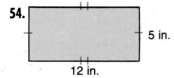 5 in. 12 in.   **55.** 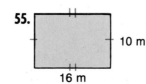 10 m 16 m   **56.** 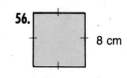 8 cm

Lesson 3-12

Find the average for each set of data. Round to the nearest tenth.

57. 1, 5, 6, 4, 3, 8, 9, 10, 7, 2   **58.** 21, 25, 16, 18, 26, 17, 19, 20, 22

59. $4.00, $5.60, $7.80, $9.35, $8.65, $9.10, $4.10

Lesson 3-13

60. Carole Jefferson pays 12 equal additional charges on cable TV service that total $27 per year. How much is each service charge?

61. A long-distance telephone call costs $0.18 for the first minute and $0.15 for each additional minute. What is the cost of a 12-minute call?

62. Darren wants to buy a radio that costs $129. He saves $6.50 each week. So far he has saved $52. How many more weeks will it take to save enough money to buy the radio?

63. Bryce Canyon National Park in Utah had 1.1 million visitors in 1989. The same year, the Canyonlands National Park in Utah had 260,000 visitors. How many more people visited Bryce Canyon than the Canyonlands?

TEST

Multiply.

1. $\begin{array}{r} 72 \\ \times\ 49 \\ \hline \end{array}$

2. $\begin{array}{r} 6{,}244 \\ \times\ \ \ \ 53 \\ \hline \end{array}$

3. 4.345×10^3

4. 7.923×10^6

Divide.

5. $7\overline{)642}$

6. $49\overline{)816}$

7. $45.03 \div 10^4$

8. Gretyl Haenes will pay $32.50 a month for 24 months for a computer system. If she made a $100 down payment, what is the credit price of the system?

Write in scientific notation.

9. 5,280

10. 392,000

Solve for d, r, or t using the formula d = rt. Round to the nearest tenth.

11. 34 kilometers, 2.6 hours

12. 450 miles per hour, 3.8 hours

Find the area of each rectangle.

13. 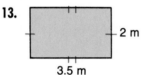 2 m

3.5 m

14. 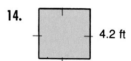 4.2 ft

Write the next three numbers in each sequence.

15. 3, 15, 75, . . .

16. 600, 300, 150 . . .

Find the average for each set of numbers.

17. 9, 8, 4, 7, 6, 11

18. 3.2, 4.6, 1.2, 3.5

19. Mr. Schultz buys numerals to put on the doors of each apartment in a 99-unit apartment building. The apartments are numbered 1 through 99. How many of each digit 0, 1, 2, 3, 4, 5, 6, 7, 8, 9 should Mr. Schultz buy?

20. Mr. Jamison bought a compact disc system on credit for a $65 down payment and a monthly payment of $38.60 for 12 months. What is the total cost of the system?

▶ BONUS: These are . . .
4×12, 6×8, 1×48

These are not . . .
3×8, 4×7, 2×13

Which of these are?
2×24, 3×16, 4×14, 5×16

NOISE POLLUTION

Heidi works in a noisy downtown office. She notices that the office seems noisier than yesterday. Then she remembers that traffic is being rerouted past the office because of construction. How much more noise does a passing car make than Heidi's office?

Sound intensity is measured in decibels.

The chart below lists the intensity of various sounds and the meaning of each measurement.

Sound	Decibels	Mathematical Language
barely heard	0	10^0
breathing	10	10^1
rustling leaves	20	10^2
talking	40	10^4
noisy office	60	10^6
vacuum cleaner	70	10^7
car	80	10^8
subway train	100	10^{10}
motorcycle	110	10^{11}
jet airplane	130	10^{13}

How many times more intense is the sound of a passing car than the sound of a noisy office?

$$\text{car} \rightarrow 80 \text{ decibels} \rightarrow 10^8$$
$$\text{noisy office} \rightarrow 60 \text{ decibels} \rightarrow 10^6$$

The sound of a passing car is 10^2 or 100 times more intense than a noisy office.

How many times more intense is the first sound than the second?

1. subway train, noisy office
2. breathing, barely heard sound
3. car, vacuum cleaner
4. jet airplane, breathing
5. At the Summerville rock concert, the sound intensity in row 20 was measured at 110 decibels. How much more intense was the sound at the concert than sound of normal talking?

Free Response

Lesson 1-10

Write an expression for each phrase.

1. 7 more than a number

2. 8 less than y

3. the product of 6 and m

4. the quotient of 4 and w

5. the sum of g and 32

6. 9 times a number

Lessons 1-11
1-12

Solve each equation. Check your solution.

7. $14y = 294$

8. $76 = 4x$

9. $17 + y = 32$

10. $13y = 52$

11. $8x = 4$

12. $3\frac{2}{3} + y = 40$

Lesson 1-13

Write an equation to solve this problem. Then solve.

13. A train travels at a constant rate of 80 miles per hour. How far will it travel in 15 hours?

Lessons 2-2
2-4

Add or subtract.

14. $\begin{array}{r} 17.4 \\ + 19.2 \end{array}$

15. $\begin{array}{r} 1.85 \\ + 1.2 \end{array}$

16. $\begin{array}{r} 6.71 \\ - 2.7 \end{array}$

17. $\begin{array}{r} 32.4 \\ - 6.27 \end{array}$

18. $7.82 - 1.3$

19. $6.45 - 0.36$

20. $20.6 + 1.01$

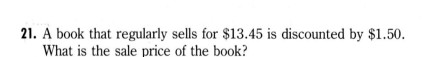

21. A book that regularly sells for $13.45 is discounted by $1.50. What is the sale price of the book?

Lesson 3-1

Multiply.

22. $\begin{array}{r} 90 \\ \times 70 \end{array}$

23. $\begin{array}{r} 280 \\ \times 0.9 \end{array}$

24. $\begin{array}{r} 24 \\ \times 1.1 \end{array}$

25. $\begin{array}{r} 42 \\ \times 22 \end{array}$

26. 263×35

27. 603×32

28. $4,010 \times 33$

Lesson 3-2

29. Ben paid $16.50 for 3 classic Elvis singles at a garage sale. He sold 2 of the records to Sheila at the same price he paid for them. How much did Sheila pay for the 2 singles?

Lessons 3-2
3-7

Multiply or divide.

30. $\begin{array}{r} 4.9 \\ \times 7 \end{array}$

31. $\begin{array}{r} 3.82 \\ \times 5.1 \end{array}$

32. $\begin{array}{r} 2.061 \\ \times 8.8 \end{array}$

33. $15\overline{)5.4}$

34. $2.7\overline{)16.74}$

35. 7.8×6.42

36. $1.08 \div 1.8$

37. $210 \div 280$

Multiple Choice

Choose the letter of the correct answer for each item.

1. The average attendance at 77 Red Sox home games was 26,026. What was the total attendance for 77 games?
 a. 338
 b. 8,796,788
 c. 563,024
 d. 2,004,002

Lesson 3-1

6. The total ticket sales for the All-Country concert are $200,824. 38 ticket outlets sold a total of 7,724 tickets. What was the price of each ticket?
 e. $31
 f. $20
 g. $26
 h. $36

Lesson 3-4

2. Subtract 7,996 from 56,489.
 e. 48,493
 f. 49,493
 g. 49,593
 h. 51,513

Lesson 2-3

7. What is the sum of 6,360 and 9,863?
 a. 15,123
 b. 15,223
 c. 16,123
 d. 16,223

Lesson 2-1

3. What is 101,740 rounded to the nearest thousand?
 a. 100,000
 b. 101,000
 c. 101,700
 d. 102,000

Lesson 1-4

8. If Sue Thomas earns $15,184 a year, what are her weekly earnings?
 e. $298
 f. $12,560
 g. $292
 h. $1,265

Lesson 3-4

4. Estimate the difference of 28,683 and 7,405.
 e. 2,000
 f. 12,000
 g. 22,000
 h. 32,000

Lesson 2-3

9. Which facts are given? Jody purchased a life insurance policy worth $50,000. Her monthly payment is $24.84.
 a. name of insurance company
 b. payment per month
 c. length of time for payments
 d. how her payments will be made

Lesson 2-9

5. What is the quotient if you divide 21,264 by 12?
 a. 17,720
 b. 1,772
 c. 255,168
 d. 1,936

Lesson 3-4

10. Mars is an average 1.4171×10^8 miles from the sun. About how many miles is Mars from the sun?
 e. 141,000
 f. 141,000,000
 g. 141,000,000,000
 h. none of the above

Lesson 3-8

			Hypatia,	
Earliest use of numbers			earliest known	
Congo (Zaire), Africa	2000 BC		woman mathematician	AD 520
6000 BC		Snow skis	AD 370-415	Decimal system
		Northern Norway		India

How does cutting down rainforests cause Earth to warm up?

It's similar to the reason a closed car becomes warm on sunny days. Sunlight passing into the car changes to heat when it is absorbed by upholstery in the car. Windows don't let heat out as easily as light. Heat is trapped inside. Similarly, Earth's atmosphere lets sunlight through, but carbon dioxide (CO_2) and other gases trap some heat. Trees and plants use CO_2 to make food. If we cut down rainforests that normally absorb tons of CO_2, CO_2 can build up in the atmosphere, trapping even more heat, and warming Earth.

MATH M·E·N·U

Pretzel Cookies

1 cup semisweet chocolate pieces
½ cup butter
1 package (10 oz) marshmallows
4 cups crisp ready-to-eat rice cereal
2 cups salted peanuts
2 cups raisins
2 cups broken pretzel sticks (about ½ inch long)

Melt chocolate, butter, and marshmallows in top of double boiler over simmering water. Stir well. Mix remaining ingredients in a large greased bowl. Pour melted mixture over dry mixture and stir until pieces are well coated. Drop by spoonfuls onto waxed paper. Let cool until set. Makes about 7 dozen.

In 1910 football teams were penalized 15 yards for each incompleted forward pass.

COMICS

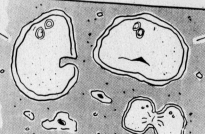

I WONDER WHY WE AMOEBAS HAVE NEVER PRODUCED A GREAT MATHEMATICIAN?

PROBABLY BECAUSE WE DIVIDE TO MULTIPLY.

THAVES 10-6

FRANK & EARNEST

First All-American made clock   AD 1843   Microwave oven   AD 1975

AD 1761   First calculating machine Sweden   AD 1945   Rubik's cube

JOKE!

Q: What kind of problem do five-foot people have?

A: They need two and a half pairs of shoes.

How does a microwave oven heat food? Molecules of food have a positive electrical charge at one end and a negative charge at the other. A pulse of microwave energy lines up all the molecules so that the positive ends point in one direction. The next pulse of microwave energy reverses the way the molecules are lined up. Heat is generated from the friction of molecules flip-flopping back and forth at an incredible 2,450,000,000 times each second.

RIDDLE

Q: Who is your father's sister's son's only uncle's only child?

A: You.

TEASER

Challenge a friend to Decimal Maze. Place a marker at Start. Enter Start number 100 in your calculators. Each player alternately moves the marker along a line segment and performs the operation indicated on his or her calculator. Proceed in any direction. Paths may be repeated, but not on consecutive plays. When the marker reaches Finish, the player with the smaller calculator display wins.

100 START

x0.9   −0.09   −0.09   +0.6   +0.7

x1.09   +2.01   x1.9   +1.9   x12

x0.5   +0.4   x0.99   +2.1

−12

−1.7   x1.09   +1.4   +0.8   +0.5

+1.2   +.87   −0.5   x1.09

x0.97   x1.01

x0.7

FINISH

QUIZ TIME

Binary numbers are written using ones and zeros. Examine the chart of binary numbers. Each *one* means the number at the top of its column is needed as an addend to total the Arabic number on the left. For 7(00111), one 4, one 2, and one 1 are needed. Make a chart of binary numbers up to 31.

Arabic	16	8	4	2	1
0	0	0	0	0	0
1	0	0	0	0	1
2	0	0	0	1	0
3	0	0	0	1	1
4	0	0	1	0	0
5	0	0	1	0	1
6	0	0	1	1	0
7	0	0	1	1	1
8	0	1	0	0	0
9	0	1	0	0	1
10	0	1	0	1	0

4

FRACTIONS: ADDING AND SUBTRACTING

▶ **Bon Appetit!**

What is your favorite kind of pizza? When was the last time you made pizza? Do you remember the steps you followed in making the pizza? Why is it important to follow the recipe or the set of directions? How did your pizza turn out?

Danny makes a pizza for his family. He cuts it into 10 pieces. If he and his sister, Emily, eat 6 of the pieces, what fraction of the pizza is left for the rest of the family?

ACTIVITY: Twenty "Decimal" Questions

The object of the game is to find two factors and their product by asking twenty questions. You may only ask questions that can be answered *yes* or *no*.

Materials: three dice or three number cubes with the digits 1, 2, 3, 4, 5, 6; paper and pencil; calculators are optional

Divide the class into two teams, Team 1 and Team 2.

Playing the Game

1. Team 2 rolls three number cubes and makes one 1-digit and one 2-digit number from the digits rolled.

2. Team 2 may place a decimal point in one or both of their numbers. For example, a 2-digit number from the digits 4 and 6 may be 0.46, 4.6, or 46. 0.046 is not allowed.

3. Team 2 multiplies the numbers and records the product.

4. Team 1 may ask up to 20 questions in order to find the numbers and their product. Some sample questions are listed below.
   - Is one number a multiple of 5?
   - Is there a 5 in the ones place?
   - Are there two decimal places in the product?
   - Is the product less than 1?

5. Keep a record of the number of questions Team 1 asks to find the answers.

6. Team 1 then takes a turn to roll the cubes and makes two numbers and finds the product.

7. The team that uses fewer questions to find the answer is the winner.

Variations of the Game

8. a. Cover the faces on one die or number cube and change the digits to: 7, 8, 9, 0, 0, 0.
   b. Roll four number cubes or four dice and make two 2-digit numbers.

Communicate Your Ideas

9. Create your own variations of the game. Adjust the rules to accommodate your changes.

4-1 FACTORS AND DIVISIBILITY

Objective

Identify factors of a whole number.

Jamie has 6 different patches to sew on his jacket. He can arrange them in a rectangular pattern. Two rectangular displays are shown at the right.

Factors of a number divide that number so that the remainder is zero. You can find the factors of a number by dividing. If the remainder is zero, the divisor and quotient are factors of the number.

What are the factors of 6? 1, 2, 3, 6

Examples

A *Find the factors of 12.*

$12 \div 1 = 12$
$12 \div 2 = 6$
$12 \div 3 = 4$
$12 \div 4$

Stop dividing since you know 4 is a factor.

The factors of 12 are 1, 2, 3, 4, 6, and 12.

B *Find the factors of 27.*

$27 \div 1 = 27$
$27 \div 2 = 13 \text{ R1}$
$27 \div 3 = 9$
$27 \div 4 = 6 \text{ R3}$
$27 \div 5 = 5 \text{ R2}$
$27 \div 6 = 4 \text{ R3}$
$27 \div 7 = 3 \text{ R6}$
$27 \div 8 = 3 \text{ R3}$

The factors of 27 are 1, 3, 9, and 27.

Stop dividing. Why?

A whole number is *divisible* by its factors. You can use these divisibility rules for 2, 3, 5, 9, and 10.

  2: A number is divisible by 2 if its ones digit is 0, 2, 4, 6, or 8.
  5: A number is divisible by 5 if its ones digit is 0 or 5.
10: A number is divisible by 10 if its ones digit is 0.
  3: A number is divisible by 3 if the sum of its digits is divisible by 3.
  9: A number is divisible by 9 if the sum of its digits is divisible by 9.

Determine whether each number is divisible by 2, 3, 5, 9, or 10.

		Ones Digit	By 2?	By 5?	By 10?	Sum of Digits		By 3?	By 9?
Example C	160	0	yes	yes	yes	$1 + 6 + 0 = 7$		no	no
Example D	255	5	no	yes	no	$2 + 5 + 5 = 12$		yes	no
Example E	3,357	7	no	no	no	$3 + 3 + 5 + 7 = 18$		yes	yes

Examples A, B

Practice

Number Sense

Applications

Critical Thinking

Using Variables

Find all the factors of each number.

1. 6   **2.** 10   **3.** 30   **4.** 55   **5.** 120

State whether each number is divisible by 2, 3, 5, 9, or 10.

6. 72   **7.** 236   **8.** 102   **9.** 957   **10.** 3,485

Find all the factors of each number.

11. 9   **12.** 16   **13.** 18   **14.** 20   **15.** 21

16. 48   **17.** 36   **18.** 25   **19.** 42   **20.** 65

State whether each number is divisible by 2, 3, 5, 9, or 10.

21. 89   **22.** 64   **23.** 125   **24.** 156   **25.** 216

26. 330   **27.** 225   **28.** 524   **29.** 1,986   **30.** 2,052

31. Write a 4-digit number that is divisible by 2, 6, 7, and 10.

32. What is the greatest 3-digit number that is divisible by 2, 3, and 5?

33. Name the first year of the twenty-first century that is divisible by only 1 and itself.

34. What number is a factor of every number?

35. Janet and two of her friends went fishing. They caught 57 fish. Can they divide the number of fish evenly among themselves? If so, how many fish would each girl receive?

36. How many trips must a six-passenger airplane make to fly 159 passengers to a summer wilderness camp?

37. Todd made 81 cookies. How many ways could he package the cookies so that there would be the same number in each package with no cookies left over?

38. Suppose you had to choose 3 more players out of 6 to be on your basketball team. Your choices are Jana, Ted, Ken, Melissa, Angie, and Lee. How many different groups of three could you choose? List all possible groups.

39. If the product of 17 and n is 51, what is the value of n?

40. If the product of 8 and b is 72, what is the value of b?

4-2 PRIME FACTORIZATION

Objective

Write the prime factorization of a whole number.

The front row bleacher seats for the cheering section at Allen High School has 5 seats. This is 1 row of 5 students or 1 × 5 students.

A **prime number** has exactly two factors, 1 and the number itself. 5 is a prime number.

5 = 1 × 5   The only factors of 5 are 1 and 5.

A **composite** number has more than two factors.

16 = 1 × 16   The factors of 16
16 = 2 × 8   are 1, 2, 4, 8,
16 = 4 × 4   and 16.

The numbers 0 and 1 are neither prime nor composite.

Every composite number can be expressed as the product of prime numbers. This is called the **prime factorization** of a number. You can use a **factor tree** to find the prime factorization of a number.

Method

1 Find any factor pair of the number.

2 Continue finding factors of each factor until all factors are prime.

Examples

Write the prime factorization for each number.

A

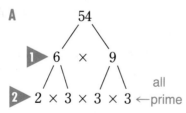

1 6 × 9

2 2 × 3 × 3 × 3 ←prime   all

The prime factorization of 54 is 2 × 3 × 3 × 3 or 2 × 3³.

B

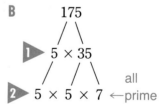

1 5 × 35

2 5 × 5 × 7 ←prime   all

The prime factorization of 175 is 5 × 5 × 7 or 5² × 7.

Guided Practice

State whether each number is prime or composite.

1. 7   **2.** 19   **3.** 15   **4.** 38   **5.** 419   **6.** 231

Examples A, B

Write the prime factorization of each number.

7. 24   **8.** 70   **9.** 102   **10.** 121   **11.** 164   **12.** 225

State whether each number is prime or composite.

13. 2   **14.** 9   **15.** 11   **16.** 24   **17.** 29   **18.** 35

19. 73   **20.** 153   **21.** 61   **22.** 291   **23.** 671   **24.** 501

Write the prime factorization of each number.

25. 20   **26.** 65   **27.** 52   **28.** 30   **29.** 28   **30.** 72

31. 155   **32.** 50   **33.** 96   **34.** 201   **35.** 1,250   **36.** 2,648

Number Sense

37. List the first ten prime numbers.

38. List the first ten composite numbers.

39. Find the least prime number that is greater than 50.

40. Find the greatest prime number that is less than 2,000.

Write Math

41. Explain why an even number greater than 2 cannot be prime.

Cooperative Groups

42. Prime numbers that differ by 2 are called *twin primes*. One such pair is 3 and 5. Discuss with your group how you would find all pairs of twin primes less than 100. Then find all the pairs.

Use a calculator to find each quotient. Then answer each question.

43. 468 ÷ 2 = ?   **44.** 3,150 ÷ 3 = ?
      468 ÷ 3 = ?   3,150 ÷ 5 = ?
      468 ÷ 13 = ?   3,150 ÷ 8 = ?
      Is 468 divisible by 2 × 3 × 13?   Is 3,150 divisible by 3 × 5 × 8?

Write a Conclusion

45. Study the results in Exercises 43 and 44. What can you say about a number that is divisible by several relatively prime numbers?

Mixed Review

Lesson 2-3

Subtract.

46.   63
   − 14

47.   819
   − 68

48.   988
   − 699

49.   4,382
   − 695

50.   62,103
   − 40,674

Lesson 3-4

51. 18)723   **52.** 22)2,210   **53.** 47)6,584   **54.** 432)6,879

Lesson 3-7

55. Alex bought a 24-can case of soda at store A for $5.29. At store B he bought two 12 packs of soda for $2.49 each. How much did he pay per can at each store? Which store had the better buy?

4-3 GREATEST COMMON FACTOR AND LEAST COMMON MULTIPLE

Objective

Find the greatest common factor and the least common multiple of two or more numbers.

Fraternal twins may have common features such as color of eyes, color of hair, or they may be the same sex. Likewise, two or more numbers have common factors and common multiples. That is, different numbers have factors or multiples that are the same.

The **greatest common factor (GCF)** of two or more numbers is the greatest number that is a factor of both numbers.

Here's one way to find the GCF of 12 and 18. List the factors of 12 and 18.

factors of 12: 1, 2, 3, 4, 6, 12
factors of 18: 1, 2, 3, 6, 9, 18

The common factors of 12 and 18, shown in red, are 1, 2, 3, and 6. The greatest common factor (GCF) of 12 and 18 is 6.

Another way to find the GCF of two or more numbers is given below.

Method

1. ▶ Write the prime factorization for each number.
2. ▶ Circle the common prime factors.
3. ▶ Find the product of the common prime factors.

Example A

Find the GCF of 16 and 20.

1. ▶ $16 = 2 \times 2 \times 2 \times 2$
   $20 = 2 \times 2 \times 5$
2. ▶ $16 = \boxed{2} \times \boxed{2} \times 2 \times 2$
   $20 = \boxed{2} \times \boxed{2} \times 5$
3. ▶ GCF: $2 \times 2 = 4$

The GCF of 16 and 20 is 4.

When you multiply a number by the whole numbers 0, 1, 2, 3, and so on, you get multiples of the number.

The **least common multiple (LCM)** of two or more numbers is the least nonzero number that is a multiple of each number.

multiples of 6: 0, 6, 12, 18, 24, 30, 36, 42, . . .
multiples of 9: 0, 9, 18, 27, 36, 45, . . .

Two common multiples of 6 and 9 are shown in red. The least nonzero common multiple (LCM) of 6 and 9 is 18.

You can also use prime factors to find the LCM of two or more numbers.

Method

1. ▶ Write the prime factorization for each number.
2. ▶ Circle the common prime factors and each remaining factor.
3. ▶ Find the product of the prime factors using each common prime factor only once.

Example B

Find the LCM of 6 and 10.

1. ▶ $6 = 2 \times 3$
   $10 = 2 \times 5$
2. ▶ $6 = \boxed{2} \times \boxed{3}$
   $10 = \boxed{2} \times \boxed{5}$
3. ▶ LCM: $2 \times 3 \times 5$
   or 30

The LCM of 6 and 10 is 30.

Guided Practice

Example A

Find the GCF of each group of numbers.

1. 9, 12   **2.** 20, 30   **3.** 15, 56   **4.** 81, 108   **5.** 6, 12, 18

Example B

Find the LCM of each group of numbers.

6. 3, 4   **7.** 12, 21   **8.** 10, 25   **9.** 12, 35   **10.** 8, 12, 24

Exercises

Practice

Find the GCF of each group of numbers.

11. 8, 18   **12.** 6, 9   **13.** 4, 12   **14.** 18, 24   **15.** 8, 24

16. 17, 51   **17.** 65, 95   **18.** 42, 48   **19.** 64, 32   **20.** 72, 144

Find the LCM of each group of numbers.

21. 5, 6   **22.** 9, 27   **23.** 12, 15   **24.** 8, 12   **25.** 5, 15

26. 13, 39   **27.** 16, 24   **28.** 18, 20   **29.** 21, 14   **30.** 25, 30

Applications

31. Mrs. Jacobs is paid $10.25 an hour as a practical nurse. To the nearest cent, how much is she paid for a 7.5-hour day?

32. Barney saws an 18-foot log into 2-foot pieces. How many cuts will he make?

33. Sweat suits usually sell for $49.99. They are on sale for $35. How much is saved if three sweat suits are bought?

Using Variables

34. If n is a prime number, what is the GCF of $12n$ and $15n$?

35. Find the prime factorization of $5c + 6d$ if $c = 4$ and $d = 5$.

36. If p is a prime number, what is the LCM of $5p$ and $15p$?

Suppose

37. Suppose your real estate taxes, car insurance, and water bill all came due in June. The real estate taxes are due every six months. The car insurance is due every four months and the water bill is due every three months. Name the next month that all three bills will come due at the same time.

38. Any odd number greater than 5 can be written as the sum of three primes. For example, $11 = 5 + 3 + 3$. Show that this statement is true for odd numbers between 20 and 30.

Research

39. Write a report about Goldbach's Conjecture.

4-4 TOTAL DEDUCTIONS AND TAKE-HOME PAY

Objective

Read an earnings statement.

Sandy Graves is an office manager for Carpenter's Lumber Company. Sandy checks her earnings statement to make sure it is correct. This earnings statement was attached to her last paycheck.

Carpenter's Lumber Company				Weekly Earnings Statement		
Sandy Graves						
Pay Period Ending 8/16		Pay Date 8/16		Check Number 43075		
Hours	Rate	Earnings	Type	Deduction		Type
40 00	8 19	327 60	Reg	1 25		Dental
				8 54		Health
				10 17		401 K
				54		AccIns
				10 00		CrUnion
				2 00		UnitedFd
This Pay	Gross Pay 327 60	Federal 52 42	F.I.C.A. 24 57	State 9 83	Local 3 28	Net Pay 205 00

Sandy adds to check her total tax deductions.

Federal Tax	$52.42
Social Security (F.I.C.A.)	24.57
State Tax	9.83
Local Tax	+ 3.28
	$90.10

Sandy also requests her employer to make these personal deductions.

Dental Insurance	$ 1.25
Health Insurance	8.54
401 K (Retirement)	10.17
Accident Insurance	0.54
Credit Union	10.00
United Way	+ 2.00
	$32.50

The *net pay* or take-home pay is found by subtracting the total tax and personal deductions from the *gross pay*, her weekly wage.

$90.10	$327.60	gross pay
+ 32.50	− 122.60	deductions
$122.60	$205.00	net pay

Understanding the Data

● How many hours a week does Sandy work?

● How much does she make an hour?

● Explain how you would check to see if her gross pay is correct. Then check to see if it is correct.

Guided Practice

1. Mr. Garcia's gross pay is $569.19. His total tax deductions are $156.54 and his personal deductions are $68.49. What are his total deductions and his net pay?

Practice

Complete the chart.

Name	Gross Pay	Total Tax Deduction	Total Personal Deduction	Take-Home Pay
L. Adams	$187.60	$25.43	$14.25	2. ■
D. Block	$193.35	$34.82	$16.70	3. ■
M. Federico	$215.70	$38.64	$10.18	4. ■
J. Mason	$243.92	$29.58	$12.50	5. ■
B. Klenk	$174.51	$31.05	$ 8.67	6. ■

Applications

| Paper/ Pencil |
| Mental Math |
| Estimate |
| Calculator |

7. Tom Poling has $36.20 deducted weekly from his paycheck for federal tax, $4.52 for state tax, and $2.05 for city tax. How much total tax does he pay yearly?

8. Jenny Spalding's gross pay is $472.36. Her total tax deductions are $129.91 and her personal deductions are $53.49. What is her net pay?

9. Todd Benson's gross pay is $394.93, and his net pay is $289.27. What are his total deductions?

10. Refer to Sandy's earnings statement on page 114. Sandy uses her credit union for a savings account. How much savings will Sandy have in one year?

Interpreting Data

11. In your own words, explain the difference between gross pay and net pay.

12. Name some ways you could make your net pay higher without getting a raise.

Suppose

13. Suppose your net pay is less than or greater than normal and you know of no changes. What would you do?

Research

14. Contact the nearest Social Security (F.I.C.A.) office to ask for a booklet that explains the laws, benefits, and current rate taken out of earnings. Determine what benefits are available. If you have a job, check the rate against your earnings statement.

Mixed Review

Lesson 3-1

Multiply.

15. 82
      × 5

16. 46
      ×15

17. 1,432
       ×   91

18. 6,423
       × 108

Lesson 3-10

Solve for d, r, or t using the formula d = rt.

19. $d = 240$ miles, $t = 6$ hours, $r = $ ___?___

20. $r = 58$ mph, $t = 8.25$ hours, $d = $ ___?___

21. $d = 48$ miles, $r = 16$ mph, $t = $ ___?___

Lesson 3-11

22. Erin is going to carpet her bedroom floor. Her bedroom is rectangular in shape and measures 12 feet by 9 feet. How many square yards of carpet does she need to cover her floor?

Carpenter's Lumber Co.
980 Maple Street
Miami, FL 33100

Pay to the order of _____ Sandy Graves _____ Da

Two Hundred and Five Dollars and No Cents----------

Memo_____

4-5 EQUIVALENT FRACTIONS

Objective

Find fractions equivalent to given fractions.

Katie tries to get 8 hours of sleep each night. If she does this the rest of her life, she will spend $\frac{8}{24}$ or $\frac{1}{3}$ of her life sleeping.

A fraction, such as $\frac{1}{3}$, may be used to name part of a whole or a group.

$\frac{1}{3}$ ← numerator
← denominator

The *denominator* tells the part of objects or equal-sized parts. The *numerator* tells the number of objects or parts being considered.

The figures at the right show that $\frac{1}{3}$ and $\frac{8}{24}$ name the same number. They are called **equivalent fractions.**

You can multiply or divide the numerator and denominator of a fraction by the same nonzero number to find an equivalent fraction.

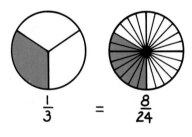

$\frac{1}{3} = \frac{8}{24}$

Examples

Replace each ▓ with a number so that the fractions are equivalent.

A $\frac{3}{7} = \frac{▓}{21}$

Multiply the numerator and the denominator by 3. This is like multiplying by what number?

$\overset{\times\ 3}{\frown}$
$\frac{3}{7} = \frac{▓}{21} \rightarrow \frac{3}{7} = \frac{9}{21}$
$\underset{\times\ 3}{\smile}$

B $\frac{12}{36} = \frac{▓}{9}$

Divide the numerator and the denominator by 4. This is like dividing by what number?

$\overset{\div\ 4}{\frown}$
$\frac{12}{36} = \frac{▓}{9} \rightarrow \frac{12}{36} = \frac{3}{9}$
$\underset{\div\ 4}{\smile}$

Name another fraction equivalent to $\frac{3}{9}$.

Guided Practice

Replace each ▓ with a number so that the fractions are equivalent.

Example A

1. $\frac{6}{7} = \frac{▓}{28}$ **2.** $\frac{1}{2} = \frac{▓}{14}$ **3.** $\frac{2}{3} = \frac{▓}{6}$ **4.** $\frac{3}{4} = \frac{▓}{8}$ **5.** $\frac{7}{9} = \frac{▓}{18}$

Replace each ■ **with a number so that the fractions are equivalent.**

6. $\frac{6}{12} = \frac{■}{2}$ **7.** $\frac{4}{20} = \frac{■}{10}$ **8.** $\frac{4}{16} = \frac{■}{4}$ **9.** $\frac{6}{15} = \frac{■}{5}$ **10.** $\frac{6}{18} = \frac{■}{3}$

Exercises

Practice

Replace each ■ **with a number so that the fractions are equivalent.**

11. $\frac{5}{9} = \frac{■}{27}$ **12.** $\frac{7}{12} = \frac{■}{36}$ **13.** $\frac{6}{1} = \frac{■}{4}$ **14.** $\frac{2}{2} = \frac{■}{12}$ **15.** $\frac{3}{16} = \frac{■}{48}$

16. $\frac{6}{16} = \frac{■}{8}$ **17.** $\frac{5}{5} = \frac{■}{1}$ **18.** $\frac{10}{25} = \frac{■}{5}$ **19.** $\frac{15}{9} = \frac{■}{3}$ **20.** $\frac{12}{18} = \frac{■}{9}$

21. $\frac{3}{4} = \frac{■}{12}$ **22.** $\frac{1}{3} = \frac{■}{6}$ **23.** $\frac{3}{15} = \frac{■}{5}$ **24.** $\frac{12}{16} = \frac{■}{4}$ **25.** $\frac{3}{7} = \frac{■}{42}$

26. $\frac{1}{25} = \frac{■}{100}$ **27.** $\frac{0}{36} = \frac{■}{3}$ **28.** $\frac{6}{32} = \frac{■}{16}$ **29.** $\frac{9}{32} = \frac{■}{64}$ **30.** $\frac{10}{5} = \frac{■}{1}$

Name a fraction equivalent to each fraction.

31. $\frac{2}{3}$ **32.** $\frac{5}{8}$ **33.** $\frac{1}{4}$ **34.** $\frac{2}{6}$ **35.** $\frac{5}{10}$ **36.** $\frac{10}{16}$

Applications

Write two equivalent fractions for each of the following. Use the circle graph.

37. What fraction of the population is from 0-19 years old?

38. What fraction of the population is from 20-34 years old?

39. What fraction of the population is 35 years old or older?

40. What fraction of the population is 34 years old or younger?

U.S. Population by Age

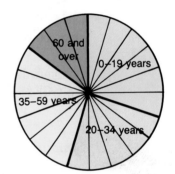

41. The trans-Alaska pipeline delivers about 2 million barrels of oil per day. The pipeline is 800 miles long. About how many barrels are delivered in 7.5 days?

Write Math

42. Write four fractions that are equivalent to three-fourths. Write each fraction in words.

Suppose

43. Suppose the denominator of a fraction is increased, what happens to the value of the fraction?

Critical Thinking

44. There is a gallon of punch in a punch bowl and a gallon of lemonade in another punch bowl. Mickey puts a ladle of punch in the lemonade. Then Mickey takes a ladle of lemonade and puts it in the punch. Now, is there more punch in the lemonade or more lemonade in the punch?

4-6 SIMPLIFYING FRACTIONS

Objective
Write fractions in simplest form.

In the 1988 presidential election, George Bush carried 40 out of 50 states. You could write the fraction of states that he carried using the fractions $\frac{40}{50}$, $\frac{8}{10}$, and $\frac{4}{5}$.

The fraction $\frac{4}{5}$ is in simplest form.

A fraction is in *simplest form* when the GCF of the numerator and denominator is 1. To write a fraction in simplest form, divide both the numerator and the denominator by common factors until their GCF is 1.

Method

1 Divide the numerator and the denominator by a common factor.

2 Continue dividing by common factors until the GCF is 1.

Examples

Simplify each fraction.

A $\frac{24}{30}$

1 **2**

$$\overset{\div 2}{\overset{\frown}{}} \overset{\div 3}{\overset{\frown}{}}$$
$$\frac{24}{30} = \frac{12}{15} = \frac{4}{5}$$
$$\underset{\div 2}{\underset{\smile}{}} \underset{\div 3}{\underset{\smile}{}}$$

The GCF of 4 and 5 is 1.

B $\frac{16}{24}$

1 **2**

$$\overset{\div 2}{\overset{\frown}{}} \overset{\div 2}{\overset{\frown}{}} \overset{\div 2}{\overset{\frown}{}}$$
$$\frac{16}{24} = \frac{8}{12} = \frac{4}{6} = \frac{2}{3}$$
$$\underset{\div 2}{\underset{\smile}{}} \underset{\div 2}{\underset{\smile}{}} \underset{\div 2}{\underset{\smile}{}}$$

How could the GCF be used to simplify $\frac{16}{24}$?

C $\frac{5}{15}$

1
$$\overset{\div 5}{\overset{\frown}{}}$$
$$\frac{5}{15} = \frac{1}{3}$$
$$\underset{\div 5}{\underset{\smile}{}}$$

D $\frac{7}{9}$

The GCF of 7 and 9 is 1 so the fraction is already in simplest form.

Guided Practice

Simplify each fraction.

Example A

1. $\frac{12}{16}$ 2. $\frac{28}{32}$ 3. $\frac{75}{100}$ 4. $\frac{8}{16}$ 5. $\frac{6}{18}$ 6. $\frac{27}{36}$

Example B

7. $\frac{16}{64}$ 8. $\frac{8}{16}$ 9. $\frac{50}{100}$ 10. $\frac{24}{40}$ 11. $\frac{32}{80}$ 12. $\frac{8}{24}$

Examples C, D

13. $\frac{20}{25}$ 14. $\frac{4}{10}$ 15. $\frac{3}{5}$ 16. $\frac{14}{19}$ 17. $\frac{9}{12}$ 18. $\frac{6}{8}$

19. $\frac{15}{18}$ 20. $\frac{9}{20}$ 21. $\frac{8}{21}$ 22. $\frac{10}{15}$ 23. $\frac{9}{24}$ 24. $\frac{6}{31}$

Exercises

Practice

Simplify each fraction.

25. $\frac{18}{32}$ 26. $\frac{15}{36}$ 27. $\frac{27}{72}$ 28. $\frac{23}{69}$ 29. $\frac{36}{48}$ 30. $\frac{6}{30}$

31. $\frac{15}{27}$ 32. $\frac{20}{24}$ 33. $\frac{6}{16}$ 34. $\frac{4}{6}$ 35. $\frac{7}{21}$ 36. $\frac{18}{27}$

37. $\frac{24}{30}$ 38. $\frac{8}{10}$ 39. $\frac{16}{18}$ 40. $\frac{12}{36}$ 41. $\frac{30}{50}$ 42. $\frac{14}{21}$

43. Write $\frac{6}{9}$ in simplest form.   44. What is the simplest form of $\frac{6}{12}$?

Applications

Paper/ Pencil
Mental Math
Estimate
Calculator

45. In the 1980 presidential election, Ronald Reagan carried 44 of the 50 states. What fraction, in simplest form, of the states did he carry?

46. On the average, 12 of the 30 days in June in Pittsburgh are rainy. What fraction, in simplest form, of the days in June are not rainy?

Mental Math

Place operation signs (+, −, ×, or ÷) and parentheses, if needed, in each equation to make a true statement.

*f***UN with MATH**

How long has the equal sign been around?
See page 175.

47. 5 __?__ 5 __?__ 5 = 50   48. 5 __?__ 5 __?__ 5 = 6

49. 5 __?__ 5 __?__ 5 = $\frac{1}{2}$ 50. 5 __?__ 5 __?__ 5 = 30

Show Math

51. Draw a rectangle and divide it into equal-sized parts in such a way that shows three-fourths is equal to six-eighths.

Critical Thinking

52. A pile of dimes is counted by 2s, 3s, 4s, and 5s but there is always one remaining. When counted by 7s, none are left over. What is the total value of the coins in the pile?

Mixed Review

Lesson 1-6

Estimate.

53. $\begin{array}{r} 65 \\ + 81 \end{array}$ 54. $\begin{array}{r} 983 \\ - 426 \end{array}$ 55. $\begin{array}{r} 6,423 \\ + 718 \end{array}$ 56. $\begin{array}{r} 7,773 \\ - 4,281 \end{array}$

Lesson 3-2

Multiply.

57. $\begin{array}{r} 6.4 \\ \times .02 \end{array}$ 58. $\begin{array}{r} 9.23 \\ \times 1.42 \end{array}$ 59. $\begin{array}{r} 72.6 \\ \times 8.3 \end{array}$ 60. $\begin{array}{r} 12.001 \\ \times 3.19 \end{array}$

Lesson 3-13

61. Marcus and his two brothers were given 8 boxes of 12 pencils each. They shared the pencils equally. How many pencils did each receive?

USES OF COMPUTERS TODAY

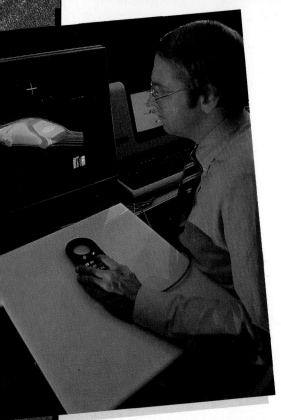

Computers are widely used in today's society. In the medical field, a computer can help a doctor make faster and more accurate diagnoses. It can analyze a patient's medical history and the results of laboratory tests. A computer can help medical researchers study the effects of drugs and procedures before they are used on patients.

In businesses, computers are largely used to help with paperwork, record keeping, and information retrieval. Word-processing software makes it possible for computers to be used in place of typewriters.

Many department stores and supermarkets are using *point-of-sale (POS) terminals* in place of cash registers. Part of a POS terminal is an optical scanning device. This device is used to read a *Universal Product Code (UPC)* that is printed on each item.

Many banks provide *automated teller machines.* These machines can be used to transfer funds from one account to another, deposit funds, withdraw funds, pay utility bills, and get a balance statement. The machines are connected directly to the bank's computer.

Drawings, music, sculpture, and poetry are being generated by computers. Computers are also being used to produce special effects in movies and animated films.

1. How are computers being used in the medical field?
2. How are businesses using computers?
3. What do POS and UPC stand for? How are they used by department stores and supermarkets?
4. **Research** How are computers being used in air and space travel?

USING CHARTS AND GRAPHS

Stacey Mills works in the advertising department of a recordings manufacturer. Her job is to increase the number of customers the company has. She puts together a report that lists the wholesale value of recordings for 1991.

To conserve space and make the numbers easy to read, Stacey writes the numbers in shortened form.

For example, the wholesale value of CD's was $4,337,700,000. Written in shortened form, 4,337,700,000 is 4,337.7 million and read *four billion, three hundred thirty-seven million, seven hundred thousand.* Many newspapers and magazines write large amounts in this way.

Manufacturers' Value of Recordings in 1991	
(in millions)	
Singles	63.9
LPs	29.4
CDs	4,337.7
Cassettes	3,019.6
Cassette singles	230.4

LP–Long Play
CD–Compact Disk

Use the appropriate table. Write your answers in standard form.

1. What was the wholesale value of singles in 1991?

2. What was the total wholesale value of recordings in the five categories in 1991?

3. How much less was the wholesale value of cassettes than CDs in 1991? (do not include cassette singles)

4. In 1970, Joe Namath earned $150,000. How much was that salary worth in 1990?

5. How much more did Joe Montana earn in 1990 than Sammy Baugh earned in 1940?

6. What is the difference in salary from what Walter Payton earned in 1980 and what he would earn in 1990?

7. Comparing the salary earned and the 1990 salary, what conclusion can you come to about the value of the dollar?

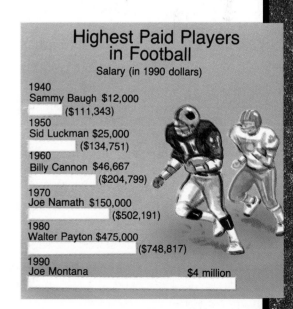

Highest Paid Players in Football
Salary (in 1990 dollars)

1940
Sammy Baugh $12,000
($111,343)
1950
Sid Luckman $25,000
($134,751)
1960
Billy Cannon $46,667
($204,799)
1970
Joe Namath $150,000
($502,191)
1980
Walter Payton $475,000
($748,817)
1990
Joe Montana $4 million

4-7 COMPARING FRACTIONS

Objective

Compare and order fractions.

Four-fifths of Earth's surface is covered by water. One-fifth of Earth's surface is covered by land. Is Earth's surface covered mostly by land or water?

To compare fractions with the same denominator, compare the numerators.

Example A

Compare $\frac{4}{5}$ and $\frac{1}{5}$.   You can use a number line to compare fractions.

Compare the numerators. $4 > 1$

Since $4 > 1$, it follows that $\frac{4}{5} > \frac{1}{5}$.

Earth's surface is covered mostly by water.

To compare fractions with different denominators, find equivalent fractions that have the same denominator. To do this, find their **least common denominator,** the LCM of the denominators. Then compare the numerators.

Method

1 Find the least common denominator (LCD).

2 Find equivalent fractions using the LCD as the denominator.

3 Compare the numerators.

Example B

Compare $\frac{4}{5}$ and $\frac{5}{6}$.

1 Find the LCD of 5 and 6. $5 = ⑤$
$6 = ② \times ③$
LCD: $5 \times 2 \times 3$ or 30

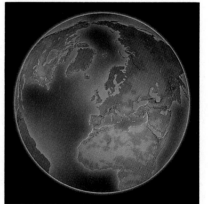

2 $\overset{\curvearrowright \times 6 \searrow}{\underset{\searrow \times 6 \nearrow}{\frac{4}{5} = \frac{24}{30}}}$ $\overset{\curvearrowright \times 5 \searrow}{\underset{\searrow \times 5 \nearrow}{\frac{5}{6} = \frac{25}{30}}}$

3 Since $24 < 25$, it follows that $\frac{24}{30} < \frac{25}{30}$ or $\frac{4}{5} < \frac{5}{6}$.

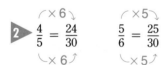 Guided Practice

Replace each ● with <, >, or = to make a true sentence.

Example A

1. $\frac{2}{3} ● \frac{1}{3}$ 2. $\frac{6}{7} ● \frac{4}{7}$ 3. $\frac{3}{8} ● \frac{7}{8}$ 4. $\frac{11}{16} ● \frac{15}{16}$ 5. $\frac{9}{11} ● \frac{3}{11}$

Example B

6. $\frac{7}{8} ● \frac{3}{4}$ 7. $\frac{1}{3} ● \frac{1}{5}$ 8. $\frac{3}{7} ● \frac{4}{5}$ 9. $\frac{4}{16} ● \frac{1}{4}$ 10. $\frac{5}{12} ● \frac{2}{3}$

Practice

Replace each ● with <, >, or = to make a true sentence.

11. $\frac{0}{5}$ ● $\frac{3}{5}$ **12.** $\frac{8}{9}$ ● $\frac{6}{9}$ **13.** $\frac{5}{6}$ ● $\frac{2}{6}$ **14.** $\frac{3}{8}$ ● $\frac{5}{8}$ **15.** $\frac{7}{10}$ ● $\frac{3}{10}$

16. $\frac{2}{9}$ ● $\frac{3}{9}$ **17.** $\frac{1}{12}$ ● $\frac{5}{12}$ **18.** $\frac{6}{7}$ ● $\frac{5}{7}$ **19.** $\frac{2}{7}$ ● $\frac{4}{14}$ **20.** $\frac{3}{4}$ ● $\frac{3}{8}$

21. $\frac{2}{3}$ ● $\frac{6}{9}$ **22.** $\frac{7}{10}$ ● $\frac{3}{5}$ **23.** $\frac{7}{11}$ ● $\frac{3}{11}$ **24.** $\frac{6}{16}$ ● $\frac{3}{8}$ **25.** $\frac{5}{7}$ ● $\frac{2}{3}$

26. $\frac{4}{9}$ ● $\frac{1}{2}$ **27.** $\frac{3}{8}$ ● $\frac{5}{6}$ **28.** $\frac{2}{5}$ ● $\frac{2}{3}$ **29.** $\frac{9}{15}$ ● $\frac{11}{15}$ **30.** $\frac{3}{5}$ ● $\frac{7}{8}$

Order the following fractions from least to greatest.

31. $\frac{1}{2}, \frac{1}{3}, \frac{1}{5}, \frac{1}{4}$ **32.** $\frac{3}{8}, \frac{1}{4}, \frac{7}{8}, \frac{3}{4}$ **33.** $\frac{1}{3}, \frac{5}{6}, \frac{3}{8}, \frac{7}{8}$

34. $\frac{5}{6}, \frac{2}{3}, \frac{3}{4}, \frac{4}{5}$ **35.** $\frac{5}{16}, \frac{1}{4}, \frac{1}{2}, \frac{11}{16}$ **36.** $\frac{3}{7}, \frac{2}{5}, \frac{4}{9}, \frac{5}{11}$

Applications

37. About $\frac{1}{12}$ of the world's population lives in North America. About $\frac{1}{8}$ of the population lives in Africa. Which continent has a greater population?

38. In a recent year, $\frac{3}{4}$ of the domestic motor vehicles sold in the U.S. were cars and $\frac{5}{24}$ were trucks. Were more cars or trucks sold in the U.S.?

39. Bananas are on sale for $0.39 a pound. About how many pounds can Mrs. Barry buy for $2.00?

Critical Thinking

40. Notice the page numbers in the open book. The product of these page numbers is 7,482. Suppose you open the book to two other page numbers whose product is 30,450. To which pages do you have the book open?

Mental Math

Justin Robinson's answers to a math quiz on adding and subtracting decimals are shown below.

Without using pencil and paper to calculate, determine whether each answer is reasonable. Write yes or no.

41. $12.1 + 0.65 = $ _14.75_   **42.** $52.1 - 0.45 = $ _51.65_

43. $8.9 + 13 + 9.5 = $ _19.7_   **44.** $5 + 5.1 + 23.41 = $ _23.97_

45. $53.7 - 36 = $ _17.7_   **46.** $35.7 - 24.83 = $ _10.87_

47. $82.6 + 15.25 = $ _97.85_   **48.** $9.36 - 2.4 = $ _9.12_

49. $10.17 + 9.16 + 21.3 = $_25.43_   **50.** $113.52 - 91.65 = $ _21.87_

4-8 MIXED NUMBERS

Objective

Change improper fractions to mixed numbers and vice versa.

A survey showed that two-thirds of the people attending an event at the Fort Wayne Coliseum were Indianapolis Colts fans.

A fraction, like $\frac{2}{3}$, that has a numerator that is less than the denominator is a **proper fraction**.

A fraction, like $\frac{7}{4}$ or $\frac{5}{5}$, that has a numerator that is greater than or equal to the denominator is an **improper fraction**.

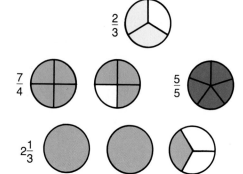

A **mixed number**, like $2\frac{1}{3}$, indicates the sum of a whole number and a fraction.

$$2\frac{1}{3} = 2 + \frac{1}{3}$$

An improper fraction may be changed to a mixed number by dividing the numerator by the denominator.

Remember that the bar separating the numerator and the denominator indicates division.

Examples

Change each fraction to a mixed number in simplest form.

A $\frac{18}{4}$

$$\begin{array}{r} 4 \text{ R2} \\ 4\overline{)18} \\ -16 \\ \hline 2 \end{array} \longrightarrow 4\frac{2}{4} = 4\frac{1}{2}$$

So $\frac{18}{4} = 4\frac{1}{2}$.

B $\frac{30}{6}$

$$\begin{array}{r} 5 \\ 6\overline{)30} \\ -30 \\ \hline 0 \end{array} \longrightarrow 5 \quad \text{The remainder is 0.}$$

So $\frac{30}{6} = 5$.

You can use the following method to change a mixed number to an improper fraction.

Method

1 Multiply the whole number by the denominator and add the numerator.

2 Write the sum over the denominator.

Example C

Change $3\frac{2}{5}$ to an improper fraction.

$$3\frac{2}{5} \xrightarrow{\;1\;} \frac{(3 \times 5) + 2}{5} \xrightarrow{\;2\;} \frac{17}{5}$$

Explain how this diagram shows that $3\frac{2}{5} = \frac{17}{5}$.

Guided Practice

Change each fraction to a mixed number in simplest form.

Example A

1. $\frac{5}{3}$ 2. $\frac{7}{6}$ 3. $\frac{9}{6}$ 4. $\frac{42}{8}$ 5. $\frac{24}{9}$ 6. $\frac{14}{5}$

Example B

7. $\frac{6}{6}$ 8. $\frac{12}{4}$ 9. $\frac{10}{5}$ 10. $\frac{36}{9}$ 11. $\frac{42}{7}$ 12. $\frac{81}{9}$

Example C

Change each mixed number to an improper fraction.

13. $1\frac{3}{8}$ 14. $1\frac{7}{12}$ 15. $3\frac{2}{3}$ 16. $4\frac{2}{5}$ 17. $3\frac{5}{7}$ 18. $11\frac{2}{3}$

Exercises

Practice

Change each fraction to a mixed number in simplest form.

19. $\frac{8}{7}$ 20. $\frac{9}{5}$ 21. $\frac{21}{7}$ 22. $\frac{7}{4}$ 23. $\frac{9}{7}$ 24. $\frac{19}{8}$

25. $\frac{11}{5}$ 26. $\frac{12}{8}$ 27. $\frac{16}{6}$ 28. $\frac{33}{9}$ 29. $\frac{76}{9}$ 30. $\frac{40}{24}$

Change each mixed number to an improper fraction.

31. $1\frac{2}{5}$ 32. $2\frac{3}{4}$ 33. $1\frac{7}{8}$ 34. $2\frac{1}{2}$ 35. $2\frac{9}{10}$ 36. $3\frac{5}{8}$

Write Math

Write the name of each fraction in words.

37. $\frac{1}{2}$ 38. $1\frac{2}{3}$ 39. $2\frac{7}{8}$ 40. $\frac{9}{5}$

Using Expressions

41. If $2 + \frac{1}{3}$ can be written as the improper fraction $\frac{7}{3}$, how would you write $n + \frac{1}{3}$ as an improper fraction?

Critical Thinking

42. Without bending or breaking the sticks, how would you place six ice cream sticks so that each of the six sticks touches each of the others?

Applications

43. Two lots for home building were sold. One lot contained $1\frac{3}{5}$ acres and the other $1\frac{7}{10}$ acres. Which lot was larger?

44. A dress pattern calls for $3\frac{5}{8}$ yards of material. Melissa bought $3\frac{3}{4}$ yards. Did she buy enough material?

45. Bart plans to set up a lawn mowing business for his summer job. He can buy a used lawn mower for $50. If he saves $4.75 a week, how long will it take him to save for the mower?

Mixed Review

Lesson 1-9

Find the value of each expression.

46. $16 \times 2 \div 8 + 20$ 47. $9 \div 3 + 14 \div 7 + 18$

Lesson 3-9

Find the common ratio. Write the next 3 terms in each sequence.

48. 5.6, 11.2, 22.4   49. 729, 243, 81

Lesson 4-6

50. Four months out of the year, Mrs. Joel lives in Florida. The rest of the year Mrs. Joel lives in Minnesota. What fraction, in simplest form, of the year does Mrs. Joel live in Florida?

4-9 ESTIMATING SUMS AND DIFFERENCES

Objective

Estimate the sum and difference of fractions.

Michael volunteers a couple of hours each Saturday at the Village Retirement Center. One Saturday he and Mr. Svenson walked $\frac{9}{10}$ mile to attend a festival in the park. On the return trip, they took a shortcut that was a $\frac{4}{5}$-mile walk. *About* how far did they walk to the park and back? (The problem is solved in Example F.)

To round fractions before estimating their sum or difference, you can use the following guidelines.
- If the numerator is much less than the denominator, round the fraction to 0.
- If the numerator is about half of the denominator, round the fraction to $\frac{1}{2}$.
- If the numerator is a little less than the denominator, round the fraction to 1.

Examples

State whether each fraction is close to 0, $\frac{1}{2}$, or 1. You can see that $\frac{1}{5}$ is close to 0, $\frac{9}{16}$ is close to $\frac{1}{2}$, and $\frac{4}{5}$ is close to 1.

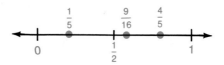

A $\frac{4}{5} \to 1$ 4 is a little less than 5.   **B** $\frac{9}{16} \to \frac{1}{2}$ 9 is about half of 16.   **C** $\frac{1}{5} \to 0$ 1 is much less than 5.

Mixed numbers are usually rounded to the nearest whole number.

Examples

Round each mixed number to the nearest whole number.

D $1\frac{5}{6} \to 2$ **E** $14\frac{5}{16} \to 14$

You can estimate sums and differences of fractions and mixed numbers.

Method

▶1 Round each fraction or mixed number.   ▶2 Add or subtract.

Examples

F *To find out about how far Michael and Mr. Svenson walked, estimate $\frac{9}{10} + \frac{4}{5}$.*

▶1 $\frac{9}{10} \to 1$ $\frac{4}{5} \to 1$

▶2 $1 + 1 = 2$

G *Estimate $2\frac{3}{4} - \frac{3}{8}$.*

▶1 $2\frac{3}{4} \to 3$ $\frac{3}{8} \to \frac{1}{2}$

▶2 $3 - \frac{1}{2} = 2\frac{1}{2}$

$2\frac{3}{4} - \frac{3}{8}$ is about $2\frac{1}{2}$.

Michael and Mr. Svenson walked about 2 miles.

State whether each fraction is close to 0, $\frac{1}{2}$, or 1. Use a number line to help.

Examples A-C

1. $\frac{11}{12}$ 2. $\frac{5}{8}$ 3. $\frac{2}{5}$ 4. $\frac{1}{10}$ 5. $\frac{1}{6}$ 6. $\frac{2}{3}$

Examples D, E

Round each mixed number to the nearest whole number.

7. $6\frac{2}{3}$ 8. $18\frac{1}{4}$ 9. $27\frac{5}{8}$ 10. $32\frac{2}{5}$ 11. $12\frac{7}{9}$ 12. $20\frac{1}{3}$

Estimate.

Example F

13. $\frac{9}{16} + \frac{13}{24}$ 14. $\frac{5}{6} + \frac{4}{9}$ 15. $3\frac{1}{4} + \frac{2}{3}$ 16. $5\frac{1}{2} + 6\frac{7}{12}$

Example G

17. $\frac{5}{12} - \frac{1}{10}$ 18. $\frac{4}{5} - \frac{1}{3}$ 19. $3\frac{2}{5} - \frac{7}{8}$ 20. $3\frac{5}{11} - 2\frac{1}{2}$

Estimate.

Practice

21. $\frac{2}{5} + \frac{1}{3}$ 22. $\frac{1}{5} + \frac{4}{10}$ 23. $\frac{1}{6} + \frac{4}{9}$ 24. $\frac{3}{4} + \frac{7}{16}$

25. $\frac{5}{6} - \frac{7}{12}$ 26. $\frac{5}{8} - \frac{1}{10}$ 27. $12\frac{3}{4} - \frac{3}{8}$ 28. $7\frac{2}{3} - \frac{3}{5}$

29. $14\frac{1}{10} - 6\frac{4}{5}$ 30. $8\frac{1}{3} + 2\frac{1}{6}$ 31. $4\frac{7}{8} + 7\frac{3}{4}$ 32. $11\frac{11}{12} - 5\frac{1}{4}$

Make Up a Problem

33. Make up a problem that has two addends with an estimated sum of 1. Use fractions and/or mixed numbers.

Applications

34. Carlos runs $1\frac{1}{2}$ miles on Monday, $2\frac{1}{3}$ miles on Tuesday, and $1\frac{5}{6}$ miles on Wednesday. About how many miles did he run in all?

***f*UN with MATH**

Ready for a game of Korean tic tac toe?
See page 175.

35. Which race is longer, a $1\frac{7}{16}$-mile race or a $1\frac{3}{8}$-mile race?

36. A pane of glass costs 99¢. What does it cost to buy glass to replace six windows that have four panes each?

Mixed Review

Write as a product and then find the number named.

Lesson 1-2

37. 4 cubed   38. 2^6 39. 5 squared   40. 8^2

Lesson 1-7

Estimate.

41. $\begin{array}{r} 16 \\ \times\ 8 \end{array}$ 42. $0.6\overline{)12}$ 43. $\begin{array}{r} 481 \\ \times\ 322 \end{array}$ 44. $22.1\overline{)437.2}$

Lesson 2-8

45. Marcia walks 5 miles each day. She walks 0.5 mile on Elm St., 0.25 mile on Oak St., 0.75 mile on Main St., and 1.00 mile on Town St. How many times does she walk this route each day?

4-10 ADDING FRACTIONS

Objective

Add proper fractions.

Barbara is making a jacket. She needs $\frac{5}{8}$ yard of backing for the front of the jacket and $\frac{2}{8}$ yard of backing for the collar. How much backing does she need to buy?

To add fractions with like denominators, add the numerators and write the sum over the denominator. Simplify if necessary.

Examples

Add.

A $\frac{5}{8} + \frac{2}{8}$ Estimate: $\frac{1}{2} + \frac{1}{2} = 1$

$\frac{5}{8} + \frac{2}{8} = \frac{7}{8}$ Compared to the estimate, is the answer reasonable?

Barbara needs $\frac{7}{8}$ yard of backing.

B $\frac{3}{18} + \frac{17}{18}$ Estimate: $0 + 1 = 1$

$\frac{3}{18} + \frac{17}{18} = \frac{20}{18} \rightarrow 18\overline{)20}^{\,1\ R2}$

$= 1\frac{2}{18}$ or $1\frac{1}{9}$

Is the answer reasonable?

Add fractions with unlike denominators as follows.

Method

1 Find the least common denominator (LCD).

2 Rename the fractions using the LCD.

3 Add. Simplify if necessary.

Example C

Add $\frac{7}{8} + \frac{5}{12}$.

1 $8 = \boxed{2} \times \boxed{2} \times \boxed{2}$
$12 = \boxed{2} \times \boxed{2} \times \boxed{3}$ LCD: $2 \times 2 \times 2 \times 3$ or 24

Estimate: $1 + \frac{1}{2} = 1\frac{1}{2}$

2
$$\begin{array}{r} \frac{7}{8} \\ + \frac{5}{12} \end{array} \rightarrow \begin{array}{r} \frac{21}{24} \\ + \frac{10}{24} \end{array}$$

3
$$\begin{array}{r} \frac{21}{24} \\ + \frac{10}{24} \\ \hline \frac{31}{24} \end{array}$$
Change to a mixed number.

$\frac{31}{24} \rightarrow 24\overline{)31}^{\,1\ R7} \rightarrow 1\frac{7}{24}$

Is the answer reasonable?

Guided Practice

Example A
Example B
Example C

Add.

1. $\frac{1}{4} + \frac{2}{4}$
2. $\frac{2}{5} + \frac{2}{5}$
3. $\frac{13}{32} + \frac{15}{32}$
4. $\frac{3}{9} + \frac{3}{9}$

5. $\frac{3}{8} + \frac{7}{8}$
6. $\frac{3}{4} + \frac{2}{4}$
7. $\frac{5}{6} + \frac{4}{6}$
8. $\frac{7}{10} + \frac{9}{10}$

9. $\frac{3}{4} + \frac{1}{8}$
10. $\frac{1}{8} + \frac{1}{6}$
11. $\frac{3}{5} + \frac{2}{7}$
12. $\frac{1}{3} + \frac{6}{7}$

Exercises

Practice

Add.

13. $\frac{1}{5} + \frac{3}{5}$
14. $\frac{2}{7} + \frac{2}{7}$
15. $\frac{1}{9} + \frac{5}{9}$
16. $\frac{7}{12} + \frac{9}{12}$
17. $\frac{11}{16} + \frac{9}{16}$

18. $\frac{2}{3} + \frac{1}{6}$
19. $\frac{3}{4} + \frac{1}{3}$
20. $\frac{2}{3} + \frac{1}{8}$
21. $\frac{7}{10} + \frac{1}{2}$
22. $\frac{1}{4} + \frac{3}{8}$

23. $\frac{7}{8} + \frac{5}{6}$
24. $\frac{3}{4} + \frac{1}{5}$
25. $\frac{9}{10} + \frac{14}{15}$
26. $\frac{2}{9} + \frac{1}{6}$
27. $\frac{2}{3} + \frac{7}{9}$

28. Find the sum of $\frac{2}{5}$, $\frac{1}{9}$, and $\frac{4}{15}$.

29. Add $\frac{1}{4}$, $\frac{1}{3}$, and $\frac{7}{9}$.

Applications

30. Hans worked on his science homework for $\frac{7}{8}$ hour and his math homework for $\frac{2}{3}$ hour. How long did he work on science and math?

31. Bev mixes $\frac{1}{3}$ can of yellow paint with $\frac{1}{4}$ can of blue paint to make green paint. How many cans of green paint does she make?

Using Algebra

Find the value for each expression if $a = \frac{1}{2}$, $b = \frac{3}{4}$, and $c = \frac{3}{8}$.

32. $a + b$
33. $b + c$
34. $a + c$
35. $a + b + c$

36. Explain how you could use prime factorization to add fractions with unlike denominators.

Critical Thinking

37. A bookworm chewed his way from Volume 1 page 1 to Volume 2, last page. The books are on a bookshelf standing next to each other. If the pages in each volume are 3 inches thick and the covers are $\frac{1}{4}$ inch thick each, how many inches did the bookworm chew?

4-11 ADDING MIXED NUMBERS

Objective

Add mixed numbers.

Ling Wu weighed $6\frac{3}{4}$ pounds at birth.

She gained $13\frac{3}{4}$ pounds in her first year. How much did she weigh on her first birthday?

You can use the following steps to add mixed numbers.

Method

1 ▶ Add the fractions. Rename first if necessary.

2 ▶ Add the whole numbers.

3 ▶ Rename and simplify if necessary.

Example A

Add $13\frac{3}{4}$ + $6\frac{3}{4}$. Estimate: 14 + 7 = 21

1 ▶
$$13\frac{3}{4}$$
$$+\ \ 6\frac{3}{4}$$
$$\overline{\ \frac{6}{4}}$$

2 ▶
$$13\frac{3}{4}$$
$$+\ \ 6\frac{3}{4}$$
$$\overline{19\frac{6}{4}}$$

3 ▶ $\frac{6}{4} = 1\frac{2}{4}$ or $1\frac{1}{2}$
$$19 + 1\frac{1}{2} = 20\frac{1}{2}$$

Ling Wu weighed $20\frac{1}{2}$ pounds on her first birthday.

Based on the estimate, explain why the answer, $20\frac{1}{2}$, is reasonable.

Example B

Add $16\frac{1}{2}$ + $5\frac{5}{8}$. Estimate: 17 + 6 = 23

1 ▶
$$16\frac{1}{2}$$
$$+\ \ 5\frac{5}{8}$$
$$\longrightarrow$$
$$16\frac{4}{8}$$
$$+\ \ 5\frac{5}{8}$$
$$\overline{\ \frac{9}{8}}$$

2 ▶
$$16\frac{4}{8}$$
$$+\ \ 5\frac{5}{8}$$
$$\overline{21\frac{9}{8}}$$

3 ▶ $\frac{9}{8} = 1\frac{1}{8}$
$$21 + 1\frac{1}{8} = 22\frac{1}{8}$$

Guided Practice — **Add.**

Example A

1. $4\frac{2}{5} + 7\frac{1}{5}$
2. $9\frac{3}{11} + 4\frac{5}{11}$
3. $9\frac{1}{8} + 6\frac{3}{8}$
4. $4\frac{3}{8} + 5\frac{3}{8}$

Example B

5. $8\frac{1}{5} + 7\frac{3}{10}$
6. $2\frac{1}{8} + 5\frac{3}{4}$
7. $5\frac{1}{2} + 3\frac{1}{3}$
8. $4\frac{1}{4} + 8\frac{5}{12}$

Practice

Add.

9. $1\frac{3}{7} + 8\frac{2}{7}$ **10.** $4\frac{3}{10} + 7\frac{3}{10}$ **11.** $4\frac{3}{8} + 3\frac{7}{8}$ **12.** $5\frac{3}{4} + 11\frac{3}{4}$

13. $7\frac{1}{6} + 9\frac{5}{8}$ **14.** $4\frac{4}{9} + 15\frac{1}{6}$ **15.** $6\frac{7}{12} + 5\frac{1}{6}$ **16.** $18\frac{3}{4} + 7\frac{1}{5}$

17. $4\frac{7}{8} + 5\frac{3}{8}$ **18.** $3\frac{3}{4} + 8\frac{7}{8}$ **19.** $4\frac{4}{5} + 6\frac{7}{10}$ **20.** $9\frac{11}{16} + 2\frac{15}{16}$

21. $8\frac{2}{3} + 6\frac{1}{6}$ **22.** $9\frac{8}{9} + 10\frac{3}{5}$ **23.** $17\frac{1}{3} + 6\frac{3}{4}$ **24.** $4\frac{5}{7} + 8\frac{1}{2}$

25. Add $1\frac{1}{12}$, $3\frac{1}{2}$, and $4\frac{1}{6}$.   **26.** Find the sum of $7\frac{8}{9}$, $2\frac{5}{6}$, and $8\frac{2}{3}$.

Using Algebra

Solve each equation.

27. $a - 2 = 3\frac{1}{4}$ **28.** $b - \frac{4}{4} = 8\frac{1}{5}$ **29.** $x - 8\frac{3}{8} = 17\frac{1}{2}$

Evaluate each expression if $a = \frac{3}{4}$, $b = 5\frac{1}{2}$, and $c = 7\frac{5}{8}$.

30. $a + b$ **31.** $c + a$ **32.** $b + c$ **33.** $a + b + c$

Applications

34. Paul has two sections of rope that he will tie together. One section is $6\frac{3}{4}$ feet long and the other is $17\frac{1}{2}$ feet long. If he loses $\frac{3}{4}$ of a foot when he ties them together, what is the total length?

35. Mrs. Baker had $5\frac{1}{2}$ yards of curtain material. She needs another $2\frac{7}{8}$ yards. How much material did she need in all?

36. Mr. James can drive his car for about 22¢ a mile. The trip home is 415 miles. Will $80 be enough to pay for the car expenses? Explain your answer.

Write Math

37. Why would it not be necessary to use the parentheses in the expression $a + (b + c)$?

Critical Thinking

38. Use all the digits 1, 2, 3, 4, and 5 and addition (in the denominator only) to write a fraction that is equal to one-half.

Mixed Review

Lesson 2-2

Add.

39. $143.6 + 18.2$ **40.** $\$82.07 + \$16.95$ **41.** $72.6 + 1.03$ **42.** $5.012 + 71.8$

Lesson 2-6

Complete each sequence.

43. 18, __, 24, __, __, 33   **44.** __, 4.6, 7.9, ___, ___, 17.8

Lesson 2-9

If the problem has the necessary facts, solve. State any missing or extra facts.

45. A newspaper ad costs $5 for the first 2 lines and $1.50 for each additional line. Bold print costs $0.50 extra per line. How much does a regular ad cost if it has 10 lines with no bold print?

NORTHEAST—SKYWAE

Spacious two BR townhomes, 1½ bath, fin. bsmt., fully equipped eat-in kitchen, patio. Close to busline, shopping, dining & freeways. ONLY $410. Open Sat. & Sun. 9–6. Weekdays by appt. 222-3333, 444-0000. Located in Strawberry Farms. ASK ABOUT FREE RENT SPECIAL. Bette Varca & Co., Realtors.

4-12 SUBTRACTING FRACTIONS

Objective

Subtract proper fractions.

Maxine is building a bird feeder as part of her science project. A peg that helps hold the feeder together needs to be $\frac{5}{8}$ inch long. The peg she has is $\frac{7}{8}$ inch long. How much does Maxine need to saw off the peg?

To subtract fractions with like denominators, subtract the numerators and write the difference over the denominator. Simplify if necessary.

Example A

Subtract $\frac{7}{8} - \frac{5}{8}$.

$$\frac{7}{8} - \frac{5}{8} = \frac{2}{8}$$

Estimate: $1 - \frac{1}{2} = \frac{1}{2}$

$$= \frac{1}{4}$$

Compared to the estimate, is the answer reasonable?

Maxine needs to saw off $\frac{1}{4}$ inch of the peg.

Subtract fractions with unlike denominators as follows.

Method

1. ▶ Find the least common denominator.
2. ▶ Rename the fractions using the LCD.
3. ▶ Subtract. Simplify if necessary.

Example B

Subtract $\frac{5}{6} - \frac{1}{4}$. Estimate: $1 - \frac{1}{2} = \frac{1}{2}$

1. ▶ $6 = \boxed{2} \times \boxed{3}$
   $4 = \boxed{2} \times \boxed{2}$
   $\Big\}$ LCD: $2 \times 3 \times 2$ or 12

2. ▶ $\begin{array}{r} \frac{5}{6} \\ -\frac{1}{4} \\ \hline \end{array} \longrightarrow \begin{array}{r} \frac{10}{12} \\ -\frac{3}{12} \\ \hline \end{array}$

3. ▶ $\begin{array}{r} \frac{10}{12} \\ -\frac{3}{12} \\ \hline \frac{7}{12} \end{array}$ Is the answer reasonable?

Guided Practice

Subtract.

Example A

1. $\frac{6}{9} - \frac{3}{9}$ 2. $\frac{7}{18} - \frac{5}{18}$ 3. $\frac{20}{21} - \frac{8}{21}$ 4. $\frac{5}{11} - \frac{1}{11}$ 5. $\frac{16}{20} - \frac{12}{20}$

Example B

6. $\frac{1}{2} - \frac{1}{3}$ 7. $\frac{2}{3} - \frac{1}{4}$ 8. $\frac{7}{10} - \frac{4}{9}$ 9. $\frac{5}{6} - \frac{3}{4}$ 10. $\frac{13}{24} - \frac{3}{8}$

Exercises

Practice

Subtract.

11. $\dfrac{7}{8} - \dfrac{3}{8}$ **12.** $\dfrac{5}{6} - \dfrac{4}{6}$ **13.** $\dfrac{15}{20} - \dfrac{8}{20}$ **14.** $\dfrac{5}{7} - \dfrac{2}{7}$ **15.** $\dfrac{17}{18} - \dfrac{5}{18}$

16. $\dfrac{4}{15} - \dfrac{1}{15}$ **17.** $\dfrac{1}{4} - \dfrac{1}{5}$ **18.** $\dfrac{2}{5} - \dfrac{1}{3}$ **19.** $\dfrac{1}{2} - \dfrac{1}{2}$ **20.** $\dfrac{5}{8} - \dfrac{1}{6}$

21. $\dfrac{7}{8} - \dfrac{5}{6}$ **22.** $\dfrac{3}{4} - \dfrac{2}{3}$ **23.** $\dfrac{23}{28} - \dfrac{9}{28}$ **24.** $\dfrac{4}{7} - \dfrac{1}{3}$ **25.** $\dfrac{5}{6} - \dfrac{4}{9}$

26. Subtract $\dfrac{1}{3}$ from $\dfrac{6}{7}$.   **27.** Find the difference of $\dfrac{7}{8}$ and $\dfrac{5}{12}$.

Using Equations

Solve each equation.

28. $a + \dfrac{3}{8} = \dfrac{5}{8}$ **29.** $x - \dfrac{1}{2} = \dfrac{1}{6}$ **30.** $c + \dfrac{3}{5} = \dfrac{4}{5}$

Applications

31. Jamie was making brownies using a recipe that called for $\dfrac{1}{4}$ cup of oil. She was talking on the phone at the same time and accidently poured $\dfrac{3}{4}$ cup of oil into the mixing bowl. How much did she need to remove from the bowl?

32. Max is making a taco salad. He needs $\dfrac{1}{4}$ cup of cheddar cheese. He already has $\dfrac{1}{2}$ cup of mozzerella in the measuring cup. If he puts the cheddar cheese in the cup on top of the mozzerella, how much cheese should be in the cup?

Critical Thinking

33. The results of a survey of 50 high school students showed that 17 liked math, 19 liked English, and the remainder liked both. How many students liked both subjects?

Estimation

Front-end estimation is one way to find the sum of mixed numbers. Consider $2\dfrac{7}{8} + 1\dfrac{1}{10} + 4\dfrac{5}{6} + 3$.

First add only the whole numbers, $2 + 1 + 4 + 3 = 10$ the "front end" of each addend.

Then mentally estimate the sum of $\dfrac{7}{8} + \dfrac{1}{10} + \dfrac{5}{6}$ is about 2. the fractions as a whole number.

So, $2\dfrac{7}{8} + 1\dfrac{1}{10} + 4\dfrac{5}{6} + 3$ is about $10 + 2$ or 12.

Estimate.

34. $6\dfrac{1}{2} + 10\dfrac{1}{5} + 4\dfrac{3}{5}$ **35.** $7\dfrac{1}{2} + 11\dfrac{5}{8} + \dfrac{1}{16}$ **36.** $3\dfrac{5}{6} + 4\dfrac{1}{3} + 4\dfrac{1}{3} + 4$

4-13 SUBTRACTING MIXED NUMBERS

Objective

Subtract mixed numbers.

Barney Lopez has to send his computer printer back to the manufacturer for repairs. Since he must pay the postage, he wants to keep the weight low without allowing damage to the printer. The printer weighs $8\frac{1}{4}$ pounds. When packaged for shipping it weighs $13\frac{3}{4}$ pounds. What is the weight of the packing materials?

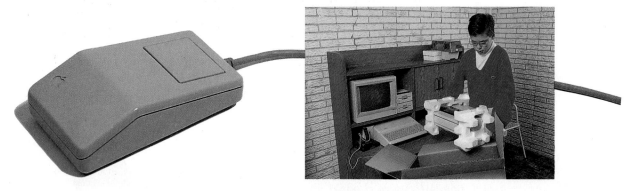

Subtracting mixed numbers is similiar to adding mixed numbers.

Method

1 Subtract the fractions. Rename first if necessary.

2 Subtract the whole numbers.

3 Rename and simplify if necessary.

Examples

A $13\frac{3}{4} - 8\frac{1}{4}$ Estimate: $14 - 8 = 6$

1
$$13\frac{3}{4}$$
$$-\ 8\frac{1}{4}$$
$$\overline{\quad \frac{2}{4}}$$

2
$$13\frac{3}{4}$$
$$-\ 8\frac{1}{4}$$
$$\overline{\ 5\frac{2}{4}}$$

3 $5\frac{2}{4} = 5\frac{1}{2}$

The weight of the packing materials is $5\frac{1}{2}$ pounds.

B $12\frac{1}{4} - 8\frac{1}{6}$ Estimate: $12 - 8 = 5$

1
$$12\frac{1}{4} \qquad 12\frac{3}{12}$$
$$-\ 8\frac{1}{6} \quad\rightarrow\quad -\ 8\frac{2}{12}$$
$$\overline{\qquad\qquad \frac{1}{12}}$$

2
$$12\frac{3}{12}$$
$$-\ 8\frac{2}{12}$$
$$\overline{\ 4\frac{1}{12}}$$

Is the answer reasonable?

Example C $9\frac{1}{4} - 6\frac{5}{8}$ Estimate: $9 - 7 = 2$

1
$$9\frac{1}{4} \qquad 9\frac{2}{8} \qquad 8\frac{10}{8}$$
$$-\ 6\frac{5}{8} \rightarrow -\ 6\frac{5}{8} \rightarrow -\ 6\frac{5}{8}$$
$$\overline{\qquad\qquad\qquad\qquad \frac{5}{8}}$$
Since $\frac{2}{8} < \frac{5}{8}$, rename $9\frac{2}{8}$ as $8\frac{10}{8}$.

2
$$8\frac{10}{8}$$
$$-\ 6\frac{5}{8}$$
$$\overline{\ 2\frac{5}{8}}$$

Based on the estimate, explain why the answer, $2\frac{5}{8}$, is reasonable.

134   CHAPTER 4   FRACTIONS: ADDING AND SUBTRACTING

Example D

$$10 - 3\frac{2}{7} \qquad \text{Estimate: } 10 - 3 = 7$$

❶
$$
\begin{array}{r}
10 \\
-\ 3\frac{2}{7}
\end{array}
\rightarrow
\begin{array}{r}
9\frac{7}{7} \\
-\ 3\frac{2}{7} \\
\hline
\frac{5}{7}
\end{array}
$$
Rename 10 as $9\frac{7}{7}$.

❷
$$
\begin{array}{r}
9\frac{7}{7} \\
-\ 3\frac{2}{7} \\
\hline
6\frac{5}{7}
\end{array}
$$
Is the answer reasonable?

Guided Practice

Subtract.

Example A

1. $3\frac{8}{9} - 1\frac{2}{9}$ 2. $5\frac{3}{4} - 4\frac{1}{4}$ 3. $7\frac{5}{8} - 2\frac{3}{8}$ 4. $9\frac{4}{5} - 6\frac{1}{5}$

Example B

5. $8\frac{4}{5} - 5\frac{3}{4}$ 6. $7\frac{1}{3} - 3\frac{1}{5}$ 7. $24\frac{3}{8} - 13\frac{1}{4}$ 8. $17\frac{5}{6} - 11\frac{7}{18}$

Example C

9. $5\frac{1}{3} - 1\frac{1}{2}$ 10. $15\frac{3}{5} - 8\frac{3}{4}$ 11. $7\frac{1}{6} - 2\frac{7}{8}$ 12. $15\frac{5}{8} - 10\frac{2}{3}$

Example D

13. $6 - 2\frac{2}{3}$ 14. $7 - 2\frac{1}{4}$ 15. $10 - \frac{5}{8}$ 16. $25 - 11\frac{4}{7}$

Exercises

Practice

Subtract.

17. $16\frac{11}{16} - 6\frac{5}{16}$ 18. $9\frac{7}{9} - 6\frac{4}{9}$ 19. $10\frac{11}{12} - 7\frac{1}{4}$ 20. $17\frac{5}{6} - 9\frac{3}{8}$

21. $17\frac{1}{6} - 8\frac{5}{6}$ 22. $4\frac{1}{7} - 2\frac{3}{4}$ 23. $3\frac{5}{8} - 1\frac{4}{5}$ 24. $9\frac{5}{8} - 7\frac{5}{6}$

25. $6\frac{1}{2} - 1\frac{2}{3}$ 26. $9\frac{3}{5} - \frac{4}{5}$ 27. $9\frac{2}{5} - 6\frac{9}{10}$ 28. $5\frac{3}{8} - 2\frac{7}{12}$

29. $14 - \frac{2}{7}$ 30. $50 - 34\frac{2}{5}$ 31. $36 - 34\frac{5}{8}$ 32. $10 - 1\frac{1}{9}$

33. $(3\frac{3}{4} + 4\frac{1}{4}) - 2\frac{1}{2}$ 34. $(16 + 14\frac{7}{8}) - 11\frac{3}{8}$ 35. $(8\frac{7}{8} + 5\frac{1}{4}) - 3\frac{1}{2}$

36. Subtract $2\frac{3}{8}$ from 20.

37. Find the difference of $9\frac{5}{8}$ and $7\frac{5}{6}$.

Applications

ƒUN with MATH

Which country invented the first wheelbarrow?
See page 174.

38. How much more energy does Europe use than it produces?

39. How much more energy do the Americas produce than China?

40. How much more energy do the Americas use than the Soviet Union?

41. How much energy is used by the Soviet Union, and Europe?

42. Which two regions produce about the same amount of energy?

43. List the regions from greatest to least based on energy use.

World's Energy Production and Usage

Region	Produced	Used
The Americas	$\frac{1}{3}$	$\frac{7}{8}$
Middle East-Africa	$\frac{3}{10}$	$\frac{1}{20}$
Soviet Union	$\frac{1}{5}$	$\frac{1}{6}$
Western Europe	$\frac{3}{25}$	$\frac{1}{4}$
China	$\frac{1}{25}$	$\frac{7}{100}$

Collect Data

44. Take a survey to find what fractional part of each day students study. Then make a table to compare study time of boys to girls according to grade level by arranging the fractions from greatest to least.

| Explore |
| Plan |
| Solve |
| Examine |

4-14   MAKE A DRAWING

Objective

Solve verbal problems by making a drawing.

Paul left his home and jogged north 2 blocks, then east 3 blocks, and then south 2 blocks. How many blocks east of his home was he when he stopped jogging?

Sometimes it helps to make a drawing when you are deciding how to solve a problem.

▶ **Explore**

What facts are given?
● The jogging path is:
 north 2 blocks
 east 3 blocks
 south 2 blocks

What fact do you need to find?
● How many blocks east of his home does Paul stop jogging?

▶ **Plan**

Draw a map that shows Paul's house (the starting point), the jogging path, and the stopping point.

▶ **Solve**

Draw a map of Paul's jogging path.

From the map, Paul stopped jogging 3 blocks east of his home.

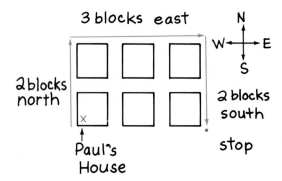

▶ **Examine**

Look at the drawing of your map. Did you draw 2 blocks north, 3 blocks east, and 2 blocks south? Yes. So you can see Paul is 3 blocks east of his home.

Guided Practice

Solve. Make a drawing.

1. Dario paints 3 sections of fence in the time it takes Marty to paint 2 sections of fence. Together they paint 40 sections of fence. How many sections does Marty paint?

2. Each page of a photo album is $8\frac{1}{2}$ inches by 11 inches. At most how many 4-inch by 5-inch pictures will fit on one page?

Solve. Use any strategy.

3. Erica plans to sew two squares of cloth together along one side to make a placemat. Then she wants to sew a braid around the outside. Each square of cloth is $8\frac{1}{2}$ inches by $8\frac{1}{2}$ inches. How many inches of braid does Ashley need?

4. It cost $0.10 to cut and weld a link. How much does it cost to make a chain out of 10 links?

5. A lizard climbs a 30-foot pole. Each day it climbs up 7 feet and each night he slips back 4 feet. How many days will it take the lizard to reach the top?

6. Luke can rake a lawn that is 60 meters by 60 meters in 4 hours. At the same rate, how long will it take him to rake a lawn that is 30 meters by 30 meters?

7. Mrs. Fern's dance class is standing evenly spaced in a circle. If the sixth person is directly opposite the sixteenth person, how many persons are in the circle?

8. Three spiders are on a 9-foot wall. The gray spider is 4 feet from the top. The brown spider is 7 feet from the bottom. The black spider is 3 feet below the brown spider. Which spider is the highest?

9. Select an assignment that shows something new you learned in this chapter.

10. Hot Hot Pizza offers a choice of 4 toppings. How many different 2-topping pizzas are possible?

11. How many different ways can you buy 4 attached postage stamps? Two possible ways are shown at the right.

Mixed Review

Lesson 2-4

Subtract.

12. 49.6
      − 18.4

13. 90.1
      − 48.6

14. 800
      − 34.5

15. 7.773
       − 4.2

Lesson 3-7

Multiply or divide.

16. 18.4 × 100   **17.** 9.6 × 10   **18.** 37.2 × 100   **19.** 62.8 × 10
20. 0.84 ÷ 10   **21.** 62.4 ÷ 100   **22.** 15.9 ÷ 1,000   **23.** 46 ÷ 10

Lesson 4-9

24. Antonio does $\frac{1}{8}$ of all his homework during study hall. He does $\frac{1}{4}$ of all his homework when he gets home from school, and he does $\frac{1}{2}$ of all his homework before he goes to bed. Does Antonio complete all of his homework? If not, how much does he still need to complete?

REVIEW

**Vocabulary/
Concepts**

Choose the word or number from the list at the right to complete each sentence.

1. The least common multiple of 10 and 15 is ___?___.

2. The number 27 is an example of a ___?___ number.

3. ___?___ is the simplest form of $\frac{16}{24}$.

4. The fraction $\frac{12}{16}$ is equivalent to the fraction ___?___.

5. 14 is the ___?___ of 28 and 42.

6. ___?___ is a factor of 675

7. A ___?___ indicates the sum of a whole number and a fraction.

8. In the fraction $\frac{23}{30}$, the numerator is ___?___.

$\frac{2}{3}$
$\frac{3}{4}$
5
10
23
30
composite
greatest common factor (GCF)
least common denominator
  (LCD)
mixed number
prime

**Exercises/
Applications**

Lesson 4-1

Find all the factors of each number.

9. 12   **10.** 18   **11.** 29   **12.** 38   **13.** 42

Lesson 4-1

State whether each number is divisible by 2, 3, 5, 9, or 10.

14. 16   **15.** 21   **16.** 41   **17.** 54   **18.** 108

Lesson 4-2

Write the prime factorization for each number.

19. 98   **20.** 78   **21.** 90   **22.** 475   **23.** 156

Lesson 4-3

Find the GCF of each group of numbers.

24. 6, 8   **25.** 32, 48   **26.** 18, 24   **27.** 54, 200   **28.** 12, 38, 62

Lesson 4-3

Find the LCM of each group of numbers.

29. 7, 4   **30.** 16, 40   **31.** 8, 12   **32.** 6, 14   **33.** 2, 6, 10

Lesson 4-4

Solve.

34. Jorge Ortega's gross pay is $523.78 a week. His total tax deductions are $142.09 and his personal deductions are $45.89. What are his total deductions and net pay?

35. Ms. Chan has $89.42 deducted weekly from her paycheck for federal tax, $13.24 for state tax, and $6.99 for city tax. How much total tax does she pay yearly?

Replace each ▓ with a number so the fractions are equivalent.

36. $\frac{4}{5} = \frac{▓}{25}$ **37.** $\frac{3}{8} = \frac{▓}{32}$ **38.** $\frac{21}{24} = \frac{▓}{8}$ **39.** $\frac{8}{36} = \frac{▓}{9}$ **40.** $\frac{9}{16} = \frac{▓}{32}$

41. $\frac{2}{15} = \frac{▓}{45}$ **42.** $\frac{27}{45} = \frac{▓}{5}$ **43.** $\frac{42}{50} = \frac{▓}{25}$ **44.** $\frac{11}{30} = \frac{▓}{60}$ **45.** $\frac{36}{90} = \frac{▓}{10}$

Simplify each fraction.

46. $\frac{4}{8}$ **47.** $\frac{12}{16}$ **48.** $\frac{21}{24}$ **49.** $\frac{6}{42}$ **50.** $\frac{32}{40}$ **51.** $\frac{38}{72}$

Replace each ● with <, >, or = to make a true sentence.

52. $\frac{3}{5}$ ● $\frac{12}{20}$ **53.** $\frac{5}{6}$ ● $\frac{7}{18}$ **54.** $\frac{4}{15}$ ● $\frac{9}{16}$ **55.** $\frac{4}{13}$ ● $\frac{4}{9}$ **56.** $\frac{7}{8}$ ● $\frac{5}{6}$

Change each fraction to a mixed number in simplest form.

57. $\frac{22}{3}$ **58.** $\frac{37}{5}$ **59.** $\frac{27}{4}$ **60.** $\frac{94}{8}$ **61.** $\frac{39}{5}$ **62.** $\frac{83}{7}$

Change each mixed number to an improper fraction.

63. $4\frac{5}{8}$ **64.** $3\frac{6}{7}$ **65.** $6\frac{2}{3}$ **66.** $5\frac{4}{5}$ **67.** $7\frac{8}{9}$ **68.** $8\frac{9}{10}$

Estimate.

69. $\frac{2}{3} + \frac{1}{6}$ **70.** $\frac{9}{20} + \frac{11}{12}$ **71.** $1\frac{3}{8} + 7\frac{1}{9}$ **72.** $\frac{7}{8} - \frac{1}{2}$ **73.** $14\frac{2}{9} - 7\frac{5}{6}$

Add.

74. $\frac{3}{5} + \frac{1}{5}$ **75.** $\frac{5}{8} + \frac{5}{8}$ **76.** $\frac{1}{3} + \frac{3}{10}$ **77.** $\frac{4}{5} + \frac{1}{4}$ **78.** $\frac{5}{6} + \frac{1}{10}$

79. $2\frac{4}{9} + 3\frac{7}{9}$ **80.** $1\frac{5}{8} + 4\frac{7}{8}$ **81.** $3\frac{5}{11} + 6\frac{4}{11}$ **82.** $14\frac{2}{3} + 4\frac{5}{12}$ **83.** $7\frac{2}{3} + 5\frac{1}{4}$

Subtract.

84. $\frac{7}{12} - \frac{1}{12}$ **85.** $\frac{9}{10} - \frac{3}{10}$ **86.** $\frac{14}{15} - \frac{8}{15}$ **87.** $\frac{5}{6} - \frac{2}{9}$ **88.** $\frac{1}{2} - \frac{1}{14}$

89. $4\frac{4}{5} - 2\frac{2}{5}$ **90.** $7\frac{1}{8} - 5\frac{7}{8}$ **91.** $6\frac{3}{4} - 3\frac{1}{3}$ **92.** $12\frac{4}{7} - 9$ **93.** $20 - 2\frac{1}{2}$

94. Janice and Julie are identical twins. At birth, Janice weighed $6\frac{5}{8}$ lb and Julie weighed 8 lb. What was the difference in their weight?

Solve. Use any strategy.

95. After a bake sale, there was one cake left. Anna cut the cake into thirds to share it with two co-workers. She then remembered Janice in the kitchen. How could Anna make one more cut to have four equal parts?

TEST

Write the prime factorization for each number.

1. 24   **2.** 54   **3.** 81

Replace each ▓ with a number so that the fractions are equivalent.

4. $\frac{1}{3} = \frac{▓}{18}$ **5.** $\frac{21}{21} = \frac{▓}{9}$ **6.** $\frac{12}{20} = \frac{▓}{5}$

Solve.

7. Tanya Smith's gross pay is $425.81. Her total tax deductions are $119.39 and her personal deductions are $58.07. What is her net pay?

Simplify each fraction.

8. $\frac{9}{12}$ **9.** $\frac{8}{16}$ **10.** $\frac{15}{20}$

Replace each ● with <, >, or = to make a true sentence.

11. $\frac{8}{9} ● \frac{7}{9}$ **12.** $\frac{2}{3} ● \frac{3}{4}$ **13.** $\frac{5}{6} ● \frac{10}{12}$

Change each mixed number to an improper fraction.

14. $2\frac{2}{3}$ **15.** $6\frac{4}{9}$ **16.** $8\frac{3}{10}$

Estimate.

17. $\frac{9}{10} + \frac{1}{8}$ **18.** $2\frac{1}{3} + 7\frac{1}{4}$ **19.** $\frac{7}{8} - \frac{1}{5}$

Add or subtract.

20. $\frac{1}{5} + \frac{2}{5}$ **21.** $\frac{2}{5} + \frac{1}{4}$ **22.** $\frac{5}{12} + \frac{7}{8}$

23. $3\frac{1}{5} + 5\frac{4}{5}$ **24.** $7\frac{1}{6} + 2\frac{2}{3}$ **25.** $13\frac{5}{7} + 6\frac{1}{4}$

26. $\frac{5}{12} - \frac{1}{12}$ **27.** $\frac{7}{9} - \frac{1}{6}$ **28.** $\frac{5}{6} - \frac{2}{3}$

29. $7\frac{5}{6} - 5\frac{1}{6}$ **30.** $7\frac{3}{8} - 2\frac{1}{4}$ **31.** $8\frac{1}{9} - 1\frac{4}{9}$

Solve. Use any strategy.

32. Andy ran on three different trails. The first was $1\frac{1}{2}$ miles long, the second $2\frac{1}{4}$, and the third $1\frac{3}{4}$. How far did he run?

33. The chess club has 6 members. If every member shakes hands with every other member, how many handshakes are there?

▶ BONUS: Explain when the statement $\frac{a}{b} + \frac{c}{b} = 1$ is true.

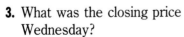

Stock		High	Low	Last	Chg.
Herculs	2.24	$32\frac{1}{4}$	$28\frac{7}{8}$	30	$-2\frac{1}{4}$
Hrshey	.90	$35\frac{1}{8}$	$32\frac{3}{4}$	$34\frac{1}{4}$	$-\frac{3}{8}$
HewlPk	.42	$35\frac{1}{8}$	31	$33\frac{3}{4}$	$-\frac{7}{8}$
Hexcel	.44	$10\frac{3}{4}$	$9\frac{3}{4}$	10	$-\frac{3}{4}$
Hibern	1.00	$14\frac{7}{8}$	$12\frac{3}{4}$	$12\frac{7}{8}$	-2
HiShear	.22	11	$7\frac{7}{8}$	$7\frac{7}{8}$	-3
Hilnco	.90	$4\frac{7}{8}$	$4\frac{3}{8}$	$4\frac{1}{2}$	$-\frac{3}{8}$
Hincll	1.08	$5\frac{1}{2}$	$4\frac{5}{8}$	$4\frac{5}{8}$	$-\frac{7}{8}$
HilnIll	1.14	$6\frac{1}{4}$	$5\frac{7}{8}$	$5\frac{7}{8}$	$-\frac{3}{8}$

STOCKS

Tim and Nadine Jackson invest some of their money in stocks. Investing in stocks can be very risky. So the Jacksons study the stock market carefully. The list at the right shows some stock trades on the New York Stock Exchange. Some company names are abbreviated.

Hrshey → Hershey

HomeSh → Home Shopping Club

The profit or loss is found by comparing the quote of the share when it is bought with the quote when it is sold. What profit or loss would Tim and Nadine have if they buy 100 shares of Hercules at $28\frac{7}{8}$ and sell it for $32\frac{1}{4}$?

The price of each share is quoted in dollars to the nearest $\frac{1}{8}$ of a dollar.

selling price → $32\frac{1}{4}$ → $32\frac{1}{4}$

buying price → $28\frac{7}{8}$ → $-\,28\frac{7}{8}$

$$\overline{3\frac{3}{8}}$$

Tim and Nadine would make a profit of $3\frac{3}{8}$ or \$3.375 on each share. The profit on 100 shares is \$3.375 × 100 or \$337.50.

1. What is the highest quote for HiShear stock this year?

2. What profit or loss would there be if 100 shares of Hercules were bought at $28\frac{1}{8}$ and sold at the last quote?

Suppose the graph at the right shows the net change in the HewlPk stock over one week.

3. What was the closing price Wednesday?

4. What was the change between Thursday and Friday?

5. What was the overall change during the week shown?

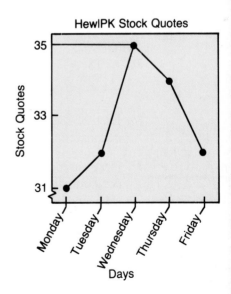

HewlPK Stock Quotes

Free Response

Lesson 1-1 | *Round each number to the underlined place-value position.*

1. 6<u>2</u>8   **2.** 5,3<u>5</u>4   **3.** <u>3</u>,890   **4.** <u>3</u>1,617   **5.** 850,<u>6</u>38

Lessons 1-6
1-7 | *Estimate.*

6. 351
    + 490

7. 6,082
    − 693

8. 495
    × 57

9. 55)‾436

10. 8,774
     − 4,876

11. 7,762 + 856   **12.** 6,493 ÷ 74   **13.** 325 × 608
14. 5,000 − 3,837   **15.** 639 × 471   **16.** 48,470 ÷ 583

Lesson 1-13 | *Write an equation to solve this problem. Then solve.*

17. Alison has a collection of 483 stamps. How many stamps does she need to add to her collection to raise the total to 625?

Lessons 2-1
2-3 | *Add or subtract.*

18. 62
    + 853

19. 540
    + 389

20. 316
    − 178

21. 506
    − 38

22. 160
    + 825

23. 600
    − 352

24. 8,706 + 537   **25.** 480 − 309   **26.** 7,800 − 279
27. 512 + 4,246   **28.** 5,123 + 5,477   **29.** 6,231 − 2,345

30. A car dealer sold 1,774 cars one year and 1,967 cars the next year. How many cars were sold in the two years?

Lesson 3-1 | *Multiply.*

31. 73
    × 2

32. 646
    × 5

33. 2,374
    × 4

34. 126
    × 58

35. 710
    × 14

36. Each of 231 persons donates $21 to the Heart Fund. How much is donated in all?

Lesson 3-13 | **37.** Jerry earns $20,800 a year. He would like to purchase a computer that costs $1,200. How long will it take him to save enough money to buy the computer if he saves $200 every two weeks?

Lesson 4-2 | *Write the prime factorization for each number.*

38. 36   **39.** 136   **40.** 720   **41.** 111   **42.** 1,800

Lesson 4-5 | *Replace each ▓ with a number so the fractions are equivalent.*

43. $\frac{7}{10} = \frac{▓}{100}$ **44.** $\frac{15}{20} = \frac{▓}{4}$ **45.** $\frac{2}{7} = \frac{▓}{14}$ **46.** $\frac{3}{6} = \frac{▓}{2}$ **47.** $\frac{8}{8} = \frac{▓}{6}$

Multiple Choice

Choose the letter of the correct answer for each item.

1. How can 6 to the third power be written?

  a. 3×6 **c.** $6 \times 6 \times 6$

  b. 6×3 **d.** $6 + 6 + 6$

2. About how much is 17×53?

  e. 350

  f. 1,000

  g. 10,000

  h. *none of the above*

Lesson 1-7

3. What is the result when 6.71 is subtracted from 3,028.5?

  a. 2,961.4

  b. 3,021.79

  c. 3,021.89

  d. *none of the above*

Lesson 2-4

4. A cheetah can run 32 meters per second. How far can the cheetah run in 2.5 seconds?

  e. 8 meters

  f. 12.8 meters

  g. 80 meters

  h. 800 meters

Lesson 3-2

5. Denny received 90, 87, 89, and 93 on his last four history exams. He needs a total of 455 points to get an A for the quarter. What must he score on the last exam?

  a. 90

  b. 95

  c. 96

  d. 93

Lesson 2-1

6. Write 17,500,000 in scientific notation.

  e. 1.75×10^5 **g.** 17.5×10^6

  f. 1.75×10^7 **h.** 17.5×10^7

Lesson 3-8

7. Find the sum of 28.5 and 6.

  a. 29.1

  b. 34.5

  c. 44.5

  d. 88.5

Lesson 2-2

8. Herb divides 1,320 marbles among 27 people. What is the result?

  e. 4 R27

  f. 48 R24

  g. 48 R34

  h. 49

Lesson 3-4

9. A book carton weighs 1 pound. What is the weight when a book carton is filled with ten books? State the missing fact.

  a. weight of address labels

  b. shipping costs

  c. weight of each book

  d. *none of the above*

Lesson 2-9

10. Janet cuts two pieces from a ribbon that is 1 foot long. One piece is $\frac{1}{2}$ foot long and the other is $\frac{1}{3}$ foot long. How long is the remaining ribbon?

  e. $\frac{1}{6}$ foot   **g.** $\frac{4}{5}$ foot

  f. $\frac{3}{5}$ foot   **h.** $\frac{5}{6}$ foot

Lesson 4-12

CUMULATIVE REVIEW/TEST **143**

FRACTIONS: MULTIPLYING AND DIVIDING

Statement Savings

Spending or Saving?

Do you have a part-time job? Are you saving money for buying something special, a stereo, a compact disc player, clothes, or a car? Are you saving money for college or career training? There are many young people who have turned part-time work into full-time businesses.

Carrie babysits for the Tremaines for $2\frac{3}{4}$ hours every Wednesday. She charges $2 an hour. How much does she earn on Wednesday?

ACTIVITY: Exploring Multiplication

In this activity you will explore multiplying fractions.

Materials: compass, paper(thin paper works better than construction paper), pencil or colored pencils, a pair of scissors

1. Use a compass to draw a circle on a piece of paper. Then cut it out.
2. Fold the circle in half, in half again, and then in half again, making sharp creases.
3. Unfold the circle.

Cooperative Groups

Work together in small groups.

4. Into how many parts do the creases separate the circle? What fraction represents one part of the circle?
5. To find the product of 5 and $\frac{1}{8}$, decide how many parts of the circle you should shade. What is the product of 5 and $\frac{1}{8}$?
6. Write a sentence describing how to find the product of any whole number and a fraction with a numerator of 1.

Extend the Concept

7. Cut a piece of paper into a 2-inch by 8-inch strip. Discuss with your group how to find $\frac{1}{2}$ of $\frac{1}{4}$, or $\frac{1}{2} \times \frac{1}{4}$ using the paper. Then find the product.
8. Cut a second strip of paper the same size as the one in Exercise 7. Discuss with your group how to show the product of $\frac{2}{3}$ and $\frac{3}{4}$. Then find the product.

Communicate Your Ideas

9. Discuss with your group how to show the product of a mixed number and a fraction. Try $1\frac{1}{2} \times \frac{1}{4}$.
10. With your group, discuss the results of Exercises 4-8 and write some ideas about the product of fractions that you have discovered.

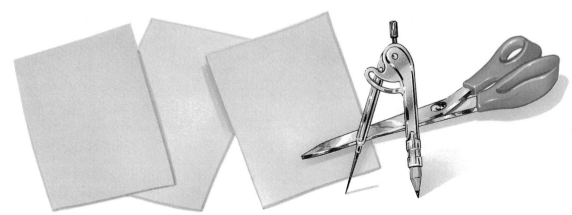

ESTIMATING PRODUCTS AND QUOTIENTS

Objective

Estimate products and quotients of fractions and mixed numbers.

Mr. Sykes's science class discussed a newspaper article on energy consumption. The article stated that an average of $2\frac{5}{6}$ quadrillion BTUs of energy were used by each person in each household in the United States during 1985. Mr. Sykes asked the class *about* how many quadrillion BTUs of energy were used by each household if each household, on average, has about $3\frac{1}{5}$ people.

As with whole numbers and decimals, round mixed numbers before estimating products.

When estimating quotients, round the divisor to a whole number. Round the dividend to the nearest multiple of the divisor.

Examples

A **Estimate $2\frac{5}{6} \times 3\frac{1}{5}$.**

$2\frac{5}{6}$ rounds to 3.

$3\frac{1}{5}$ rounds to 3.

$3 \times 3 = 9$

An average household used about 9 quadrillion BTUs.

B **Estimate $14\frac{5}{8} \div 2\frac{1}{4}$.**

$2\frac{1}{4}$ rounds to 2.

Round $14\frac{5}{8}$ to 14 since 14 is the nearest multiple of 2.

Since $14 \div 2$ is 7, the estimate is 7.

To estimate the product of fractions or a fraction and a mixed number, note that multiplying by a proper fraction produces a product *less* than the other factor.

When you divide by a proper fraction, note that the quotient is always *greater* than the dividend.

Examples

C **Is $\frac{1}{4}$ of $\frac{2}{3}$ less than or greater than $\frac{2}{3}$?**

Because you are finding $\frac{1}{4}$ of $\frac{2}{3}$, or only a *part* of $\frac{2}{3}$, the result will be less than $\frac{2}{3}$.

D **Is $4 \div \frac{1}{3}$ less than or greater than 4?**

Dividing 4 by 1 gives 4 parts or a quotient of 4. Dividing 4 by a number less than 1 such as $\frac{1}{3}$ gives *more* than 4 parts or a quotient greater than 4.

Guided Practice

Estimate each product.

Example A

1. $2\frac{3}{4} \times 7$

2. $1\frac{1}{4} \times 1\frac{3}{5}$

3. $3\frac{3}{4} \times 2\frac{2}{3}$

4. $3\frac{1}{8} \times 3\frac{2}{5}$

Example B

Examples C, D

Estimate each quotient.

5. $5 \div 3\frac{1}{3}$ **6.** $10\frac{1}{2} \div 2\frac{1}{4}$ **7.** $4\frac{2}{3} \div 2$ **8.** $8\frac{1}{3} \div 1\frac{5}{6}$

9. Is $\frac{3}{4}$ of 2 less than or greater than 2?

10. Is $18 \div \frac{1}{3}$ less than or greater than 18?

Exercises

Practice

Estimate.

11. $4\frac{1}{2} \times 2\frac{3}{4}$ **12.** $5\frac{1}{3} \times 4\frac{1}{2}$ **13.** $9\frac{3}{5} \div 4\frac{1}{5}$ **14.** $8\frac{2}{3} \div 1\frac{3}{5}$

15. $20\frac{5}{8} \div 3\frac{2}{5}$ **16.** $17\frac{7}{8} \div 2\frac{3}{16}$ **17.** $2\frac{5}{6} \times 1\frac{7}{8}$ **18.** $1\frac{9}{16} \times 4\frac{4}{5}$

19. Is $\frac{3}{4}$ of 12 less than or greater than 12?

20. Is $15 \div \frac{1}{4}$ less than or greater than 15?

21. Is $\frac{1}{2}$ of $\frac{1}{2}$ less than, greater than, or equal to $\frac{1}{2}$?

22. Is $6\frac{2}{3} \div \frac{3}{5}$ less than or greater than 6?

Applications

23. Alec's car used 7 gallons of gasoline last week. If he paid $1.29 a gallon for gasoline, *about* how much did he spend on gasoline for the week?

24. A $3\frac{1}{2}$-ounce ice cream bar contains 5 grams of fat per ounce. *About* how many grams of fat are contained in the whole bar?

f UN with MATH

What does your refrigerator do to the ozone layer?
See page 174.

25. Linda breaks a seed-treat stick into 3 pieces for her guinea pig. If she feeds the guinea pig $1\frac{2}{3}$ sticks per week, *about* how long will a supply of $8\frac{1}{3}$ sticks last?

26. If the Smith's fuel bill for last year was $1,180, *about* how much was their average monthly fuel bill?

Using Algebra

JOURNAL ENTRY

Solve each equation.

27. $x + 6 = 11$ **28.** $h - 9 = 4$ **29.** $\frac{t}{4} = 9$

30. Describe a situation in your life where you need to use estimation.

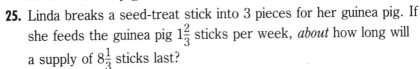

Mixed Review

Lesson 3-7

Divide. Round to the nearest hundredth.

31. $6.3\overline{)43.7}$ **32.** $1.8\overline{)9.76}$ **33.** $22.4\overline{)484.3}$

Lesson 3-8

Write in standard form.

34. 4.8×10^4 **35.** 6.23×10^3 **36.** 9.01×10^5

Lesson 4-10

37. Lindsey uses $\frac{3}{4}$ cup of sugar in her cookies. She uses $\frac{2}{3}$ cup of sugar for her cake. How much sugar does she use for both?

5-2 MULTIPLYING FRACTIONS

Objective
Multiply fractions.

Lisa is a baker for Mm-Mm Good Caterers. Her recipe for filling 100 cream puffs calls for $\frac{1}{2}$ cup of vanilla flavoring. If Lisa needs to make $\frac{3}{4}$ of a recipe, how much vanilla flavoring does she need? Lisa needs to find $\frac{3}{4}$ of $\frac{1}{2}$.

$\frac{1}{2}$ cup of vanilla

You are making $\frac{3}{4}$ of the recipe.

$\frac{3}{4}$ of the vanilla is shaded blue.

You can see that $\frac{3}{4}$ of $\frac{1}{2}$ is $\frac{3}{8}$.

Method

1 Multiply the numerators. Multiply the denominators.

2 Write the product in simplest form.

Example A

Multiply $\frac{3}{4} \times \frac{1}{2}$.

1 $\frac{3}{4} \times \frac{1}{2} = \frac{3 \times 1}{4 \times 2}$

2 $= \frac{3}{8}$

Lisa needs $\frac{3}{8}$ cup of vanilla flavoring.

Example B

Multiply $8 \times \frac{2}{3}$. Explain why the product should be a little more than 4.

1 Write 8 as a fraction. $8 = \frac{8}{1}$ $\qquad \frac{8}{1} \times \frac{2}{3} = \frac{8 \times 2}{1 \times 3}$

2 $= \frac{16}{3}$ or $5\frac{1}{3}$

You can use the following shortcut to simplify a product.

$$\frac{3}{5} \times \frac{10}{21} = \frac{3 \times 10}{5 \times 21}$$

$$= \frac{\overset{1}{\cancel{3}} \times \overset{2}{\cancel{10}}}{\underset{1}{\cancel{5}} \times \underset{7}{\cancel{21}}}$$

Divide both the numerator and the denominator by 3 and by 5.
3 is the GCF of 3 and 21.
5 is the GCF of 10 and 5.

$$= \frac{2}{7} \quad \leftarrow \text{simplest form}$$

Guided Practice

Multiply.

· Example A

1. $\frac{3}{4} \times \frac{1}{4}$ **2.** $\frac{2}{3} \times \frac{5}{6}$ **3.** $\frac{2}{3} \times \frac{4}{5}$ **4.** $\frac{3}{8} \times \frac{4}{5}$ **5.** $\frac{4}{9} \times \frac{3}{4}$

6. $\frac{1}{7} \times 3$ **7.** $3 \times \frac{7}{16}$ **8.** $\frac{5}{14} \times 42$ **9.** $7 \times \frac{5}{8}$ **10.** $\frac{2}{3} \times 9$

Exercises

Practice

Multiply.

11. $\frac{2}{9} \times \frac{2}{7}$ **12.** $\frac{2}{3} \times \frac{1}{4}$ **13.** $\frac{5}{7} \times \frac{8}{9}$ **14.** $\frac{4}{5} \times \frac{1}{2}$

15. $\frac{1}{2} \times \frac{8}{9}$ **16.** $\frac{4}{5} \times \frac{1}{9}$ **17.** $\frac{7}{8} \times \frac{8}{9}$ **18.** $\frac{7}{9} \times \frac{2}{21}$

19. $\frac{3}{4} \times 5$ **20.** $\frac{2}{3} \times 5$ **21.** $16 \times \frac{5}{8}$ **22.** $64 \times \frac{5}{16}$

23. $\frac{3}{4} \times \frac{3}{5} \times \frac{5}{6}$ **24.** $\frac{6}{7} \times \frac{1}{3} \times \frac{1}{5}$ **25.** $\frac{7}{8} \times \frac{2}{7} \times \frac{9}{10}$

26. Find the product of $\frac{3}{7}$ and $\frac{21}{33}$.

27. *True* or *false:* $\frac{2}{3} \times \frac{12}{30} = \frac{4}{15}$.

Applications

28. Bev wants to wallpaper one-half of her bedroom in a pastel print and the other half in the coordinating stripe. If each roll is the same size and it takes $5\frac{1}{2}$ rolls of wallpaper to do the whole bedroom, how many rolls of each pattern does Bev need?

29. A rod and reel usually sell for $65. They are on sale for three-fourths of the regular price. What is the sale price?

Critical Thinking

30. Move one straight line segment to make this true statement different, but true, also.

$$77 \boxed{+} 1 \boxed{=} 78$$

Using Variables

Evaluate each expression if $a = 4$, $b = \frac{3}{4}$, and $c = \frac{2}{3}$.

31. $b + c$ **32.** $a - (b + c)$ **33.** ac

Make Up a Problem

34. Make up a problem using two proper fractions. The answer must be $3\frac{3}{4}$.

Mixed Review

Lesson 2-2

Add.

35.   48.5   **36.**   6.23   **37.**   1.07
      + 16.7   + 10.8   + 2.98

Lesson 3-1

Multiply.

38.   13   **39.**   45   **40.**   51
      × 14   × 8   × 63

Lesson 3-13

41. Kisha is building a fence around her rectangular garden. The fencing she wants to use costs $0.75 per foot. The garden measures 12 feet by 10 feet. How much will fencing for her garden cost?

5-3   MULTIPLYING MIXED NUMBERS

Objective

Multiply mixed numbers.

Jeff made eight sections of a silver necklace and decorated them with hand-hammered designs and turquoise stones. Each section is $2\frac{3}{4}$ inches long. How long will the necklace be when all eight sections are joined? Assume that no length is lost when the sections are joined.

You need to multiply $2\frac{3}{4}$ by 8. Multiply mixed numbers as follows.

Method

1▶ Rename each mixed number as an improper fraction.

2▶ Multiply the fractions.

3▶ Write the product in simplest form.

Example A

Multiply $8 \times 2\frac{3}{4}$.   Estimate: $8 \times 3 = 24$

1▶ $8 = \frac{8}{1}$

$2\frac{3}{4} = \frac{(2 \times 4) + 3}{4}$ or $\frac{11}{4}$

2▶ $8 \times 2\frac{3}{4} = \frac{8}{1} \times \frac{11}{4}$

$= \frac{\overset{2}{\cancel{8}} \times 11}{1 \times \cancel{4}}$
$\phantom{= \frac{8 \times 11}{1 \times}}_{1}$

Compared to the estimate, is the answer reasonable?

3▶ $= \frac{22}{1}$ or 22

The necklace will be 22 inches long.

Example B

Multiply $2\frac{5}{8} \times 1\frac{3}{7}$.   Estimate: $3 \times 1 = 3$

1▶ $2\frac{5}{8} = \frac{(2 \times 8) + 5}{8}$ or $\frac{21}{8}$

$1\frac{3}{7} = \frac{(1 \times 7) + 3}{7}$ or $\frac{10}{7}$

2▶ $2\frac{5}{8} \times 1\frac{3}{7} = \frac{21}{8} \times \frac{10}{7}$

$= \frac{\overset{3}{21} \times \overset{5}{10}}{\underset{4}{\cancel{8}} \times \underset{1}{\cancel{7}}}$

Is the answer reasonable?

3▶ $= \frac{15}{4}$ or $3\frac{3}{4}$

Guided Practice

Example A

Multiply.

1. $3 \times 1\frac{3}{8}$

2. $2\frac{1}{6} \times 4$

3. $2\frac{5}{8} \times 3$

4. $7 \times 2\frac{2}{3}$

5. $12 \times 3\frac{1}{4}$

6. $5\frac{1}{5} \times 25$

Example B

7. $1\frac{1}{6} \times 1\frac{2}{5}$

8. $2\frac{2}{9} \times 1\frac{4}{7}$

9. $2\frac{2}{3} \times \frac{2}{3}$

10. $9\frac{3}{8} \times 3\frac{1}{5}$

11. $3\frac{1}{3} \times 3\frac{4}{5}$

12. $2\frac{4}{5} \times 2\frac{6}{7}$

Multiply.

13. $2\frac{1}{4} \times 2\frac{1}{7}$ 14. $2\frac{2}{3} \times 5\frac{1}{3}$ 15. $3\frac{1}{5} \times 1\frac{1}{8}$ 16. $1\frac{1}{6} \times \frac{7}{8}$

17. $7\frac{1}{2} \times 3\frac{1}{3}$ 18. $2\frac{6}{7} \times 2\frac{4}{5}$ 19. $1\frac{5}{6} \times 2\frac{1}{2}$ 20. $8\frac{1}{3} \times 3\frac{2}{5}$

21. $\frac{3}{8} \times 2\frac{2}{3}$ 22. $3\frac{1}{2} \times 1\frac{1}{2}$ 23. $6\frac{1}{4} \times 8\frac{1}{5}$ 24. $1\frac{1}{6} \times 1\frac{3}{5}$

25. $7\frac{1}{2} \times \frac{2}{5} \times 15$ 26. $5\frac{1}{2} \times \frac{3}{5} \times 8\frac{1}{3}$ 27. $3\frac{4}{5} \times 3\frac{1}{3} \times 1\frac{1}{5}$

28. Multiply $2\frac{1}{2}$ and $4\frac{1}{3}$.

29. What is the product of 6 and $7\frac{1}{24}$?

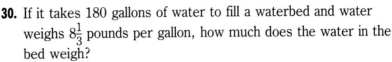

Applications

30. If it takes 180 gallons of water to fill a waterbed and water weighs $8\frac{1}{3}$ pounds per gallon, how much does the water in the bed weigh?

f **UN with MATH**

Suppose you were to cross a horse with a kangaroo?
See page 174.

31. A manager spends $\frac{1}{3}$ of her $7\frac{1}{2}$-hour day in meetings and the rest in directing her staff. How many hours a day does she spend directing her staff?

32. Carla's car travels $22\frac{1}{2}$ miles on one gallon of gasoline. How far can it travel on $4\frac{1}{2}$ gallons of gasoline?

Write Math

33. Write a letter to a sixth-grade student explaining how to multiply mixed numbers. Be sure to use an example that shows the shortcut on page 148.

Critical Thinking

34. A horse, rider, and saddle weigh 1,210 pounds. The horse weighs 9 times as much as the rider. The rider weighs 12 times as much as the saddle. How much does the horse weigh?

Suppose

35. Suppose you are driving 800 miles from Ohio to Massachusetts. Your car gets $23\frac{1}{2}$ miles to a gallon of gasoline and has a 12-gallon gas tank. Can you make the trip with 3 tankfuls?

36. Suppose, in Exercise 35, you had a 15-gallon gas tank. Could you make the trip in 2 tankfuls?

Mental Math

A **unit fraction** is a fraction that has a numerator of 1. Multiplying by a unit fraction is the same as dividing by the denominator of the unit fraction.

$$\frac{1}{4} \text{ of } 360 \to 360 \div 4 \text{ or } 90$$

Suppose you need to find $\frac{3}{4}$ of 360. Note that $\frac{3}{4}$ of 360 is 3 times $\frac{1}{4}$ of 360. So $\frac{3}{4}$ of 360 is 3×90 or 270.

Multiply. Write only the answer.

37. $\frac{1}{2} \times 70$ 38. $\frac{1}{3} \times 240$ 39. $\frac{1}{4} \times 160$ 40. $\frac{1}{5} \times 250$

5-4   ENERGY AND COAL

Objective

Use fractions in a real world application.

Energy and its uses are an important concern for everyone. The manufacturers of appliances put labels on their products to show the consumer the savings that can be expected by using their energy efficient appliances.

Coal is often used to produce energy in the form of electricity. How much coal does it take to produce electricity to run a color television for one year? The chart at the right lists the energy equivalents in tons of coal for one year of operation for several appliances. The electricity needed to run a color television for one year is produced by $\frac{1}{4}$ ton of coal.

Appliance	Energy Equivalent In Tons of Coal
Range	$\frac{1}{2}$
Microwave oven	$\frac{1}{10}$
Water heater	2
Lighting a 6-room house	$\frac{1}{3}$
Refrigerator	1
Radio	$\frac{1}{20}$
Dishwasher	$\frac{1}{5}$
Color TV	$\frac{1}{4}$

Understanding the Data

● Which appliance, a dishwasher or a color TV, takes more electricity to operate for one year?

● What types of appliances, ones that heat and cool or ones that produce light and sound, take more electricity to operate?

Exercises

Use the chart to find the energy equivalent in tons of coal for using each appliance.

1. a range and a radio for 1 year

2. a microwave oven for 2 years

3. a color TV for $\frac{1}{2}$ year

4. a dishwasher for 5 years

5. a refrigerator for $\frac{1}{4}$ year

6. lighting a 6-room house for $\frac{3}{4}$ year

7. a color TV for 2 years

8. a water heater for 1 month

9. a radio for 6 months

10. List the appliances in the chart according to the amount of energy they use. List in order from greatest to least energy use.

Coal is found in many areas of the United States. There are coal mines in 26 states. The circle graph at the right shows areas where the coal deposits are located.

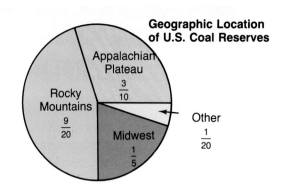

Geographic Location of U.S. Coal Reserves

11. What part of the coal deposits are found in the Midwest and the Rocky Mountains?

12. Which area has the most coal deposits?

13. Which area has the greater coal deposits, the Appalachian Plateau or the Midwest?

14. About $\frac{4}{5}$ of the coal in the Appalachian Plateau is mined underground. What part of the total U.S. coal deposits is this?

15. About $\frac{2}{5}$ of the coal in the Rocky Mountains can be mined on the surface. What part of the total U.S. coal reserves is this?

16. Recently, there were about 470 billion tons of coal in deposits. About how much of this coal was in the Midwest?

17. Chin's family has a monthly income of $1,940. They spend about $\frac{1}{30}$ of their income for electricity. About how much do they spend each month for electricity?

Suppose

18. Suppose Mrs. Crewe installed a solar heating system in her home for $5,720. If she saves $40 a month on her heating bill, in *about* how many years will the savings equal the cost of the solar heating system?

19. The U.S. produces about 740 million tons of coal in one year. If Canada produces $\frac{1}{25}$ of that amount, *about* how many tons of coal does Canada produce?

Mixed Review

Lesson 1-2

Write each number as a power of another number in two different ways.

20. 64   21. 81   22. 1   23. 10,000

Lesson 4-1

Find all the factors of each number.

24. 32   25. 45   26. 24   27. 54

Lesson 5-2

28. One winter day only $\frac{5}{8}$ of the entire student body of Central High School was in school. How many students were in school that day if the total number of students who attend Central is 1,120?

> Explore
> Plan
> Solve
> Examine

5-5   SIMPLIFYING THE PROBLEM

Objective

Solve problems by simplifying the problem.

Two art students are making posters for the school carnival. If they make two posters in two hours, how many posters can six students make in ten hours? Assume all the students work at the same rate.

▶ **Explore**

What is given?

● Two students make two posters in two hours.

What is asked?

● How many posters can six students make in ten hours?

▶ **Plan**

Simplify the problem by finding how long it takes each student to make one poster. Then find how many posters each can make in ten hours and multiply by the number of students, or six.

▶ **Solve**

$2 \div 2 = 1$ Each student makes 1 poster in 2 hours.

How many posters can one student make in 10 hours?

Divide 10 by 2 since each poster takes 2 hours.

$10 \div 2 = 5$ Each student can make 5 posters in 10 hours.

How many posters can 6 students make in 10 hours? $6 \times 5 = 30$

Six students can make 30 posters in ten hours.

▶ **Examine**

Check your solution by thinking another way. If two students can make two posters in two hours, six students can make three times as many posters in two hours or 6 posters in two hours. In ten hours the students can make 5 times as many or 30 posters. The solution is 30 posters.

Guided Practice

Solve. Use simplifying the problem.

1. Two boys can mow four lawns in two hours. How many lawns can six boys mow in eight hours? Assume all the boys work at the same rate.

2. Sixteen people are sharing a super-sized submarine sandwich. How many cuts must be made to divide the sandwich equally among the sixteen people? Explain your answer.

Solve. Use any strategy.

3. A plane flying at an altitude of 33,700 feet is directly above a submarine at a depth of 400 feet. Find the vertical distance between them.

Paper/ Pencil
Mental Math
Estimate
Calculator

4. A rectangular field is fenced on two adjacent sides by a stone wall. The length of the field is 63 meters. The area of the field is 1,323 square meters. Determine how much fencing is needed to enclose the two sides not fenced in by stone walls.

5. Find the number of seconds in the month of July. Write the method of computation used: estimation, mental math, calculator, or paper and pencil.

6. Chris, Peter, Maria, Danielle, Sarah, and Mike have tickets to a rock concert. Their tickets are seats 1-6 in row L. The lowest numbered seat is on the aisle. Chris is sitting between Maria and Sarah. Sarah is sitting on the aisle. Peter is in seat 4 and is between two girls. Find the number of each person's seat.

7. Five different colored pairs of socks are mixed up in a drawer. If socks are pulled out at random, what is the greatest number of socks that could be chosen before you get a matching pair?

8. A jet can seat 525 passengers. Find the number of passengers on the plane if four seats are occupied for every empty seat.

Research

9. In the library, find information about the history of calculators. See the Fun Pages, 104 and 105, in Chapter 3 to get started.

Critical Thinking

10. How many different routes can Peter take from Room 101 to Room 205 if the hallway from Room 104 to Room 203 does not intersect with the hallway from Room 103 to Room 204?

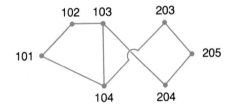

Mixed Review

Find the average for each set of numbers. Round to nearest tenth.

Lesson 3-12

11. 18, 24, 25   **12.** 4.5, 6.2, 5.8   **13.** 0.83, 0.74, 0.78

Lesson 4-2

Write the prime factorization of each number.

14. 18   **15.** 20   **16.** 28   **17.** 42   **18.** 102

Lesson 4-11

19. Miranda works for $3\frac{1}{2}$ hours on Saturday and $4\frac{1}{3}$ hours on Sunday. How many hours does she work on both days?

CHEF

Diane Seriani is the head chef in a hotel restaurant. Chef Seriani coordinates the work of the kitchen staff and directs preparation of special dishes. One of the house specialties is Grilled Lemon Chicken.

Grilled Lemon Chicken

6 pieces chicken
(leg, thigh, or split
breast)

$\frac{1}{4}$ pound corn oil margarine

$\frac{1}{2}$ cup lemon juice

1 tbsp poultry seasoning
1 tsp dried parsley, crumbled

$1\frac{1}{2}$ cups long grain and wild
rice, uncooked

$\frac{1}{2}$ tsp salt

Cook rice according to package directions. In a small saucepan, combine lemon juice, poultry seasoning, margarine, and parsley. Bring mixture to a boil and remove from heat. Baste chicken with lemon mixture. Broil 4 inches from heat, placing breasts with rib side facing heat, 20 to 25 minutes. Turn chicken and baste again. Broil an additional 10 minutes until brown. Serve on the rice. Pour remaining lemon mixture on chicken. Serves 4.

Suppose Chef Seriani doubles the recipe. How much rice does she need?

cups of rice needed
for one recipe ⟶ $\frac{3}{2} \times 2 = \frac{6}{2}$ or 3 ← cups of rice needed
for double recipe

change $1\frac{1}{2}$ to an
improper fraction

How much of each item is needed to double the recipe?

1. margarine   **2.** lemon juice   **3.** poultry seasoning

How many ounces of each item is needed for $\frac{3}{4}$ of the recipe?

4. salt   **5.** parsley   **6.** rice

7. How many pieces of chicken are needed to serve 16 people?

8. If 2 cups of uncooked rice is equivalent to one pound, how many pounds of rice are needed to serve 16 people?

9. Tomorrow Chef Seriani needs 2 extra cooks. Each will work 4 extra hours to prepare for a banquet. The rate for cooks is $10.25. How much will this overtime help cost?

ESTIMATING UTILITY BILLS

Pam and Mike Spiker live in an apartment. They have to pay three utility bills. Each month they receive an electric bill and a gas bill. Every three months—four times a year—they receive a water bill.

The amount of electricity, natural gas, and water they use is measured using the following units. The cost per unit is also given below.

Utility	Electricity	Natural Gas	Water
Unit	Kilowatt hour (kWh)	100 cubic feet (CCF)	cubic feet (ft^3)
Cost per unit	5.6¢	56.5¢	7.85¢

Mrs. Spiker estimates each utility bill to see if the amount charged is reasonable.

Estimate the cost of 906 kWh of electricity at 5.6¢ per kWh.

$$\begin{array}{cc} 906 & 900 \\ \underline{\times\ 5.6¢} & \underline{\times\quad 6¢} \\ & 5400¢ \end{array}$$

Round 906 to 900.
Round 5.6¢ to 6¢.
Cost in cents

Write 5400¢ as $54. Mrs. Spiker estimates the electric bill to be $54.

Estimate the amount of each electric bill. The number of kilowatt-hours (kWh) and the cost per kWh are given.

1. 563 kWh, 8.23¢   **2.** 795 kWh, 7.6¢   **3.** 241 kWh, 5.5¢

Estimate the amount of each natural gas bill. The amount of gas used (CCF) and the cost per CCF are given.

4. 64 CCF, 56.6¢   **5.** 125 CCF, 48.9¢
6. 187 CCF, 59¢

7. Last month the Jones family used 945 kWh of electricity at 7.6¢ per kWh and 57 CCF of gas at 58.2¢ per CCF. Estimate the total of these two utility bills.

8. Estimate each utility bill. Then estimate the total. 894 kWh of electricity at 8.9¢ per kWh, 94 CCF of gas at 49.9¢ per CCF, and 726 ft^3 of water at 5.7¢ per ft^3.

5-6 DIVIDING FRACTIONS

Objective

Find the reciprocal of a number. Divide fractions.

The Student Council at Jefferson High is planning a cookout. Harry found the best price on hamburger was $16.50 for a 10-pound package. How many $\frac{1}{4}$-pound patties can Harry make from 10 pounds of hamburger? Harry divided 10 by $\frac{1}{4}$ to find the answer. He made this drawing to show 10 pounds of hamburger.

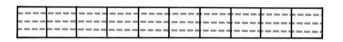

The dashed lines show that each pound makes four $\frac{1}{4}$-pound hamburger patties. Ten pounds makes 10 times 4 or 40 hamburger patties. Example C shows another way to solve this problem.

Two numbers whose product is 1, such as 4 and $\frac{1}{4}$, are **reciprocals**.

Examples

A *What is the reciprocal of $\frac{5}{12}$?*

Since $\frac{5}{12} \times \frac{12}{5} = 1$, the reciprocal of $\frac{5}{12}$ is $\frac{12}{5}$.

B *What is the reciprocal of 3?*

Since $3 \times \frac{1}{3} = 1$, the reciprocal of 3 is $\frac{1}{3}$.

To divide by a fraction, multiply by its reciprocal.

Method

1 ▶ Rewrite the problem as a product of the dividend and the reciprocal of the divisor.

2 ▶ Multiply.

3 ▶ Write the result in simplest form if necessary.

Examples

C $10 \div \frac{1}{4}$ The quotient will be greater than 10.

1 ▶ The reciprocal of $\frac{1}{4}$ is 4.

$10 \div \frac{1}{4} = \frac{10}{1} \times \frac{4}{1}$

2 ▶ $= \frac{40}{1}$

3 ▶ $= 40$

Harry can make 40 hamburger patties from 10 pounds of hamburger.

D $\frac{5}{9} \div \frac{5}{12}$

1 ▶ The reciprocal of $\frac{5}{12}$ is $\frac{12}{5}$.

$\frac{5}{9} \div \frac{5}{12} = \frac{5}{9} \times \frac{12}{5}$

2 ▶ $= \frac{\overset{1}{\cancel{5}}}{\underset{3}{\cancel{9}}} \times \frac{\overset{4}{\cancel{12}}}{\underset{1}{\cancel{5}}}$

3 ▶ $= \frac{4}{3}$ or $1\frac{1}{3}$

Example E

$\frac{14}{15} \div 7$ Will the quotient be less than or greater than 7?

1 ▶ The reciprocal of 7 is $\frac{1}{7}$.

$\frac{14}{15} \div 7 = \frac{14}{15} \times \frac{1}{7}$

2 ▶ $= \frac{\overset{2}{\cancel{14}}}{15} \times \frac{1}{\underset{1}{\cancel{7}}}$ or $\frac{2}{15}$

Examples A, B

Name the reciprocal of each number.

1. $\frac{8}{9}$ **2.** $\frac{2}{3}$ **3.** $\frac{3}{20}$ **4.** $\frac{5}{7}$ **5.** 5 **6.** 12

Divide.

Example C

7. $4 \div \frac{3}{4}$ **8.** $10 \div \frac{5}{8}$ **9.** $14 \div \frac{7}{9}$ **10.** $6 \div \frac{3}{7}$

Examples D, E

11. $\frac{2}{3} \div \frac{1}{4}$ **12.** $\frac{7}{15} \div \frac{5}{6}$ **13.** $\frac{3}{8} \div 5$ **14.** $\frac{11}{12} \div 11$

Practice

Name the reciprocal of each number.

15. $\frac{19}{20}$ **16.** 24 **17.** $\frac{11}{16}$ **18.** $\frac{1}{6}$

Divide.

19. $\frac{3}{4} \div \frac{2}{5}$ **20.** $\frac{7}{8} \div \frac{3}{5}$ **21.** $\frac{5}{12} \div \frac{2}{3}$ **22.** $\frac{2}{3} \div \frac{2}{3}$

23. $6 \div \frac{3}{8}$ **24.** $3 \div \frac{2}{3}$ **25.** $2 \div \frac{1}{2}$ **26.** $8 \div \frac{2}{3}$

27. $\frac{1}{4} \div 9$ **28.** $\frac{7}{8} \div 3$ **29.** $\frac{3}{5} \div 12$ **30.** $\frac{3}{8} \div 3$

31. Divide $\frac{7}{8}$ by $\frac{1}{2}$.

32. What is the quotient if $\frac{3}{10}$ is divided by 2?

Applications

33. Angie has $\frac{7}{8}$ yard of material left over after making chair pads. If it takes $\frac{1}{6}$ yard to make a napkin ring, does Angie have enough material left to make 6 matching napkin rings?

34. How many $\frac{5}{6}$-foot vinyl tiles will fit along the edge of a 15-foot room? Make no allowance for seams.

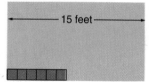

Talk Math

35. Tell how this drawing shows $6 \div \frac{1}{3} = 18$.

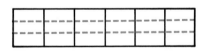

Using Variables

Evaluate each expression if $p = \frac{1}{2}$, $q = \frac{3}{4}$, and $r = \frac{1}{3}$.

36. $q \div r$ **37.** $p \div q$ **38.** $p \times q$ **39.** $q \times r$

Critical Thinking

40. A knitting pattern for the body of a sweater is given below.

    Row 1 Knit 1, Purl 1, . . .   Row 2 Knit 1, Purl 1, . . .
    Row 3 Purl 1, Knit 1, . . .   Row 4 Purl 1, Knit 1, . . .
    Row 5 Knit 1, Purl 1, . . .   Row 6 Knit 1, Purl 1, . . .
    Row 7 Purl 1, Knit 1, . . .   and so on.
    Does Row 35 begin with *Knit 1* or *Purl 1*?

5-7 DIVIDING MIXED NUMBERS

Objective

Divide mixed numbers.

Justin jogs $12\frac{1}{2}$ miles a day when training for a wrestling event. If the course he jogs is $2\frac{1}{2}$ miles long, how many times does he jog the course each day?

Method

1. ▶ Rename each mixed number as an improper fraction.
2. ▶ Divide the fractions.
3. ▶ Write the result in simplest form, if necessary.

Example A

$12\frac{1}{2} \div 2\frac{1}{2}$ Estimate: $12 \div 3 = 4$

1. ▶ $12\frac{1}{2} = \frac{(12 \times 2) + 1}{2}$ or $\frac{25}{2}$

   $2\frac{1}{2} = \frac{(2 \times 2) + 1}{2}$ or $\frac{5}{2}$

2. ▶ $12\frac{1}{2} \div 2\frac{1}{2} = \frac{25}{2} \div \frac{5}{2}$

   $= \frac{\overset{5}{\cancel{25}}}{\cancel{2}} \times \frac{\cancel{2}}{\cancel{5}}$

   $= \frac{5}{1}$

   The reciprocal of $\frac{5}{2}$ is $\frac{2}{5}$.

3. ▶ $= 5$

Subtract $2\frac{1}{2}$ five times to check the answer.

$12\frac{1}{2} - 2\frac{1}{2} - 2\frac{1}{2} - 2\frac{1}{2} - 2\frac{1}{2} - 2\frac{1}{2} = 0$

Based on the estimate, explain why the answer is reasonable.

Justin jogs around the course 5 times.

Example B

$2\frac{2}{3} \div 4$ Estimate: $2 \div 4 = \frac{2}{4}$ or $\frac{1}{2}$

1. ▶ $2\frac{2}{3} = \frac{(2 \times 3) + 2}{3}$ or $\frac{8}{3}$

2. ▶ $2\frac{2}{3} \div 4 = \frac{8}{3} \div 4$

   $= \frac{\overset{2}{\cancel{8}}}{3} \times \frac{1}{\underset{1}{\cancel{4}}}$

   $= \frac{2}{3}$ ← simplest form

   The reciprocal of 4 is $\frac{1}{4}$.

   Is the answer reasonable?

Guided Practice

Divide.

Example A

1. $1\frac{2}{3} \div 1\frac{1}{5}$
2. $8\frac{1}{4} \div 5\frac{1}{6}$
3. $2\frac{2}{5} \div \frac{2}{5}$
4. $\frac{1}{3} \div 1\frac{5}{6}$

5. $\frac{2}{5} \div 5\frac{4}{5}$
6. $3\frac{3}{8} \div 2\frac{1}{5}$
7. $5\frac{1}{4} \div 8\frac{1}{6}$
8. $1\frac{1}{8} \div \frac{9}{10}$

Example B

9. $3\frac{1}{3} \div 4$
10. $1\frac{3}{4} \div 2$
11. $1\frac{1}{4} \div 15$
12. $9 \div 1\frac{4}{5}$

13. $6 \div 3\frac{1}{5}$
14. $12 \div 1\frac{3}{5}$
15. $10 \div 13\frac{3}{4}$
16. $10\frac{1}{2} \div 9$

Practice

Divide.

17. $1\frac{1}{8} \div 3$ **18.** $2\frac{1}{2} \div 3$ **19.** $6 \div 2\frac{2}{5}$ **20.** $6 \div 9\frac{3}{5}$

21. $14 \div 2\frac{1}{10}$ **22.** $6 \div 3\frac{1}{2}$ **23.** $3\frac{3}{5} \div 12$ **24.** $13\frac{3}{4} \div 5$

25. $1\frac{3}{4} \div 4\frac{3}{8}$ **26.** $8\frac{2}{5} \div 1\frac{2}{5}$ **27.** $6\frac{3}{5} \div 6$ **28.** $1\frac{2}{3} \div \frac{2}{3}$

29. $5 \div 2\frac{1}{2}$ **30.** $\frac{1}{4} \div 2\frac{1}{3}$ **31.** $12\frac{3}{4} \div 4\frac{7}{8}$ **32.** $1\frac{7}{8} \div 5$

33. Divide $3\frac{1}{8}$ by $\frac{5}{8}$.

34. What is the quotient if $\frac{1}{2}$ is divided by $1\frac{1}{2}$?

Applications

35. A $3\frac{3}{4}$-foot board is cut into pieces that are $\frac{1}{8}$ of a foot long. How many pieces are cut from the board?

36. Don has two stacks of $\frac{3}{8}$-inch plywood sheets. One stack is 30 inches high. The other is 24 inches high. How many sheets of plywood are there in both stacks?

37. How many boards of 6-inch siding does it take to cover the side of a house if each board covers $5\frac{1}{4}$ inches and the height of the wall is 105 inches?

Critical Thinking

38. A football team can score 6 points for a touchdown, 1 point for a point after a touchdown, 2 points for a safety, and 3 points for a field goal. Write 5 different ways a team can score 34 points.

Cooperative Groups

39. In groups of three or four, use a newspaper ad for a building supply company and make up three problems. In one problem you must use multiplication of fractions to solve it. In another problem you must use division of fractions to solve it. In the third problem you must use mixed numbers to write the problem.

Problem Solving

40. The product of two fractions is $15\frac{1}{8}$. One of the fractions is $2\frac{3}{4}$. What is the other fraction?

Estimation

Most department stores have sales several times a year. You can estimate the savings mentally to be sure you are charged the correct price.

About how much are a pair of shoes that regularly cost $32.99?

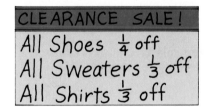

CLEARANCE SALE!
All Shoes $\frac{1}{4}$ off
All Sweaters $\frac{1}{3}$ off
All Shirts $\frac{1}{3}$ off

Think: $\frac{1}{4}$ of $32.99 \rightarrow \frac{1}{4}$ of $32 or $8 $\leftarrow$ estimated savings
$32.99 $-$ $8 \rightarrow $33 $-$ $8 or $25 $\leftarrow$ estimated sale price

Estimate the sale price of each item. Write only your answer.

41. a $27.50 sweater   **42.** a $14.99 shirt   **43.** a $39 pair of shoes

44. an $18.95 shirt   **45.** a $45 pair of shoes   **46.** a $38.99 sweater

5-8 CHANGING FRACTIONS TO DECIMALS

Objective

Change fractions to decimals.

Joy needs to add $\frac{1}{4}$ liter of a salt solution to the mixture in her beaker. The graduated cylinder is marked in decimal parts of a liter, so Joy needs to change the fraction to a decimal. She divides 1 by 4 to change $\frac{1}{4}$ to a decimal.

To change a fraction to a decimal, divide the numerator by the denominator. If the division ends or terminates with a remainder of zero, the quotient is called a **terminating decimal.** If a remainder of zero cannot be obtained, the digits in the quotient repeat. The quotient is called a **repeating decimal.**

Method

1 ▶ Divide the numerator by the denominator.

2 ▶ Stop when a remainder of zero is obtained or a pattern develops in the quotient.

Examples

A *Change $\frac{1}{4}$ to a decimal.*

1 ▶ $1 \div 4 = 0.25$

Joy measures to the line marked 0.25 liter.

B *Change $\frac{2}{3}$ to a decimal.*

1 ▶ $2 \div 3 =$

0.6666667
Your calculator rounds.
0.6666666

Your calculator carries another digit in memory.

$\frac{2}{3} = 0.\overline{6}$ The bar is used to show digits that repeat.

Fractions that are equivalent to repeating decimals may also be written as mixed decimals.

Example C

Change $\frac{4}{7}$ to a mixed decimal.

$$7\overline{)4.00}^{\quad 0.57\frac{1}{7}}$$
$$\underline{-3\ 5}$$
$$50$$
$$\underline{-\ 49}$$
$$1$$

$\frac{4}{7} = 0.57\frac{1}{7}.$

Guided Practice

Example A

Change each fraction to a decimal.

1. $\frac{2}{5}$ 2. $\frac{3}{8}$ 3. $\frac{3}{4}$ 4. $\frac{5}{16}$

Change each fraction to a decimal. Use bar notation.

5. $\frac{1}{9}$ **6.** $\frac{5}{33}$ **7.** $\frac{5}{6}$ **8.** $\frac{7}{12}$

Example C

Change each fraction to a mixed decimal.

9. $\frac{2}{3}$ **10.** $\frac{7}{13}$ **11.** $\frac{8}{9}$ **12.** $\frac{13}{24}$

Exercises

Practice

Change each fraction to a decimal. Use bar notation to show a repeating decimal.

13. $\frac{7}{10}$ **14.** $\frac{12}{25}$ **15.** $\frac{11}{20}$ **16.** $\frac{5}{8}$ **17.** $\frac{3}{11}$

18. $\frac{7}{16}$ **19.** $\frac{2}{3}$ **20.** $\frac{5}{9}$ **21.** $\frac{7}{9}$ **22.** $\frac{10}{33}$

23. $\frac{28}{45}$ **24.** $\frac{9}{16}$ **25.** $\frac{3}{16}$ **26.** $\frac{31}{40}$ **27.** $\frac{11}{12}$

Change each fraction to a mixed decimal.

28. $\frac{2}{7}$ **29.** $\frac{19}{30}$ **30.** $\frac{6}{7}$ **31.** $\frac{1}{3}$ **32.** $\frac{13}{15}$

33. Write $\frac{1}{125}$ as a decimal.

34. Write $\frac{5}{11}$ as a decimal.

Applications

f **UN with MATH**

Ever thought about how a VCR works?
See page 175.

35. The length of one bolt is $\frac{5}{6}$ inch. The length of another is 0.875 inch. Which bolt is longer?

36. A box of electrical connectors weighs $6\frac{1}{4}$ pounds. If the box weighs $\frac{1}{4}$ pound and each connector weighs $\frac{1}{16}$ pound, how many connectors are in the box?

Cooperative Groups

37. In groups of three or four, use a calculator to list examples of fractions that can be expressed as repeating decimals. Examine your list to discover what kind of numbers in the denominator result in a repeating decimal.

Critical Thinking

38. If 🌻 represents $3\frac{1}{2}$, then 🌻 represents $\frac{?}{}$.

Using Data

Solve. Use the chart.

39. If 1,181 students were surveyed, how many considered themselves competent in computer games?

Collect Data

40. Conduct your own survey of the students in your class or in your school. Compare your results to the results of the survey.

Computer Know How
Percent of Students with Computer Skills

Programming 39%
Database Use 55%
Word Processor Use 81%
Data Entry 81%
Computer Games 89%

CHANGING DECIMALS TO FRACTIONS

Objective

Change decimals to fractions.

Missy is shopping for pizza supplies. She asks the clerk in the meat department for $\frac{7}{8}$ pound of ground sausage. The package the clerk handed to Missy was labeled 0.875 pound. Missy changed 0.875 to a fraction to make sure she had $\frac{7}{8}$ pound of ground sausage.

To change a terminating decimal to a fraction, write the decimal as a fraction and simplify.

Method

▷**1** Write the digits of the decimal as the numerator. Use the appropriate power of ten (10, 100, 1,000, and so on) as the denominator.

▷**2** Simplify.

Examples

A **Change 0.875 to a fraction.**

▷**1** $0.875 = \dfrac{\overset{7}{\cancel{875}}}{\underset{8}{\cancel{1,000}}}$ The GCF is 125.

▷**2** $\qquad = \dfrac{7}{8}$

Missy found she had $\frac{7}{8}$ pound of sausage.

B **Change 0.32 to a fraction.**

▷**1** $0.32 = \dfrac{\overset{8}{\cancel{32}}}{\underset{25}{\cancel{100}}}$ The GCF is 4.

▷**2** $\qquad = \dfrac{8}{25}$

To change a mixed decimal to a fraction, remember that a fraction indicates division.

Example C

Change $0.16\frac{2}{3}$ to a fraction.

▷**1** $0.16\frac{2}{3} = \dfrac{16\frac{2}{3}}{100}$

▷**2** $\dfrac{16\frac{2}{3}}{100} = 16\frac{2}{3} \div 100$ Write the fraction as a division problem.

$\qquad = \dfrac{50}{3} \div 100$ Write $16\frac{2}{3}$ as an improper

$\qquad = \dfrac{\overset{1}{\cancel{50}}}{3} \times \dfrac{1}{\underset{2}{\cancel{100}}}$ fraction: $\dfrac{(16 \times 3) + 2}{3}$

$\qquad = \dfrac{1}{6}$

Guided Practice

Change each decimal to a fraction.

Example A

1. 0.006   **2.** 0.084   **3.** 0.125   **4.** 0.650   **5.** 0.408

Example B

6. 0.98   **7.** 0.53   **8.** 0.31   **9.** 0.64   **10.** 0.05

Example C

Change each mixed decimal to a fraction.

11. $0.14\frac{2}{7}$ **12.** $0.66\frac{2}{3}$ **13.** $0.22\frac{2}{9}$ **14.** $0.85\frac{5}{7}$ **15.** $0.71\frac{3}{7}$

Change each decimal to a fraction.

16. 0.5   **17.** 0.8   **18.** 0.32   **19.** 0.75

20. 0.54   **21.** 0.38   **22.** 0.744   **23.** 0.101

24. 0.303   **25.** 0.486   **26.** 0.626   **27.** 0.448

28. 0.074   **29.** 0.008   **30.** 9.36   **31.** 10.18

Change each mixed decimal to a fraction.

32. $0.11\frac{1}{9}$ **33.** $0.33\frac{1}{3}$ **34.** $0.83\frac{1}{3}$ **35.** $0.09\frac{1}{11}$

Replace each ● with =, <, or > to make a true sentence.

36. 0.25 ● $\frac{1}{4}$ **37.** $0.\overline{1}$ ● $\frac{1}{9}$ **38.** $\frac{2}{7}$ ● 0.286   **39.** $\frac{9}{10}$ ● $0.\overline{9}$

Applications

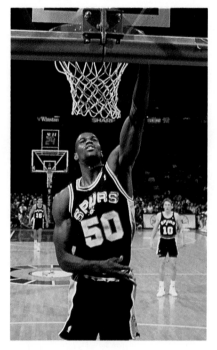

What, besides pizza, originated in Italy?
See page 175.

40. Scott's meatloaf recipe calls for $1\frac{3}{4}$ pounds of ground beef. The price label on a ground beef package indicates that it contains 1.89 pounds of ground beef. Does the package contain enough ground beef to make the meatloaf?

41. A 10-pound spool of solder for making a stained glass window contains 9.375 feet of solder. How many $\frac{5}{8}$-foot pieces of solder can be cut from the spool?

42. Suppose David Robinson's field-goal average was 0.501 in 1993. We say he makes 5 out of 10 field goals. Suppose your field-goal average is 0.40. You can say you make $\underline{\ ?\ }$ out of $\underline{\ ?\ }$ field goals.

Number Sense

A method for finding a fraction that is halfway between a given pair of fractions is to find their average or one-half their sum.

43. Find a fraction halfway between $\frac{7}{9}$ and $\frac{8}{9}$.

44. Find a fraction halfway between $\frac{6}{11}$ and $\frac{7}{11}$.

45. Find a decimal that is halfway between 0.67 and 0.68.

Using Algebra

Evaluate each expression if $a = 0.25$, $b = \frac{3}{8}$, and $c = 1.375$.

46. $a + b$ **47.** $b - a$ **48.** ac **49.** bc

> Explore
> Plan
> Solve
> Examine

5-10 USING FRACTIONS

Objective

Solve verbal problems involving fractions.

Janie bought an aquarium. She uses a $2\frac{1}{4}$-gallon bucket to fill it with $22\frac{1}{2}$ gallons of water. How many buckets of water does it take?

What facts are known?
- Janie fills the aquarium with $22\frac{1}{2}$ gallons of water.
- She uses a bucket that holds $2\frac{1}{4}$ gallons of water.

What do you need to find?
- the number of buckets of water it will take to fill the aquarium

Divide the total number of gallons by the number of gallons in one bucket.

Estimate $22 \div 2 = 11$ It takes between 7 and 11
              $21 \div 3 = 7$ buckets to fill the aquarium.

$$22\frac{1}{2} \div 2\frac{1}{4} = \frac{45}{2} \div \frac{9}{4}$$

$$= \frac{\overset{5}{\cancel{45}}}{\underset{1}{\cancel{2}}} \times \frac{\overset{2}{\cancel{4}}}{\underset{1}{\cancel{9}}}$$

$$= 10$$

It takes 10 buckets of water to fill the aquarium.

Compared to the estimate the answer is reasonable. You can use repeated subtraction to show the answer is correct.

$$22\frac{1}{2} - \left(2\frac{1}{4} + 2\frac{1}{4} + 2\frac{1}{4} + 2\frac{1}{4} + 2\frac{1}{4} + 2\frac{1}{4} + 2\frac{1}{4} + 2\frac{1}{4} + 2\frac{1}{4} + 2\frac{1}{4}\right) = 0$$

— Guided Practice —

1. Boyd buys $8\frac{3}{4}$ yards of blue fabric and $6\frac{7}{8}$ yards of plaid fabric to make a sleeping bag. How many yards of fabric does he buy in all?

2. To hang some of her paintings, Heather needs six pieces of wire each $1\frac{3}{4}$ feet long. At least how many feet of wire should she buy?

3. A survey indicated that of the 660 students at Harry Johnson High School, $\frac{2}{5}$ participate in the sports programs. How many students participate in sports?

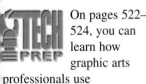

Paper/ Pencil
Mental Math
Estimate
Calculator

4. Rita and three friends divide $\frac{2}{3}$ of a pizza evenly among themselves. How much pizza does each person get?

5. Horace walks one block in $1\frac{3}{4}$ minutes. At that rate, about how many minutes does it take him to walk $7\frac{1}{2}$ blocks to school?

6. Ben works $5\frac{1}{4}$ hours overtime on Saturday and $3\frac{1}{2}$ hours overtime on Sunday. How many hours of overtime does he work on the weekend?

7. The Mississippi River is about $2\frac{1}{3}$ times as long as the Ohio River. If the Mississippi River is about 2,350 miles long, about how long is the Ohio River?

8. On a recent trip Tom averaged $48\frac{1}{2}$ miles per hour. How far did he travel in $3\frac{3}{4}$ hours?

9. Joan drove $62\frac{1}{2}$ miles in $1\frac{1}{3}$ hours on Thursday. How many miles per hour did she average?

10. There are $5\frac{1}{2}$ cups of sugar in a canister. Clare uses $2\frac{3}{4}$ cups for one recipe and $1\frac{3}{4}$ cups for another recipe. How much sugar is left in the canister?

11. Mr. Acker has $8\frac{1}{2}$ bushels of apples. He sells $\frac{3}{4}$ of them for $0.79 a pound. How many bushels of apples are left?

12. Juan buys $2\frac{1}{2}$ pounds of ground beef at $1.90 a pound and half as much ground pork at $2.80 a pound. He gives the cashier $10. How much change does he receive?

On pages 522–524, you can learn how graphic arts professionals use mathematics in their jobs.

Critical Thinking

13. Four monkeys can eat 4 sacks of peanuts in 3 minutes. How many monkeys, eating at the same rate, will it take to eat 100 sacks of peanuts in one hour?

14. Review the items in your portfolio. Make a table of contents of the items, noting why each item was chosen. Replace any items that are no longer appropriate.

Mixed Review

Subtract.

Lesson 2-3

15. 82	**16.** 95	**17.** 436	**18.** 204
− 16	− 73	− 58	− 196

Lesson 4-3

Find the GCF of each group of numbers.

19. 21, 35   **20.** 15, 45   **21.** 81, 99   **22.** 34, 102

Lesson 4-8

23. Albert ran $4\frac{7}{10}$ miles and Leonard ran $4\frac{5}{8}$ miles. Which runner ran farther?

REVIEW

Vocabulary/ Concepts

Choose the letter of the number, word, or phrase at the right that best matches each description.

1. a decimal whose digits end, or terminate, like 0.25

2. an estimate of $16\frac{1}{2} \div 1\frac{3}{4}$

3. a number that consists of a decimal and a fraction, like $0.83\frac{1}{3}$

4. two numbers whose product is 1

5. an estimate of $2\frac{1}{4} \times 5\frac{2}{3}$

6. a decimal whose digits repeat in groups of one or more, like 0.525252 ...

a. $\frac{1}{2}$
b. 2
c. 8
d. 12
e. mixed decimal
f. mixed numeral
g. reciprocals
h. repeating decimal
i. terminating decimal

Exercises/ Applications

Lesson 5-1

Estimate.

7. $4\frac{2}{3} \times 9$

8. $4\frac{1}{2} \times 2\frac{1}{8}$

9. $7\frac{1}{8} \div 3\frac{4}{5}$

10. $12\frac{2}{5} \div 3\frac{3}{4}$

11. Is $\frac{2}{3}$ of 5 less than or greater than 5?

12. Is $\frac{5}{6} \div \frac{3}{4}$ less than or greater than $\frac{5}{6}$?

Lesson 5-2

Multiply.

13. $\frac{4}{9} \times \frac{1}{4}$

14. $\frac{3}{5} \times \frac{5}{8}$

15. $\frac{5}{6} \times 8$

16. $\frac{7}{12} \times \frac{4}{5}$

17. $\frac{2}{5} \times \frac{9}{10}$

18. $7 \times \frac{3}{14}$

19. $\frac{4}{5} \times \frac{5}{9}$

20. $\frac{4}{7} \times \frac{7}{8}$

21. $\frac{3}{4} \times \frac{8}{9}$

22. $\frac{3}{5} \times 25$

Lesson 5-3

Multiply.

23. $\frac{3}{4} \times 1\frac{2}{5}$

24. $7\frac{1}{2} \times 3\frac{3}{5}$

25. $6 \times 4\frac{3}{5}$

26. $2\frac{1}{2} \times \frac{2}{3}$

27. $6\frac{1}{4} \times 2\frac{2}{3}$

28. $9\frac{1}{4} \times 12$

29. $\frac{3}{11} \times 3\frac{2}{3}$

30. $2\frac{2}{9} \times 2\frac{4}{7}$

Lesson 5-4

Use the table on page 152 to find energy equivalents for each appliance.

31. a dishwasher for 6 months

32. a color TV for 8 months

33. a radio and TV for 2 years

34. a refrigerator for 4 years

Lesson 5-5

Solve. Use simplifying the problem.

35. Three students can make 6 floral centerpieces for the prom in 1 hour. How many centerpieces can the whole prom committee make in 2 hours if there are 9 people on the committee and they all work at the same rate?

Lesson 5-6

Name the reciprocal of each number.

36. 7   **37.** $\frac{3}{4}$ **38.** $\frac{7}{12}$ **39.** 6   **40.** $\frac{7}{5}$ **41.** $\frac{9}{2}$

Lesson 5-6

Divide.

42. $\frac{3}{8} \div \frac{1}{2}$ **43.** $\frac{7}{10} \div \frac{2}{5}$ **44.** $\frac{8}{9} \div 4$ **45.** $\frac{5}{8} \div \frac{5}{6}$ **46.** $9 \div \frac{3}{8}$

47. $\frac{9}{10} \div 6$ **48.** $\frac{5}{9} \div \frac{9}{10}$ **49.** $\frac{11}{12} \div \frac{3}{4}$ **50.** $15 \div \frac{5}{7}$ **51.** $\frac{5}{12} \div \frac{5}{7}$

Lesson 5-7

Divide.

52. $10 \div 3\frac{3}{4}$ **53.** $1\frac{2}{5} \div 2\frac{2}{3}$ **54.** $3\frac{1}{5} \div 1\frac{1}{3}$ **55.** $2\frac{2}{5} \div \frac{3}{10}$

56. $3\frac{1}{2} \div 5\frac{1}{2}$ **57.** $\frac{8}{9} \div 2\frac{2}{5}$ **58.** $1\frac{5}{8} \div 2$ **59.** $5\frac{1}{2} \div 2\frac{3}{4}$

Lesson 5-8

Change each fraction to a decimal. Use bar notation to show a repeating decimal.

60. $\frac{2}{3}$ **61.** $\frac{1}{10}$ **62.** $\frac{11}{15}$ **63.** $\frac{13}{25}$ **64.** $\frac{7}{22}$ **65.** $\frac{1}{99}$

Lesson 5-9

Change each decimal to a fraction.

66. 0.20   **67.** 0.034   **68.** 0.36   **69.** 0.25   **70.** 0.425

Lesson 5-10

Solve.

71. A sugar scoop holds $\frac{2}{3}$ cup. How many scoops will equal 2 cups of sugar?

72. Leonard has $5\frac{1}{2}$ rows of green beans in his garden. He has $\frac{2}{3}$ as many rows of peas as green beans. How many rows of peas does he have?

73. Alice saves $\frac{1}{5}$ of her salary and spends $\frac{1}{6}$ of her salary for transportation. What part of her salary does she have left?

74. Mike walked $12\frac{1}{4}$ miles in $2\frac{1}{2}$ hours to raise money for Central City's Fund for the Homeless. What was the average distance he walked each hour?

75. If $2\frac{3}{5}$ is multiplied by a number, the product is $17\frac{1}{3}$. What is the number?

TEST

Estimate.

1. $3\frac{1}{7} \times 1\frac{7}{9}$ **2.** $8\frac{1}{3} \times 4\frac{4}{5}$ **3.** $2\frac{1}{3} \div 1\frac{1}{3}$ **4.** $5\frac{1}{2} \div 1\frac{3}{4}$

5. Is $10 \div \frac{1}{5}$ less than or greater than 10?

6. Is $\frac{1}{5}$ of 5 less than, equal to, or greater than 5?

Multiply.

7. $2 \times \frac{2}{3}$ **8.** $\frac{5}{8} \times 6$ **9.** $\frac{1}{3} \times \frac{3}{4}$ **10.** $\frac{7}{18} \times \frac{9}{14}$

11. $1\frac{1}{4} \times 2\frac{1}{4}$ **12.** $1\frac{5}{6} \times 8\frac{2}{11}$ **13.** $2\frac{2}{3} \times 5\frac{1}{7}$ **14.** $1\frac{5}{16} \times 3\frac{3}{7}$

Divide.

15. $6 \div \frac{1}{3}$ **16.** $\frac{5}{8} \div 5$ **17.** $\frac{7}{8} \div \frac{3}{4}$ **18.** $\frac{3}{7} \div \frac{4}{5}$

19. $5\frac{3}{4} \div 1\frac{1}{3}$ **20.** $4\frac{1}{2} \div 1\frac{1}{2}$ **21.** $26 \div 1\frac{3}{10}$ **22.** $4\frac{1}{8} \div 1\frac{3}{4}$

23. Gary and Suki can make 3 dozen cupcakes for the bakesale in a half-hour. How many dozens of cupcakes can they make in 1 hour if Peter and Jillian help? Assume Peter and Jillian work at the same rate.

Change each fraction to a decimal. Use bar notation to show a repeating decimal.

24. $\frac{2}{5}$ **25.** $\frac{5}{16}$ **26.** $\frac{5}{11}$ **27.** $\frac{7}{12}$

Change each decimal to a fraction.

28. 0.6   **29.** 0.35   **30.** 0.875   **31.** 0.92

Solve.

32. Janet's car will travel $22\frac{1}{2}$ miles on one gallon of gas. At this rate, how far will it travel on $6\frac{1}{2}$ gallons of gas?

33. Betty has $6\frac{1}{4}$ bags of plant food. She uses $1\frac{1}{4}$ bags each month. How many months will her supply of plant food last?

▶ BONUS: If $\frac{a}{b} = \frac{c}{d}$, explain in your own words why $ad = bc$.

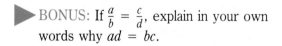

UNIT PRICE

Careful shopping can lead to considerable savings. This is especially true for food shopping. Many grocery stores now display the unit price so that comparison shopping is easier. It is a good idea to compare different brands for quality and price. To tell which is the better buy of two items equal in quality but different in size, find the cost of one unit of the product. This is the **unit price.**

Which is the better buy, $1.59 for 12 oz of peanut butter, or $2.94 for 28 oz of peanut butter?

Small Jar of Peanut Butter

$1.59 for 12 ounces

Find the cost for 1 ounce.

$1.59 \div 12 = 0.1325$

Large Jar of Peanut Butter

$2.94 for 28 ounces

Find the cost for 1 ounce.

$2.94 \div 28 = 0.105$

Since 10.5¢ < 13.25¢, the large jar is the better buy. $0.1325 = 13.25¢
$0.105 = 10.5¢

Assume the quality in each case is the same. Determine which size is the better buy for each item.

1. cereal   10 oz for $1.79   15 oz for $2.39

2. ketchup   14 oz for $0.89   28 oz for $1.29

3. peaches   16 oz for $0.77   29 oz for $1.52

4. bread   16 oz for $0.85   24 oz for $1.39

5. Which is the better buy, a 6.5-ounce bag of potato chips for $1.69 or a 14.5-ounce bag for $2.99?

6. Brittany buys a 10-ounce bottle of shampoo for $2.79. She has a coupon for 20¢ off the regular price. What is the cost per ounce of the shampoo with the coupon?

7. At the Corner Market, one quart of milk costs 63¢, one-half gallon costs $1.25, and one gallon costs $2.53. Which is the best buy?

8. George is in charge of buying paper plates, napkins, and cups for the sophomore dance. Third Street Market sells 100 plates for $1.39. Joy's Supermarket sells 50 plates for $0.74. Which store has the better buy?

Free Response

Lesson 1-2

Write as a product and then find the product.

1. 3^4 **2.** 7^3 **3.** 10^5 **4.** 2^4

5. 1^{10} **6.** 6^2 **7.** 8^2 **8.** 5^3

Lesson 1-9

Find the value of each expression.

9. $6 + 18 \div 3$ **10.** $30 - 5 \times 4$ **11.** $(9 + 21) \div 3$

12. $4.6 \times 2 \div 4$ **13.** $7.8 \times (15 + 5)$ **14.** $37.8 - 25 + 3.6$

15. $16.4 - (8.5 + 0.1)$ **16.** $3 \times 10^3 + 2 \times 10^3$ **17.** $1.6 \times 10^2 + 2.7$

Lesson 2-3

18. The Big Time Amusement Park had an average daily attendance of 23,760 during July. This was 5,807 more than during June. Find the average daily attendance in June.

Lesson 3-1

Multiply.

19. 54×72 **20.** 87×37

21. 20×384 **22.** 86×4.709

23. $3,000 \times 21$ **24.** $50 \times 2,000$

Lesson 3-4

Divide.

25. $10\overline{)40}$ **26.** $60\overline{)300}$

27. $23\overline{)92}$ **28.** $60\overline{)11,895}$

29. $30\overline{)12,000}$ **30.** $48\overline{)600}$

Lesson 3-13

31. A hotel room costs $29 a day plus $2.24 tax. What is the charge on a room for four days?

Lesson 4-1

Find all the factors of each number.

32. 36   **33.** 89   **34.** 45   **35.** 39   **36.** 144

Lesson 4-3

Find the GCF and the LCM of each group of numbers.

37. 36, 40   **38.** 15, 50   **39.** 9, 25   **40.** 4, 28

Lesson 4-12

41. A recipe calls for $2\frac{3}{4}$ cups of flour. Mark has measured 1 cup. How much more flour does he need?

Lessons 5-2
5-3

Multiply.

42. $\frac{3}{4} \times \frac{5}{8}$ **43.** $\frac{2}{7} \times \frac{1}{6}$ **44.** $\frac{2}{3} \times \frac{9}{10}$ **45.** $\frac{7}{9} \times 20$

46. $1\frac{3}{4} \times \frac{4}{5}$ **47.** $\frac{11}{18} \times 2\frac{2}{3}$ **48.** $7\frac{5}{6} \times 18$ **49.** $3\frac{3}{5} \times 10\frac{1}{9}$

Multiple Choice

Choose the letter of the correct answer for each item.

1. What is 238.675 rounded to the nearest hundredth?
- **a.** 200
- **b.** 238.67
- **c.** 238.68
- **d.** 238.7

Lesson 1-4

2. Elia spends $4.50 at the record store and $36.79 at the clothing store. She arrives home with $12.61. Which question can you answer from the information above?
- **e.** At what store did she buy clothes?
- **f.** How many miles between stores?
- **g.** How much money did she have to begin with?
- **h.** How much change she received at the record store?

Lesson 2-2

3. John withdraws $25.68 from his savings of $810.50. What is the amount left in his account?
- **a.** $553.70
- **b.** $784.82
- **c.** $784.98
- **d.** $794.98

Lesson 2-4

4. What is the product of 0.05 and 1,000?
- **e.** 0.005
- **f.** 5
- **g.** 50
- **h.** 5,000

Lesson 3-2

5. In one game, Pat kicks field goals from 43, 38, and 41 yards. To the nearest tenth of a yard, what is the average length of the field goals?
- **a.** 40.6 yards
- **b.** 40.7 yards
- **c.** 46.6 yards
- **d.** 46.7 yards

Lesson 3-12

6. Linda Fada earns $16,500 per year. What is her weekly wage?
- **e.** $317.31
- **f.** $317.37
- **g.** $3,173
- **h.** $3,173.07

Lesson 3-4

7. Order from least to greatest.
$$\frac{1}{2}, \frac{5}{8}, \frac{6}{5}, \frac{7}{16}$$
- **a.** $\frac{7}{16}, \frac{1}{2}, \frac{5}{8}, \frac{6}{5}$
- **b.** $\frac{1}{2}, \frac{5}{8}, \frac{6}{5}, \frac{7}{16}$
- **c.** $\frac{1}{2}, \frac{7}{16}, \frac{5}{8}, \frac{6}{5}$
- **d.** $\frac{7}{16}, \frac{5}{8}, \frac{6}{5}, \frac{1}{2}$

Lesson 4-7

8. If you combine $5\frac{3}{4}$ cups of flour with $4\frac{7}{8}$ cups of flour, how much flour will you have?
- **e.** $9\frac{1}{8}$ cups
- **f.** $10\frac{5}{8}$ cups
- **g.** $9\frac{5}{6}$ cups
- **h.** $9\frac{7}{8}$ cups

Lesson 4-11

9. *About* how much is 1,180 divided 375?
- **a.** 1
- **b.** 3
- **c.** 30
- **d.** 40

Lesson 1-7

10. If $4\frac{1}{2}$ dozen doughnuts are divided between 2 tenth-grade classes, how many doughnuts does each person receive? Which fact is missing?
- **e.** the number of classes
- **f.** the price of the doughnuts
- **g.** the number of people in each class
- **h.** none of the above

Lesson 5-7

fun with MATH

CFCs could outlive you!

If you have a refrigerator in your home or an air conditioner in your car, its cooling fluid probably contains CFCs, or chlorofluorocarbons. Ozone is a form of oxygen with three oxygen atoms, O_3. The ozone layer protects us from dangerous ultraviolet radiation from the sun. CFCs rise to the upper atmosphere, releasing chlorine atoms that combine with ozone molecules, in effect, slowly destroying the ozone layer. One CFC molecule can remain in the upper atmosphere for more than 100 years.

MATH M·E·N·U

Carrot Bars

4 eggs (beaten)
2 tsp baking soda
2 tsp salt
2½ cups flour
2 cups sugar
2 tsp cinnamon
1½ cups vegetable oil
½ cup nuts (large pieces)
3 small jars strained carrots (baby food)

Combine and mix ingredients in order given. Bake 30-40 minutes at 350° on greased 10″ × 15″ jelly-roll pan or cookie sheet.

Frosting: 1/2 tsp vanilla
1/2 cup margarine
3 1/2 cups powdered sugar
1 8-oz pkg of cream cheese

Combine ingredients and beat until smooth. Spread on cooled bars.

COMICS

BROOM-HILDA

When playing golf, Mary Queen of Scots called the boys who fetched her golf balls "cadets," pronounced "cadday-"; hence "caddie." Today we use the term caddy."

The equal sign England	AD 1718-1799	The "Real McCoy" Ypsilanti, Michigan	AD 1956
AD 1545	The "Witch" of Agnesi Bologna, Italy	AD 1872	The VTR California

JOKE!

Q: What professional football team has a name which is the same as 7^2?

A: San Francisco Forty-Niners.

RIDDLE

Q: To what number can you add a letter and make it three less?

A: IX (9), prefixed by an 'S' makes it SIX.

How does a VCR work? When recording, the VCR converts incoming TV signals into electric current. The current travels to the head, an electromagnet. The head magnetizes the particles on the tape into patterns that represent video signals, audio signals, and control signals. When playing the tape back, the head reads the magnetic patterns on the tape and creates an electric current. The current is sent to the TV and transformed into sounds and pictures.

TEASER

Challenge a friend to a Korean form of tic-tac-toe, called *ko-no*. You will need two red and two blue playing pieces, or two types of coins. Place the red pieces on the circles marked A and B, the blue pieces on C and D. The player on A and B goes first. The player attempts to move either playing piece to the empty circle along a connecting segment. *Jumping a piece is not allowed.* The object of the game is to block your opponent from moving his/her playing pieces.

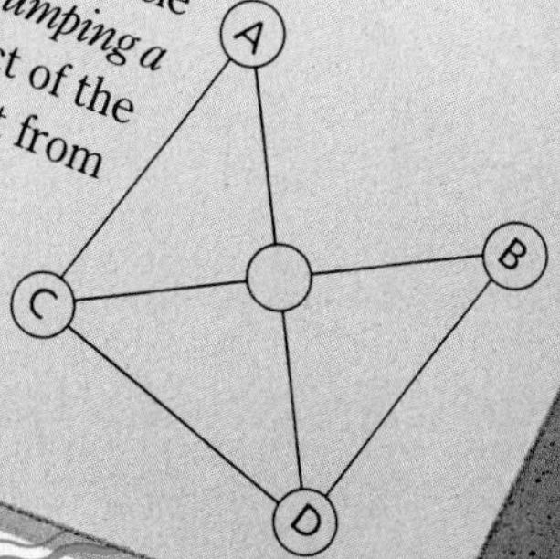

QUIZ TIME

Why is a Black Hole black?

$\frac{1}{2} + \frac{2}{4}$ = ____ of

$\frac{3}{6} - \frac{1}{3}$ = ____ a

$\frac{1}{5} \times \frac{2}{4}$ = ____ is

$\frac{1}{12} \div \frac{1}{3}$ = ____ even

$\frac{6}{16} - \frac{3}{8}$ = ____ escape

$\frac{5}{6} \times \frac{2}{3}$ = ____ gravity

$\frac{3}{4} \div \frac{1}{8}$ = ____ Black

$\frac{5}{6} - \frac{1}{3}$ = ____ so

$\frac{4}{9} \times \frac{7}{8}$ = ____ light

$\frac{1}{2} \div \frac{6}{10}$ = ____ the

$\frac{1}{9} + \frac{5}{9}$ = ____ in

$\frac{11}{15} - \frac{2}{5}$ = ____ Hole

$\frac{1}{4} \times \frac{6}{12}$ = ____ great

$\frac{15}{16} \div \frac{3}{4}$ = ____ can't

$\frac{3}{5} + \frac{2}{6}$ = ____ force

ANSWER

____ $\frac{5}{6}$	____ $\frac{14}{15}$	____ 1
____ $\frac{5}{9}$	____ $\frac{2}{3}$	____ $\frac{1}{6}$
____ 6	____ $\frac{1}{3}$	____ $\frac{1}{10}$
____ $\frac{1}{2}$	____ $\frac{1}{8}$	____ $\frac{1}{4}$
____ $\frac{7}{18}$	____ $\frac{5}{4}$	____ 0

6

MEASUREMENT

How much is that doggie in the window?

Do you have a pet? What kind of pet do you have and what makes your pet special to you? If you don't have a pet, what kind would you like to have? Pets are a responsibility and need feeding, care, and attention every day.

Julio's German shepherd puppy eats 400 grams of puppy food a day. How many days will a 6-kilogram bag of puppy food last?

ACTIVITY: Create Your Own Measuring System

In this activity, you will make your own measuring device and create your own system for measuring capacity. Take some time to read the activity with your group before preparing any materials and planning your system.

This chapter is about the standard units of measure that everyone uses. You can have some fun creating your own system.

Materials: large glass (straight sides are the easiest to use); spoon, tablespoon, or coffee measure; small container of water; masking tape; felt-tip pen; other containers like paper cups, coffee cans, small bottles, and so on.

Cooperative Groups

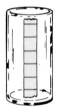

1. Prepare the large glass by putting a piece of tape vertically on the outside from top to bottom.

2. Decide with your group how many spoonfuls you will use as the next smallest amount after one spoonful. Use the spoon to add the number of spoonfuls of water that you have chosen and mark the tape with a line. Write the total number of spoonfuls next to the line.

3. Now add the same number of spoonfuls as you did in Step 2, and mark and number the tape again. Continue to add and mark until you have the number of spoonfuls that will equal **one** of the largest unit in your system.

4. Think up a name that will represent half of the largest unit and also a name for the whole unit.

5. Use your measuring device to find the capacity of other containers, some larger than your device and some smaller, such as paper cups, small bottles, and so on.

6. Set up a table or chart for changing from spoonfuls to the unit for one half, from spoonfuls to the whole unit, and from the half unit to the whole unit.

Communicate Your Ideas

7. Is your system easy to use? Is it easy to change from one unit to another?

8. Does your system have units that are small enough to measure a small capacity and units that are large enough to measure a larger capacity?

9. Write about a system of length or weight that you would like to create. Include the names of the units you would use and a chart for changing from one unit to another.

6-1 THE METRIC SYSTEM

Objective

Identify and use metric prefixes.

Tom wanted to know why his teacher said the metric system is an easier system to use and remember than the customary system. He found there are fewer terms to remember. Also, changing from one unit to another is very similar to changing money; ten pennies equal one dime, ten dimes equal one dollar.

The basic units in the metric system are the **meter** (length), **gram** (mass), and **liter** (capacity).

Prefixes and **basic units** are used to name units larger and smaller than the basic units. The chart at the right shows how the prefixes relate to decimal place values and powers of ten.

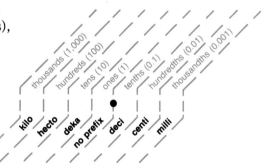

Some examples of metric units and their meanings are given below.

prefix basic unit	prefix basic unit	prefix basic unit
centimeter	kilogram	milliliter
0.01 meter	1,000 grams	0.001 liter

Look at the chart at the right for the basic unit of a meter. Each unit is equal to ten times the unit to its right. For example, 1 kilometer equals 10 hectometers. Each unit is equal to one-tenth of the unit to its left. For example, 1 meter equals 0.1 dekameter.

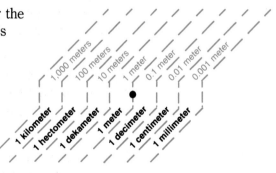

The system is the same for changing units of mass and capacity.

Guided Practice **Name the place value related to each prefix.**

1. milli **2.** kilo **3.** centi

4. deka **5.** deci **6.** hecto

Complete. Use the place-value chart.

7. 1 kilogram = $\underline{?}$ hectogram(s) **8.** 1 dekagram = $\underline{?}$ gram(s)

9. 1 liter = $\underline{?}$ milliliter(s) **10.** 1 kiloliter = $\underline{?}$ liter(s)

Name the metric unit for each measurement.

11. 10 grams **12.** 0.01 gram

13. 0.001 meter **14.** 0.1 meter

Practice

Complete. Use the place-value chart.

15. 1 kilometer = $\underline{?}$ meter(s)
16. 1 deciliter = $\underline{?}$ liter(s)
17. 1 milligram = $\underline{?}$ gram(s)
18. 1 dekameter = $\underline{?}$ meter(s)
19. 1 centiliter = $\underline{?}$ liter(s)
20. 1 hectogram = $\underline{?}$ gram(s)

fUN with MATH

*Where and when
did the metric
system originate?
See page 241.*

Name the metric unit for each measurement.

21. 0.001 meter   **22.** 1,000 liters   **23.** 100 grams
24. 0.1 meter   **25.** 10 liters   **26.** 0.01 gram

Name the larger unit.

27. 1 meter or 1 kilometer   **28.** 1 centigram or 1 milligram
29. 1 milliliter or 1 liter   **30.** 1 kilometer or 1 centimeter
31. Change 1,000 centimeters to dekameters.
32. Is 100 millimeters equivalent to 1 meter or 1 decimeter?

Applications

33. Eric is the school play sound technician. He needs 100 meters of cable to extend the ceiling microphones. How many 15-meter spools of cable should he buy?

34. In science class Sharyn needs to add 1 dekagram of silver nitrate to the water in her beaker. How many 1-gram packets should she add?

Write Math

35. If the U.S. adopts the metric system, what are some of the problems you expect to develop?

36. Name three areas involving measure that would be improved by using the metric system.

Research

37. Who developed the metric system? What is the policy in the U.S. for using metric measure?

38. In a grocery or hardware store, find three items you could buy that are labeled in metric units.

Mixed Review

Lesson 3-7

Multiply.

39. 4.63×100 **40.** 5.82×10 **41.** $16.4 \times 1,000$ **42.** 6.97×10^2

Lesson 4-13

Subtract.

43. $8\frac{5}{6} - 3\frac{1}{6}$ **44.** $16\frac{4}{7} - 8\frac{3}{7}$ **45.** $6\frac{4}{9} - 3\frac{1}{6}$ **46.** $5\frac{3}{5} - 2\frac{1}{2}$

Lesson 5-8

47. Kevin purchased a $6\frac{3}{8}$ inch length of pipe to replace a pipe that was to be 6.425 inches long. Did Kevin purchase the right length pipe? If not, is the pipe he purchased too short or too long?

6-2 MEASURING LENGTH

Objective

Estimate and measure metric units of length.

In the 1992 Summer Olympics, Yang Wenyi swam the 50-meter freestyle and won in 24.76 seconds. She set a winning record time for the gold medal.

The most commonly used metric units of length are the **meter (m), centimeter (cm), millimeter (mm),** and **kilometer (km).**

One meter (1 m) is about the height of a car fender.

One centimeter (1 cm) is about the width of a large paper clip.

One millimeter (1 mm) is about the thickness of a dime.

One kilometer (1 km) is about the length of 5 city blocks.

You can use a metric ruler to measure objects in centimeters or millimeters.

Example A

Use the portion of a metric ruler shown below to find the length of the safety pin in centimeters and millimeters.

The length of the safety pin is between 3 and 4 centimeters or 30 and 40 millimeters. Since each small unit represents 1 millimeter, the safety pin is about 32 millimeters long. This can also be read as 3.2 centimeters.

Guided Practice

Name the most reasonable unit for measuring each item. Use kilometer, meter, centimeter, or millimeter.

1. height of a tree

2. length of a pencil

3. thickness of an electrical wire

4. width of a classroom

5. distance from Dallas to Tampa

6. width of a staple

Choose the most reasonable measurement.

7.	diameter of a bicycle wheel	66 cm	66 m	66 km
8.	length of a bolt	21 cm	21 m	21 mm
9.	width of notebook paper	22 mm	22 cm	22 km
10.	width of school hallway	5 mm	5 cm	5 m
11.	width of a door	120 mm	120 km	120 cm

Use a metric ruler to measure each item. Give the measurement in centimeters and also in millimeters.

12.   **13.**   **14.**

Applications

15. Sing Lu makes flower arrangements for Caufield's Flowers. If it takes 1.2 m to make a bow for an arrangement, how many bows can Sing Lu make if he has 30 m of ribbon?

16. A pencil, measured to the nearest centimeter, is 13 centimeters long. Between what two measurements does the pencil have to be?

17. If the air distance from San Francisco to New York is 4,139 kilometers and the air distance from San Francisco to Chicago is 2,990 kilometers, how much farther is it from San Francisco to New York than to Chicago?

Critical Thinking

18. The Big K ranch brought six horses to the fair for the calf-roping contest. They have six adjacent stalls for the horses. Your job is to assign each horse to a stall. Where will you put them if Lola cannot be next to Lady, Trudy cannot be next to King, Lucky cannot be next to Jim, and Lady behaves well when she is next to Lucky.

Number Sense

19. Locate three items in your classroom that are about 1 centimeter, about 1 millimeter, and about 1 meter in length or height.

Estimation

You can estimate the length or height of an object by comparing it to an object with which you are familiar.

about one-half of person's height or 80 cm

160 cm

20. Noelle is about half as tall as her father. If she is 92 cm tall, about how tall is her father?

21. A classroom wall is 24 blocks high. If each block is about 20 centimeters high and there is about 1 centimeter of mortar between blocks, about how many meters high is the wall?

6-3 CHANGING METRIC UNITS

Objective

Change metric units of length.

Mr. Santos wants to run at least 4,000 meters a day while training for the 5,000-meter run at Boyce Park. His office is 4.3 kilometers from his home. If Mr. Santos runs to his office in the morning instead of taking the bus, will he run at least 4,000 meters?

To change metric units, multiply or divide by powers of ten. When changing a larger unit to a smaller unit, you should *multiply*. When changing a smaller unit to a larger unit, you should *divide*.

Method

1 Determine if you are changing from smaller to larger units or vice versa.

2 Divide to change from smaller to larger units. Multiply to change from larger to smaller units.

3 Determine the multiple of ten by which to multiply or divide. Use this diagram.

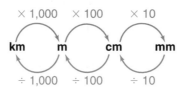

Example A

Change 4.3 kilometers to meters.

1 kilometers → meters
You are changing from larger to smaller units.

2 There will be more units after the change. Multiply.

3 There are 1,000 meters in a kilometer.
Multiply by 1,000.

$$4.3 \text{ km} = \underline{\ ?\ } \text{ m} \quad \times 1{,}000$$
$$4.3 \text{ km} = 4{,}300 \text{ m}$$

Mr. Santos will run 4,300 meters.

Example B

Change 375 centimeters to meters.

1 centimeters → meters
You are changing from smaller to larger units.

2 Divide.

3 There are 100 centimeters in a meter. Divide by 100.

$$375 \text{ cm} = \underline{\ ?\ } \text{ m} \quad \div 100$$
$$375 \text{ cm} = 3.75 \text{ m}$$

Guided Practice

Example A

Complete.

1. $3 \text{ m} = \underline{\ ?\ } \text{ cm}$

2. $8 \text{ km} = \underline{\ ?\ } \text{ m}$

3. $2 \text{ cm} = \underline{\ ?\ } \text{ mm}$

4. $1.2 \text{ m} = \underline{\ ?\ } \text{ cm}$

5. $0.4 \text{ cm} = \underline{\ ?\ } \text{ mm}$

6. $0.96 \text{ km} = \underline{\ ?\ } \text{ m}$

Complete.

7. 20 mm = ? cm   **8.** 400 cm = ? m   **9.** 4.6 m = ? km

10. 27.5 cm = ? m   **11.** 9.78 m = ? km   **12.** 5,670 mm = ? cm

Exercises

Practice

Complete.

13. 700 mm = ? cm   **14.** 0.8 m = ? cm   **15.** 350 cm = ? m

16. 3 km = ? m   **17.** 0.4 km = ? m   **18.** 40 mm = ? cm

19. 5,000 m = ? km   **20.** 0.9 cm = ? mm   **21.** 3 mm = ? cm

22. 2,600 m = ? km   **23.** 4.2 m = ? cm   **24.** 40 cm = ? m

25. 8.7 m = ? km   **26.** 15 mm = ? cm   **27.** 3.6 cm = ? mm

28. How many meters are in 38.2 kilometers?

Applications

f **UN with MATH**

Why do we only see one side of the moon?
See page 240.

29. In air, sound at a particular frequency travels at about 343 meters per second. In water, sound at the same frequency travels at about 1.435 kilometers per second. How many meters per second faster does this sound wave travel in water than in air?

30. Kim walks every day for 1 hour as part of her aerobic program. If she walks at a rate of 1 kilometer in 15 minutes, how many meters will she walk in 1 hour?

31. Suppose Dr. Seuss' book *Green Eggs and Ham* has sold 3.8 million copies. Imagine that the books are stacked on top of each other. If each book is 13 mm thick, how many kilometers high would the stack be?

JOURNAL ENTRY

32. The Andromeda Galaxy is about 1.4×10^{22} meters from Earth. How far is this in kilometers?

33. Write a few sentences about something new you learned in this lesson. Give examples.

Critical Thinking

34. Nick has 4 pictures to put in a grouping frame. The largest picture will only fit in the top left-hand corner. The other three pictures will fit in any of the other corners. How many different ways can Nick put the pictures in the frame?

Mixed Review

Lesson 3-8

Write in scientific notation.

35. 7,010,000   **36.** 830,000   **37.** 48,000,000   **38.** 53,700

Lesson 4-1

State whether each number is divisible by 2, 3, 5, 9, or 10.

39. 14   **40.** 24   **41.** 33   **42.** 45   **43.** 90

Lesson 4-7

44. At the school cafeteria $\frac{8}{15}$ of the students ate pizza and $\frac{9}{20}$ of the students ate hamburgers. Which food did more students eat?

6-4 MEASURING MASS

Objective

Estimate and use metric units of mass.

In Chang's chemistry experiment, the water dissolved 523 milligrams of sodium chloride (table salt). How many grams of sodium chloride did the water dissolve?

The most commonly used metric units of mass are the **gram (g)**, **kilogram (kg)**, and **milligram (mg)**.

The mass of a large paper clip is about one gram (1 g).

The mass of a baseball bat is about one kilogram (1 kg).

The mass of a grain of salt is about one milligram (1 mg)

Method

Changing metric units of mass is similar to changing metric units of length.

1 ▸ Determine if you are changing from smaller to larger units or vice versa.

2 ▸ Divide to change from smaller to larger units. Multiply to change from larger to smaller units.

3 ▸ Determine the multiple of ten by which to multiply or divide.

$$\times 1{,}000 \qquad \times 1{,}000$$

kg   **g**   **mg**

$$\div 1{,}000 \qquad \div 1{,}000$$

Examples

A *Change 523 milligrams to grams.*

1 ▸ milligrams → grams
smaller → larger

2 ▸ Divide.

3 ▸ There are 1,000 milligrams in a gram.

523 mg = $\underline{\quad ? \quad}$ g ÷ 1,000

523 mg = 0.523 g

The water dissolved 0.523 g of sodium chloride.

B *Change 28 kilograms to grams.*

1 ▸ kilograms → grams
larger → smaller

2 ▸ Multiply.

3 ▸ There are 1,000 grams in a kilogram.

28 kg = $\underline{\quad ? \quad}$ g × 1,000

28 kg = 28,000 g

─── **Guided Practice** ───

Choose the better estimate for the mass of each item.

1. 1 vitamin tablet, 1 g or 1 kg

2. a feather, 17 mg or 17 g

3. automobile, 1,000 g or 1,000 kg

4. pencil, 10 g or 10 mg

5. grocery bag, 80 g or 80 mg

6. bowling pin, 1.4 g or 1.4 kg

Complete.

Example A   **7.** 4.9 mg = $\underline{\ ?\ }$ g   **8.** 5,862 g = $\underline{\ ?\ }$ kg   **9.** 8,754 mg = $\underline{\ ?\ }$ g

Example B   **10.** 67 g = $\underline{\ ?\ }$ mg   **11.** 73 kg = $\underline{\ ?\ }$ g   **12.** 9.7 g = $\underline{\ ?\ }$ mg

Exercises

Practice

Complete.

13. 8 kg = $\underline{\ ?\ }$ g
14. 62 g = $\underline{\ ?\ }$ mg
15. 752 mg = $\underline{\ ?\ }$ g
16. 2,197 g = $\underline{\ ?\ }$ kg
17. 9.4 g = $\underline{\ ?\ }$ mg
18. 6.8 kg = $\underline{\ ?\ }$ g
19. 912 g = $\underline{\ ?\ }$ kg
20. 1,775 mg = $\underline{\ ?\ }$ g
21. 24 kg = $\underline{\ ?\ }$ g
22. 17.8 g = $\underline{\ ?\ }$ mg
23. 5 mg = $\underline{\ ?\ }$ g
24. 40 g = $\underline{\ ?\ }$ kg

25. Change 25.8 kilograms to grams.
26. How many kilograms are in 260,000 milligrams?

Applications

27. Ace Athletic Club has 4 different sizes of hand weights in the workout room. The lightest weight is 0.5 kg. If each size increases by 500 g, how many kilograms does the heaviest size weigh?

28. Kurt has a jar full of nickels and a jar full of pennies. He wants to know how much money he has in nickels. Kurt knows that a nickel has a mass of 5 grams. The nickels from his jar weigh 2.5 kilograms. How many nickels does Kurt have?

29. A case of cans of vegetable soup weighs 4 kilograms. If the weight of the packaging is 40 grams and 1 can contains 330 grams of soup, how many cans of soup are there in the case?

Number Sense

30. The following paragraph has four errors in units of mass or length. Find and correct the errors.

The mass of a hummingbird is about 11 mg and is light compared to the mass of an orange, about 330 g. The bill of a hummingbird is about 1 m long, so it can get nectar from flowers. Another insect that gets nectar is the honeybee that has a total mass of about 200 g. A honeybee is about 1 cm long with a wingspan of about 1 m so it can fly easily.

Mixed Review

Lesson 2-1

Add.

31. 42
    + 87

32. 631
    + 825

33. 7,340
    + 283

34. 4,522
    + 8,766

35. 10,487
    + 68,313

Lesson 3-4

36. 16$\overline{)483}$
37. 42$\overline{)723}$
38. 123$\overline{)6,458}$
39. 381$\overline{)19,482}$

Lesson 5-10

40. Janelle noted that over the last 45 days, $\frac{4}{5}$ of the days were sunny. How many days were sunny?

6-5 MEASURING CAPACITY

Objective

Estimate and use metric units of capacity.

Mary and Jake each spend $0.75 a day for a drink at lunch. Jake buys 1 carton of fruit juice from the vending machine in the cafeteria for $0.75. The carton contains 300 milliliters of juice. Mary brings a 0.5-liter carton of juice that she buys in the grocery store for $0.75. Who gets more juice for the same amount of money? Mary changes 0.5 liter to milliliters to find out.

The most commonly used metric units of capacity are the **liter (L)**, **kiloliter (kL),** and **milliliter (mL).**

The capacity of the pitcher is about one liter (1 L).

The capacity of the average above-ground swimming pool is 15 kiloliters (15 kL).

The capacity of the eyedropper is about one milliliter (1 mL)

Changing metric units of capacity is similar to changing metric units of length and mass.

Method

1. Determine if you are changing from smaller units to larger units or vice versa.

2. Divide to change from smaller to larger units. Multiply to change from larger to smaller units.

3. Determine the multiple of ten by which to multiply or divide.

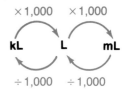

Examples

A *Change 0.5 liters to milliliters.*

1. liters → milliliters
   larger → smaller

2. Multiply.

3. There are 1,000 milliliters in a liter.
   0.5 L = _?_ mL × 1,000
   0.5 L = 500 mL
   Mary gets more juice than Jake for the same amount of money.

B *Change 4,560 liters to kiloliters.*

1. liters → kiloliters
   smaller → larger

2. Divide.

3. There are 1,000 liters in a kiloliter.
   4,560 L = _?_ kL ÷ 1,000
   4,560 L = 4.560 kL

Guided Practice

Name the most reasonable unit for measuring the capacity of each item. Use kiloliter, liter, or milliliter.

1. aquarium

2. large coffee pot

3. jar of mayonnaise

4. bottle of perfume

Example A

Example B

Complete.

5. 42 L = <u>?</u> mL   **6.** 0.9 L = <u>?</u> mL   **7.** 36 kL = <u>?</u> L

8. 791 L = <u>?</u> kL   **9.** 235.6 mL = <u>?</u> L   **10.** 4 L = <u>?</u> kL

Exercises

Practice

Complete.

11. 6 L = <u>?</u> mL   **12.** 3 kL = <u>?</u> L   **13.** 7.5 kL = <u>?</u> L

14. 0.03 L = <u>?</u> mL   **15.** 234 mL = <u>?</u> L   **16.** 58 L = <u>?</u> kL

17. 127 L = <u>?</u> kL   **18.** 42 mL = <u>?</u> L   **19.** 28 L = <u>?</u> mL

20. 0.22 kL = <u>?</u> L   **21.** 262 mL = <u>?</u> L   **22.** 500 L = <u>?</u> kL

23. 17 kL = <u>?</u> L   **24.** 34.2 L = <u>?</u> mL   **25.** 96.8 L = <u>?</u> kL

26. Change 34.2 L to milliliters.

27. How many milliliters are in 0.78 kiloliters?

Applications

28. When it was Charlene's turn to bring the drink for the photography club meeting, she brought three 2-liter bottles of punch. How many milliliters did each person have to drink if there were fifteen people at the meeting and they shared equally?

29. A pharmacist fills a prescription for pediatric antibiotic liquid. The dosage prescribed by the doctor is 2 mL 3 times a day for 5 days. How many milliliters of antibiotic should the pharmacist prepare?

f **UN with MATH**

Lasers are used in medicine. Find out what a laser is. See page 241.

30. The Boosters Club buys cases of lemonade to sell at the track meets for $4.25. There are 12 cans of lemonade in a case. Each can contains 180 mL of lemonade. If the club wants to make a profit of $0.20 a can, how much should it charge for a can of lemonade?

Suppose

31. Suppose Benney's Sweet Shoppe has a drink dispenser that holds 75 L. Mr. Benney blended a punch that he wants to sell in a 300-mL cup. How many cups can he sell from a full dispenser?

Research

32. Interview a pharmacist, nurse, doctor, or lab technician about the use of the metric system in medicine. Write a paragraph about your interview.

Mixed Review

Lesson 2-4

Subtract.

33.   18.5   **34.**   9.03   **35.**   25.4

       − 6.2   − 4.67   − 6.73

Lesson 4-3

Find the LCM of each group of numbers.

36. 5,7   **37.** 9, 12   **38.** 17, 51

Lesson 5-2

39. A leather jacket sells for $185. It is on sale for $\frac{1}{5}$ off the regular price. How much money is saved by buying the jacket on sale?

▶ Explore
▶ Plan
▶ Solve
▶ Examine

6-6 ACTING IT OUT

Objective

Solve problems by acting out a solution.

George collects baseball cards and posters. He decides to sell the poster collection so he can buy some special baseball cards. On Saturday, George sells every third poster. The next Saturday, he sells every third poster that he has left, and so on, for four Saturdays. How many posters does George have left if he started with twenty posters?

▶ **Explore**

What is given?
● George had twenty posters.
● George sells every third poster, four times.
What is asked?
● How many posters are left after George sells every third poster four times?

▶ **Plan**

The problem could be solved by having twenty students act out the sale of the posters.

▶ **Solve**

Twenty students stand in a line and count off by ones to twenty. Each student whose number is a multiple of three sits down. This indicates that the poster they represent has been sold.

Students count off again by ones. Then every student whose number is a multiple of three sits down. Students repeat the procedure two more times.

There are five students left standing, so the solution is five posters.

▶ **Examine**

Look at the diagram below. The numbers represent the students. The blue circles represent the first group to sit. The green circles represent the second group to sit. The red circles represent the third group, and the yellow circles, the fourth group.

1	2	③	④	⑤	⑥	⑦
⑧	⑨	10	⑪	⑫	⑬	14
⑮	16	⑰	⑱	⑲	20	

There are five numbers left so the solution is correct.

Guided Practice

Solve. Use acting it out.

1. At Savory Snacks, pretzels are moved from the ovens to packaging on a conveyor belt in a line of 25. Every fifth pretzel is tested for weight. Every third pretzel is tested for size. How many of the 25 pretzels are not checked or tested?

2. Dominique bought a classic rock album for $5. She sold it for $8.00. Then she bought it back for $10.50, and sold it again for $13.00. How much money did Dominique lose or gain in buying and selling the album?

Paper/ Pencil
Mental Math
Estimate
Calculator

Solve. Use any strategy.

3. Leah initially ordered 250 flowers to sell at the homecoming game. After checking people's preferences, she changes her order for 50 carnations to 20 roses, changes 40 roses to 50 orchids, and changes 10 orchids to 20 mums. How many flowers will Leah have for sale?

4. Ms. Dorman is planning to drive her daughter to college, a 500-mile trip. If her car gets 32 miles per gallon and has an 18-gallon gas tank, can she make the trip without stopping for more gas?

f UN with MATH

What country was the first to use a compass? When?
See page 240.

5. Main Street runs north and south, and Third Avenue runs east and west. From the intersection of Main and Third, John jogs 5 blocks south, 6 blocks west, 3 blocks north, and 2 blocks east. How many blocks west of Main Street is John?

6. Jeanne, Mary, and some friends order a pizza that is cut into 8 pieces. Mary takes the first piece and passes the pizza around the table. If each friend has one piece and there is one piece left when the pizza returns to Mary, how many friends joined Jeanne and Mary?

7. What whole number between 50 and 70 is a multiple of 3, 4, and 5?

8. Jody starts baking her cherry pie at 450° F for $\frac{1}{4}$ of an hour. Then she bakes the pie at 350° for $\frac{5}{6}$ of an hour. How long is the pie in the oven altogether?

9. Juan and Susan are painting opposite sides of the fence. If Susan paints one-fourth of her side in an hour and Juan paints one-half of his side in the same time, how much longer will it take Susan than Juan to paint her side of the fence?

Mixed Review

Lesson 5-1

Estimate.

10. $6\frac{1}{4} \times 3\frac{4}{5}$

11. $8\frac{5}{8} \times 6\frac{3}{4}$

12. $10\frac{1}{6} \div 1\frac{7}{8}$

13. $16\frac{3}{8} \div 4\frac{1}{7}$

Lesson 5-6

Divide.

14. $\frac{6}{7} \div \frac{3}{8}$

15. $\frac{5}{9} \div \frac{10}{9}$

16. $\frac{6}{11} \div 3$

17. $\frac{7}{12} \div \frac{2}{3}$

Lesson 6-2

18. Mario bought 50 meters of rope to make a rope ladder to use in a tree house. If each rung of the ladder uses $2\frac{1}{2}$ meters, how many rungs will there be?

6-7 CUSTOMARY UNITS OF LENGTH

Objective

Change from one customary unit of length to another.

Henry James High School is voting for student council members on Thursday. Molly bought 8 yards each of red, white, and blue bunting to decorate the tables. How many 2-foot lengths of each color can Molly cut? She needs to change the number of yards to feet and divide.

The most commonly used customary units of length are the **inch (in. or ″), foot (ft or ′), yard (yd), and mile (mi).**

To change customary units, multiply or divide by the appropriate number.

1 foot	= 12 inches
1 yard	= 3 feet or 36 inches
1 mile	= 5,280 feet or 1,760 yards

Method

1 ▶ Determine if you are changing from smaller to larger units or vice versa.

2 ▶ Divide to change from smaller to larger units. Multiply to change from larger to smaller units.

3 ▶ Determine the number by which to multiply or divide.

Example A

How many 2-foot lengths can Molly cut from 8 yards of bunting.

1 ▶ yards → feet
You are changing from larger to smaller units.

2 ▶ There will be more units after the change.
Multiply.   8 yards = $\frac{?}{}$ feet

3 ▶ There are 3 feet in one yard.
Multiply by 3.   8 yards = 24 feet

Molly divides 24 feet by 2 to find the number of 2-foot lengths in 8 yards.   24 ÷ 2 = 12

Molly can cut 12 2-foot lengths from 8 yards of bunting.

Example B

Change 50 inches to feet and inches.

1 ▶ inches → feet   smaller → larger
2 ▶ Divide.   50 in. = $\frac{?}{}$ ft __ in.
3 ▶ Divide by 12.   ÷ 12

$$\begin{array}{r} 4 \text{ R2} \\ 12\overline{)50} \\ -48 \\ \hline 2 \end{array}$$

50 in. = 4 ft 2 in.

Guided Practice

Complete.

Example A

1. 3 ft = $\frac{?}{}$ in.   **2.** 2 yd = $\frac{?}{}$ ft   **3.** 2 mi = $\frac{?}{}$ yd

4. 3 yd = $\frac{?}{}$ in.   **5.** 2 mi = $\frac{?}{}$ ft   **6.** 15 yd = $\frac{?}{}$ ft

Example B

7. 24 in. = $\frac{?}{}$ ft   **8.** 72 in. = $\frac{?}{}$ yd   **9.** 15 ft = $\frac{?}{}$ yd

10. 30 in. = $\frac{?}{}$ ft $\frac{?}{}$ in.   **11.** 23 ft = $\frac{?}{}$ yd $\frac{?}{}$ ft   **12.** 76 in. = $\frac{?}{}$ ft $\frac{?}{}$ in.

Complete.

13. 5 ft = $\underline{\ ?\ }$ in. **14.** 4 yd = $\underline{\ ?\ }$ ft **15.** 2 yd = $\underline{\ ?\ }$ in.

16. 67 in. = $\underline{\ ?\ }$ ft $\underline{\ ?\ }$ in. **17.** 86 in. = $\underline{\ ?\ }$ ft $\underline{\ ?\ }$ in. **18.** 36 in. = $\underline{\ ?\ }$ ft

19. 27 ft = $\underline{\ ?\ }$ yd **20.** 10 ft = $\underline{\ ?\ }$ in. **21.** 180 in. = $\underline{\ ?\ }$ yd

22. $\frac{1}{2}$ mi = $\underline{\ ?\ }$ yd **23.** $1\frac{1}{2}$ ft = $\underline{\ ?\ }$ in. **24.** $2\frac{1}{4}$ yd = $\underline{\ ?\ }$ in.

25. 96 in. = $\underline{\ ?\ }$ yd $\underline{\ ?\ }$ ft **26.** 18 in. = $\underline{\ ?\ }$ ft **27.** 54 in. = $\underline{\ ?\ }$ yd

Applications

28. Romona walks $\frac{1}{2}$ mile to school on Mondays, Wednesdays, and Fridays. On Tuesdays and Thursdays she walks $2\frac{1}{4}$ miles to the athletic club. How much farther is the walk to the athletic club?

29. In the high jump, Allan cleared 5 ft 3 in. David jumped 13 inches higher. How high was David's jump?

30. Nan walked an average of $2\frac{1}{2}$ miles a day for three days. The first day she walked 3 miles, the second day she walked $1\frac{3}{4}$ miles. How far did she walk the third day?

Cooperative Groups

31. In groups of three or four, measure each person's height with a yardstick and a meterstick. Find the average height of the group in customary units and in metric units. Which system is easier to use?

Talk Math

32. In your own words, explain how to find the average of 5 ft 2 in., 4 ft 10 in., 5 ft 3 in., 4 ft 5 in., and 3 ft 11 in.

Using Equations

33. The formula that relates shoe size s and foot length f (in inches) for men is $s = 3f - 24$. What is the shoe size for a man whose foot is 11 inches long?

Estimation

34. While shopping, Celia found a bin of different-sized candles on sale. She knows a dollar bill is about 6 inches long, so she used a bill to measure the unmarked candles. If the white candles measure about $2\frac{1}{3}$ dollar bills long and the red candles measure about $1\frac{1}{2}$ dollar bills long, *about* how long is each candle?

DATA PROCESSING

Peter Soanes works for the credit card division of a bank. He enters data for individual accounts into the computer system. He must type quickly and accurately.

Peter had to take a typing test to get his job in data processing. In five minutes he typed 325 words and made 5 errors. How many words did he type per minute? The formula used to determine typing speed is $\frac{\text{number of words}}{\text{number of minutes}}$ − number of errors.

$$\text{number of} \rightarrow 5\overline{)325}^{\,65} \leftarrow \text{number of}$$
minutes   words

Subtract one for each error made.

$$65 - 5 = 60$$

Peter typed 60 words per minute. At this rate, how many words can he type in one hour?

$$60 \times 60 = 3,600$$

Peter can type 3,600 words in one hour.

Find the number of words typed per minute.

1. 300 words in 5 minutes with 6 errors

2. 210 words in 3 minutes with 4 errors

Find the number of words typed in one hour.

3. 48 words per minute

4. 63 words per minute

5. Richard plans to type his 8-page term paper on Wednesday. He can fit about 250 words on each page. If Richard types 50 words per minute, about how long will it take him to type 8 pages?

6. Anne can type 50 words per minute. If she works for 4 hours and takes one 15-minute break, how many words can she type?

7. Kevin can type 63 words per minute with 2 errors. If he works for 7 hours and takes two 20-minute breaks, how many words can he type?

NONSTANDARD UNITS OF LENGTH AND AREA

Linear measure probably developed around 10,000 to 8,000 BC before the development of measures of weight and capacity. Measurements were made by comparing distances with available units. People used their fingers, hands, arms, and feet as measuring tools. Since these differ from person to person, the units were not standard.

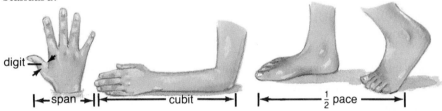

digit

|←span→||←———— cubit ————→| |←————— ½ pace —————→|

In early England, an acre of land was measured in different (nonstandard) ways. An acre might be the amount of land that could be plowed in one year, or it might be several lengths of a stick or rod. One old English system that may have been used through the Middle Ages is shown at the right.

1 hide	=	120 acres
1 virgate	=	30 acres
		or
		$\frac{1}{4}$ hide

Name the most reasonable unit for measuring each item.
Use digit, span, cubit, or pace.

1. length of a car   **2.** width of a paper clip

3. width of a skateboard   **4.** height of a candle

Choose the best estimate for each of the following.

5. height of a desk   **a.** 2 cubits   **b.** 2 paces

6. width of a book   **a.** 1 cubit   **b.** 1 span

7. length of a chalkboard   **a.** 4 paces   **b.** 4 digits

Complete.

8. 1 hide = $\underline{\ ?\ }$ virgate(s)

9. $\frac{1}{2}$ hide = $\underline{\ ?\ }$ acres

10. 8 virgates = $\underline{\ ?\ }$ hide(s)

11. Could one person's span equal another person's cubit?

6-8 CUSTOMARY UNITS OF WEIGHT AND CAPACITY

Objective

Change customary units of weight and capacity.

Melinda and Jeff are making ice cream. The recipe calls for 8 cups of heavy cream. At the store, heavy cream is available in pints. They change cups to pints to find the amount to buy.

The most commonly used customary units of weight are the **ounce (oz), pound (lb),** and **ton (T).**

1 pound = 16 ounces
1 ton = 2,000 pounds

The most commonly used customary units of capacity are the **fluid ounce (fl oz), cup (c), pint (pt), quart (qt),** and **gallon (gal).**

1 cup = 8 fluid ounces
1 pint = 2 cups
1 quart = 2 pints
1 gallon = 4 quarts

Changing customary units of weight and capacity is similar to changing customary units of length.

Method

1. Determine if you are changing from smaller to larger units or vice versa.

2. Divide to change from smaller to larger units.
   Multiply to change from larger to smaller units.

3. Determine the number by which to multiply or divide.

Examples

A *Change 8 cups to pints.*

1. cups → pints
   smaller → larger
2. Divide.
3. There are 2 cups in a pint. So divide by 2.

$8 \text{ c} = \underline{\overset{?}{}} \text{ pt} \quad \div 2$

$8 \text{ c} = 4 \text{ pt}$

Melinda and Jeff buy 4 pints of heavy cream.

B *Change 2 pounds to ounces.*

1. pounds → ounces
   larger → smaller
2. Multiply.
3. There are 16 ounces in a pound. So multiply by 16.

$2 \text{ lb} = \underline{\overset{?}{}} \text{ oz} \quad \times 16$

$2 \text{ lb} = 32 \text{ oz}$

Guided Practice

Example A

Complete.

1. $2 \text{ c} = \underline{\overset{?}{}} \text{ fl oz}$
2. $2 \text{ gal} = \underline{\overset{?}{}} \text{ qt}$
3. $40 \text{ fl oz} = \underline{\overset{?}{}} \text{ c}$
4. $12 \text{ qt} = \underline{\overset{?}{}} \text{ gal}$
5. $10 \text{ c} = \underline{\overset{?}{}} \text{ pt}$
6. $3 \text{ pt} = \underline{\overset{?}{}} \text{ qt}$

Example B

7. $2 \text{ T} = \underline{\overset{?}{}} \text{ lb}$
8. $48 \text{ oz} = \underline{\overset{?}{}} \text{ lb}$
9. $96 \text{ oz} = \underline{\overset{?}{}} \text{ lb}$
10. $5 \text{ lb} = \underline{\overset{?}{}} \text{ oz}$
11. $3 \text{ lb} = \underline{\overset{?}{}} \text{ oz}$
12. $128 \text{ oz} = \underline{\overset{?}{}} \text{ lb}$

Practice

Complete.

13. 8 gal = $\underline{\ ?\ }$ qt

14. 14 pt = $\underline{\ ?\ }$ qt

15. 4 c = $\underline{\ ?\ }$ pt

16. 8 qt = $\underline{\ ?\ }$ gal

17. 5 pt = $\underline{\ ?\ }$ c

18. $2\frac{1}{2}$ pt = $\underline{\ ?\ }$ c

19. $5\frac{1}{2}$ qt = $\underline{\ ?\ }$ pt

20. $\frac{1}{2}$ lb = $\underline{\ ?\ }$ oz

21. 96 fl oz = $\underline{\ ?\ }$ c

22. 3 T = $\underline{\ ?\ }$ lb

23. 1 gal = $\underline{\ ?\ }$ pt

24. 1 qt = $\underline{\ ?\ }$ fl oz

25. 1 gal = $\underline{\ ?\ }$ c

26. 500 lb = $\underline{\ ?\ }$ T

27. 1 pt = $\underline{\ ?\ }$ fl oz

28. 4 gal = $\underline{\ ?\ }$ c

29. 3 qt = $\underline{\ ?\ }$ fl oz

30. 6 gal = $\underline{\ ?\ }$ pt

31. How many pounds are in 112 ounces?

32. Change 2.5 quarts to cups.

Number Sense

*Give the most appropriate unit of capacity. Use **ounces, cups, pints, quarts,** and **gallons.***

33. bath tub

34. soup spoon

35. coffee cup

36. apple cider

Applications

37. Bernie jogs 3 hours a week. If she loses an average of 8 ounces a week by jogging, how many weeks will it take her to lose 10 pounds?

38. Rich swims 6 hours a week in order to lose 10 ounces each week. How many weeks will it take him to lose $5\frac{1}{2}$ pounds?

39. A newborn hooded seal pup weighs about 45 pounds at birth. In the first 4 days of life the pup gains about 3.5 pounds a day. How much will the average pup weigh in 4 days?

40. Ken's recipe for chili calls for $2\frac{1}{2}$ cups of tomato sauce. Ken doubles the recipe for his party Friday night. If Ken buys three 14-ounce cans, will he have enough tomato sauce?

Using Variables

Evaluate each expression if c = 8, q = 32, and p = 16.

41. $4c$

42. $q \div c$

43. $p \div c$

44. $q \div p$

Talk Math

45. Explain in your own words how to change 8 gallons to the equivalent number of ounces.

Critical Thinking

46. You are arranging the games for the volleyball intramural finals. There are 8 teams ranked 1 to 8 in the tournament. Draw a schedule tree like the one at the right and assign the teams so that the two top ranked teams might play the final match against each other.

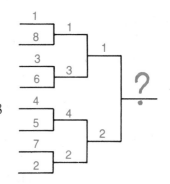

6-9 FORMULAS AND TEMPERATURE

Objective

Estimate temperature and use formulas to change Celsius to Fahrenheit and vice versa.

Jonine received a letter from her pen pal Bryan in Australia. He told her that he had been on vacation in Sidney, and the temperature had been about 25°C every day. Jonine changed the Celsius temperature to Fahrenheit.

Temperature is commonly measured in degrees Celsius (°C) or degrees Fahrenheit (°F). A Celsius thermometer and a Fahrenheit thermometer are shown at the right.
The formulas for changing temperature scales are given below.

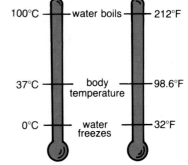

Celsius to Fahrenheit: $F = \frac{9}{5} \times C + 32$

Fahrenheit to Celsius: $C = \frac{5}{9} \times (F - 32)$

Method

1 ▶ Determine if you are changing from Celsius to Fahrenheit units or vice versa.

2 ▶ Choose the correct formula.

3 ▶ Perform the required calculations.

Example A

Change 25°C to degrees Fahrenheit.

1 ▶ Celsius → Fahrenheit   **3** ▶ $F = \frac{9}{5} \times 25 + 32$ Replace C with 25.

2 ▶ Use $F = \frac{9}{5} \times C + 32$. $F = 45 + 32$ $\frac{9}{5} \times 25 = 45$

$F = 77$

The temperature in Sidney was about 77°F every day.

Example B

Change 68°F to degrees Celsius.

1 ▶ Fahrenheit → Celsius   **3** ▶ $C = \frac{5}{9} \times (68 - 32)$ Replace F with 68.

2 ▶ Use $C = \frac{5}{9} \times (F - 32)$. $C = \frac{5}{9} \times (36)$ $\frac{5}{\underset{1}{\cancel{9}}} \times \frac{\overset{4}{\cancel{36}}}{1} = 20$

$C = 20$ 68°F is the same as 20°C.

Guided Practice

Choose the better temperature for each activity.

1. ice hockey, 120°F or 20°F   **2.** swimming, 30°C or 10°C

3. snow skiing, 59°F or 23°F   **4.** jogging, 70°F or 10°F

Example A

Find the equivalent Fahrenheit temperature to the nearest degree.

5. 60°C   **6.** 15°C   **7.** 50°C   **8.** 85°C   **9.** 30°C

Example B

Find the equivalent Celsius temperature to the nearest degree.

10. 41°F   **11.** 59°F   **12.** 149°F   **13.** 239°F   **14.** 194°F

Practice

Find the equivalent Fahrenheit temperature to the nearest degree.

15. 100°C   **16.** 5°C   **17.** 53°C   **18.** 47°C   **19.** 122°C

20. 27°C   **21.** 13°C   **22.** 18°C   **23.** 63°C   **24.** 33°C

Find the equivalent Celsius temperature to the nearest degree.

25. 95°F   **26.** 104°F   **27.** 100°F   **28.** 77°F   **29.** 212°F

30. 32°F   **31.** 90°F   **32.** 80°F   **33.** 34°F   **34.** 87°F

Applications

35. The high temperature on March 18 in Omaha, Nebraska, was 2°C. What was the temperature in degrees Fahrenheit?

36. Jennifer likes to practice in the outdoor pool for the 25-meter backstroke if the temperature is between 20°C and 30°C. If the Fahrenheit temperature is 76°, what is the Celsius temperature?

37. The Greenhouse Crisis Foundation predicts that the global warming effect may increase average temperatures between 4° and 9°F during the next 60 to 70 years. This means that the average monthly temperature for Jacksonville for the month of January could increase to between 69° and 74°F instead of the current 65°F. What is the predicted range for Jacksonville in degrees Celsius?

Cooperative Groups

38. In groups of three or four, collect pictures from magazines of the group's favorite activities or sports and place each picture on a separate piece of paper. On index cards, write appropriate Celsius temperatures for each picture. See if other groups can match the temperature with the activity.

Critical Thinking

39. Hector was leading his flock of geese south for the winter. There were 5 geese behind him in each wing of the V-formation. Hector dropped back after 1 hour to the tail of the right-hand section. Alternating one leader from each side of the formation, each new leader heads the flock for one hour. In how many hours will Hector have to lead again?

Calculator

Find the equivalent Fahrenheit temperature for 60°C.

$9 \div 5 \times 60 + 32 = 140$

60°C is equivalent to 140°F.

Find the equivalent Celsius temperature for 140°F.

$140 - 32 = 108 \times 5 \div 9 = 60$

140°F is equivalent to 60°C.

Complete. Use a calculator.

40. 32°C = $\underline{?}$ F   **41.** 70°C = $\underline{?}$ F

42. 99°F = $\underline{?}$ C   **43.** 212°F = $\underline{?}$ C

6-10 MEASURING TIME

Objective

Change from one unit of time to another.

As part of her weight reducing program, Pat plays tennis for one hour three times a week. In three weeks Pat can lose about one pound. How many minutes a week does Pat play tennis?

Units of time are the same in the customary and metric systems. Common units of time are shown at the right.

1 day (d) = 24 hours (h)
1 hour (h) = 60 minutes (min)
1 minute (min) = 60 seconds (s)

Method

1 ▶ Determine if you are changing from smaller to larger units or vice versa.

2 ▶ Divide to change from smaller to larger units.
Multiply to change from larger to smaller units.

3 ▶ Determine the number by which to multiply or divide.

Example A

Change 3 hours to minutes.

1 ▶ hours → minutes
2 ▶ larger → smaller

3 ▶ Multiply by 60.

$3 h = \underline{\ ?\ } min$

$3 \times 60 = 180$

$3 h = 180 min$

Pat plays tennis for 180 minutes a week.

Example B

Change 325 seconds to minutes.

1 ▶ seconds → minutes
2 ▶ smaller → larger

3 ▶ Divide by 60.

$325 s = \underline{\ ?\ } min \quad \div 60$

$325 s = 5 min 25 s$

$$\begin{array}{r} 5 \ R25 \\ 60\overline{)325} \\ -300 \\ \hline 25 \end{array}$$

Elapsed time is the amount of time that has passed from one time to another. Remember to watch for A.M. and P.M. changes when you determine elapsed time.

Example C

Find the elapsed time from 11:00 A.M. to 3:30 P.M.

Beginning
11:00 A.M.

11:00 A.M. to
12:00 noon is
1 hour

12:00 noon to
3:30 P.M. is
$3\frac{1}{2}$ hours.

Ending
4:30 P.M.

The elapsed time is $1 + 3\frac{1}{2}$ or $4\frac{1}{2}$ hours.

Guided Practice

Name the larger measurement.

1. 60 min or 1 h 5 min
2. 3 days or 60 h
3. 6 h or 350 min
4. 8 min or 500 s
5. 30 h or 2 days
6. 4 h 10 min or 300 min

Example A
Example B

Complete.

7. 5 min = _?_ s   **8.** 8 h = _?_ min   **9.** 4 d = _?_ h

10. 48 h = _?_ d   **11.** 360 s = _?_ min   **12.** 1,440 min = _?_ d

Example C

Find the elapsed time.

13. from 4:10 P.M. to 7:35 P.M.   **14.** from 11:25 P.M. to 1:48 A.M.

Exercises

Practice

Complete.

15. 10 d = _?_ h   **16.** 345,600 s = _?_ d   **17.** 72 h = _?_ d

18. $\frac{1}{2}$ h = _?_ min   **19.** 45 min = _?_ h   **20.** $\frac{1}{3}$ h = _?_ min

21. 1 day = _?_ min   **22.** 1 h 30 min = _?_ min   **23.** $4\frac{1}{2}$ min = _?_ s

Find the elapsed time.

24. from 3:30 A.M. to 6:12 A.M.   **25.** from 10:30 A.M. to midnight

Applications

26. Nina wants to get to the airport in plenty of time so she won't miss her plane. If it takes about 35 minutes to drive to the airport, about 20 minutes to park the car and ride the shuttle, and she plans to check in 45 minutes early for her 8:20 A.M. flight, what time should she leave home for the airport?

27. Janet Evans won the 400-meter freestyle in the 1988 Summer Olympics with a time of 4 minutes 3.85 seconds. If you disregard the hundredths of a second, how many seconds is this?

28. Brandon likes to sleep about 7 hours a night. If he went to bed at 10:45 P.M. and got up at 6:15 A.M., how much more or less than 7 hours did he sleep?

Make Up a Problem

29. Using your school schedule, make up two problems that relate to time. One problem must use elapsed time. One problem must involve changing units of time.

Critical Thinking

30. On a digital clock during a 12-hour period, which numeral appears more often, *1* or *0*?

31. If a standard chiming clock became a 24-hour clock, it would strike 24 times at midnight. How many times would the clock strike during one whole day?

Mental Math

Solve Exercise 32 by counting. Count the hours, then count the minutes. For example, Jamie checked out a video that has a running time of 97 minutes. If she begins watching it at 8:30 P.M., what time will the video be over?

Think: 97 minutes is 7 minutes more than an hour and a half. An hour and a half from 8:30 P.M. is 10:00 P.M., so 7 more minutes will be 10:07 P.M.

32. Larry begins watching a video movie at 7:20 P.M. His brother wants to know if he can watch his favorite sitcom on the same T.V. If the video is 128 minutes long, when will it be over?

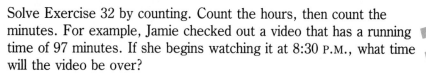
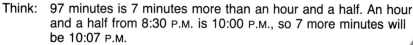
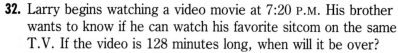

6-11 TIME CARDS

Objective

Compute the number of hours worked using 24-hour notation.

April punches a time clock when she begins and ends her work shift. On Wednesday her time card showed IN times of 7:45 and 12:30 and OUT times of 12:00 and 14:30. How many hours did she work on Wednesday?

Finding the number of hours worked from time cards is similar to finding elapsed time. Many time clocks print the time in 24-hour notation. It is easier to find elapsed time using 24-hour notation because you do not have to adjust for A.M. to P.M. changes in the same 24-hour day.

The A.M. times are 00:01 to 12:00. The P.M. times are 12:01 to 24:00. For example, 14:30 means the same as 2:30 P.M.

April subtracts to find the amount of time she worked on Wednesday.

Before Lunch		After Lunch		Total Time	
OUT	12:00 Rename 1 h	OUT	14:30	4 h 15 min	before lunch
IN	− 7:45 as 60 min.	IN	− 12:30	2 h 00 min	after lunch
	4:15		2:00	6 h 15 min	

(OUT 12:00 shown with renaming marks "1 6 0")

April worked 6 hours and 15 minutes on Wednesday.

Understanding the Data

● What time did April punch in Wednesday morning?

● What P.M. time corresponds to 17:00 in 24-hour notation?

● What P.M. time corresponds to 21:30 in 24-hour notation?

Check for Understanding

Change each 12-hour time to 24-hour notation.

1. 1:35 P.M.   **2.** 7:20 P.M.   **3.** 2:10 A.M.

4. 6:40 A.M.   **5.** 4:50 P.M.   **6.** 12:10 A.M.

Compute the working hours for each day.

7.	IN	OUT	8.	IN	OUT	9.	IN	OUT
	8:00	12:00		8:30	12:00		7:45	11:45
	13:00	17:00		13:00	17:30		13:15	17:15

Compute the working hours for each day.

10. IN	OUT		11. IN	OUT		12. IN	OUT
7:15	12:00		9:30	13:00		9:15	12:15
13:30	18:00		13:30	16:30		13:45	18:15

13. IN	OUT		14. IN	OUT		15. IN	OUT
13:30	18:30		10:50	14:15		8:45	11:10
19:00	22:10		15:00	19:20		11:40	16:25

Applications

16. Sherri's time card showed IN times of 09:00 and 13:15 and OUT times of 12:15 and 16:30 for Thursday. If she makes $4.75 an hour, how much will Sherri earn Thursday?

17. Howard was scheduled to work from 2:00 P.M. on Saturday until 7:00 P.M. If his time card shows he punched in at 14:20, was Howard on time? If not, how late was he?

Critical Thinking

18. A jeweler has 6 lengths of gold chain. Each length has 4 links. To join any 2 lengths, a link must first be cut, looped through a closed link, and soldered together. It costs $5 to cut a link and $10 to solder it together. What is the least expensive way to rejoin the six lengths into a 24-link length? How much will it cost?

Suppose

Use the table for Exercises 19–21.

19. Suppose the number of households in Medford, Oregon, is 675. *About* how many households own pets? (41.2% = 0.412)

20. Suppose *about* one-sixth of the households that own pets in Medford own more than one pet. How many households is this?

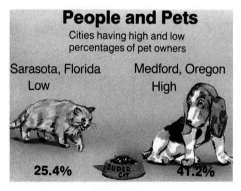

People and Pets

Cities having high and low percentages of pet owners

Sarasota, Florida — Low   Medford, Oregon — High

25.4%   SUPER CAT   41.2%

Collect Data

21. Survey the members of your class to find how many own pets. Find the number of pet owners in the class compared to the total number in the class. Is the portion of the class that owns pets closer to the portion of Medford, Oregon (0.412), or Sarasota, Florida (0.254)?

Mixed Review

Lesson 4-11

Add.

22. $6\frac{1}{8} + 4\frac{2}{8}$ **23.** $15\frac{1}{4} + 6\frac{2}{3}$ **24.** $5\frac{2}{5} + 3\frac{8}{9}$ **25.** $9\frac{2}{7} + 8\frac{2}{3}$

Lesson 5-3

Multiply.

26. $4\frac{1}{5} \times 3\frac{2}{3}$ **27.** $6\frac{2}{5} \times 5$ **28.** $2\frac{1}{4} \times 3\frac{1}{3}$ **29.** $1\frac{1}{8} \times 3\frac{2}{3}$

Lesson 6-3

30. At the print shop they print posters that are 1.2 meters long. If 300 posters are printed end to end, what is the length of paper, in km, needed to print the posters?

► Explore
► Plan
► Solve
► Examine

6-12 USING MEASUREMENTS

Objective

Solve problems involving measurements.

State marching band competition will be held in 1 week and José wants 8 hours of extra practice. If José practices the trumpet for 35 minutes every day before school and on Saturday and Sunday, will he practice for 8 hours?

► **Explore**

- What is given?
  José practices 35 minutes for 7 days a week.

- What is asked?
  At 35 minutes a day, will José practice for 8 hours in a week?

► **Plan**

Find the number of *minutes* José practices each week by multiplying 35 minutes by 7. Change the number of minutes to hours by dividing by 60.

► **Solve**

Estimate: José practices a little more than $\frac{1}{2}$ hour each day. In 7 days, he practices a little more than 7 times $\frac{1}{2}$ or $3\frac{1}{2}$ hours.

$$\begin{array}{r} 35 \text{ minutes} \\ \times\ \ 7 \text{ days} \\ \hline 245 \text{ minutes} \end{array}$$

$$\begin{array}{r} 4 \ \text{R5} \\ 60\overline{)245} \\ -\underline{240} \\ 5 \end{array}$$

José practices 4 hours 5 minutes. He will not practice an extra 8 hours.

► **Examine**

To solve another way, think how many days are in one week. There are 7. If José practices one whole hour a day for 7 days, he will practice an extra 7 hours. Since one whole hour a day is not enough, 35 minutes a day is not enough.

Guided Practice

1. Kyle has 2 boards for shelving. One is 18 inches longer than the other. If the shorter board is 6 feet 3 inches long, how long is the other board?

2. Karen brought 3 quarts of orange juice in her thermal jug for the break at hockey practice. How many 4-ounce glasses can she serve to the team?

3. Kurt added walking 5 days a week to his reducing program. If he walks 40 minutes a day for 5 days, how many hours will he walk?

4. A rectangular garden measures 55 feet wide by 83 feet long. Jed has 100 yards of fencing to enclose the garden. How many yards of fencing will Jed have left after enclosing the garden?

Paper/ Pencil
Mental Math
Estimate
Calculator

5. Linda's blouse pattern calls for buttons that are 9 millimeters in diameter. Linda chose buttons that are 1 centimeter in diameter. By how many millimeters are the buttons too large or too small?

6. Kent wants to measure the top of his desk for a new blotter, but he didn't want to walk downstairs for the yardstick. Kent knows that a sheet of typing paper is 8.5 inches wide and 11 inches long. He measures the top of his desk with a sheet of typing paper using only the side that measures 11 inches. If the desk measures $2\frac{1}{2}$ sheets by 4 sheets, what are the approximate dimensions of the desk?

7. A dime has a mass of 2.27 grams. What is the mass of a roll of dimes that is worth $5?

8. The Wangler's dog Gracie had three puppies. The first one weighed 1 pound 3 ounces, the next one 1 pound 1 ounce, and the last one 1 pound 5 ounces. What was the average weight of the puppies?

9. Select your favorite word problem from this chapter. Attach a note explaining why it is your favorite.

Mixed Review

Lesson 1-6

Estimate.

10. 586
      +381

11. 4,621
      + 314

12. 8.12
        3.98
      +4.76

13. 3,512
      −2,621

Lesson 3-1

Multiply.

14. 62
       × 3

15. 143
       × 12

16. 280
       × 45

17. 651
       ×231

Lesson 6-7

18. Max lives 8 blocks from the school. On his way home from drama club, he alternately walks two blocks, then jogs two blocks. If each block is 210 yards long, how many blocks does Max jog?

6

REVIEW

**Vocabulary/
Concepts**

*Choose a word or numeral from the list at the right to
complete each sentence.*

1. There are __?__ centimeters in one meter.
2. The prefix __?__ means 1,000.
3. The basic unit for capacity in the metric
   system is the __?__.
4. The prefix milli means __?__.
5. Water freezes at 0 degrees __?__.
6. The basic unit for mass in the metric
   system is the __?__.
7. Units of __?__ are the same in the
   customary and metric systems.
8. There are __?__ grams in one kilogram.
9. The most commonly used __?__ units of
   weight are the ounce, pound, and ton.
10. There are __?__ quarts in one gallon.
11. There are 12 inches in one __?__.

0.001
4
100
1,000
Celsius
customary
Fahrenheit
foot
gram
kilo
liter
metric
time

**Exercises/
Applications**

Lesson 6-1

Name the metric unit for each measurement.

12. 1,000 grams
13. 0.01 meter
14. 100 centimeters
15. 0.001 liter
16. 1,000 meters
17. 0.001 gram

Lesson 6-2

*Use a metric ruler to measure each segment. Give the
measurement in centimeters and also in millimeters.*

18. _____
19. _____

20. _____
21. _____

Lesson 6-3

Complete.

22. 400 mm = __?__ cm
23. 0.5 m = __?__ cm
24. 5 km = __?__ m
25. 100 cm = __?__ m
26. 2,372 m = __?__ km
27. 4 mm = __?__ cm

**Lesson 6-4
6-5**

Complete.

28. 4 kg = __?__ g
29. 34 g = __?__ mg
30. 346 mg = __?__ g
31. 4,536 g = __?__ kg
32. 4.9 g = __?__ mg
33. 8.5 kg = __?__ g

Lesson 6-6

34. The Mustangs were on their own 22-yard line. The quarterback
   for the Mustangs threw the football 22 yards forward from 6
   yards behind the line of scrimmage. The wide receiver caught the
   ball but decided he couldn't advance so he threw a lateral (no
   forward or backward distance) to another receiver. The second
   receiver ran 11 yards forward before he was tackled. Where did
   the referee spot the ball?

Lessons 6-7
6-8

Complete.

35. 3 ft = $\underline{\text{?}}$ in.
36. 7 yd = $\underline{\text{?}}$ ft
37. 120 in. = $\underline{\text{?}}$ ft
38. 2 lb 2 oz = $\underline{\text{?}}$ oz
39. 36 ft = $\underline{\text{?}}$ yd
40. 7 lb 10 oz = $\underline{\text{?}}$ oz
41. $2\frac{1}{2}$ ft = $\underline{\text{?}}$ in.
42. 3 lb = $\underline{\text{?}}$ oz
43. 2 T = $\underline{\text{?}}$ lb
44. 5 c = $\underline{\text{?}}$ fl oz
45. $2\frac{1}{2}$ lb = $\underline{\text{?}}$ oz
46. 64 fl oz = $\underline{\text{?}}$ c

Lesson 6-9

Find the equivalent Fahrenheit temperature or Celsius temperature to the nearest degree.

47. 30°C
48. 120°C
49. 100°F
50. 36°F

Choose the best temperature.

51. a summer day   100°C   42°C   42°F
52. an ice cube   26°F   62°F   26°C
53. winter day   2°C   75°C   75°F
54. swimming outdoors   25°C   0°C   25°F

Lesson 6-10

Complete.

55. 3 min = $\underline{\text{?}}$ s
56. 72 h = $\underline{\text{?}}$ d
57. 7 d = $\underline{\text{?}}$ h
58. 4 h 30 min = $\underline{\text{?}}$ min
59. 5 min 2 s = $\underline{\text{?}}$ s
60. $\frac{1}{2}$ h = $\underline{\text{?}}$ min

Lesson 6-11

Change each 12-hour time to 24-hour notation.

61. 2:33 A.M.
62. 5:56 P.M.
63. 11:29 P.M.

Compute the working hours for each day.

64.

IN	OUT
9:32	11:50
12:20	3:40

65.

IN	OUT
8:21	12:06
12:56	4:10

66.

IN	OUT
6:45	11:00
11:30	2:45

Lesson 6-12

67. For a Friday evening flight to Boston from Newark, a Boeing 737 carried 120 passengers. If the average weight of a passenger was 64 kg and the average weight of each passenger's luggage was 16 kg, what was the total weight of the passengers and their luggage?

68. Melinda needs 3 cups of apricot nectar for a gelatin salad. If each can of nectar contains 14 ounces, how many cans should Melinda buy?

69. The longest Wimbledon tennis match was in 1982 between Jimmy Connors and John McEnroe. It lasted 4 hours 16 minutes. How many minutes did the match last?

TEST

Complete.

1. 400 cm = ___?___ mm
2. 0.9 m = ___?___ cm
3. 2 km = ___?___ m
4. 260 cm = ___?___ m
5. 9,000 m = ___?___ km
6. 17 mm = ___?___ cm
7. 30 g = ___?___ kg
8. 3.42 g = ___?___ mg
9. 3.6 g = ___?___ kg
10. 10 L = ___?___ mL
11. 705 mL = ___?___ L
12. 45.5 L = ___?___ mL
13. 9 ft = ___?___ yd
14. 32 oz = ___?___ lb
15. 7 ft = ___?___ in.
16. $\frac{1}{2}$ gal = ___?___ qt
17. 10 pt = ___?___ c
18. 60 oz = ___?___ lb ___?___ oz
19. 3 pt = ___?___ c
20. 381 min = ___?___ h ___?___ min
21. 2 h 45 min = ___?___ min

Solve.

22. Smythe's tree farm has 30 maple trees planted in each row. The first year Smythe sold every fourth tree from each row. The second year they sold every third tree in a row, and the third year they sold every other tree in the row. How many trees are left in each row?

23. Rosa made 5 pounds of fudge to sell at the Photography Club bake sale. She packaged it in $\frac{1}{4}$-pound boxes. How many boxes of fudge does Rosa have to sell?

24. Coach Thompson uses blue ribbon armbands in physical education class to separate players into teams. She cuts 8 yards of ribbon into 1.5-foot pieces. How many armbands does she have?

25. Kevin went to football practice at 3:45 P.M. If practice lasted two and a half hours, what time did practice end?

▶ BONUS: Which one does not belong? Explain why.

2,000 mg   20 cg   2 g   0.02 hg

NONSTANDARD UNITS OF WEIGHT AND VOLUME

In ancient Egypt the Qedet system of weights was based on a unit of grain. The type of a grain was probably barley. The city of Heliopolis used a system that was based on 140 grains that equaled one qedet. Since the value of a qedet varied between cities, this unit was *not* standard.

Cities kept their own standards for weights in their temples. These were often decorated shapes of stone marked with the weight they represented.

The Qedet System

70 grains $= \frac{1}{2}$ qedet (or 1 drachm)

140 grains $=$ 1 qedet

10 qedet $=$ 1 deben

10 deben $=$ 1 sep

The animal figure is a 10-qedet weight carved from malachite.

The ancient Romans used a system of volume based on containers of a certain size. They kept standards of each in their temples.

Approximate Roman Volume Equivalents

2 acetabulum $=$ 1 quartarius

2 quartarius $=$ 1 hemina

25 quartarius $=$ 1 congius

4 congius $=$ 1 urna

2 urna $=$ 1 amphora

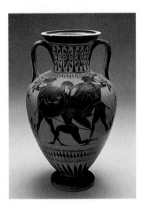

Complete.

1. 420 grains $= \underline{\ ?\ }$ qedet

2. 50 qedet $= \underline{\ ?\ }$ deben

3. 60 deben $= \underline{\ ?\ }$ sep

4. 3 sep $= \underline{\ ?\ }$ grains

5. 6 qedet $= \underline{\ ?\ }$ drachm

6. 2 sep $= \underline{\ ?\ }$ qedet

7. 630 grains $= \underline{\ ?\ }$ drachm

8. 60 drachm $= \underline{\ ?\ }$ deben

9. 28,000 grains $= \underline{\ ?\ }$ sep

10. 1 congius $= \underline{\ ?\ }$ acetabulum

11. 1 urna $= \underline{\ ?\ }$ quartarius

12. 1 amphora $= \underline{\ ?\ }$ congius

13. 75 quartarius $= \underline{\ ?\ }$ congius

14. 6 congius $= \underline{\ ?\ }$ hemina

15. 4 urna $= \underline{\ ?\ }$ acetabulum

16. Make up your own system of nonstandard weights or volume. Use at least three different units.

Free Response

Lesson 1-6   *Estimate.*

1. 4,327 + 2,294   **2.** 7,468 − 4,923   **3.** 750,290 − 61,499

Lesson 2-4   **4.** Jovita adds 12.5 gallons of gasoline to fill her car. The tank holds 15 gallons. How much gasoline was in the car's gas tank before she added any?

Lesson 2-6   **5.** Write the next three numbers in the arithmetic sequence 8.7, 7.5, 6.3.

Lesson 3-4   **6.** Dave wants to earn $76 this week. He can make $4 an hour at Colony's Mini-Golf. How many hours does Dave need to work?

Lesson 4-1   *State whether each number is divisible by 2, 3, 5, 9, or 10.*

7. 36   **8.** 136   **9.** 180

Lesson 4-7   *Replace each ● with <, >, or = to make a true sentence.*

10. $\frac{4}{5}$ ● $\frac{3}{10}$ **11.** $\frac{6}{7}$ ● $\frac{3}{14}$ **12.** $\frac{1}{2}$ ● $\frac{1}{10}$ **13.** $\frac{2}{3}$ ● $\frac{13}{14}$

Lesson 4-9   *Estimate.*

14. $\frac{4}{4} + \frac{3}{10}$ **15.** $\frac{7}{15} + \frac{1}{10}$ **16.** $\frac{43}{45} - \frac{2}{5}$ **17.** $\frac{11}{12} - \frac{7}{8}$

Lesson 5-3   *Multiply.*

18. $3 \times 19\frac{1}{3}$ **19.** $2\frac{1}{2} \times \frac{1}{5}$ **20.** $15 \times 2\frac{4}{5}$ **21.** $3\frac{1}{2} \times 6\frac{3}{4}$

22. Alberto has 230 albums. He sells $\frac{2}{5}$ of them at $1.50 each. How much money does he make?

Lessons 5-6   *Divide.*
5-7

23. $\frac{3}{4} \div \frac{1}{3}$ **24.** $\frac{5}{16} \div \frac{3}{8}$ **25.** $\frac{3}{10} \div 10$ **26.** $\frac{7}{8} \div 4$

27. $5\frac{1}{2} \div \frac{1}{2}$ **28.** $7\frac{3}{4} \div 1\frac{1}{2}$ **29.** $\frac{1}{5} \div \frac{4}{5}$ **30.** $\frac{3}{4} \div \frac{9}{10}$

Lesson 6-3   *Complete.*

31. 5,000 g = __?__ kg   **32.** 5,000 g = __?__ mg   **33.** 75 L = __?__ mL

34. 1,500 m = __?__ km   **35.** 32 km = __?__ m   **36.** 7 m = __?__ cm

Lesson 6-10   **37.** Jack has 8 minutes of blank tape left. How much tape will be left blank after he records a song that lasts 3 minutes 17 seconds?

Multiple Choice

Choose the letter of the correct answer for each item.

1. What is 6,785,201 rounded to the nearest ten thousand?
   a. 6,790,000
   b. 6,800,000
   c. 6,890,000
   d. 7,000,000

   Lesson 1-4

6. What is the difference of 5,421 and 372?
   e. 1,701
   f. 4,049
   g. 5,049
   h. 5,059

   Lesson 2-3

2. Find the quotient when 79,272 is divided by 8.
   e. 899   g. 9,909
   f. 999   h. 9,999   *Lesson 3-4*

7. What is the GCF of 12 and 8?
   a. 2   c. 24
   b. 4   d. 96

   Lesson 4-3

3. Three fourths of a pizza added to seven twelfths of a pizza gives you how much pizza?
   a. $\frac{10}{16}$ c. $1\frac{1}{4}$
   b. $\frac{5}{6}$ d. $1\frac{1}{3}$

   Lesson 4-10

8. Find the result of dividing $7\frac{3}{4}$ by $\frac{1}{2}$.
   e. $3\frac{7}{8}$ g. $7\frac{3}{4}$
   f. $7\frac{3}{8}$ h. $15\frac{1}{2}$

   Lesson 5-7

4. Lewis pole-vaulted $16\frac{3}{4}$ ft, $17\frac{3}{4}$ ft, and $17\frac{1}{4}$ ft. Find the average height of his vaults.
   e. $16\frac{31}{36}$ ft
   f. 17 ft
   g. $17\frac{1}{4}$ ft
   h. $50\frac{7}{4}$ ft

   Lessons 3-12, 4-11

9. A recipe calls for $\frac{3}{4}$ cup of sugar for one batch of cookies. How much sugar would half of a batch of cookies require?
   a. $\frac{3}{8}$ cup   c. $\frac{6}{8}$ cup
   b. $\frac{1}{2}$ cup   d. $\frac{3}{2}$ cup

   Lesson 5-3

5. In 1988, the USA's population was said to be increasing at a rate of 100,000 people a day. What is the rate of increase per minute?
   a. 15
   b. 12
   c. 69
   d. 24

   Lesson 6-10

10. Choose the most reasonable unit for measuring the mass of a science textbook.
    e. centimeter
    f. gram
    g. kilogram
    h. meter

    Lesson 6-4

7

GEOMETRY

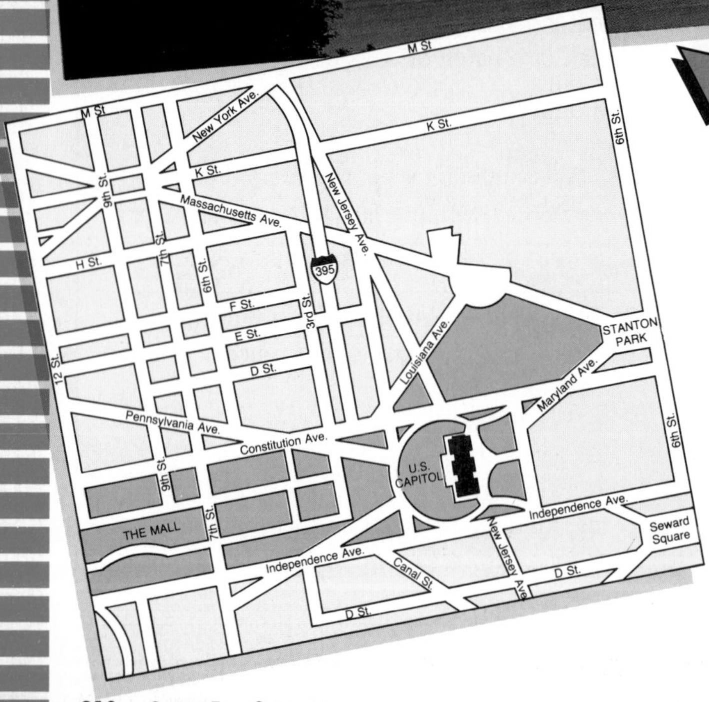

How do I get there from here?

How is your city laid out? Was it a planned city? Washington, D.C., and Savannah, Georgia, are both planned cities. A partial map of Washington, D.C., is shown at the left. Notice how most streets run east and west or north and south.

Can you find out how Savannah is laid out?

- Using the map, name two streets that will never cross. Do you know what these represent in geometry?
- Name two streets that form square corners. Do you know what these represent in geometry?
- Does Maryland Avenue represent a line or a line segment? Explain.

ACTIVITY: Exploring Constructions

In geometry the compass and straightedge are two important tools. A **straightedge** is any object that can be used to draw a straight line. A **compass** is used to draw circles or parts of circles, called **arcs**.

Constructions are drawings made with only a compass and a straightedge.

Materials: compass, straightedge, and paper

Work together in groups of two. Make your own drawing and compare it to your partner's.

1. Follow these steps to draw a six-sided figure.
   a. Use the compass to draw a circle on your paper. Do not change the compass setting.
   b. Using the same compass setting, put the metal point on the circle and draw a small arc across the circle.
   c. Move the compass point to the arc and then draw another arc across the circle. Continue doing this until you have six arcs.
   d. Use a straightedge to connect the intersection points in order.

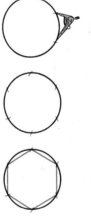

2. What is the name of this six-sided figure?

3. Explain how you would compare the distance between each arc using a compass. What do you notice about the distance between the arcs?

4. Discuss with your partner how you could complete steps **a–d** above and then draw a triangle that has sides of equal measure.

5. Discuss with your partner how to draw the six-petaled flower shown at the right. Then draw the figure using your compass.

6. Why is a construction different from some other types of drawings?

7. Make up your own design using a compass and straightedge.

BASIC TERMS OF GEOMETRY

Objective

Identify, name, and draw points, lines, and planes.

Erin is helping her father build a short brick wall around the end of their patio. Erin notices that parts of the wall illustrate a point, a line, a line segment, a ray, and a plane.

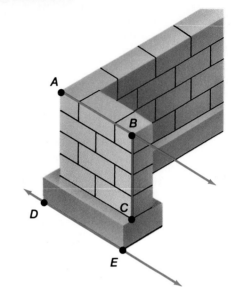

A **point** is an exact location in space. A point is named by a capital letter and is represented by a dot. Point *A* and point *B* are two points shown in the drawing at the right.

A **line** is formed by all the points on a never-ending straight path. A line is named by two points on the line. Line *DE* or line *ED,* shown in red, is written as $\overleftrightarrow{DE}$ or $\overleftrightarrow{ED}$.

A **line segment** is formed by two endpoints and the straight path between them. Line segment *BC* or line segment *CB,* shown in green, is written as $\overline{BC}$ or $\overline{CB}$.

A **ray** is formed by all the points in a never-ending straight path extending in only one direction. A ray is named by its endpoint (*always* given first) and a point on the ray. Ray *AB,* shown in blue, is written as $\overrightarrow{AB}$.

A **plane** is formed by all the points in a never-ending flat surface. A plane has no boundaries. A plane is named by three points not on a line. Part of plane *ABC* is shown in gray.

Name three points, two line segments, and two rays in the drawing of the wall that were not named above.

Guided Practice

Use symbols to name each figure.

1. *C* ●——● *D*

2. *B* ●——● *D* →

3. *R* ●——● *S*

4. *M* ←●——● *N* →

5. *A* ●——● *B*

6. *T* ●

Name two other real-life models for each geometric term.

7. point: pencil tip

8. line segment: ruler

9. ray: radar beam

10. part of a plane: sheet of paper

Exercises

Practice

f **UN with MATH**

Where and when did geometry originate?
See page 240.

Use words and symbols to name each figure in as many ways as possible.

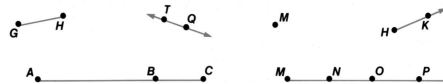

Make a drawing of each figure.

line *AB*   point *K*   line segment *MN*

ray *DE* $\overleftrightarrow{AD}$ $\overrightarrow{RS}$

Critical Thinking

How many line segments can you draw through two points?

How many lines can you draw through one point?

Draw models to illustrate your answers in Exercises

How many rays can have the same endpoint?

How many planes can contain the same line?

How many points, not in a line, determine a plane?

Show Math

Make a drawing of a desk using a ruler and pencil. Label the corner points with letters. Name a point, a line segment, a ray, and a plane.

Estimation

Estimate the length of each line segment to the nearest $\frac{1}{4}$ inch.

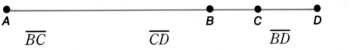

$\overline{BC}$ $\overline{CD}$ $\overline{BD}$ $\overline{AD}$

Interpret Data

Draw a straight line. Display the data from the table on the line using points, line segments, and rays.

SIT-UPS (MEN)	
Number	Fitness Level
50+	very high
40–49	high
30–39	average
20–29	low
0–19	very low

Mixed Review

Lesson 2-3

Subtract.

$$\begin{array}{r} 89 \\ -37 \\ \hline \end{array} \qquad \begin{array}{r} 223 \\ -147 \\ \hline \end{array} \qquad \begin{array}{r} 4{,}621 \\ -\ \ 858 \\ \hline \end{array}$$

Lesson 4-2

State whether each number is prime or composite.

18   23   31   49

Lesson 4-14

Brandon has 15 gallons of gasoline in his car. A trip to his aunt's house uses ten gallons. Then Brandon puts three gallons more in the tank. A trip to a ballgame uses four gallons and then Brandon puts another eight gallons in the tank. How many gallons of gasoline are in the tank now?

7-2 MEASURING ANGLES

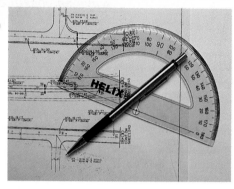

Objective

Name, draw, and measure angles.

Tom Andres is a drafter. He is preparing a detailed drawing of a bridge from an engineer's rough sketch. One tool Tom uses is a protractor. Using a protractor helps Tom measure and draw angles to the correct size.

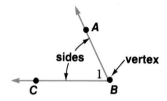

Two rays with a common endpoint form an **angle.** The common endpoint is called the **vertex.** The rays form the **sides** of the angle. In the angle shown at the left, Point B is the vertex. $\overrightarrow{BA}$ and $\overrightarrow{BC}$ are the sides of the angle.

Angle ABC is symbolized as $\angle ABC$. The letter naming the vertex must be in the middle. The angle can also be named as $\angle CBA$, $\angle B$, or $\angle 1$.

The most common unit used in measuring angles is the **degree.** Imagine a circle separated into 360 equal-sized angles with their vertex at the center. Each angle would measure one degree (1°) as shown at the left.

A **protractor** is used to measure angles.

Method

Use a protractor to measure angle DEF.

1. ▶ Place the center of the protractor on the vertex *(E)* and the 0° mark on one side of the angle *(ED)*.

2. ▶ Use the scale that begins with 0 at $\overrightarrow{ED}$. Read where the other ray $(\overrightarrow{EF})$ crosses this scale.

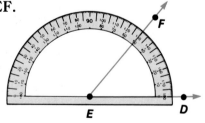

$\angle DEF$ measures 50°. This is symbolized m$\angle DEF$ = 50°.

Guided Practice

Use symbols to name each angle in three ways.

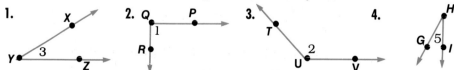

1.
2.
3.
4.

Practice

Trace each angle and extend the sides. Then find the measure of each angle.

5. 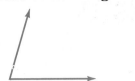   **6.** 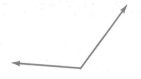   **7.**

8. 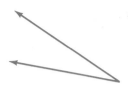   **9.** 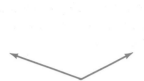   **10.**

Talk Math

Use the figure at the right.

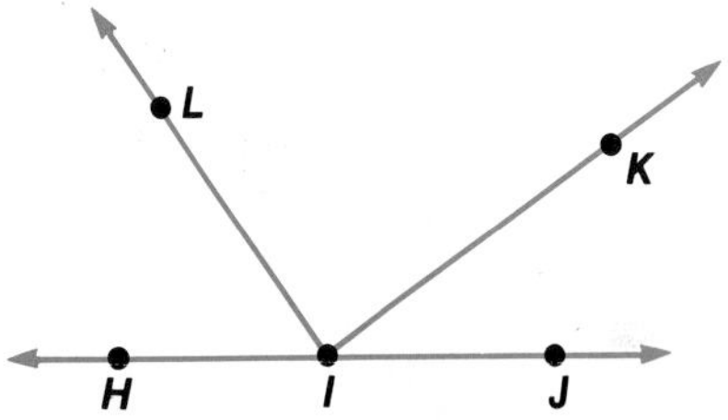

11. Name five angles. Explain why none of the angles can be named ∠*I*.

Estimation

12. Use the figure in Exercise 11. Estimate the degree measure of ∠*HIL*, ∠*LIK*, and ∠*KIJ*. Then use a protractor to measure ∠*HIL*, ∠*LIK*, and ∠*KIJ*. How close are your estimates to the actual measures?

Applications

13. Use the figure in Exercise 11. What is m∠*HIL* + m∠*LIK* + m∠*KIJ?*

14. When you do sit-ups in an exercise program, describe the angle or angles your body makes.

15. When you do push-ups in an exercise program describe the angle or angles your body makes with the floor.

Show Math

16. Draw two lines as shown at the right. Measure each of the four angles. What do you notice about the sum of the four measures?

Mixed Review

Find the common ratio and write the next three terms in each sequence.

Lesson 3-9

17. 112, 56, 28, . . .   **18.** 192, 48, 12, . . .

Lesson 4-5

Replace each ■ with a number so that the fractions are equivalent.

19. $\frac{3}{5} = \frac{■}{15}$ **20.** $\frac{8}{9} = \frac{■}{27}$ **21.** $\frac{4}{7} = \frac{■}{35}$ **22.** $\frac{1}{12} = \frac{■}{48}$

Lesson 5-5

23. Six employees can make 24 pizzas in 30 minutes. How many pizzas can 10 employees make in $1\frac{1}{2}$ hours?

7-3 CLASSIFYING ANGLES

Objective
Classify angles.

Bridges are classified according to the way they are constructed. The Golden Gate Bridge in San Francisco is a suspension bridge.
Likewise, an angle can be classified according to its degree measure.

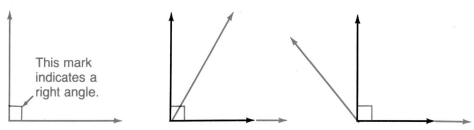

This mark indicates a right angle.

A **right angle** measures 90°.

An **acute angle** measures between 0° and 90°.

An **obtuse angle** measures between 90° and 180°.

Method

1 Measure the angle.

2 Determine if the measure is greater than 90°, less than 90°, or equal to 90°.

3 Classify the angle. If the measure is greater than 90° it is an obtuse angle. If less than 90°, and less than 180°, it is an acute angle. If equal to 90°, it is a right angle.

Example A

Classify ∠PQR.

1 m∠PQR = 75°

2 75° < 90°

3 Since 75° is less than 90°, ∠PQR is an acute angle.

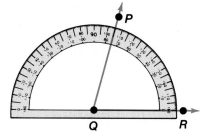

Guided Practice

Example A

Classify each angle.

1.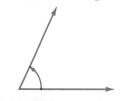

2.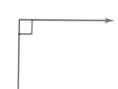

3.

Exercises

Practice

Classify each angle as right, acute, or obtuse.

4.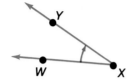

5.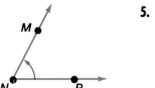

6.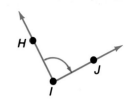

7.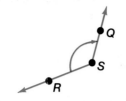

8.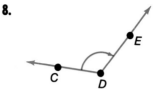

9.

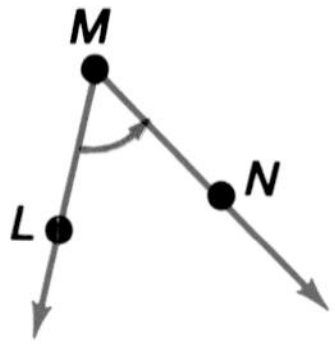

Use the figure at the right to classify each angle.

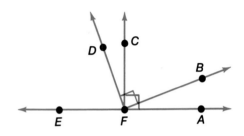

10. ∠*BFA*   **11.** ∠*AFC*

12. ∠*BFD*   **13.** ∠*DFA*

14. ∠*DFE*   **15.** ∠*BFE*

Talk Math

16. In the figure at the right, explain why none of the angles are named ∠*F*.

Make Up a Problem

17. Make up a problem using the measures of two acute angles that, when added, equal 90°.

Show Math

18. Make a drawing that shows that the measure of one acute angle plus the measure of one obtuse angle can equal exactly 180°.

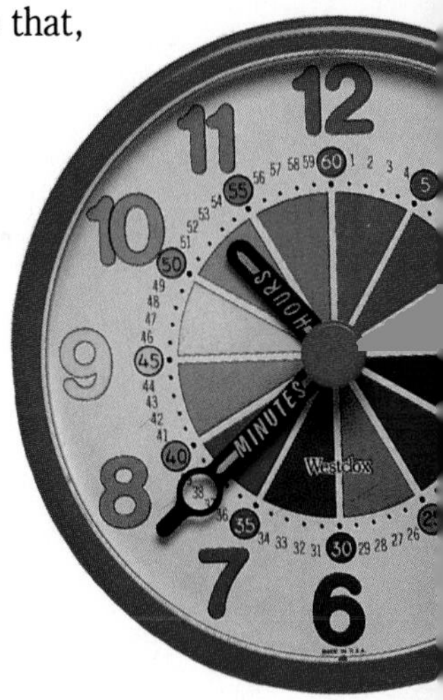

Problem Solving

Paper/ Pencil
Mental Math
Estimate
Calculator

19. Two angles are **complementary** if the sum of their measures is 90°. If the measure of an angle is half of its complement, what is the measure of each angle?

20. From 12:00 noon to 3:00 P.M., how many times do the hands of a standard clock form a right angle?

Critical Thinking

21. Each figure below shows rays with a common endpoint. Count the number of angles in each figure.

f **UN with MATH**

Try a few magic squares.
See page 241.

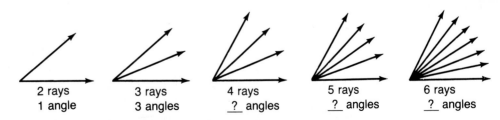

Do you see a pattern? Predict the number of angles that are formed by 7 rays.

7-4 CONGRUENT FIGURES AND CONSTRUCTIONS

Objective

Make constructions using a compass and a straightedge.

Mika Han is an architect. She is working on a design for a new building. Sometimes she uses a compass and straightedge to draw geometric constructions.

Congruent line segments and congruent angles can be constructed using a compass and a straightedge.

Congruent line segments have the same measure.
Congruent angles have the same measure.

Construction A

Construct a line segment congruent to a given line segment.

Given:

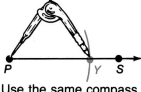

1. Use a straightedge to draw $\overrightarrow{PS}$ longer than $\overline{AB}$.

2. Open the compass to match $\overline{AB}$.

3. Use the same compass setting. With the compass point at P, draw an arc that intersects $\overrightarrow{PS}$ at Y, $\overline{PY} \cong \overline{AB}$.

Construction B

Construct an angle congruent to a given angle.

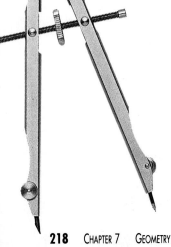

Given:

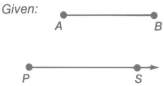

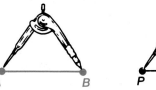

1. Use the straightedge to draw $\overrightarrow{PS}$.

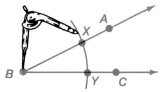

2. With the compass point at B, draw an arc that intersects $\angle ABC$ at X and Y.

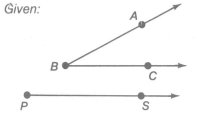

3. Use the same compass setting. With the compass point as P, draw an arc that intersects $\overrightarrow{PS}$ at N.

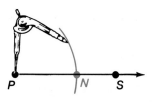

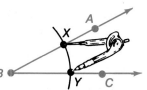

4. Open the compass to match XY. With the compass point at N, draw an arc. Label the intersection of the arcs L. Use a straightedge to draw $\overrightarrow{PL}$. $\angle LPS \cong \angle ABC$.

You can also use a compass and a straightedge to bisect a line segment. **Bisect** means to separate into two congruent parts.

Construction C

Bisect a given line segment.

Given:

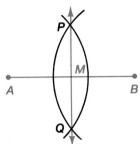

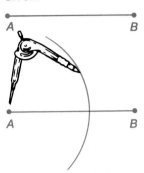
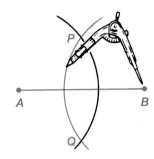

1. Open the compass to more than half the length of $\overline{AB}$. With the compass point at A, draw an arc.

2. Use the same compass setting. With the compass point at B, draw another arc. Label the intersection points P and Q.

3. Draw $\overleftrightarrow{PQ}$. $\overleftrightarrow{PQ}$ bisects $\overline{AB}$. Point M is called the midpoint of $\overline{AB}$. $\overline{AM} \cong \overline{MB}$

Exercises

Practice

Trace each figure. Construct a figure congruent to each figure given.

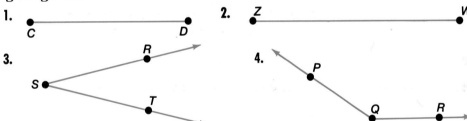

1.
2.
3.
4.

Trace each line segment. Bisect each line segment.

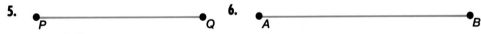

5.
6.

7. Trace $\overline{AB}$ from page 218 and perform Construction A.
8. Trace $\overline{AB}$ from page 218 and perform Construction C.

Problem Solving

9. Draw an obtuse angle. Separate it into four congruent angles by construction.

10. Suppose you bisect an acute angle. What type of angle is each of the two angles formed?

11. If you bisect a right angle, what type of angle is each of the two angles formed?

12. Draw a line segment. Separate it into four congruent parts by construction.

Talk Math

13. Explain how to construct a line segment that is twice as long as $\overline{CD}$ in exercise 1. Then construct the line segment.

7-5 PARALLEL AND PERPENDICULAR LINES

Objective

Identify and name parallel and perpendicular lines.

Katie was reading an office building floor plan for industrial technology class, when she noticed that many of the hallways were examples of parallel and perpendicular lines.

Lines in the same plane that *do not* intersect are called **parallel lines.**

Example A

Line *MN* is parallel to line *OP*.

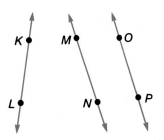

You can write this using symbols.

$\overleftrightarrow{MN} \parallel \overleftrightarrow{OP}$ Read as: Line *MN* is *parallel* to line *OP*.

Imagine that lines *KL* and *MN* are extended. They do intersect, so they are not parallel.

Two lines that intersect to form right angles are called **perpendicular lines.**

Example B

Angles 1, 2, 3, and 4 are right angles.
So, line *XZ* is perpendicular to line *YW*.

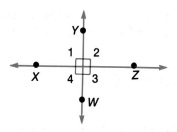

You can write this using symbols.

$\overleftrightarrow{XZ} \perp \overleftrightarrow{YW}$ Read as: Line *XZ* is *perpendicular* to line *YW*.

Lines that *do not* intersect and *are not* parallel are called **skew lines.**

Example C

In the figure at the right, line *FJ* and line *HK* are skew lines.

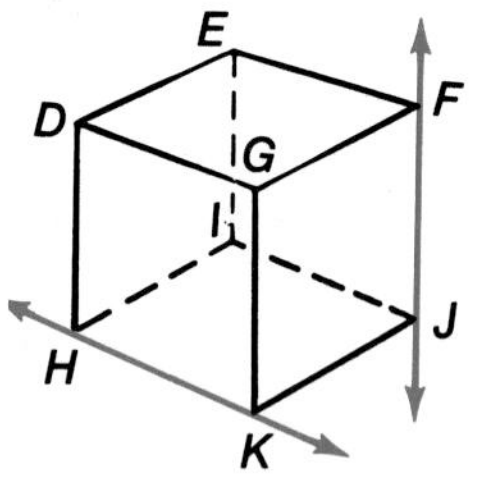

Guided Practice

State whether each pair of blue lines is parallel, perpendicular, or skew. Use symbols to name all parallel and perpendicular lines.

Examples A–C

1.

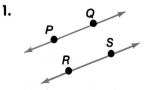

2.

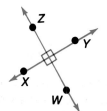

3.

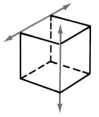

State whether each pair of blue lines is parallel, perpendicular, *or* skew. *Use symbols to name all parallel and perpendicular lines.*

4.

5.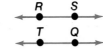

6.

7.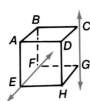

8.

9.

 On pages 525–527, you can learn how robotics technicians use mathematics in their jobs.

Complete. Use the cube at the right.

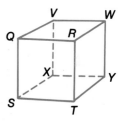

10. Name all the line segments that are perpendicular to line segment QS.

11. Name all the line segments that are parallel to line segment RW.

12. Name all the line segments that are skew to line segment QS.

13. Use a compass and a straightedge to draw two line segments that are perpendicular. Hint: Think about what happens when you bisect a line.

14. Find out how to use a compass and a protractor to draw two line segments that are parallel.

15. Suppose you are in the starting position for doing push-ups. How is your body positioned relative to the floor?

Critical Thinking

16. Explain how you would compare the lengths of $\overline{DE}$ and $\overline{EF}$ if you didn't have a ruler. Which looks longer $\overline{DE}$ or $\overline{EF}$?

Mixed Review

Lesson 1-4

Round each number to the nearest hundredth.

17. 0.378   **18.** 0.844   **19.** 4.0949   **20.** 9.982   **21.** 13.004

Lesson 5-7

Divide.

22. $4\frac{1}{4} \div 1\frac{1}{2}$ **23.** $3\frac{4}{5} \div 5$ **24.** $2\frac{3}{8} \div 1\frac{2}{3}$ **25.** $8 \div 1\frac{1}{2}$

Lesson 6-9

26. The temperature at 5 A.M. was 2°F. At 3:00 P.M. the temperature reached a high of 44°F. What was the change in temperature?

INTRODUCTION TO BASIC

BASIC (Beginner's All-purpose Symbolic Instruction Code) is one of the most widely-used languages for microcomputers. This language uses the operational symbols shown at the right.

Operation	Mathematical Expression	BASIC Expression
addition	$a + b$	A + B
subtraction	$a - b$	A − B
multiplication	$a \times b$, $a \cdot b$, or ab	A * B
division	$a \div b$ or $\frac{a}{b}$	A / B
raising to a power	n^2	N ↑ 2

BASIC expressions are evaluated using the standard order of operations.

All variables are represented by capital letters in BASIC.

Write each expression in BASIC.

Mathematical	BASIC
$\frac{1}{2}h \times (a + b)$	1/2*H*(A+B)
$4^3 - 2 \times 5 \div (2 + 3)$	4↑3−2*5/(2+3)

Find the value of 6*4+(2+3)↑2/5−7.

6*4+(2+3)↑2/5−7 = 6*4+5↑2/5−7

Do operations in parentheses.
Evaluate all powers. = 6*4+25/5−7

Do all multiplications and divisions from left to right. = 24+25/5−7

= 24+5−7

Do all additions and subtractions from left to right. = 29−7

= 22

Write each expression in BASIC.

1. $\frac{1}{4}y + 5$ **2.** z^3 **3.** $(4 - p) \times 6$

Find the value of each expression.

4. 2*(4+8)/6   **5.** (2*3)↑2   **6.** 3+4+2↑2−7

7. Write an expression in BASIC that will give your age in five years.

8. Beth found the value of the expression $59 - 4 \times 3^2$ as 495. Write the expression in BASIC. Then determine the answer given by a computer. Is Beth's answer correct?

PHYSICAL FITNESS

Chris is a member of the high school soccer team. In order to stay in shape, she stays active and counts the calories in the food she eats. A calorie is a heat-producing or energy-producing unit.

To remain at her present weight, Chris must use the same number of calories that her meals contain.

The chart below gives the number of calories used per minute for each activity.

Activity	Calories Used per Minute
Bicycling	8.2
Running	13.0
Skiing	9.5
Swimming	8.7
Walking	5.2

Chris runs for 45 minutes each Saturday. How many calories does she use?

Multiply.   13   calories per minute
          ×45   minutes
          ───
          585   Chris uses 585 calories.

Determine the number of calories used in each activity.

1. 90 minutes of skiing

2. 1 hour of swimming

3. 2 hours of bicycling

4. $1\frac{1}{2}$ hours of running

5. Jennifer walked for 40 minutes and ran for 20 minutes. How many calories were used in these two activities?

6. How many minutes must Jill ski to use the calories in a 532-calorie hamburger?

7. How many calories are used by Corey if he swims for 45 minutes, bicycles for 45 minutes, and walks for 45 minutes?

8. How long would Shawn have to run in order to use the same number of calories that are used in 40 minutes of walking?

9. Michael had the following for lunch: a hot dog with cheese, 330 calories; a pear, 121 calories; a glass of milk, 164 calories. How many minutes must he bicycle to use these calories?

10. If Jason swims $1\frac{1}{2}$ hours every day, how many extra calories a day must he consume in order to maintain his weight?

7-6 POLYGONS

Objective

Identify and name various types of polygons.

Street signs have many shapes. Simple, closed figures were used in the design of these signs.

Polygons are closed plane figures formed by line segments called **sides.** The line segments intersect at their endpoints, but do not cross. The intersection points are called **vertices** (plural of vertex).

Polygons are named by the number of sides they have. Some common polygons are listed in the chart at the right and examples are shown below.

Polygon	Sides
Triangle	3
Quadrilateral	4
Pentagon	5
Hexagon	6
Octagon	8
Decagon	10

Triangle   Quadrilateral   Pentagon   Hexagon   Octagon   Decagon

A polygon in which all sides are congruent and all angles are congruent is called a **regular polygon.**

3 congruent angles	4 congruent angles	5 congruent angles
3 congruent sides	4 congruent sides	5 congruent sides

The matching red marks indicate congruent parts.

Regular Triangle or
Equilateral Triangle

Regular Quadrilateral
or Square

Regular Pentagon

Name each polygon by the number of sides. Then state whether it is regular or not regular.

1. **2.** **3.** **4.**

Practice

Name each polygon by the number of sides. Then state whether it is regular or not regular.

5.
6.
7.
8.

Explain why each figure is not a polygon.

9. 10. 11.

Draw an example of each polygon.

12. regular triangle   13. regular quadrilateral   14. regular hexagon
15. not regular triangle   16. not regular pentagon   17. not regular octagon

18. Name two conditions that are necessary for a hexagon to be called a regular hexagon.

19. Explain why the words *closed* and *plane* are used to define a polygon.

Applications

20. Cal Walker's patio is in the shape of a decagon. Each side is $2\frac{1}{2}$ feet long. What is the perimeter of his patio?

21. Janet Guthrie drove a lap at the Indianapolis 500 in 47 seconds. One lap is 13,200 feet. Find the speed to the nearest tenth of a foot per second.

22. What were the top five oil producing states in the United States in 1991?

23. Which state produced the most oil? the least oil?

24. How much more oil did Texas produce than Louisiana?

25. Which two states produced *most closely* the same amount of oil?

26. *About* how much oil did the top oil producing states in the United States produce?

27. If the average price per barrel was $21, what was the value of the oil produced in Alaska?

Top Oil States in 1991

In 1991, the United States produced 2.7 billion barrels of crude oil.

In millions of barrels

Texas	683
Alaska	656
California	320
Louisiana	147
Oklahoma	108

7-7 TRIANGLES

Objective

Identify and name various types of triangles.

The fastest roller coaster in the United States is the Steel Phantom in West Mifflin, Pennsylvania. Triangular braces are used in the construction of roller coasters. The triangular braces add rigidity to the structure.

Triangles can be classified by the number of congruent sides.

Example A

An **equilateral triangle** has 3 congruent sides.

An **isosceles triangle** has at least 2 congruent sides.

A **scalene triangle** has no congruent sides.

Triangles can also be classified by angles. All triangles have at least two acute angles. The third angle is used to classify the triangle.

Example B

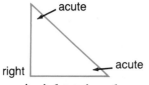
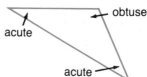
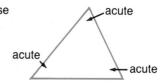

A **right triangle** has one right angle.

An **obtuse triangle** has one obtuse angle.

An **acute triangle** has all acute angles.

Guided Practice

Example A

Classify each triangle by its sides.

1.
2.
3.
4.

Example B

Classify each triangle by its angles.

5.
6.
7.
8.

Classify each triangle by its sides and then by its angles.

9.   10.   11.   12.   13.

Draw an example of each triangle.

14. scalene triangle
15. isosceles triangle
16. right triangle
17. equilateral triangle
18. acute triangle
19. obtuse triangle

Using Reasoning

f **UN with MATH**

What did the acorn say when he finally grew up?
See page 240.

State whether each statement is **true** *or* **false.** *Draw an example of each figure when the statement is true.*

20. An equilateral triangle must be an acute triangle.
21. An isosceles triangle can have a right angle.
22. A scalene triangle may also be a right triangle.
23. An obtuse triangle may also be an equilateral triangle.
24. An equilateral triangle is a regular triangle.
25. An acute triangle may also be an isosceles triangle.

Use the figure at the right to name each of the following. Use the symbol △ for triangle.

26. five right triangles.
27. two acute triangles
28. three obtuse triangles
29. one isosceles triangle
30. seven scalene triangles

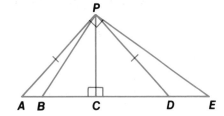

Collect Data

31. List the road signs you see on the way to school that are shaped like triangles.

Cooperative Groups

32. Draw and cut out any triangle. Label the angles 1, 2, and 3. Tear off each angle. Draw a line and place point *P* on the line. Place angles 1, 2, and 3 on the line so that their vertices are on Point *P*. Explain what you discover. What is the sum of the measures of the angle?

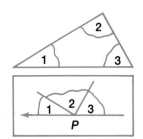

33. Suppose that a right triangle has a 42° angle. Find the measure of the other angles.

34. What is the measure of each of the angles in an equilateral triangle?

Write Math

35. Explain why you cannot draw a triangle with two right angles.

7-8 QUADRILATERALS

Objective

Identify and name quadrilaterals.

A tennis court is in the shape of a rectangle. A net and a white line divide the court in half. The two halves are also rectangles. Another white line is drawn lengthwise to allow for four players. This line divides the court into four congruent rectangles.

A rectangle is a quadrilateral because it has four sides. *Quadrilaterals* are classified according to the following.

- parallel sides
- congruent angles
- congruent sides

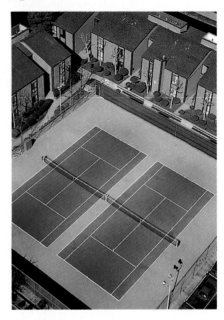

A **trapezoid** is a quadrilateral with only one pair of parallel sides.

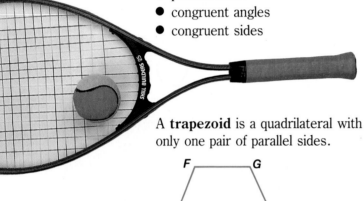

A **parallelogram** is a quadrilateral with two pairs of parallel sides.

Opposite sides are congruent.

Opposite angles are congruent.

A **rectangle** is a parallelogram with all angles congruent.

Opposite sides are congruent.

A **square** is a parallelogram with all sides and angles congruent.

A square is a regular quadrilateral.

A **rhombus** is a parallelogram with all sides congruent.

Opposite angles are congruent.

Guided Practice — *Classify each quadrilateral.*

1.

2.

3.

4.

Practice

Classify each quadrilateral.

5. 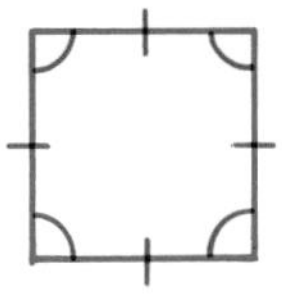   6. 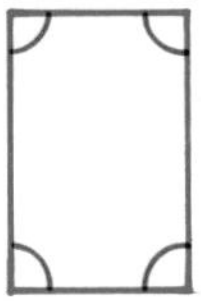   7. 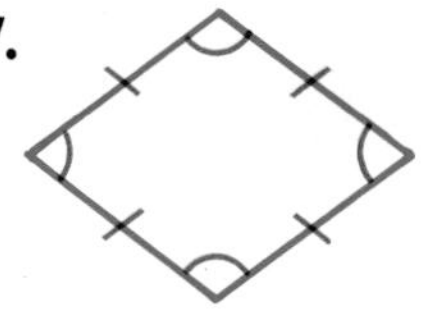   8.

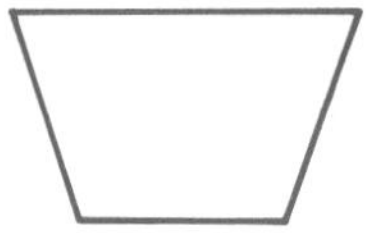

Show Math

Draw an example of each quadrilateral.

9. square   10. parallelogram   11. rhombus   12. trapezoid

Using Reasoning

State whether each statement is true *or* false. *Draw an example of each figure when the statement is true.*

13. All rectangles are squares.

14. All squares are rectangles.

15. All parallelograms are squares.

16. A rectangle is a quadrilateral.

17. A rhombus is a rectangle.

18. Not every parallelogram is a rectangle.

19. A square is a rhombus.

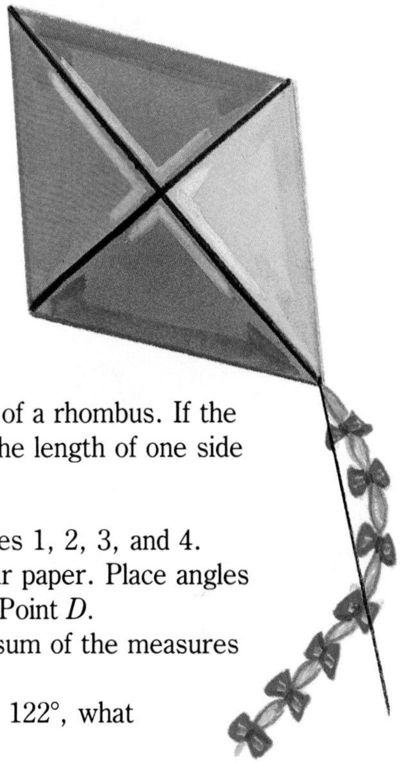

Problem Solving

20. Is this kite shape a parallelogram? Explain in your own words why or why not.

21. The wood trim on a cabinet is in the shape of a rhombus. If the total length of the trim is 148 cm, what is the length of one side of the trim?

Cooperative Groups

22. Draw and cut any rectangle. Label the angles 1, 2, 3, and 4. Tear off each angle. Place a point D on your paper. Place angles 1, 2, 3, and 4 so that their vertices are on Point D.
    a. Explain what you discover. What is the sum of the measures of the angles?
    b. If a rhombus has an angle that measures 122°, what are the measures of the other angles?

Critical Thinking

23. How many squares are in the figure at the right? Count only those squares that are outlined.

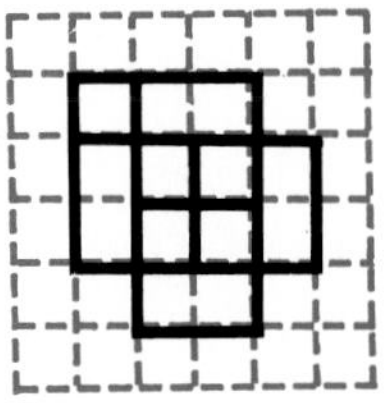

Mixed Review

Change each fraction to a mixed number in simplest form.

Lesson 4-8   24. $\frac{6}{5}$ 25. $\frac{9}{5}$ 26. $\frac{18}{7}$

Lesson 5-9

Change each decimal to a fraction.

27. 0.6   28. 0.42   29. 0.325   30. 0.638

Lesson 6-10

31. Alisa has to be at school at 8:00 A.M. It takes her 35 minutes to shower and get dressed. She needs 20 minutes to eat breakfast and 15 minutes to drive to school. What time should she get up in order to be at school on time?

THREE-DIMENSIONAL FIGURES

Objective

Identify and name three-dimensional figures.

Sally Cooper and her family camp in a tent that looks like a rectangular pyramid. The sides of her tent are triangular shaped and the floor of her tent has a rectangular shape.

Three-dimensional figures are solid figures that enclose part of space.

A solid figure with flat surfaces is a **polyhedron.** The flat surfaces are called **faces.** The faces intersect to form **edges.** The edges intersect to form **vertices.**

A **prism** is a polyhedron with two parallel congruent **bases** that are shaped like polygons. The other faces of a prism are shaped like parallelograms. A prism is named by the shape of its bases.

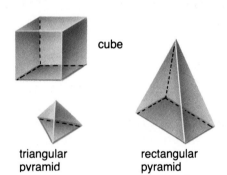

cube

triangular pyramid

rectangular pyramid

A **cube** is a rectangular prism that has six square faces. All edges of a cube are the same length. How many edges are there?

A **pyramid** is a polyhedron with a single base shaped like a polygon. The faces of a pyramid are triangular and meet at a point. A pyramid is named by the shape of its base.

Some three-dimensional figures have curved surfaces.

A **cone** has a circular base and one vertex.

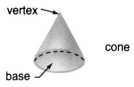

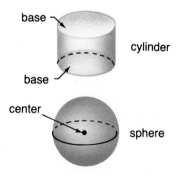

A **cylinder** has two parallel congruent circular bases.

A **sphere** is a solid with all points the same distance from a given point called the **center.**

Name each shape.

1.

2.

3.

4.

Practice

5. Copy and complete.

Polyhedron	Number of Faces *(F)*	Number of Vertices *(V)*	Number of Edges *(E)*
Rectangular Prism	6	8	12
Triangular Pyramid	▨	▨	▨
Pentagonal Prism	▨	▨	▨
Triangular Prism	▨	▨	▨
Rectangular Pyramid	▨	▨	▨
Pentagonal Pyramid	▨	▨	▨

6. Select an assignment from this chapter that shows photos or sketches of physical models you made.

Using Algebra

7. Write a formula called Euler's formula to show the relationship among the number of faces *(F)*, vertices *(V)*, and edges *(E)* of a polyhedron.

Problem Solving

8. Name a figure that has exactly one base and one vertex.

9. If a prism has 100 sides and two of them are bases, how many vertices would it have?

10. The running track at the high school is $\frac{1}{4}$-mile long. *About* how many feet is this?

Cooperative Groups

Materials: gumdrops, straws, paper, pencil

11. Make a model of a triangular prism, and a rectangular pyramid. Use the figures at the right as a guide. Then see if you can make the rest of the prisms and pyramids on page 230. Determine how each figure got its name.

Materials: paper towel tubes, paper, pencil, scissors, tape

12. Discuss with your group how to make a cylinder using a tube, paper, and tape. Then make the cylinder.

13. How many bases does a cylinder have? Does the cylinder have vertices? Why or why not?

Critical Thinking

14. Copy and enlarge the figure shown at the right. Cut it out and fold it to make a cube. Three faces meet at each corner. What is the greatest sum of three numbers whose faces meet at a corner?

	1		
6	2	4	5
	3		

> ▶ Explore
> ▶ Plan
> ▶ Solve
> ▶ Examine

7-10 USE LOGICAL REASONING

Objective

Solve problems using logical reasoning.

Sean must work one more problem to complete his geometry assignment. He must decide which one of the following polygons does not belong in the group.

A   B   C   D

▶ **Explore**

What is given?
● four different polygons

What is asked?
● Which polygon does not belong?

▶ **Plan**

Look at the figures to decide what attributes they have in common, and what attributes distinguish one from the other three. Use logical reasoning.

▶ **Solve**

● All the figures are about the same size.
● All the figures are the same color.
● All the figures are polygons.
● Figures A, C, and D are regular polygons.
● Figure B is not a regular polygon.

So figure B does not belong in the group.

▶ **Examine**

By definition, a regular polygon is one in which all sides are congruent and all angles are congruent. Figure B is the only polygon that does not meet these requirements. So the answer is reasonable.

Guided Practice

Which one does not belong? Explain your answer.

1.

   a.   b.   c.   d.

2.

   a.   b.   c.   d.

3.

   a.   b.   c.   d.

4.

   a.   b.   c.   d.

Choose the letter of the best answer. Explain your answer.

5. shoe is to foot as right angle is to
    a. square   **b.** circle   **c.** pentagon   **d.** triangular prism

6. hot is to cold as quadrilateral is to
    a. square   **b.** decagon   **c.** parallelogram   **d.** circle

7.

 is to as is to

  a.   **b.**   **c.**   **d.**

Solve. Use any strategy.

8. Two adjacent elevators are on the same floor. The first elevator goes up 4 floors and down 7. The second elevator goes down 11 floors and up 9. How many floors apart are the elevators when they last stopped?

9. Mr. Harding tells his students that he is thinking of a number. If you multiply the number by 5, subtract 15, and add 4, the result is 44. Find the number.

f UN with MATH
When and where was the wheel first used?
See page 240.

10. Chad wants to take a vacation after he graduates. In order to visit three friends, he drives 300 miles east, 100 miles northwest, 300 miles west, and then returns home. Name the geometric figure that represents his path.

11. What value of b makes $6^2 + b^2 = 10^2$ true?

12. The number 16 is to 256 as 35 is to $\underline{\ ?\ }$.

13. These are. . . .   These are not. . . .   Which of these are?

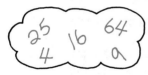      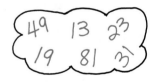

Critical Thinking

14. Symmetry means that a line can be drawn on an object so that the shape on one side of the line matches the shape on the other or that the shapes are congruent. Print all the letters of the alphabet as capital letters. How many letters have a vertical line of symmetry?

Mixed Review

Lesson 4-10

Add.

15. $\frac{3}{5} + \frac{4}{5}$ **16.** $\frac{2}{3} + \frac{3}{4}$ **17.** $\frac{3}{5} + \frac{4}{7}$ **18.** $\frac{6}{8} + \frac{1}{4}$

Lesson 6-4

Complete.

19. $42g = \underline{\ ?\ } kg$ **20.** $9.5 g = \underline{\ ?\ } mg$ **21.** $0.25 kg = \underline{\ ?\ } g$

Lesson 7-3

22. Recall that two angles are complementary if the sum of their measures is 90°. Find the measure of an angle if it is 50° less than that of its complement.

REVIEW

Vocabulary/ Concepts

Choose the word or number from the list at the right to complete each sentence.

1. A ? is formed by two endpoints and the straight path between them.

2. Two rays with a common endpoint form an ?.

3. An ? angle measures between 0° and 90°.

4. Lines in the same plane that do not intersect are ?.

5. An ? triangle has at least two congruent sides.

6. A ? is an exact location in space.

7. An equilateral triangle has three ? sides.

8. A ? is a quadrilateral with four congruent sides and four congruent angles.

9. The faces of a polyhedron intersect to form ?.

acute
angle
congruent
edges
isosceles
line segment
parallel
plane
point
scalene
square

Exercises/ Applications

**Lessons 7-1
7-2
7-3**

Make a drawing of each figure.

10. ray *AB*

11. line *MN*

12. line segment *RS*

13. $\overleftrightarrow{BC}$

14. two skew lines

15. $\angle DEF$

16. $\overline{HJ}$

17. $\overleftrightarrow{OP} \parallel \overleftrightarrow{XY}$

18. $\overleftrightarrow{RD} \perp \overleftrightarrow{MN}$

**Lessons 7-2
7-3**

Find the measure of each angle. Then classify each angle as right, acute, *or* obtuse.

19.

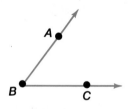

20.

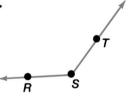

21.

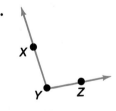

Lesson 7-4

Trace each figure. Construct a figure congruent to each figure given.

22.

23.

Lesson 7-5 *State whether each pair of lines is parallel, perpendicular, or skew. Use symbols to name all parallel and perpendicular lines.*

24.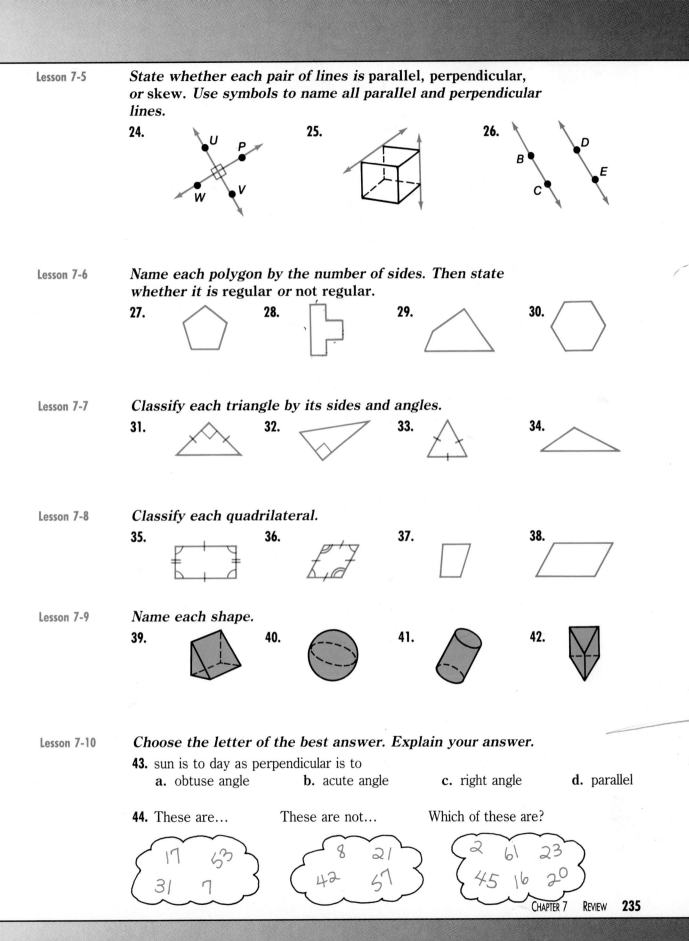

25.

26.

Lesson 7-6 *Name each polygon by the number of sides. Then state whether it is regular or not regular.*

27.   **28.**   **29.**   **30.**

Lesson 7-7 *Classify each triangle by its sides and angles.*

31.   **32.**   **33.**   **34.**

Lesson 7-8 *Classify each quadrilateral.*

35.   **36.**   **37.**   **38.**

Lesson 7-9 *Name each shape.*

39.   **40.**   **41.**   **42.**

Lesson 7-10 *Choose the letter of the best answer. Explain your answer.*

43. sun is to day as perpendicular is to
    a. obtuse angle   **b.** acute angle   **c.** right angle   **d.** parallel

44. These are...   These are not...   Which of these are?

   17   53   8   21   2   61   23

   31   7   42   57   45   16   20

TEST

Name each figure. Use symbols if possible. Be as specific as possible.

1.

2.

3.

4.

5.

Find the measure of each angle. Then classify each angle as right, acute, or obtuse.

6.

7.

8.

Trace each figure. Construct a figure congruent to each figure given.

9.

10.

Trace each line segment. Bisect each line segment.

11.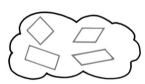

12.

State whether each pair of lines is parallel, perpendicular, or skew. Use symbols to name all parallel and perpendicular lines.

13.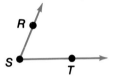

14.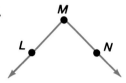

15.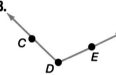

Classify each triangle by its sides and then by its angles.

16.

17.

18.

Solve. Use any strategy.

19. The number 3,000,000 is to 3×10^6 as 4×10^5 is to ___?___ ?

20. These are . . .   These are not . . .   Which of these are?

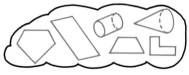

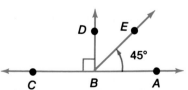

▶ BONUS: Use the figure at the right to explain why m∠DBE is 45°.

CONSTRUCTING TRIANGLES

You can use a compass and straightedge to construct triangles that are congruent to other triangles.

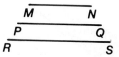

Construct a triangle congruent to a given triangle.

Given:

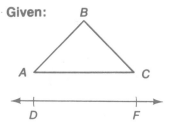

1. Draw a line. Construct a line segment congruent to $\overline{AC}$. Label it $\overline{DF}$.

2. With compass open to the length of $\overline{AB}$, draw an arc from point D. With compass open to the length of $\overline{BC}$, draw an arc from point F.

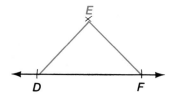

3. Label the intersection of the arcs E. Draw $\overline{DE}$ and $\overline{EF}$.

1. Trace △ABC. Construct a triangle congruent to △ABC.

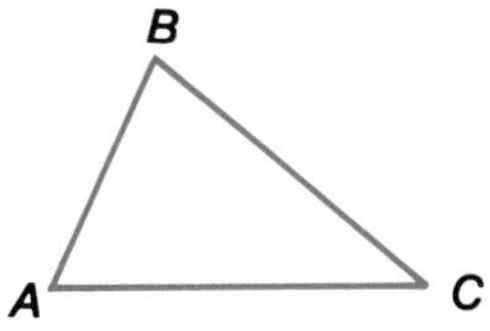

2. Trace △JKL. Construct a triangle congruent to △JKL.

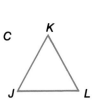

Use a straightedge to draw each triangle. Then construct a triangle congruent to each triangle you drew.

3. scalene triangle

4. isosceles triangle

5. Trace each line segment at the right. Then construct a triangle having sides the length of line segments *MN*, *PQ*, and *RS*.

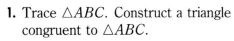

Free Response

Lesson 1-6

Estimate.

1. $3.568 + 9.237$ **2.** $86.2 + 39.92$ **3.** $0.42 + 1.7$

4. $17.325 - 11.758$ **5.** $541.2 - 38.7$ **6.** $9.6 - 0.12$

Lessons 2-1 2-3

Add or subtract.

7. $\begin{array}{r} 563 \\ +\,103 \end{array}$ **8.** $\begin{array}{r} 375 \\ -\ 25 \end{array}$ **9.** $\begin{array}{r} 251 \\ -\ 34 \end{array}$ **10.** $\begin{array}{r} 551 \\ +\,359 \end{array}$

Lesson 3-1

11. The cost of a 30-second commercial during Super Bowl XX was \$500,000. What was the cost of three 30-second commercials?

Lesson 3-4

12. Mr. Benitez receives a bonus of \$125 for each life insurance policy he sells. Last month he received a bonus of \$1,000. How many policies did he sell?

Lesson 4-6

Simplify each fraction.

13. $\frac{36}{40}$ **14.** $\frac{4}{28}$ **15.** $\frac{45}{105}$ **16.** $\frac{160}{224}$ **17.** $\frac{1,000}{40}$ **18.** $\frac{95}{38}$

Lessons 5-3 5-7

Multiply or divide.

19. $\frac{1}{2} \times \frac{1}{4}$ **20.** $\frac{7}{8} \times \frac{4}{3}$ **21.** $\frac{36}{40} \div \frac{45}{100}$ **22.** $\frac{15}{18} \div 2$

23. $20 \div \frac{1}{5}$ **24.** $7\frac{1}{5} \times 8\frac{1}{3}$ **25.** $12\frac{5}{6} \div 1\frac{3}{8}$

Lesson 5-8

Change each fraction to a decimal.

26. $\frac{6}{8}$ **27.** $\frac{17}{20}$ **28.** $\frac{1}{5}$ **29.** $\frac{3}{10}$ **30.** $\frac{60}{10}$ **31.** $\frac{18}{12}$

Lesson 6-6

32. In two throws, a discus thrower hurls a discus 218 feet and 226 feet. What is the total length of the two throws in yards?

Lesson 7-7

Classify each triangle by its sides and angles.

33.   **34.**   **35.**

Multiple Choice

Choose the letter of the correct answer for each item.

1. Round 8,079 to the nearest hundred.
  a. 8,000
  b. 8,070
  c. 8,080
  d. 8,100

Lesson 1-4

2. What is the product of 870 and 12?
  e. 2,610 **g.** 9,440
  f. 8,874 **h.** 10,440

Lesson 3-1

3. Find all the factors of 20.
  a. 2, 5
  b. 1, 2, 5
  c. 1, 2, 4, 5, 10, 20
  d. 0, 20, 40, 60, 80, . . .

Lesson 4-1

4. Akai ran a marathon in 3 h 48 min 10 s. Ross had a time of 4 h 27 min 35 s. How much faster than Ross was Akai?
  e. 21 min 25 s
  f. 39 min 25 s
  g. 1 h 21 min 25 s
  h. 1 h 39 min 25 s

Lesson 6-10

5. An average envelope weighs about 4 grams and a sheet of typing paper weighs 3 grams. How many sheets of paper can you mail in an envelope and keep the total weight below 30 grams?
  a. 7
  b. 6
  c. 8
  d. 5

Lesson 6-4

6. What is the result of dividing 22 by $\frac{9}{10}$?
  e. $\frac{9}{220}$
  f. $19\frac{4}{5}$
  g. $22\frac{4}{9}$
  h. $24\frac{4}{9}$

Lesson 5-7

7. What is the prime factorization of 20?
  a. 1×20 **c.** 2×5
  b. 4×5 **d.** none of these

Lesson 4-2

8. Estimate the sum of $\frac{7}{8}$ and $\frac{3}{4}$.
  e. 0
  f. 1
  g. 2
  h. 3

Lesson 4-10

9. What word describes two lines that intersect to form right angles?
  a. congruent
  b. parallel
  c. perpendicular
  d. similar

Lesson 7-5

10. Brian wanted to travel the shortest distance between each of his classes. Which of the following would not help him find the shortest distance?
  e. a map of his school
  f. the room numbers of his classes
  g. the school bell schedule
  h. the location of stairs

Lesson 6-7

fun with MATH

The wheel Sumer & Syria (Babylon/Assyria)	1800 BC	First compass China	AD 132
3500 BC	Geometry Babylon	· 300 BC	First seismograph China

What is a Bar Code?

A bar code is a set of binary numbers represented by black bars and white spaces. A wide bar or space stands for 1 and a thin bar or space stands for 0. The bar code below uses five elements (three bars and two spaces) for numbers only. The bar code below also begins with a five-element start code and ends with a stop code. This allows a scanner to read the whole code either forward or backward.

Start   1   5   0   Stop

MATH M·E·N·U

Egg-in-the-Toast

1 slice white bread   1 egg
3 tablespoons butter

Cut a round piece from the center of the bread with a 2½ inch cookie cutter. Melt half the butter in a skillet and put in the bread with the center cut out. Break the egg into a cup and slide it carefully into the hole in the bread. Cook over medium heat until egg is set and bread is nicely brown. Add remaining butter. Carefully turn over the egg-in-the toast. Brown the second side and serve.

We only see one side of the moon because it simultaneously makes one rotation on its axis and one revolution about Earth, every 28 days.

COMICS

CALVIN & HOBBES

JOKE!

Q: What did the acorn say when he finally grew up?

A: Geometry!

What is a carat anyway?

The carat is a unit of measure. It is equivalent to 200 milligrams. It is based on the weight of the carob seed which was once a weighing standard used by jewelers in Africa and the Middle East. The word *carat* is believed to come from an Arabic word meaning *bean* or *seed*.

TEASER

Building roads is expensive, so civil engineers try to make them as short as possible while maintaining safety. A new interstate highway passes by the small towns of Apple Junction and Hermit Cove. One access ramp is proposed with two-lane roads connecting it to both towns. Where should the access ramp be placed so the total length of road A to R to H is as short as possible? (Hint: The shortest distance between two points is a straight line.)

A — Apple Junction · 1 km · 4 km · R · Access Ramp · 2 km · H — Hermit Cove

(Flip Flop) Palindrome Riddles

1. Thigh ointment
   A: *LEG GEL*
2. A drink for tiny flies
   A: *GNAT TANG*
3. A hobo who tells untruths
   A: *RAIL LIAR*
4. Curves on a thread cylinder
   A: *SPOOL LOOPS*
5. Bedsheets
   A: *SLEEP PEELS*

QUIZ TIME

3 × 3 Magic Squares

A magic square is a square of numbers in which every row, column, and diagonal add up to the same total. In example A the magic number is 24. Copy and complete the remaining magic squares. Then try creating your own magic square and challenge a friend to complete it.

A
11	3	10
7	8	9
6	13	5

B
6	—	—
7	5	3
—	—	—

C
—	—	10
—	7	—
4	—	5

D
14	3	—
—	—	13
8	15	—

AREA AND VOLUME

Chapter 8

How big? How many?

Have you ever watched a house or an office building being built? Have you ever wondered who decides how large the building should be? Many factors determine the size of a building, such as the cost of construction or the size of the property the building occupies. Can you think of any other factors that affect the size of a building?

The Theatre Troupe at Central High is painting three walls for a backdrop for the fall play. The walls are 10 feet high. Two of the walls are 16 feet wide, and one wall is 22 feet wide. If one can of paint covers 250 square feet, how many cans of paint does the group need to paint the three walls?

ACTIVITY: Area of Irregular Figures

In this activity you will find the area of figures and shapes that do not have straight edges and square corners.

Materials: quarter-inch grid paper, pencil, straightedge, different colored pencils (optional)

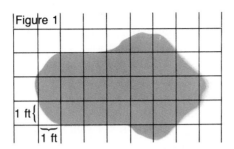
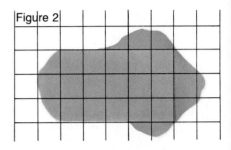

1. To find the area of an irregular shape, like the sand trap in Figure 1, separate it into regular and irregular parts.

2. In Figure 2, the brown rectangle encloses a regular part. What is the area of the rectangle in brown?

3. Estimate the area of the part of the sand trap shaded in orange by counting whole squares and partial squares. Estimate the area of the part of the sand trap shaded in blue.

4. Add the areas of the rectangle shaded in brown and the parts of the sand trap shaded in orange and blue. What is the approximate area of the whole sand trap?

Cooperative Groups

Work in groups of three or four.

5. Estimate the area of your hand in square inches. Then outline your hand on quarter-inch grid paper. Work in groups of three or four and find the area of each person's hand by separating the outline of the hand into regular and irregular shapes. Find the difference between your estimate and your result using the drawing.

Variations You Can Try

6. Have a contest to see which group can best estimate areas of irregularly shaped objects that have been selected by the groups. Use objects already in the classroom, such as bookends, a calculator, a paperweight, and so on.

7. Try estimating the area of some circular objects, such as a clock or a plate. Outline the object and find an approximate area.

Communicate Your Ideas

8. Discuss your methods for separating the outline of a figure into regular and irregular parts with other groups.

8-1 CIRCUMFERENCE OF CIRCLES

Objective

Find the circumference of circles.

Did you ever wonder how far you travel on a Ferris wheel ride? Katie is riding on a Ferris wheel and is sitting 19 feet from the center. If the wheel goes around nine times for one ride, how far does Katie travel?

First we need to find the circumference of the Ferris wheel. The distance around a circle is called the **circumference**. To find circumference, the diameter or the radius of the circle must be known.

A line segment through the center of a circle with endpoints on the circle is called a *diameter*.

The ratio of the measure of the circumference *(C)* of a circle to the measure of its diameter *(d)* is the same for all circles. The Greek letter π (pi) stands for this ratio. An approximation for π is 3.14.

$$\frac{C}{d} = \pi \rightarrow C = \pi \times d \text{ or } C = \pi d$$

The diameter *(d)* of a circle is twice the length of the radius *(r)*. Another formula for circumference can be found by substituting *2r* for *d*.

$$C = \pi \times 2 \times r \text{ or } C = 2\pi r$$

d = 2r

A line segment from the center of a circle to any point on the circle is called a *radius*.

Method

1. Determine if the diameter or the radius is given.

2. If the diameter is given, use $C = \pi d$.
   If the radius is given, use $C = 2\pi r$.

Example A

Find the distance that Katie travels in 9 rotations of the Ferris wheel. Use 3.14 for π. Round the answer to the nearest tenth.

1. The radius is given.

2. Use $C = 2\pi r$.
   $C \approx 2 \times 3.14 \times 19$
   $C \approx 119.32$
       or 119.3 feet

$\approx$ means is approximately equal to

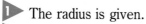
19 feet

Katie travels about 119.3 feet in one rotation. Multiply by 9 to find the total distance that Katie travels in one ride.
119.3 × 9 = 1,073.88 Katie travels *about* 1,073.9 feet in one ride.

Example B

Find the circumference of a circle whose diameter is 16 inches. Use 3.14 for π. Round the answer to the nearest tenth.

1. The diameter is given.

2. Use $C = \pi d$.
   $C \approx 3.14 \times 16$
   $C \approx 50.24$

16 in.

The circumference is about 50.2 inches.

The radius of a circle is given. Find the circumference. Use 3.14 for π. Round decimal answers to the nearest tenth.

Example A

1. 4 cm   **2.** 3.5 m   **3.** 13 in.   **4.** 240 mm

The diameter of a circle is given. Find the circumference. Use 3.14 for π. Round decimal answers to the nearest tenth.

Example B

5. 1 m   **6.** 6.7 ft   **7.** 51 cm   **8.** 23 in.

Find the circumference of each circle. Use 3.14 for π. Round decimal answers to the nearest tenth.

Practice

9. 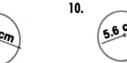 58 cm

10. 5.6 cm

11. 18 in.

12. 12 yd

Problem Solving

13. Shirley is making a tablecloth with lace trimming for a round table that has a diameter of 4 feet. She wants the tablecloth to extend 1 foot over the edge of the table all the way around. How much trimming does Shirley need to buy if she sews the trimming to the outside edge of the tablecloth?

14. Stephanie is painting a 3-foot tall strip of blue around a silo for background for advertising. One gallon of paint will cover 250 square feet. If the diameter of the silo is 55 feet, how many gallons of paint does Stephanie need?

3 feet

Mixed Review

Lesson 3-12

Find the average for each set of data. Round to the nearest tenth.

15. 16.2, 18.4, 19.1   **16.** 130.3, 132.7, 131.6, 128.5, 131.0

Lesson 7-8

Classify each quadrilateral.

17.   **18.**   **19.**   **20.**

Lesson 5-3

21. On a trip to Erie, Marshall used $\frac{3}{4}$ of the gasoline in the tank. If he started with $14\frac{1}{2}$ gallons, how many gallons did he have left?

8-2 AREA OF PARALLELOGRAMS

Objective

Find the area of parallelograms.

Four streets border an area of downtown Tulsa that is being renovated as a park. George Sands needs to find out the area of the new park to order sod. Main Street is 220 feet from Broad Street and Carver Street is 300 feet from Front Street. How many square feet of sod should Mr. Sands order?

You need to find the area of the park. You already know how to find the area of a rectangle. You can change a rectangle into a **parallelogram** that is not a rectangle by moving the triangle as shown below.

RECTANGLE

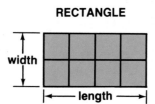

PARALLELOGRAM

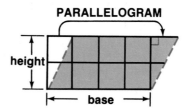

A *parallelogram* is a quadrilateral with two pairs of parallel sides. Explain how you know that the area of both figures is 8 square units.

Notice that the area of the rectangle is the same as the area of the parallelogram. So the area of any parallelogram can be found by multiplying the measures of the base *(b)* and the height *(h)*. The formula is $A = b \times h$ or $A = bh$.

Example A

Find the area of the new park.

$A = bh$ $b = 300, h = 220$

$A = 300 \times 220$

$A = 66,000$

The area of the park is 66,000 square feet. Mr. Sands should order 66,000 square feet of sod.

Guided Practice

Example A

Find the area of each parallelogram.

1.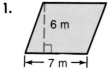
   6 m
   7 m

2. 5 in.
   12 in.

3.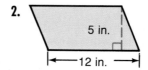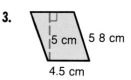
   5 cm   5 8 cm
   4.5 cm

4. base = 10 m
   height = 4.5 m

5. base = 8.5 cm
   height = 35 cm

6. base = 4 ft
   height = 12 ft

Practice

Find the area of each parallelogram.

7.

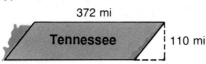

372 mi

Tennessee

110 mi

8.

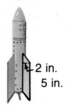

2 in.

5 in.

9.

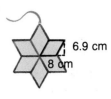

6.9 cm

8 cm

10. base = 23 m
height = 14 m

11. base = 1.5 yd
height = 3.7 yd

12. base = 10 m
height = 18.5 m

13. What is the area of a parallelogram with a base of 4.5 m and a height of 6.7 m?

14. If the height of a parallelogram is 43 inches and the base is 91 inches, what is its area?

Applications

15. The blue strip around the edge of Joan's quilt takes 48 parallelogram-shaped pieces. If the width of the strip is 8 in. and the length is 10 in., how many square inches of blue calico will Joan need? Ignore the extra material it will take to cut the pieces on the diagonal.

8 in.

10 in.

16. Which parallelogram has the larger area, parallelogram A or parallelogram B?

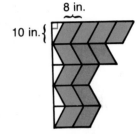

2 ft

6 ft   3 ft   A   8 ft

B

17. The receivers of a communications satellite are 12 meters deep and 14.6 meters long. If there are 16 receivers on the satellite, how many square meters of receiver area does the satellite have?

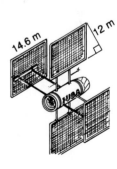

14.6 m   12 m

Show Math

18. Cut 6 one-inch squares from a piece of paper. Use the squares to show two different rectangles with an area of 6 square inches.

19. Use your squares from Exercise 18 to show as many different designs as possible. The squares must be adjacent.

Using Formulas

Solve each problem using the formula for the area of a parallelogram.

20. If the area of a parallelogram is 450 square centimeters and the base is 12.5 centimeters, what is the height of the parallelogram?

21. The height of a parallelogram is 7.4 feet. If the area of the parallelogram is 59.2 square feet, find the length of the base.

8-3 AREA OF TRIANGLES

Objective

Find the area of triangles.

Chou Lin needs 250 square feet or less of mainsail to qualify for the class C Hampshire races. The base of the sail is 11 feet, and the height of the sail is 21 feet. Does Chou Lin qualify for the class C race?

We can find the formula for the area of a triangle from the formula for the area of a parallelogram. Two **congruent** triangles form a parallelogram. The area of each triangle is one-half the area of the parallelogram.

The formula for the area of a parallelogram is $A = bh$. So the formula $A = \frac{1}{2} \times bh$ or $A = \frac{1}{2}bh$ can be used to find the area of a triangle.

Congruent triangles have the same size and shape. So they have the same area.

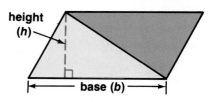

Example A

Find the area of Chou Lin's mainsail.

$A = \frac{1}{2}bh \qquad b = 11, h = 21$

$A = \frac{1}{2} \times 11 \times 21$

$A = 115.5$

To multiply by $\frac{1}{2}$ on a calculator, you can divide by 2.

The area of the mainsail is 115.5 square feet. Chou Lin qualifies.

Guided Practice

Example A

Find the area of each triangle.

1.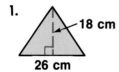
26 cm, 18 cm

2. 9 ft, 8 ft

3.
8 cm, 10 cm, 15 cm   8.

4. base = 7.5 ft
   height = 3 ft

5. base = 6.4 in.
   height = 3.5 in.

6. base = 2.1 m
   height = 1.5 m

7. base = 22 yd
   height = 10 yd

8. base = 14 mm
   height = 11 mm

9. base = 3.9 cm
   height = 0.5 cm

Find the area of each triangle.

10.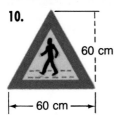
60 cm

60 cm

11.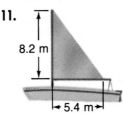
8.2 m

5.4 m

12.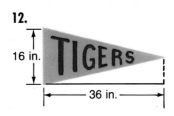
16 in.

36 in.

13. base = 17 cm
height = 6 cm

14. base = 7.4 m
height = 12 m

15. base = 3.5 in.
height = 2.6 in.

16. base = $2\frac{1}{2}$ ft

height = $1\frac{1}{4}$ ft

17. base = 21 mm
height = 32 mm

18. base = $1\frac{1}{3}$ yd

height = $2\frac{2}{3}$ yd

19. Find the area of a triangle whose base is 4.5 meters and whose height is 7 meters.

20. If the base of a triangle is 52 feet and the height is 21 feet, what is its area?

Applications

UN with MATH
Has a kite been used for military reconnaissance? See page 308.

21. Jake is replacing the siding on a triangular portion of his house that was damaged by high winds. How many square feet does Jake need to replace?

8 ft

26 ft

22. Tuygen is building a kite for Sunday's competition. The cross-shaped supports are 6 feet long and 4 feet long. What will be the total area of the finished kite?

2 ft   2 ft

Show Math

23. Draw three triangles, one with an area of 8 square inches, one with an area of 16 square inches, and one with an area of 32 square inches.

24. What is the greatest number of right triangles you can cut from a 4-foot by 8-foot sheet of plywood if the triangle has a base of 2 feet and a height of 2 feet? Use grid paper to show your answer.

Mixed Review

Lesson 6-12

Solve each equation. Check each solution.

25. $x + 7 = 23$ **26.** $y - 14 = 89$ **27.** $t + 13.6 = 59.1$

Lesson 3-5

Multiply or divide.
28. 0.613×10^5 **29.** $426.5 \div 10^3$ **30.** 0.21×10^4

Lesson 6-12

31. As a weather front passed Wheeling, West Virginia on Friday, the temperature fell 12°F, then rose 9°F. If the temperature before the front passed was 36°F, what was the temperature after the front passed?

8-4 AREA OF CIRCLES

Objective

Find the area of circles.

As station manager at KCTY, Elena Simon tells potential advertisers that the listening area covers 10,560 square miles. If the tower can broadcast 58 miles in any direction, is Ms. Simon correct in her statement of the area covered by the broadcast? In order to answer this question, you need to find the area of a circle.

You can find a formula for the approximate area of a circle by starting with the formula for the area of a parallelogram. Suppose the circle is separated into parts. Then the parts are put together to form a figure that looks like a parallelogram. The circumference of the circle can be represented by C units, so the base of the parallelogram is $\frac{1}{2}C$ units. The height is r.

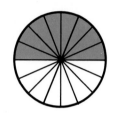

$A = b \times h$ area of parallelogram

$A = (\frac{1}{2}C) \times r$ Replace b with $\frac{1}{2}C$. Replace h with r.

$A = \frac{1}{2}(2\pi r) \times r$ Replace C with $2\pi r$.

$A = \pi \times r \times r$ Simplify.

$A = \pi r^2$

Method

1 If the diameter is given, multiply it by $\frac{1}{2}$ to find the radius.

2 Use $A = \pi r^2$ to find the area.

Examples

A *Find the area of a circle with a radius of 58 miles, KCTY radio's broadcast area.*

1 The radius is given.

2 $A = \pi r^2$ $r = 58$

$A \approx 3.14 \times 58 \times 58$

$A \approx 10,562.96$

Ms. Simon is correct that the broadcast covers 10,560 square miles.

B *Find the area of a circle with a diameter 32 inches long. Use 3.14 for π.*

1 The radius is $\frac{1}{2} \times 32$ or 16 in.

2 $A = \pi r^2$ $r = 16$

$A \approx 3.14 \times 16 \times 16$

$A \approx 803.84$

The area of the circle is *about* 804 in².

Guided Practice

Find the area of each circle whose radius is given. Use 3.14 for π. Round decimal answers to the nearest tenth.

Example A

1. 5 cm   **2.** 9 m   **3.** 20 yd   **4.** 30 mm

Example B

Find the area of each circle whose diameter is given. Use 3.14 for π. Round decimal answers to the nearest tenth.

5. 8 cm   **6.** 10 in.   **7.** 30 yd   **8.** 22 m

Find the area of each circle described below. Use 3.14 for π.
Round decimal answers to the nearest tenth.

9.
2 cm

10.
5 ft

11.
13 cm

12. radius 3.2 in.   **13.** radius, 9 yd   **14.** diameter, 6 cm

15. What is the area of a circle with a diameter of 19 cm?

16. *True* or *False*   The area of a circle with a diameter of 6.4 inches is 32.2 in².

Applications

17. Each lawn sprinkler at Markos, Inc. can reach 18 feet in any direction as it rotates. What is the area one sprinkler can water?

Problem Solving

18. Meg wanted to give her dog King as much space as possible in which to play. She has enough fencing to make a square pen that would have 6 feet on each side. She can also put King on a 6-foot chain. Will King have more room to play in the pen or on the chain? How much more?

19. The circular flower garden in Ascott Gardens, Buckinghamshire, has a radius of 23 feet. If half of the area of the garden is planted with evergreens and the other half is planted with annuals, how many square feet has been planted with evergreens?

20. If the area of a compact disc is *about* 19.6 square inches, what is the diameter of the disc?

Critical Thinking

21. Draw as many four-sided figures as possible using any of the five points on the circle as vertices.

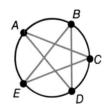

Calculator

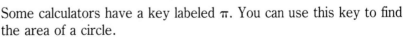

Some calculators have a key labeled π. You can use this key to find the area of a circle.

Find the area of a circle with a radius of 3 m.

 3 2 28.274334

The area is *about* 28.3 m².

Use a calculator to find the area of each circle whose radius is given. Round decimal answers to the nearest tenth.

22. 6 mm   **23.** 56 in.   **24.** 49.8 cm   **25.** 76.83 yd

8-5 INSTALLING CARPET

Objective

Find the cost of carpeting a room.

Diana Shultz is a designer and sales associate for The Carpetman. Mrs. Shultz arranges for installation, and computes the number of square yards and the cost of the order.

To compute the cost of carpet for a room that measures 4 yards by 6 yards, Mrs. Shultz finds the area of the floor in square yards and multiplies by the price per square yard.

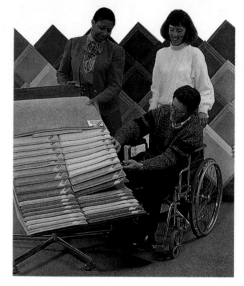

4 yd

6 yd

$A = \ell w$

$4 \times 6 = 24$

The area is 24 square yards.

Then multiply the area by the price per square yard.

$$24 \boxed{\times} 16.95 \boxed{=} 406.8$$

The cost of the carpet is $406.80.

Understanding the Data

- What are the dimensions of the room?
- What is the cost of the carpet per square yard?

Guided Practice

Find the total cost of carpeting each room pictured below. The cost per square yard of the carpeting is given.

1.

3 yd

6 yd

$13.50 per square yard

2.

12 ft

15 ft

$15.75 per square yard

(Hint: Change the length and width to yards before using $A = \ell w$

3.

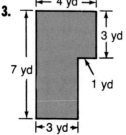

|← 4 yd →|

3 yd

7 yd

1 yd

|←3 yd→|

$19.00 per square yard

(Hint: Find the area of each rectangle and add to find the total area.)

4.

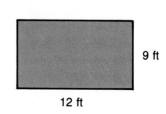

9 ft

12 ft

$22.00 per square yard

Mrs. Shultz is ordering the carpet for the Hernandez's new house. The floor plan is shown at the right. *Use the floor plan to answer Exercises 5-9.*

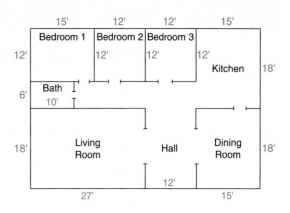

5. Mrs. Hernandez is carpeting the living room, dining room, and hall in the same color and style. The beige plush carpet costs $23.95 per square yard. What is the cost of the carpet?

6. Mrs. Hernandez has chosen the same sculptured style carpet for the three bedrooms, each in a different color, but at the same price. How much will it cost to carpet the three bedrooms if the carpet costs $21.95 per square yard?

7. The kitchen carpet is an indoor-outdoor carpet that costs $14.95 per square yard. How much will the kitchen carpet cost?

8. The bathroom carpet is a washable shag that comes on a roll 6 feet wide. The cost of the carpet is $13.95 for a length of carpet 1 yard long. How much will the carpet for the bathroom cost?

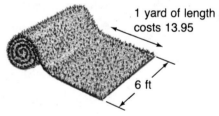

1 yard of length costs 13.95

6 ft

9. What is the total cost of carpeting the Hernandez's new house using the dollar amounts from Exercises 5–8?

Suppose

10. Suppose The Carpetman pays a wholesale price of $12.95 per square yard and charges $23.95 per square yard for plush carpet. How much profit does the store make when carpeting a room 4 yards by 6 yards?

Mixed Review

Lesson 1-7

Estimate.

11. 45×9 **12.** 29.7×4.5 **13.** 33×23.6 **14.** 509×1.8

Lesson 7-5

15. Name all the line segments that are parallel to segment AB in the cube shown at the right.

Lesson 6-6

16. Mr. Harold has a grandfather clock that chimes the number of the hour at the beginning of every hour and once every half hour. Ten minutes after Mr. Harold comes home, he hears one chime, a half hour later he hears one chime, and a half hour later he hears one chime. Between what two times did Mr. Harold come home?

▶ Explore
▶ Plan
▶ Solve
▶ Examine

8-6 MAKING A DIAGRAM

Objective
Solve problems by making a diagram.

Three trains run on different schedules. The Express leaves the Fortieth Street station every 10 minutes, the Local leaves every 15 minutes, and the Commuter leaves every 20 minutes. If all three trains left the Fortieth Street station at noon, when will all three trains leave the station together again?

▶ **Explore**

What is given?

● One train runs every 10 minutes.
● One train runs every 15 minutes.
● One train runs every 20 minutes.
● Three trains left at noon from the same station.

What is asked?

● When will all three trains again leave Fortieth Street station at the same time?

▶ **Plan**

Make a diagram to represent the three trains and the times they leave the station. The solution will be the time when all three trains are represented at the same time.

▶ **Solve**

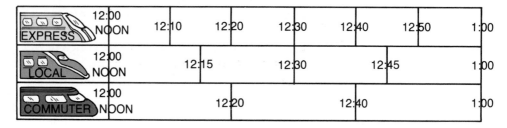

The trains will leave Fortieth Street station together at 1:00 P.M.

▶ **Examine**

The three trains leave together on the hour. The LCM of 10, 15, and 20 is sixty.

━━ **Guided Practice** ━━

Solve by making a diagram.

1. The advisor of the Freshman Class informs three students about a meeting. It takes 1 minute for her to tell three students and one minute for each of those three students to tell three other students, and so on. How many students will know about the meeting in three minutes?

Paper/Pencil
Mental Math
Estimate
Calculator

2. Joan and Sally share a dorm room at college and both alarms are set for 6:30 A.M. Suppose Joan's snooze alarm is set to ring every 4 minutes while Sally's rings every 10 minutes. If they both forget to turn off their alarms and leave at 6:45 A.M. how many times will the alarms ring at the same time before they return at 9:00 A.M.?

3. Tim divides his entire stamp collection equally among Sarah, Scott, Dekker, and Akiko. Akiko gives her brother and sister each one third of her stamps. If they each received 5 stamps, how many stamps did Tim have in his collection?

4. Sixteen students in a math class are standing. If one fourth of these students sit down, one third of those still standing sit down, and 3 rise from their seats, how many are standing?

*f*UN with MATH

Who was the first known woman doctor?
See page 308.

5. If three out of five doctors recommend walking instead of jogging, how many doctors out of 600 would probably recommend jogging?

6. Julie packs a red blouse, white blouse, black slacks, gray slacks, a blue blazer, and a black blazer for the weekend. Make a list to find how many different outfits she can make consisting of a blazer, slacks, and a blouse.

7. Kevin's beginning checking account balance is $197.53. If he writes checks for $21.35, $15.87, and $36.22, how much must he deposit in his account to bring the balance up to $200? Name the method of computation used.

Critical Thinking

8. Name a year in the twentieth century that reads the same backward and forward. Numbers such as this are called palindromes.

Mixed Review

Lesson 2-4

Subtract.

9. $28 − $4.52

10. 8.97 − 3.66

11. 43.087 − 21.98

Lesson 7-1

Use symbols to name each figure.

12.

13.

14.

15.

Lesson 8-1

16. Jackie rides her horse around a circular pen. The distance across the pen is 50 feet. How many feet will she ride her horse if she rides around the pen twenty times?

BASIC PROGRAMS

A **computer program** is a list of numbered statements that gives instructions to a computer. The number given to each statement is called a **line number.** Multiples of 10 are often used for line numbers so statements can be inserted later if necessary.

```
10 READ A,B
20 LET C=2*A+B↑3
30 PRINT C
40 DATA 5,2
50 END
RUN
18
```

Consider the program, RUN command, and output for the program shown at the right.

The **READ statement** (line 10) assigns values from the **DATA statement** (line 40) to the variables A and B. After line 10 is performed, the value of A is 5 and the value of B is 2. Every program that has a READ statement must have a DATA statement.

The **LET statement** (line 20) tells the computer to assign the value of the expression on the right to the variable on the left. For example, in the program above, C is assigned the value of $2*5+2↑3$ or 18.

The **PRINT statement** (line 30) tells the computer to print the value assigned to C. The output for the program above is 18.

The **END statement** (line 50) tells the computer that the program is complete. The **RUN command** tells the computer to execute the program. It has no line number.

For each READ and DATA statement, state the value that the computer will assign to each variable.

1. 10 READ M
    20 DATA 3

2. 10 DATA 1.2,7
    20 READ X, Y

3. 30 READ A,B,C
    95 DATA 3.2,
        0.6, 5

Determine the output for each PRINT statement.

4. 10 LET X=3.8
    20 LET Y=6.81
    30 PRINT X+Y

5. 10 READ N
    20 LET S=5*N−6
    30 PRINT S
    40 DATA 27.3

6. 10 LET X=3+5
    20 LET Y=2
    30 LET Z=X↑Y
    40 PRINT Z

PYTHAGOREAN THEOREM

The longest side of a right triangle is called the **hypotenuse.** To find the length of the hypotenuse when you know the lengths of the other two sides, you can use the **Pythagorean Theorem.**

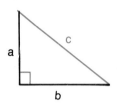

$c^2 = a^2 + b^2$, where c is the measure of the hypotenuse, and a and b are the measures of the other two sides

Find the length of the hypotenuse of the triangle at the right.

$c^2 = a^2 + b^2 \qquad a = 9, b = 12$

Enter: $9 \; \boxed{x^2} \; \boxed{+} \; 1 \; 2 \; \boxed{x^2} \; \boxed{=} \; 225$

The display shows the sum of a^2 and b^2 or c^2.

Use the square root key to find the **square root** of 225.

Enter: $\boxed{\sqrt{x}} \; 15$

The display shows the square root of 225.

The hypotenuse is 15 m long.

Find the length of the hypotenuse.

1.
4 in.
3 in.

2.
15 yd
8 yd

Solve.

3. If the foot of a ladder is 10 feet from the wall, how high up on the wall does a 50-foot ladder reach?

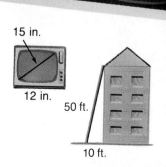
15 in.
12 in.
50 ft.
10 ft.

4. If a television screen measures 15 inches diagonally and 12 inches across the bottom, how high is the screen?

8-7 SURFACE AREA OF RECTANGULAR PRISMS

Objective

Find the surface area of rectangular prisms.

The Home Economics club is covering tissue boxes with quilted fabric to sell at their craft fair. How much fabric does Sheila need to cover a tissue box if the box is $9\frac{1}{2}$ inches long, $4\frac{1}{2}$ inches wide, and 4 inches high?

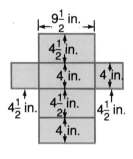

Sheila finds the **surface area** of the box by adding the areas of the faces.

Method

▶1 Use the formula $A = \ell w$ to find the area of each face.

▶2 Add the areas to find the surface area.

Example A

Find the surface area of the tissue box.

▶1
Front	$9\frac{1}{2} \times 4$	$=$	38
Back	$9\frac{1}{2} \times 4$	$=$	38
Top	$9\frac{1}{2} \times 4\frac{1}{2}$	$=$	$42\frac{3}{4}$
Bottom	$9\frac{1}{2} \times 4\frac{1}{2}$	$=$	$42\frac{3}{4}$
Right side	$4 \times 4\frac{1}{2}$	$=$	18
Left side	$4 \times 4\frac{1}{2}$	$=$	$+ 18$

▶2 $\overline{197\frac{1}{2}}$

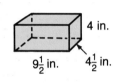

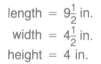

length = $9\frac{1}{2}$ in.
width = $4\frac{1}{2}$ in.
height = 4 in.

Sheila needs $197\frac{1}{2}$ square inches of fabric to cover a box.

Guided Practice

Find the surface area of each figure by adding the areas.

Example A

1.
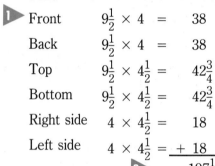

2.
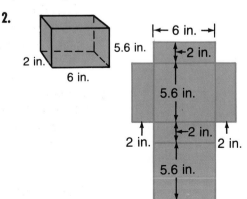

Practice

Find the surface area of each rectangular prism.

3.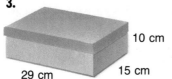
10 cm
29 cm   15 cm

4.
25 cm
15 cm   2 cm

5.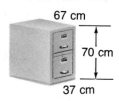
67 cm
70 cm
37 cm

6. length = 8 cm
   width = 4 cm
   height = 3 cm

7. length = 20 in.
   width = 8.3 in.
   height = 15 in.

8. length = 2.7 ft
   width = 6.5 ft
   height = 12 ft

9. What is the surface area of a carton with a length of 45 cm, a width of 22 cm, and a height of 25 cm?

Applications

10. One box is 10 in. long, 4 in. wide, and 8 in. high. Another box is 10 in. long, 6 in. wide, and 6 in. high. Which box has the greater surface area?

11. Frank is building three wooden storage cubes for his stereo system. Each cube will be $2\frac{1}{2}$ feet long, $1\frac{1}{2}$ wide, and 1 foot high. How many square feet of plywood does Frank need for three cubes if the front of each cube is left open?

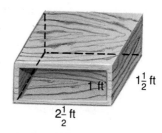
1 ft
$1\frac{1}{2}$ ft
$2\frac{1}{2}$ ft

Write Math

12. In your own words, explain the change in the surface area of a rectangular prism if you double the length, double the width, and double the height.

Cooperative Groups

13. In groups of three or four, use quarter-inch grid paper and draw a flat pattern that when cut out, folded, and taped will form a rectangular prism with a surface area of 88 square inches. Use 4 squares on the grid to represent 1 square inch.

Critical Thinking

14. A fruitbowl at the salad bar is full of apples and oranges. After 8 apples are eaten, 2 oranges remain for each apple. Then 12 oranges are eaten and 2 apples remain for each orange. How many apples were in the bowl before any were eaten?

Using Variables

Find the value of each expression if V = 24, ℓ = 2, w = 3, and h = 4.

15. $V \div (\ell \times w)$

16. $(\ell \times w \times h) \div V$

8-8   SURFACE AREA OF CYLINDERS

Objective

Find the surface area of cylinders.

Mr. Brooke's fourth grade class is making banks to sell at the school fair. The students cover clean, empty coffee cans with fabric. Mr. Brooke cuts the slot in the top. How much fabric is needed to cover one can if the radius is 2 inches and the height is 5 inches? You need to find the surface area of the coffee can.

The surface area of a cylinder is the sum of the area of its two circular bases and the rectangle that forms its curved surface. The length of the rectangle is equal to the circumference of each circle.

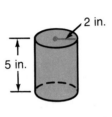
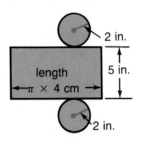

The radius of each circle is 2 in. So the diameter is 4 in. The circumference is πd or $\pi \times 4$.

The curved surface, when flat, forms a rectangle.

Method

1 ▶ Use the formula $A = \pi r^2$ to find the area of each circular base.

2 ▶ Use the formula $A = \ell w$ to find the area of the curved surface.

3 ▶ Add the areas to find the surface area.

Example A

Find the surface area of the coffee can. Use 3.14 for π.

1 ▶
$$A = \pi r^2$$
$$A \approx 3.14 \times 2^2$$
$$A \approx 3.14 \times 4$$
$$A \approx 12.56$$

2 ▶
$$A = \ell w$$
$$A = \pi \times 4 \times 5$$
$$A \approx 3.14 \times 20$$
$$A \approx 62.8$$

3 ▶
$$\begin{array}{r} 12.56 \\ 12.56 \\ + \; 62.80 \\ \hline 87.92 \text{ in}^2 \end{array}$$

The surface area of the coffee can is about 88 square inches.

Guided Practice

Find the surface area of each cylinder by adding the areas of the curved surface and each base. Use 3.14 for π. Round decimal answers to the nearest tenth.

Example A

1.

2.

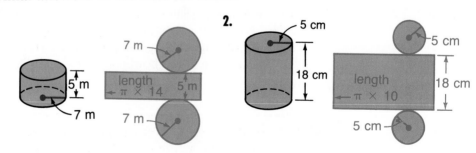

Practice

Find the surface area of each cylinder. Use 3.14 for π.
Round decimal answers to the nearest tenth.

3.

6 cm   20 cm

4.

4 cm

4 cm

5.

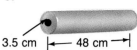

3.5 cm   48 cm

6. radius = 3.3 cm
height = 10 cm

7. radius = 6 in.
height = 16 in.

8. diameter = 19 m
height = 10.5 m

9. One can has a radius of 5 cm and a height of 20 cm. Another can has a radius of 10 cm and height of 5 cm. Which can has more surface area? Use 3.14 for π.

Applications

10. The curved surface and top of each of Hawley's 3 cylindrical water towers are being painted. If the diameter of a tower is 30 feet and the height is 70 feet, how many gallons of paint are needed to paint 3 towers? One gallon covers 250 square feet.

11. Jenny orders the aluminum sheets from which Canco stamps the parts of aluminum cans for soft drinks. If the cans are 12 cm high and have a diameter of 6 cm, how many cans can be stamped from a sheet of aluminum that is 1,000 cm long and 200 cm wide? Use 3.14 for π. Subtract 1 can from your answer for scrap area around tops and bases of cans.

Critical Thinking

12. Mike's archery target has the scoring circles shown at the right. If Mike's score is 65 with four shots, write three different ways he could have scored 65.

5
10
20
30
50

Show Math

13. Draw and label a diagram of a cylinder with a height of 11 feet that has a surface area of *about* 395 square feet.

VOLUME OF RECTANGULAR PRISMS

Objective

Find the volume of rectangular prisms.

The inner dimensions of Hank's freezer are 2 feet long, 2 feet wide, and 3 feet high. How many cubic feet of space does Hank have in the freezer?

Hank needs to find the volume inside the freezer. *Volume* is the amount of space that a solid contains. Volume is measured in cubic units.

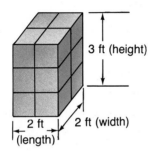

The volume *(V)* of a rectangular prism can be found by multiplying the measures of the length *(ℓ)*, width *(w)*, and height *(h)*. The formula is $V = \ell \times w \times h$ or $V = \ell wh$.

Example A

Find the volume of Hank's freezer.

$V = \ell wh$ $\ell = 2, w = 2, h = 3$
$V = 2 \times 2 \times 3$
$V = 12$ or 12 ft³

The volume of the freezer is 12 cubic feet.

Example B

Find the volume of the rectangular prism.

$V = \ell wh$ $\ell = 4, w = 3, h = 1.5$
$V = 4 \times 3 \times 1.5$
$V = 18$ or 18 cm³

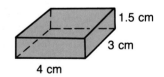

Guided Practice

Find the volume of each rectangular prism.

Example A

1.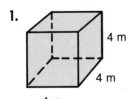
4 m
4 m
4 m

2.
35 cm
50 cm
60 cm

3.
11 in.
3 in.
9 in.

4. $\ell = 7$ in.
    $w = 2.5$ in.
    $h = 11$ in.

5. $\ell = 14$ ft
    $w = 15$ ft
    $h = 0.5$ ft

6. $\ell = 60$ in.
    $w = 5$ in.
    $h = 30$ in.

Practice

Find the volume of each rectangular prism described below.

7.

12 cm
45 cm
35 cm

8.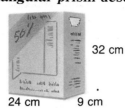

32 cm
24 cm   9 cm

9.

20 ft
54 ft   22 ft

fUN with MATH

*How well can you
concentrate on a
moving cube?*
See page 309.

Applications

10. $\ell = 8$ cm
    $w = 5$ cm
    $h = 6$ cm

11. $\ell = 8.5$ m
    $w = 3$ m
    $h = 9$ m

12. $\ell = 6$ in.
    $w = 2$ in.
    $h = 5.6$ in.

13. Each edge of a cube is 15 m long. What is the volume of the cube?

14. The dimensions of an aquarium are 20 inches long, 14 inches wide,
    and 18 inches high. How many cubic inches of water does it hold?

15. A water bed measures $6\frac{1}{2}$ feet long, 4 feet wide, and $\frac{3}{4}$ foot deep.
    How many cubic feet of water does it take to fill the bed?

16. If water weighs *about* 62 pounds per cubic foot, how much does
    the water for the bed in Exercise 15 weigh?

17. A rectangular greenhouse has a square base 64 feet on each side.
    The height of the ceiling is 12 feet. It takes a 6 horsepower
    pump to humidify 25,000 cubic feet and a 10 horsepower pump to
    humidify 50,000 cubic feet. Which pump is needed to humidify the
    greenhouse?

18. A rectangular carton holds 800 cubic inches of grass seed. It has
    a height of 10 inches and a width of 10 inches. How long is it?

Collect Data

19. In your classroom or home, find three different rectangular prisms
    and measure the length, width, and height. Then find their volume.

Using Data

Use the chart for Exercises 20-23.

20. The space shuttle *Discovery* has
    been used for several different
    missions. How many missions has
    Discovery completed?

21. How many defense and
    commercial missions has
    Discovery had?

22. What is the most frequent purpose
    for *Discovery's* missions?

Research

23. In the library find some of the
    actual missions on which
    Discovery has been sent and the
    results of those missions.

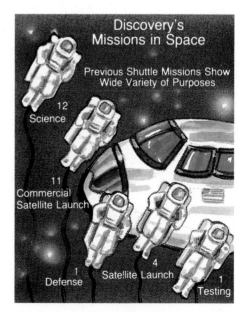

Discovery's
Missions in Space

Previous Shuttle Missions Show
Wide Variety of Purposes

12
Science

11
Commercial
Satellite Launch

1
Defense

4
Satellite Launch

1
Testing

8-10 VOLUME OF PYRAMIDS

Objective

Find the volume of pyramids.

Bob's tent has a square base 7 feet on each side. If the central (and only) tent pole is 6 feet high, how many cubic feet of space are there in the tent?

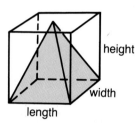

Bob's tent is shaped like a pyramid. The volume of a pyramid is one-third the volume of a prism with the same base and height as the pyramid. The formula is $V = \frac{1}{3}\ell wh$.

Example A

Find the volume of Bob's tent.

$V = \frac{1}{3}\ell wh$ $\ell = 7, w = 7, h = 6$

$V = \frac{1}{3} \times 7 \times 7 \times 6$ To multiply by $\frac{1}{3}$

$V = 98$ or $98\ ft^3$ on a calculator, you can divide by 3.

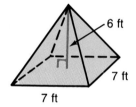

The volume of Bob's tent is about 98 cubic feet.

Example B

Find the volume of a pyramid with length 200 meters, width 150 meters, and height 200 meters.

$V = \frac{1}{3}\ell wh$ $\ell = 200, w = 150, h = 200$

$V = \frac{1}{3} \times 200 \times 150 \times 200$

$V = 2,000,000$ or $2,000,000\ m^3$

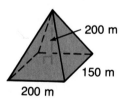

Guided Practice

Find the volume of each pyramid. Round decimal answers to the nearest tenth.

Example A

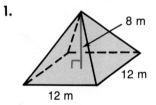

1. 8 m, 12 m, 12 m

2. 9 cm, 10 cm, 8 cm

3. 16 in., 15 in., 18 in.

Example B

4. $\ell = 11$ m
$w = 6$ m
$h = 13$ m

5. $\ell = 4$ cm
$w = 3$ cm
$h = 2.5$ cm

6. $\ell = 11$ yd
$w = 10.5$ yd
$h = 9.5$ yd

Exercises

Practice

Find the volume of each pyramid.

7.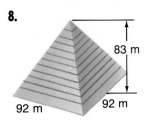

6 m

5 m   4 m

8.

83 m

92 m   92 m

9.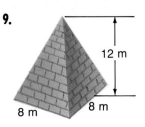

12 m

8 m   8 m

10. $\ell = 12$ m
$w = 6$ m
$h = 7$ m

11. $\ell = 4$ cm
$w = 5$ cm
$h = 2.4$ cm

12. $\ell = 9$ in.
$w = 10$ in.
$h = 12.3$ in.

Applications

$\int$ **UN with MATH**

Would you taste honey found in an Egyptian Pyramid?
See page 308.

13. The pyramid at the top of the Washington Monument has a square base 34 feet 5.5 inches on each side. The height of the pyramid is 55 feet. What is the volume of the pyramid to the nearest cubic foot?

14. Right Time Corporation produces a clock that is set in a clear acrylic pyramid. The clock has a square base 5 inches on each side and a height of 5 inches. It is shipped in a cube-shaped carton with inside dimensions of 5 inches by 5 inches by 5 inches. How many cubic inches of plastic filler are needed to fill the carton after the clock is placed inside?

Critical Thinking

15. If each dimension of a pyramid is doubled, how is the volume changed?

Using Equations

Solve for x, y, or z using the equation x = yz.

16. $x = 21$, $y = 7$, $z = \underline{\ ?\ }$

17. $y = 5$, $z = 3$, $x = \underline{\ ?\ }$

18. $z = 45$, $x = 180$, $y = \underline{\ ?\ }$

19. $x = 1.2$, $z = 6$, $y = \underline{\ ?\ }$

8-10   VOLUME OF PYRAMIDS   **265**

8-11 VOLUME OF CYLINDERS

Objective

Find the volume of cylinders.

Party Potato Chips packs their potato chips in cylindrical cartons. The new dip-style chip needs a container with *at least* 112 cubic inches of volume. Ginger Gomez's container design has a diameter of 4 inches and a height of 9 inches. Does the new container meet the requirement for volume?

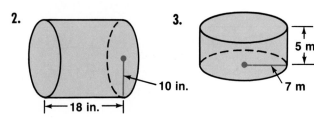

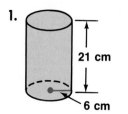

The volume of a cylinder is found by multiplying the measure of the area of the base by the height. Since the base is a circle, the area of the base is πr^2. So the formula is $V = \pi r^2 \times h$ or $V = \pi r^2 h$.

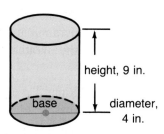

height, 9 in.

base

diameter, 4 in.

Example A

Find the volume of the container Ms. Gomez designed. Use 3.14 for π.

$$V = \pi r^2 h \qquad r = \frac{d}{2} \text{ or } 2, h = 9$$
$$V \approx 3.14 \times 2 \times 2 \times 9$$
$$V \approx 113.04 \text{ or } 113.04 \text{ in}^3$$

The volume of the cylinder is 113.0 cubic inches. Ms. Gomez's design meets the requirement for at least 112 cubic inches of volume.

Guided Practice

Practice

Find the volume of each cylinder. Use 3.14 for π. Round decimal answers to the nearest tenth.

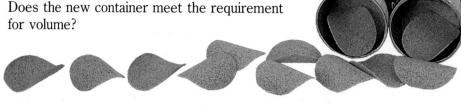

1.

21 cm

6 cm

2.

10 in.

18 in.

3.

5 m

7 m

4. radius = 2 in.
   height = 5 in.

5. diameter = 12 cm
   height = 12 cm

6. radius = 9 ft
   height = 5 ft

Practice

Find the volume of each cylinder. Use 3.14 for π. Round decimal answers to the nearest tenth.

7.
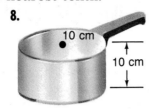
4 cm
12 cm

8.
10 cm
10 cm

9.
5 m
2 m

10. radius = 20 mm
height = 20 mm

11. radius = 19 in.
height = 10 in.

12. diameter = 14 cm
height = 17 cm

Applications

13. A storage silo on a farm has the shape of a cylinder with a diameter of 15 meters and a height of 20 meters. What is the volume of the silo?

14. Dottie's swimming pool has a diameter of 12 feet and a depth of 4 feet. If water costs $0.08 a cubic foot, by how much should Dottie expect her normal water bill to increase when she fills the pool?

Critical Thinking

15. At Mystic Seaquarium, the starfish moves up its rock 2 inches each day and down 1 inch at night. At this rate how many days will it take the starfish to reach the top if it is 15 inches tall?

16. Draw a diagram of a cylinder with a height of 4 inches and a radius of 1 inch. Use the diagram to show the change in the volume if the height is cut in half. Describe the volume if the height is cut in half.

JOURNAL ENTRY

17. Explain why it is always important to examine your answers when working with word problems.

Mixed Review

Trace each angle and extend the sides. Then find the measure of each angle.

Lesson 7-2

18.

19.

20.

Lesson 6-5

Complete.

21. 18 kL = $\underline{\ ?\ }$ L

22. 9.5 L = $\underline{\ ?\ }$ mL

23. 28 mL = $\underline{\ ?\ }$ kL

Lesson 8-2

24. The walkway to Samuel's house is in the shape of a parallelogram. The width of the walkway near the house is 3 feet. The distance from the house to the sidewalk is 12 feet. What is the area of the walkway?

3 ft
12 ft

8-12 VOLUME OF CONES

Objective

Find the volume of cones.

Steve works at the snow cone stand at Wyandot Park. The cone he fills with shaved ice is 6 inches tall and has a radius of 2 inches at the top. How many cubic inches of ice are needed to fill the cone?

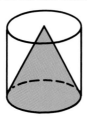

The volume of a cone is one-third the volume of a cylinder with the same radius and height as the cone. The formula is $V = \frac{1}{3}\pi r^2 h$.

Example A

Find the volume of the snow cone that Steve fills with shaved ice. Use 3.14 for π.

$V = \frac{1}{3}\pi r^2 h$ $r = 2, h = 6$

$V \approx \frac{1}{3} \times 3.14 \times 2 \times 2 \times 6$

$V \approx 25.12$ or 25.12 in³

To multiply by $\frac{1}{3}$ on a calculator, you can divide by 3.

The volume of the cone is *about* 25 cubic inches.

Guided Practice

Example A

Find the volume of each cone. Use 3.14 for π. Round decimal answers to the nearest tenth.

1.

12 cm

2 cm

2.

24 yd

20 yd

3.

9 m

4 m

4. radius = 8 m
    height = 6 m

5. radius = 2 mm
    height = 24 mm

6. diameter = 12 in.
    height = 10 in.

Find the volume of each cone. Use 3.14 for π. Round decimal answers to the nearest tenth.

Practice

7.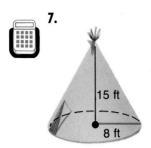

15 ft

8 ft

8.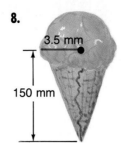

3.5 mm

150 mm

9.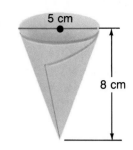

5 cm

8 cm

10. radius = 6 m
    height = 6 m

11. radius = 4 mm
    height = 24 mm

12. diameter = 5 in.
    height = 10 in.

Applications

13. If the top of a silo is cone-shaped and has a radius of 12 feet and a height of 18 feet, what is the volume of the cone-shaped portion of the silo?

14. Goody's Ice Cream Novelties makes an ice cream cone from a sugar cone filled with vanilla ice cream. Then the cone is dipped in chocolate. Each cone is 5 inches high and has a diameter of 2 inches across the top. Find the volume of the cone.

Problem Solving

15. A cone has a volume of 376.8 cubic centimeters. If the radius of the cone is 6 centimeters, what is the height of the cone?

Talk Math

16. How does the volume of a cone change if the height is cut in half? What happens to the volume of a cone if the height is doubled?

Critical Thinking

17. If four days before tomorrow is Sunday, what is three days after yesterday?

Cooperative Groups

18. In groups of three or four, use a clean, empty soup can or a cylindrical drinking glass, a piece of construction paper, and some sand or salt to prove that the volume of a cone is one-third the volume of a cylinder with the same height and base. Roll the construction paper into a cone shape with a sharp point. Place the point down into the can or glass. The base of the cone must have the same diameter as the top of the soup can. Mark the paper cone on the inside at the height of the soup can. Fill the cone with sand or salt and pour it into the can. How many conefuls does it take to fill the can? Explain your findings.

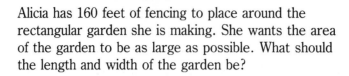

► Explore
► Plan
► Solve
► Examine

8-13 USING AREA AND VOLUME

Objective

Solve problems using area and volume.

Alicia has 160 feet of fencing to place around the rectangular garden she is making. She wants the area of the garden to be as large as possible. What should the length and width of the garden be?

 Explore

What is given?

● the perimeter of the garden, 160 feet

What is asked?

● With a perimeter of 160 feet, what width and length will make the largest area?

► **Plan**

Draw various rectangles with a perimeter of 160 feet. Then find the area of each rectangle.

If the rectangles have different areas, look for a pattern to determine the greatest area.

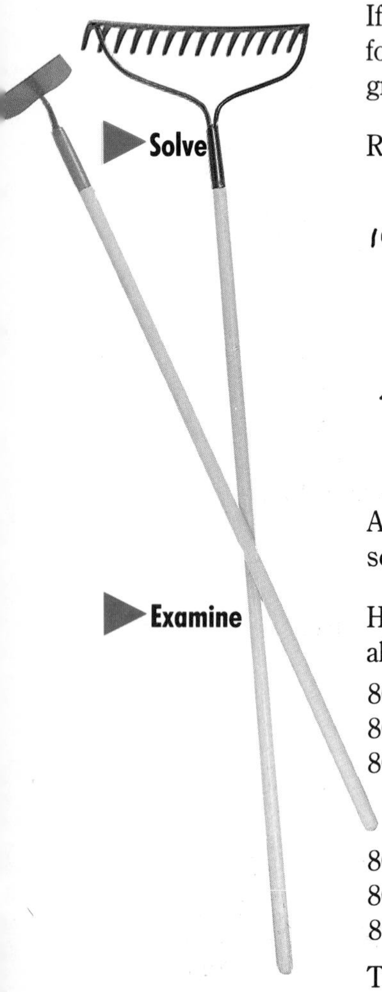

►**Solve**

Rectangles with a perimeter of 160 feet include:

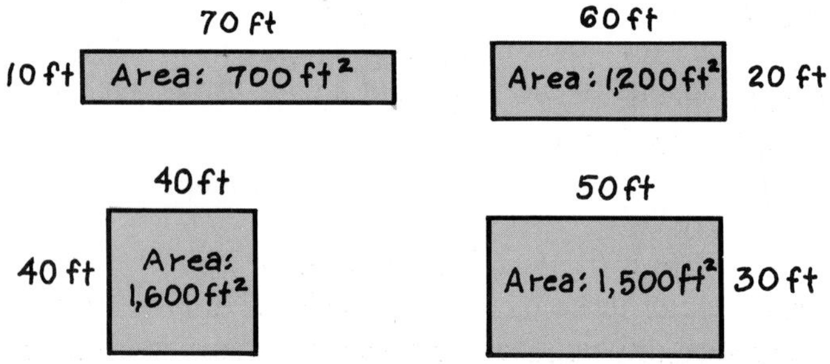

70 ft
10 ft | Area: 700 ft²

60 ft
Area: 1,200 ft² | 20 ft

40 ft
40 ft | Area: 1,600 ft²

50 ft
Area: 1,500 ft² | 30 ft

A 40-ft × 40-ft garden with 1,600 square feet of garden space seems to have the greatest area for 160 feet of fence.

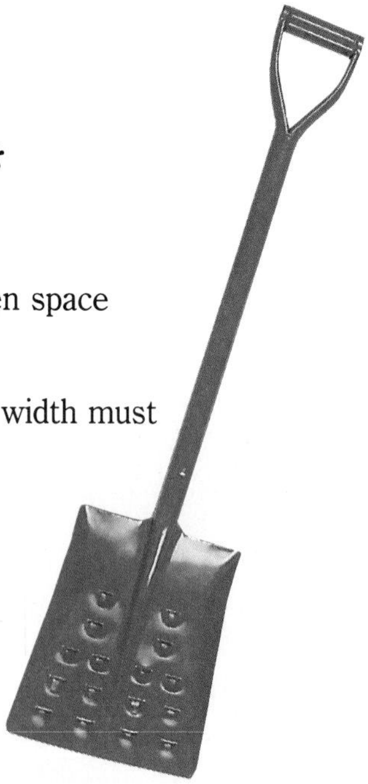

►**Examine**

Here's another way to solve the problem. Length plus width must always be 80 feet. But length times width varies.

$80 = 1 + 79$ $1 \times 79 = 79$
$80 = 2 + 78$ $2 \times 78 = 156$
$80 = 3 + 77$ $3 \times 77 = 231$
      ⋮   ⋮
$80 = 39 + 41$ $39 \times 41 = 1,599$
$80 = 40 + 40$ $40 \times 40 = 1,600$
$80 = 41 + 39$ $41 \times 39 = 1,599$

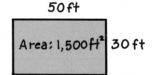

The greatest product is 1,600. So the answer is correct.

Solve. Round decimal answers to the nearest tenth.

1. Kayray's department store provides free delivery within a 60-mile radius of the store. What is the area covered by free delivery?

Solve. Round decimal answers to the nearest tenth.

2. Neil plants 12 tomato plants. Each plant will have 1 square foot of area. If the bed for the tomatoes is 2 feet wide, how long should the bed be?

Paper/Pencil
Mental Math
Estimate
Calculator

3. Find the area of the figure shown on the geoboard.

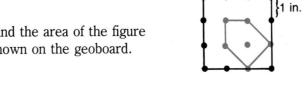
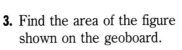

4. If it takes 25 centimeters of speaker wire to go around a spool once, what is the diameter of the spool?

5. Kate is making a circular strawberry patch three tiers high. The radius of the center circle is 1.5 feet, the radius of the inner circle is 3 feet, and the radius of the outer circle is 6 feet. If each tier is 0.5 feet high, how many cubic feet of garden soil should Kate buy to fill the tiers?

6. Draw a diagram of a 4-foot by 6-foot rectangle on a piece of paper. Choose 5 different vegetables or flowers to plant and separate the garden into 5 areas for each vegetable or flower. Find the area of each section of your garden and make sure that the five areas total the area of the whole garden.

PORTFOLIO

7. Select an assignment from this chapter that shows your favorite mathematical artwork.

Mixed Review

Lesson 4-9

Estimate.

8. $\frac{4}{5} + \frac{7}{8}$

9. $\frac{1}{6} + \frac{3}{7}$

10. $\frac{7}{8} - \frac{1}{4}$

11. $\frac{8}{15} - \frac{2}{9}$

Lesson 8-3

12. Eli has a garden that is triangular in shape. Eli wants to fertilize the garden with fertilizer that comes in bags that cover 500 square feet. How many bags of fertilizer does he need to buy?

49 ft

39 ft

REVIEW

Write the letter of the term that best completes each sentence.

1. The distance around a circle is the _?_.

2. Volume is expressed in _?_ units.

3. The ratio of the measure of the circumference of a circle to the measure of its diameter is _?_.

4. To find the area of a parallelogram you multiply the _?_ times the _?_.

5. Two _?_ triangles form a parallelogram.

6. To find the area of a circle you multiply _?_ times the _?_.

7. The surface area of a rectangular prism is the sum of the area of its _?_.

8. The volume of a rectangular prism is found by multiplying the measures of the _?_, _?_ and _?_.

9. The volume of a _?_ is one-third the volume of a rectangular prism with the same base and height.

10. The volume of a cylinder is found by _?_ the measure of the area of the base times the height.

a. base
b. circumference
c. congruent
d. cubic
e. cylinder
f. faces
g. height
h. length
i. multiplying
j. π
k. pyramid
l. radius
m. radius squared
n. rectangular prism
o. width

Exercises/
Applications

Lesson 8-1

Find the circumference of each circle. Use 3.14 for π. Round decimal answers to the nearest tenth.

11.
24 cm

12.
14 ft

13.
8 in.

Lessons 8-2
8-3
8-4

Find the area of each figure. Use 3.14 for π.

14.
6 cm
17 cm

15.
4 in.
6 in.

16.
9 cm
12 cm

17.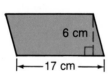
6 in.

18.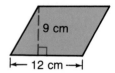
9 ft
8 ft

19.
64 cm

20. Find the total cost of carpet for a room that is 15 feet long and 12 feet wide if the carpet includng padding costs $22.95 a square yard?

21. Chang is reading a choose-your-own mystery book. At the end of the first chapter, there are two different chapter 2s from which to choose. At the end of each of the two different chapter 2s, there are two different chapter 3s from which to choose, and so on. How many different chapter 4s are there in the book?

Find the surface area of each rectangular prism or cylinder. Use 3.14 for π. Round decimal answers to the nearest tenth.

22.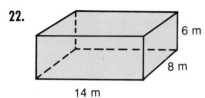
6 m
8 m
14 m

23.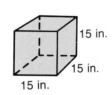
15 in.
15 in.
15 in.

24.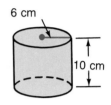
6 cm
10 cm

Find the volume of each rectangular prism, pyramid, cylinder, or cone. Use 3.14 for π. Round decimal answers to the nearest tenth.

25. rectangular prism
$\ell = 7$ cm, $w = 4$ cm, $h = 3$ cm

26. pyramid
$\ell = 8$ in., $w = 10$ in., $h = 9$ in.

27. cylinder
$r = 3$ ft, $h = 7$ ft

28. cone
$r = 11$ cm, $h = 12$ cm

29. rectangular prism
$\ell = 10$ ft, $w = 8$ ft, $h = 5$ ft

30. cylinder
$d = 4$ m, $h = 6$ m

31. Applegate Farm ships milk to market daily in a tanker truck that is 23 feet long and has a diameter of 10 feet. How many gallons of milk do they ship in a full tanker if there are about 7.5 gallons in a cubic foot?

32. Find the area of the piece of wood shown at the right.

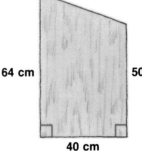

64 cm
50 cm
40 cm

TEST

Find the circumference of each circle. Use 3.14 for π. Round decimal answers to the nearest tenth.

1. $r = 4$ cm   **2.** $r = 15$ ft   **3.** $d = 22$ in.   **4.** $d = 6$ m   **5.** $r = 2.2$ m

Find the area of each figure. Use 3.14 for π. Round decimal answers to the nearest tenth.

6.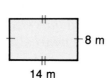
7.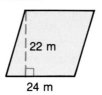
8.
9.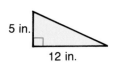

10. Jill is carpeting the kitchen with carpet that costs $15.95 per square yard. The kitchen is 10 feet long and 9 feet wide. How much does Jill pay for the carpet?

Find the surface area of each solid. Use 3.14 for π. Round decimal answers to the nearest tenth.

11.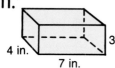
12.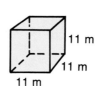
13.
14.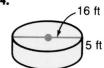

Find the volume of each solid. Use 3.14 for π. Round decimal answers to the nearest tenth.

15.
16.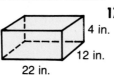
17.
18.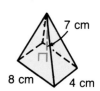

19. If Jane sends a letter to 4 cousins and the cousins each send letters to 4 friends, how many letters will have been sent in all when the 4 friends each send letters to 4 more friends?

20. How many cubic feet of concrete are needed for a driveway that is 14 feet wide, 50 feet long, and 4 inches deep?

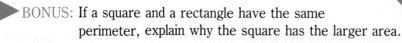

▶ BONUS: If a square and a rectangle have the same perimeter, explain why the square has the larger area.

TELEPHONE RATES

The phone company charges different rates for phone calls. The rates listed below are sample rates from a long distance company.

Dial-direct	Weekday full rate 8 A.M. to		Evening 35% discount 5 P.M. to		Night & weekend 60% discount 11 P.M. to 8 A.M.		Operator-assisted	
Sample Rates from City of Columbus, Oh. to:	First minute	Each additional minute	First minute	Each additional minute	First minute	Each additional minute	First three minutes	Each additional minute
Atlanta, Ga.	.53	.36	.34	.24	.21	.15	2.15	See dial-direct tables at left under time period that applies (weekday, evening or night & weekend).
Boston, Mass.	.53	.36	.34	.24	.21	.15	2.15	
Chicago, Ill.	.48	.34	.31	.23	.19	.14	2.05	
Denver, Colo.	.55	.38	.35	.25	.22	.16	2.25	
Houston, Tex.	.55	.38	.35	.25	.22	.16	2.25	
Philadelphia, Pa.	.50	.36	.32	.24	.20	.15	2.10	
St. Louis, Mo.	.50	.36	.32	.24	.20	.15	2.10	
Washington, D.C.	.50	.36	.32	.24	.20	.15	2.10	

Shannon lives in Columbus. On Thursday she calls her sister in Boston and talks from 8:00 P.M. to 8:15 P.M. What is she charged for this 15-minute call?

She finds the rate for an evening call to Boston.

Then she calculates the charge for the 15-minute call.

Dial-direct	Evening 35% discount	
Chicago, Ill.	.31	.23
Denver, Colo.	.35	.25
Detroit, Mich.	.30	.21

first minute
$0.34

next 14 minutes (15 − 1)
.24 ⊠ 14 ▭ 3.36

total charge
⊞ .34 ▭ 3.70

Shannon is charged $3.70 for the call.

Use the rate table above to find the charge for each call described below. Assume each call is from Columbus.

1. direct-dial to Houston, 11:00 P.M. to 11:09 P.M.

2. operator-assisted to Atlanta, 9:00 A.M. to 9:05 A.M.

3. Arlanda calls her brother in Philadelphia on Monday and talks from 5:30 P.M. to 6:15 P.M. What is she charged for the call?

4. Darrell calls his parents in Denver on Sunday. They talk from 10:04 A.M. to 10:25 A.M. How much is Darrell charged for the call?

Free Response

Lessons 1–11
1–12

Solve each equation. Check your solution.

1. $8n = 48$ **2.** $26y = 13$ **3.** $17 + m = 103$ **4.** $n - 3.5 = 2$

Lesson 2–3

Subtract.

5. $5,476 - 1,362$ **6.** $46 - 39$ **7.** $3,762 - 553$

Lesson 3–12

8. Bridget received the following test scores: 84, 93, 77, 89, 97, and 82. What is her average test score?

Lesson 4–8

Change each fraction to a mixed number in simplest form.

9. $\frac{118}{12}$ **10.** $\frac{250}{12}$ **11.** $\frac{94}{24}$

Lesson 5–9

Change each decimal to a fraction.

12. 0.07   **13.** 0.7   **14.** 0.55

Lesson 6–3

Complete.

15. $1,000 \text{ m} = \underline{?} \text{ km}$ **16.** $1 \text{ m} = \underline{?} \text{ cm}$

Lessons 6–7
6–8

Complete.

17. $3 \text{ ft} = \underline{?} \text{ in.}$ **18.** $8 \text{ ft} = \underline{?} \text{ yd}$

Lesson 7–5

State whether each pair of lines is parallel, perpendicular, or skew. Use symbols to name all parallel and perpendicular lines.

19. 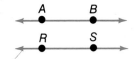   **20.** 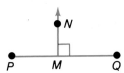   **21.**

Lesson 7–8

Classify each quadrilateral.

22.   **23.**   **24.**

Lesson 7–10

25. A certain number is increased by 8, multiplied by 3, divided by 9, and decreased by 6. The result is 4. What is the original number?

Lesson 8–3

26. Find the area of a triangle whose base is 8 cm and whose height is 4.5 cm.

Multiple Choice

Choose the letter of the correct answer for each item.

1. Reba has 100 magazines in her collection. Each magazine is 0.343 cm thick. If Reba stacks the magazines in a pile, how thick will the pile be?

  a. 0.00343 cm
  b. 3.43 cm
  c. 34.3 cm
  d. *none of the above*

Lesson 3-7

2. *About* how much is 8.982 divided by 2.3?

  e. 0.4
  f. 4
  g. 40
  h. 400

Lesson 1-7

3. Change $2\frac{5}{6}$ to an improper fraction.

  a. $\frac{16}{6}$ **c.** $\frac{16}{5}$
  b. $\frac{17}{6}$ **d.** $\frac{17}{5}$

Lesson 4-8

4. How many different outfits can be made from 2 pairs of pants, 3 shirts, and 5 ties?

  e. 5
  f. 10
  g. 25
  h. 30

Lesson 5-5

5. Estimate the quotient when 5 is divided by $\frac{3}{4}$.

  a. 1   **c.** 6
  b. 2   **d.** 20

Lesson 5-6

6. Cars were stalled bumper to bumper in two lanes for 4 miles as they approached the bridge. If cars occupy *about* 20 feet of length, *about* how many cars were stalled in both lanes?

  e. 2,640   **g.** 1,584
  f. 3,680   **h.** 2,112

Lesson 6-7

7. Do not use a protractor. Estimate the measure of angle *ABC*.

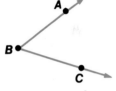

  a. *about* 10°
  b. *about* 50°
  c. *about* 100°
  d. *about* 150°

Lesson 7-2

8. Complete. 7 qt = $\frac{?}{}$ gal

  e. 1   **g.** 7
  f. $1\frac{3}{4}$ **h.** $2\frac{1}{2}$

Lesson 6-8

9. The formula of the area of a circle is $A = \pi r^2$. Find the area of the circle. Use 3.14 for π.

  a. 9.42 cm²
  b. 18.84 cm²
  c. 28.26 cm²
  d. 113.04 cm²

Lesson 8

10. Which of the following is equivalent to 1 kilogram?

  e. 100 mg   **g.** 100 g
  f. 1,000 mg   **h.** 1,000 g

Lesson 6-1

9 RATIO, PROPORTION, AND PERCENT

▶ **Win, Team, Win!**

What sports do you enjoy? Have you ever been on a team? Team standings are determined by comparing the number of games the team has won to the total number of games the team has played. Have you ever participated in a sports activity in which you kept track of your team's wins and losses?

The Jets have won 9 games and lost 7. The Rockets have won 6 games and lost 4. Which team has the better record?

ACTIVITY: Similar Figures

Suppose you don't draw very well but you want to enlarge a picture of your favorite cartoon. There is a way you can draw it and make it look like the original.

Materials: pencil, half-inch grid paper, ruler

$\frac{1}{4}$-inch grid

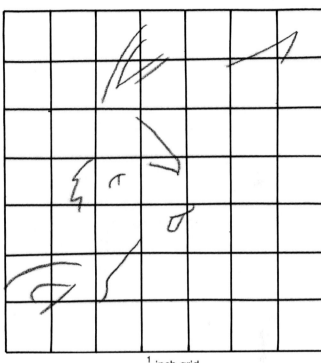

$\frac{1}{2}$-inch grid

Procedures

1. A quarter-inch grid has been placed over the cartoon. How many squares long and how many squares wide is the grid?

2. Make sure the half-inch grid has the same number of squares.

3. Study the drawings above and explain how the grids are used to enlarge the cartoon.

On Your Own

4. Now trace the larger grid and all the lines within each square. Then complete the drawing.

Cooperative Groups

Materials: a favorite cartoon, pencil, half-inch grid paper

5. Make an enlargement of your favorite cartoon. Decide what you need to do to get started.

6. Does your cartoon look like the original? If so, congratulations!

Communicate Your Ideas

7. What is the same about your drawing and the original? How is your drawing different from the original?

9-1 RATIO

Objective
Write ratios.

The Texas Rangers won 19 baseball games and lost 25 games.

You can compare their wins and losses by writing a ratio. A **ratio** is a comparison of two numbers by division. Ratios can be written three ways.

American League Western Division		
Club	**W**	**L**
California	22	26
Seattle	20	25
Texas	19	25
Oakland	13	33

19 to 25   19:25 $\frac{19}{25}$ Each ratio is read 19 to 25.
The Rangers' win-loss ratio is 19:25.

Method

▶1 Write the two numbers in the order you want to compare them.
▶2 Write ratios as fractions in simplest form.

Example A

Write the win-loss ratio for California as a fraction in simplest form.

▶1 22 wins, 26 losses
$\frac{22}{26}$ ← wins
       ← losses

▶2
$\frac{22}{26} = \frac{11}{13}$ (÷ 2)

California has an $\frac{11}{13}$ win-loss ratio.

Example B

Write the ratio for games won by Seattle to total games played.

▶1 20 wins, 20 + 25 or 45 total games
$\frac{20}{45}$ ← wins
       ← total games

▶2
$\frac{20}{45} = \frac{4}{9}$ (÷ 5)   Seattle has a $\frac{4}{9}$ ratio for games won to total games.

Example C

Phil drives about 300 miles with a full 15-gallon tank of gas. Write the ratio for miles Phil drives to gallons of gas used.

▶1 300 miles, 15 gallons   300 to 15 = $\frac{300}{15}$

▶2
$\frac{300}{15} = \frac{20}{1}$ (÷ 15)   Phil has a miles-to-gallons ratio of about $\frac{20}{1}$.

Keep the 1 as the denominator.

Guided Practice

Write each ratio as a fraction in simplest form.

Example A
1. New York: wins to losses
2. Philadelphia: losses to wins

Example B
3. Boston: wins to total games played
4. 6 absences in 180 school days

Example C
5. 300 students to 20 teachers
6. 76 heartbeats per 60 seconds

NBA Eastern Conference Atlantic Division		
Club	**W**	**L**
New York	60	22
Boston	48	34
New Jersey	43	39
Orlando	41	41
Miami	36	46
Philadelphia	26	56
Washington	22	60

Practice

Write each ratio as a fraction in simplest form.

7. 20 wins to 15 losses

8. 24 losses to 32 wins

9. 35 wins in a total of 55 games

10. 8 tickets for $48

11. 4 inches of rain in 30 days

12. 2 pounds of bananas for 88¢

13. 156 kilometers in 3 hours

14. $30.72 interest for 6 months

15. 4 students representing a class of 28 students

16. 102 passengers on 3 buses

17. 27 losses to 36 wins

18. 42 wins to 18 losses

Interpreting Data

19. Which Central Division team has a win-loss ratio of $\frac{5}{11}$?

20. Which Central Division team has a loss-win ratio of $\frac{1}{3}$?

21. Suppose Chicago had won two more of the 16 games they played. What would their win-loss ratio have been?

NFL National Conference Central Division			
Club	**W**	**L**	**T**
Chicago	12	4	0
Minnesota	11	5	0
Tampa Bay	5	11	0
Detroit	4	12	0
Green Bay	4	12	0

Applications

22. Carla's team has a 3:2 win-loss ratio. They lost four games. How many games did they win?

23. In 1988, Minnesota Viking's quarterback Wade Wilson completed 204 passes of 332 attempts. What was his ratio of completed passes to incomplete passes?

24. Quan puts 2 quarts of antifreeze and 5 quarts of water into her radiator. What is the ratio of antifreeze to water that she added?

Write Math

25. Explain in your own words what the ratio 10 fingers to 2 hands means.

Collect Data

26. Survey members of your class to see how many walk to school. Write a ratio to show walkers as a part of all students in your class.

Mixed Review

Lesson 4-3

Find the LCM of each group of numbers.

27. 4, 6

28. 7, 13

29. 9, 12

30. 16, 20

Lesson 5-6

Name the reciprocal of each number.

31. $\frac{5}{9}$

32. $\frac{15}{31}$

33. $\frac{8}{13}$

34. $\frac{7}{3}$

Lesson 6-4

35. A $10 roll of quarters has a mass of 240 grams. What is the mass of one quarter in kilograms?

9-2 AN INTRODUCTION TO PROBABILITY

Objective

Find the probability of an event.

Without looking, Jerry takes a marble out of a bag containing 3 red, 4 green, and 1 blue marble. What are the chances that the marble will be red?

The **probability** of an event is a ratio that describes how likely it is that the event will occur.

An **event** is a specific outcome or type of outcome.

Method

1 ▶ Determine the number of ways the desired event can occur.

2 ▶ Determine the number of ways all possible outcomes can occur.

3 ▶ Write the ratio the number of ways the desired event can occur to the number of ways all possible outcomes can occur.

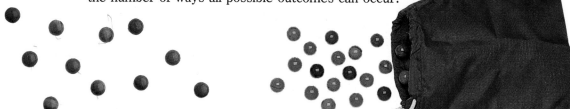

Example A

What is the probability that Jerry will take out a red marble?

1 ▶ The event is choosing a red marble. There are 3 red marbles. So, there are 3 ways to choose a red marble.

2 ▶ Since there are 8 marbles, there are 8 possible outcomes.

3 ▶ The probability of choosing a red marble is $\frac{3}{8}$. $P(\text{red}) = \frac{3}{8}$.

Example B

What is the probability of taking out a green or blue marble?

1 ▶ The event is choosing a green or a blue marble. There are 4 green marbles and 1 blue marble. So, there are 5 ways to choose either a green or a blue marble.

2 ▶ There are 8 possible outcomes.   3 ▶ $P(\text{green or blue}) = \frac{5}{8}$.

Example C

What is the probability that Jerry will take out a white marble?

1 ▶ The event is choosing a white marble. Since there are 0 white marbles, the event cannot occur.

2 ▶ There are 8 possible outcomes.   3 ▶ $P(\text{white}) = \frac{0}{8}$ or 0.

Guided Practice

Find the probability of choosing each marble from the bag described above. Write the probability in simplest form.

Examples A, B

1. green marble   **2.** blue marble   **3.** green or red marble

Examples B, C

A drawer contains 3 black socks, 4 navy socks, and 5 white socks. One sock is chosen. Find the probability of each event.

4. white sock   **5.** black sock   **6.** navy sock

7. white or navy sock   **8.** not a navy sock   **9.** brown sock

Practice

A die is rolled once. Find the probability of rolling each of the following.

10. a 4   **11.** a 2 or 4   **12.** a 2, 3, or 5   **13.** *not* a 4

14. a number less than 3   **15.** a number greater than 0

A box contains two red pencils, four yellow pencils, and three green pencils. One pencil is chosen. Find the probability of each event.

16. red pencil   **17.** yellow pencil   **18.** black pencil

19. not a red pencil   **20.** red, yellow, or green pencil

Interpreting Data

The possible outcomes for a roll of two dice are listed at the right. One die is red and the other is blue. Suppose the dice are rolled once. Find each probability.

6,1	6,2	6,3	6,4	6,5	6,6
5,1	5,2	5,3	5,4	5,5	5,6
4,1	4,2	4,3	4,4	4,5	4,6
3,1	3,2	3,3	3,4	3,5	3,6
2,1	2,2	2,3	2,4	2,5	2,6
1,1	1,2	1,3	1,4	1,5	1,6

21. $P(6,1)$ **22.** $P(2,3)$

23. $P(\text{even, odd})$ **24.** $P(\text{a sum of 3})$

25. $P(\text{both numbers even})$ **26.** $P(not \text{ a sum of 7})$

27. $P(\text{a sum of 5 or 11})$ **28.** $P(\text{a sum of at least 10})$

Applications

29. A box contains 3 red pencils and 2 blue pencils. Brent chooses a red pencil and keeps it. Then Seth chooses a pencil. What is the probability that Seth's pencil is blue?

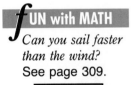
***f*UN with MATH**
Can you sail faster than the wind?
See page 309.

30. A test has three true-false questions. Becky guesses on each question. What is the probability that Becky answers all of the questions correctly? Hint: List all possible sets of three guesses.

31. A tour boat can hold 12 passengers. If 80 people are waiting for the tour, how many boats are needed?

Critical Thinking

32. There are seven basketball teams in the Central Hoosier League. Every year each team plays every other team in the league. How many league games must be scheduled every year?

Suppose

33. Suppose you have three types of outcomes: *a*, *b* and *c*. If the probability of either *a* or *b* is $\frac{7}{12}$, how many outcomes of each type are there? Can you find more than one answer?

Cooperative Groups

34. Predict how many times a coin will land with heads showing if it is tossed 4 times. Then toss a coin 4 times and record the results. Do the same for 8 tosses, 16 tosses, and 32 tosses. Is tossing a coin close to your prediction when you toss a lesser or greater number of times? Why?

9-3 PROPORTION

Objective

Determine if two ratios form a proportion.

A Tampa High School basketball team won 9 of the first 12 games they played. During the entire season the team won 18 of 24 games. Do the mid-season and end-of-season records form a proportion?

If two ratios are equivalent, they form a **proportion**. The method below shows how to determine if two ratios form a proportion.

Method

1. Write each ratio as a fraction in simplest form.
2. Compare the fractions. Are they equivalent?

Example A

Determine if the ratios 9 to 12 and 18 to 24 form a proportion.

$$\overset{\div\ 3}{\frac{9}{12}} = \frac{3}{4} \qquad \overset{\div\ 6}{\frac{18}{24}} = \frac{3}{4}$$

$$\frac{3}{4} = \frac{3}{4} \text{ or } \frac{9}{12} = \frac{18}{24}$$

The ratios 9 to 12 and 18 to 24 are equivalent. So, the records form a proportion.

Example B

Determine if the ratios 20:8 and 16:4 form a proportion.

$$\overset{5}{\underset{2}{\frac{20}{8}}} = \frac{5}{2} \qquad \overset{4}{\underset{1}{\frac{16}{4}}} = \frac{4}{1}$$

$$\frac{5}{2} \neq \frac{4}{1} \qquad \text{So, it follows that } \frac{20}{8} \neq \frac{16}{4}$$

The ratios 20:8 and 16:4 are not equivalent. They do not form a proportion.

Guided Practice

Determine if each pair of ratios forms a proportion.

Example A

1. 3 to 6, 4 to 8   **2.** 12 to 18, 20 to 24

Example B

3. 9 to 3, 6 to 2   **4.** 15:45, 7:28

5. 20:16, 35:28   **6.** 16:12, 24:20

Determine if each pair of ratios forms a proportion.

7. 3 to 4, 9 to 12   **8.** 4 to 5, 5 to 4   **9.** 8:10, 16:20

10. 5 to 8, 2 to 3   **11.** 3:7, 21:49   **12.** 5:24, 10:12

13. 14 to 8, 8 to 2   **14.** 2:50, 4:100   **15.** 18:27, 18:36

16. 80 to 100, 8 to 10   **17.** 3 to $\frac{1}{3}$, 2 to $\frac{1}{2}$ **18.** 1 to $\frac{2}{3}$, 9 to 6

19. Explain why the ratios 8:20 and 40:100 form a proportion.

20. Write *true* or *false*. 12:18 and 21:14 form a proportion. Defend your answer.

Mental Math

21. Barb adds 2 quarts of antifreeze to the 5 quarts of water in her radiator. Which of these is an equivalent ratio? 5 to 2, 7 to 10, 8 to 20, 10 to 28.

22. Brian is 16 years old and his brother is 12. Steve is 20 years old and his brother is 15. Len is 24 years old and his brother is 19. Which ratio is not equivalent to the others?

23. Office workers often use 3 in. by 5 in. cards, 4 in. by 6 in. cards, and 5 in. by 8 in. cards. The dimensions of each card form a ratio. Are the ratios for any two cards equivalent?

Applications

24. Explain whether the 1988 win-loss ratios for the Buffalo Bills and the Cincinnati Bengals were equivalent.

25. Compare the ratio of games lost to total games played for the Indianapolis Colts with the ratio for the New York Jets. Are they equivalent?

NFL American Conference Final 1988 Standings			
Club	**W**	**L**	**T**
Eastern Division			
Buffalo	12	4	0
Indianapolis	9	7	0
New England	9	7	0
New York Jets	8	7	1
Miami	6	10	0
Central Division			
Cincinnati	12	4	0
Cleveland	10	6	0
Houston	10	6	0
Pittsburgh	5	11	0

26. The Eastland High School football team won 3 of the 8 games they played. Which NFL team had an equivalent win-loss ratio for 1988?

Science Connections

27. Light travels almost 1,900,000 miles in 10 seconds. Would it travel the 93,000,000 miles from the sun to Earth in 5 minutes?

Using Equations

Solve each equation. Check your solution.

28. $12p = 120$ **29.** $m - 56 = 32$ **30.** $24 = 11 + x$

31. $68 \div t = 17$ **32.** $r + 234 = 823$ **33.** $8f = 2$

Critical Thinking

34. While cleaning the attic, Mrs. Stamos finds ten boxes. Five contain letters, four contain postcards, two contain both, and the others are empty. How many boxes are empty?

9-4 SOLVING PROPORTIONS

Objective
Solve proportions by using cross products.

Valerie is making pudding using a recipe that calls for 2 cups of milk for 5 servings. Will 6 cups of milk be enough for 15 servings? To find out, see if $\frac{2}{5}$ and $\frac{6}{15}$ form a proportion.

Besides using fractions in simplest form, another way to determine if two ratios form a proportion is to use **cross products.** If the cross products of two ratios are equal, then the ratios are equivalent and form a proportion.

Method
1 ▶ Find the cross products.
2 ▶ If the cross products are equal, the ratios form a proportion.

Example A

Use cross products to determine if $\frac{2}{5}$ and $\frac{6}{15}$ form a proportion.

1 ▶
$$\frac{2}{5} \overset{?}{=} \frac{6}{15}$$
$$2 \times 15 \overset{?}{=} 5 \times 6$$
The cross products are 2×15 and 5×6.
$$30 = 30$$
Both equal 30.

2 ▶ The cross products are equal.
$\frac{2}{5}$ and $\frac{6}{15}$ form a proportion.

Note that $\frac{6}{15}$ in simplest form is $\frac{2}{5}$.

Six cups of milk will be enough for 15 servings.

Some proportions contain variables. To solve a proportion, use cross products. Then solve the resulting equation.

Method
1 ▶ Find the cross products.
2 ▶ Solve the equation.

Example B

Solve the proportion $\frac{7}{8} = \frac{p}{32}$.

1 ▶
$$\frac{7}{8} = \frac{p}{32}$$
$$7 \times 32 = 8 \times p$$
The cross products are 7×32 and $8 \times p$.

2 ▶
$$224 = 8 \times p$$
Multiply.
$$\frac{224}{8} = \frac{8 \times p}{8}$$
Divide each side by 8.
$$28 = p$$
Check: $7 \times 32 \overset{?}{=} 8 \times 28$
$$224 = 224 \checkmark$$
Replace p with 28 in the equation.

Guided Practice

Use cross products to determine if each pair of ratios forms a proportion.

Example A
1. $\frac{4}{6}, \frac{2}{3}$ 2. $\frac{4}{2}, \frac{3}{6}$ 3. $\frac{16}{4}, \frac{20}{8}$ 4. $\frac{15}{21}, \frac{5}{7}$

Solve each proportion.

5. $\frac{6}{8} = \frac{n}{24}$ **6.** $\frac{6}{7} = \frac{36}{r}$ **7.** $\frac{t}{5} = \frac{9}{15}$ **8.** $\frac{7}{a} = \frac{21}{28}$

Use cross products to determine if each pair of ratios forms a proportion.

9. $\frac{8}{12}, \frac{9}{15}$ **10.** $\frac{4}{7}, \frac{36}{70}$ **11.** $\frac{21}{28}, \frac{3}{4}$ **12.** $\frac{4}{15}, \frac{80}{300}$

Solve each proportion.   *even*

13. $\frac{1}{2} = \frac{a}{14}$ **14.** $\frac{3}{4} = \frac{c}{16}$ **15.** $\frac{f}{15}, \frac{4}{5}$ **16.** $\frac{g}{32} = \frac{3}{8}$

17. $\frac{5}{9} = \frac{40}{a}$ **18.** $\frac{14}{16} = \frac{7}{c}$ **19.** $\frac{3}{5} = \frac{y}{42}$ **20.** $\frac{1}{2} = \frac{x}{15}$

21. $\frac{5}{4} = \frac{u}{100}$ **22.** $\frac{6}{33} = \frac{2}{a}$ **23.** $\frac{2}{3} = \frac{x}{100}$ **24.** $\frac{1}{8} = \frac{n}{100}$

25. Tell if the solution is correct. If it is not correct, explain why not.

$$\frac{5}{8} = \frac{n}{24}$$

$$5 \times 24 = 8 \times n$$
$$120 = 8 \times n$$
$$120 \times 8 = n$$
$$960 = n$$

f UN with MATH
*Try our recipe for
Mexican cocoa.*
See page 308.

26. Harry uses a recipe that requires 32 ounces of fruit juice to make 2 gallons of punch. He wants to make 6 gallons, so he estimates that he'll need almost 100 ounces of fruit juice. Is this amount reasonable? Why?

27. A park ranger stocks a pond with 4 sunfish for every 3 perch. Suppose 296 sunfish are put in the pond. How many perch should be stocked?

28. The ratio of land area to total area of Earth's surface is about 3 to 10. If the area of Earth's surface is about 197,000,000 square miles, *about* how many square miles of land are on Earth's surface?

29. For the three variables a, b, and c, $\frac{a}{b} = \frac{b}{c}$. Find a set of values for a, b, and c. Then find at least one other set of values.

30. Make up your own problem that requires solving a proportion. Use three of these numbers: 3, 4, 9, and 15.

Subtract.

31. $2\frac{3}{4} - 1\frac{1}{4}$ **32.** $5\frac{8}{9} - 3\frac{2}{3}$ **33.** $6\frac{5}{8} - 2\frac{1}{6}$

34. A furniture store bought 8 identical sofas for $4,000. How much did each sofa cost?

9-5 SCALE DRAWINGS

Objective

Interpret scale drawings.

Mr. Upton uses a scale drawing as he lays out a baseball field for the new City Park. He determines the actual distances by measuring on the drawing and then solving a proportion.

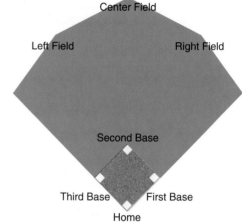

The scale drawing shows the new baseball field. On the drawing 1 inch represents 180 feet.

To determine the actual distance, measure the distance on the drawing. Then write and solve the proportion.

Method

1 ▶ Write the scale as a ratio.

2 ▶ Set up a proportion.

3 ▶ Solve the proportion.

Example A

Find the actual length of the right field line.

1 ▶ $\dfrac{1}{180}$ ← in.
← ft   1 in. represents 180 feet.

2 ▶ On the drawing, the right field line measures $1\frac{3}{4}$ or 1.75 in.

$$\frac{1}{180} = \frac{1.75}{x} \quad \begin{array}{l} \leftarrow \text{in.} \\ \leftarrow \text{ft} \end{array}$$

3 ▶ $1 \times x = 180 \times 1.75$ Use cross products to solve.

$x = 315$ The actual distance is 315 feet.

Example B

Ricardo makes a scale drawing of his house. On his drawing 1.5 centimeters represents 5 meters. If the actual dimensions of the living room are 8 m by 6 m, what should be the dimensions on the drawing?

1 ▶ $\dfrac{1.5}{5} \quad \begin{array}{l} \leftarrow \text{cm} \\ \leftarrow \text{m} \end{array}$

2 ▶ $\dfrac{1.5}{5} = \dfrac{x}{8} \quad \begin{array}{l} \leftarrow \text{cm} \\ \leftarrow \text{m} \end{array}$ ⟵ (Two proportions must be solved.) ⟶ $\dfrac{1.5}{5} = \dfrac{y}{6} \quad \begin{array}{l} \leftarrow \text{cm} \\ \leftarrow \text{m} \end{array}$

3 ▶ $1.5 \times 8 = 5 \times x$ Use cross products. $1.5 \times 6 = 5 \times y$

$12 = 5 \times x$ $9 = 5 \times y$

$\dfrac{12}{5} = \dfrac{\overset{1}{\cancel{5}} \times x}{\underset{1}{\cancel{5}}}$ Divide each side by 5. $\dfrac{9}{5} = \dfrac{\overset{1}{\cancel{5}} \times y}{\underset{1}{\cancel{5}}}$

$2.4 = x$ $1.8 = y$

On the drawing, the dimensions should be 2.4 cm by 1.8 cm.

Find the actual distance. Use the scale drawing on page 288.

Example A

1. the distance from home plate to the center field fence

2. the distance between first base and second base

3. the distance from home plate to second base

Example B

Find the distance on a scale drawing for each actual distance. The scale is 1 in.:50 ft. (1 in.:50 ft means 1 in. represents 50 ft.)

4. 150 ft   **5.** 125 ft   **6.** 75 ft   **7.** 20 ft   **8.** $6\frac{1}{4}$ ft

Exercises

Find the actual distances. Use the scale drawing at the right.

Practice

9. What are the dimensions of the master bedroom?

10. What are the dimensions of the dining room?

11. What are the dimensions of bedroom 2?

12. How many feet does $\frac{1}{4}$ inch represent on the scale drawing?

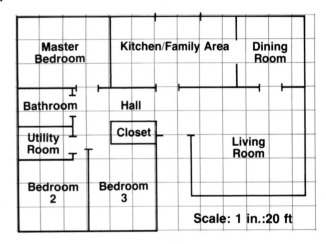

Find the distance on a scale drawing for each actual distance. The scale is 1 in.: 200 ft.

13. 300 feet   **14.** 1,000 feet   **15.** 250 feet   **16.** 175 feet

17. 50 feet   **18.** 1 foot   **19.** 50 yards   **20.** 200 yards

Applications

21. Use the scale drawing shown above. Tile is sold in pieces that measure 1 foot by 1 foot. How many pieces of tile are needed to cover the bathroom floor?

22. A car travels 144 miles on 4 gallons of gasoline. At this rate, how many gallons are needed to drive 450 miles?

23. Select an exercise set from this chapter that you feel shows your best work. Explain why it is your best work.

24. To make a scale drawing as large as possible of your home on an $8\frac{1}{2}$ in. by 11 in. sheet of paper, how would you decide what ratio to use for a scale?

LIFE INSURANCE

Life insurance protects against a financial loss due to death. Three of the four types of life insurance described build up a cash value. **Cash value** is money that the policy accumulates as the premiums are paid. This money may be returned to the policyholder upon retirement age or paid to the **beneficiary** if the policyholder dies before retirement age.

Insurance rates vary according to the age of the insured person and the type of policy. The younger the insured person is when the policy is purchased, the smaller the premium.

Denise is 20 years old. Use the table below to find her annual premium for a $15,000 straight life policy.

$17.60	premium per $1,000 of insurance
× 15	15 × $1,000 = $15,000 insurance
$264.00	Denise's annual premium is $264.

> **Term insurance** gives temporary protection. The beneficiary, or beneficiaries, receive the money only if the insured person dies during a certain period of time.
>
> **Straight life insurance** gives permanent lifetime protection. It builds up a cash value.
>
> **Limited pay-life insurance** gives permanent lifetime protection. It builds up a cash value. The premiums are paid only for a specific number of years.
>
> **Endowment insurance** builds up a cash value. The premiums are paid only for a specific number of years. After all the premiums are paid, the policyholder receives the money.

Annual Premiums Per $1,000 of Insurance							
Age at Issue	**Term**		**Straight Life**	**Limited Payment**		**Endowment**	
	5-Year	**10-Year**		**20-Year**	**30-Year**	**20-Year**	**30-Year**
15			$15.75	$28.10	$22.10	$49.50	$31.90
20	$ 7.75	$ 8.10	17.60	30.60	24.00	49.80	32.40
25	8.50	9.10	20.00	33.30	26.25	50.30	33.20
30	9.70	10.60	22.80	36.50	28.90	51.10	34.40
35	11.50	12.80	26.40	40.15	32.15	52.30	36.20
40	14.30	16.30	31.00	44.50	36.25	54.15	39.00

1. What is Denise's annual premium for a $15,000, 10-year term policy?
2. What is Denise's annual premium for a $20,000, 20-year endowment policy?
3. At the age of 25, Mr. Gerard bought a $25,000, 30-year limited payment policy. What is the total amount he will pay at the end of 30 years? Will the beneficiary receive more or less than Mr. Gerard paid? How much?
4. At age 30, how much 10-year term insurance could you buy for approximately the same annual premium as the straight life insurance?

AERIAL PHOTOGRAPHER

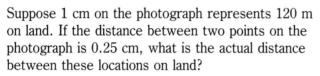

Tony Cecil is an aerial photographer. He uses special equipment to take photographs from the air. The photographs are a scale model of what is actually on land. They are used to make maps for travel, to study soil erosion and water pollution, or for military purposes.

Suppose 1 cm on the photograph represents 120 m on land. If the distance between two points on the photograph is 0.25 cm, what is the actual distance between these locations on land?

Write the scale as a ratio.	$\frac{1}{120}$ $\leftarrow$ cm $\leftarrow$ m
Set up a proportion.	$\frac{1}{120} = \frac{0.25}{n}$ 0.25 cm represents n meters.
Solve the proportion.	$1 \times n = 120 \times 0.25$
The actual distance is 30 m.	$n = 30$

The distance on an aerial photograph is given. Find the actual distance if the scale is 1 cm:360 m.

1. 1.5 cm   **2.** 0.6 cm   **3.** 2.8 cm

The actual distance is given. Find the distance on an aerial photograph if the scale is 1 cm:240 m.

4. 156 m   **5.** 864 m   **6.** 1,200 m

The distance on an aerial photograph and the actual distance are given. Find the scale.

7. 100 cm on photograph, 400 m actual   **8.** 12 cm on photograph, 360 m actual

9. On a radar screen the distance between two planes is $3\frac{1}{2}$ inches. If the scale is 1 inch on the screen to 2 miles in the air, what is the actual distance between the two planes?

10. Two planes in flight are 16 km apart. How far apart are they on a radar screen if the scale is 1 cm:1.6 km?

9-6 SIMILAR FIGURES

Objective

Find the measure of a side of a similar figure.

Alissa is a film processor for Speedy Photo Labs. She often enlarges photos during processing. If a figure is enlarged or reduced, the new figure is *similar* to the original.

The measures of corresponding sides of similar figures are proportional. That is, the ratios formed using pairs of corresponding sides are equivalent.

The triangles formed by the pole, tree, and shadows are similar.

Method

1. ▶ Identify two corresponding ratios.
2. ▶ Set up a proportion.
3. ▶ Solve the proportion.

5-ft pole

◀—8-ft—▶
shadow

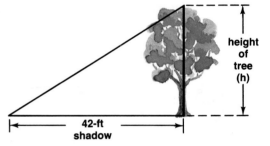

42-ft
shadow

height of tree (h)

Example A

Find the height (h) of the tree.

1. ▶ ratio of shadows: 8 to 42   ratio of heights: 5 to h

2. ▶ shadow of pole → $\dfrac{8}{42}$ = $\dfrac{5}{h}$ ← height of pole
   shadow of tree → ← height of tree

3. ▶

$$8 \times h = 42 \times 5 \qquad \text{Find cross products.}$$

$$8 \times h = 210 \qquad \text{Solve for } h.$$

$$\frac{\overset{1}{\cancel{8}} \times h}{\underset{1}{\cancel{8}}} = \frac{210}{8} \qquad \text{Divide each side by 8.}$$

$$h = 26.25 \qquad \begin{array}{l}\text{The height of the tree is 26.25 feet}\\\text{or 26 feet 3 inches.}\end{array}$$

Guided Practice

Example A

Find the missing length for each pair of similar figures.

1.

5 cm

n

3 cm   6 cm

2.

12 in.

9 in.

m

12 in.

3.

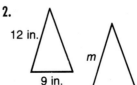

16 cm

8 cm

25 cm

p

Practice

Find the missing length for each pair of similar figures.

4. 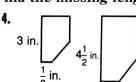 3 in. $4\frac{1}{2}$ in. $\frac{1}{2}$ in. n

5. 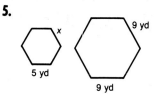 x 9 yd 5 yd 9 yd

6. 80 cm 60 cm p 36 cm

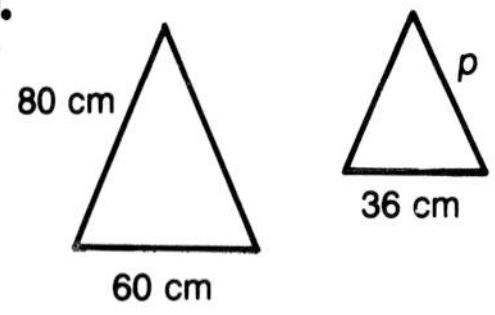

7. 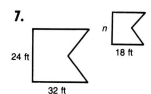 n 24 ft 18 ft 32 ft

8. 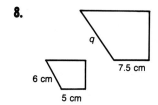 q 7.5 cm 6 cm 5 cm

9. 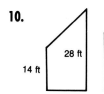 15 m 7.3 m x 7.3 m

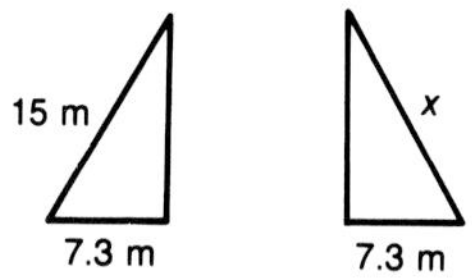

10. 28 ft 14 ft 18 ft z

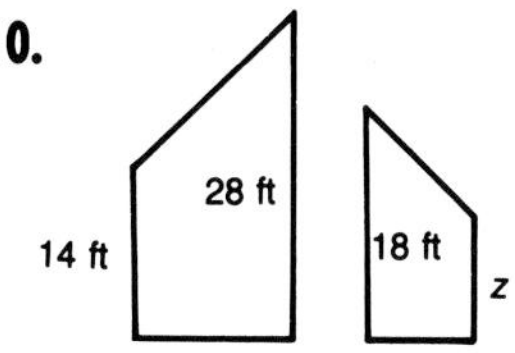

11. 7 cm 4 cm 3 cm 6 cm x

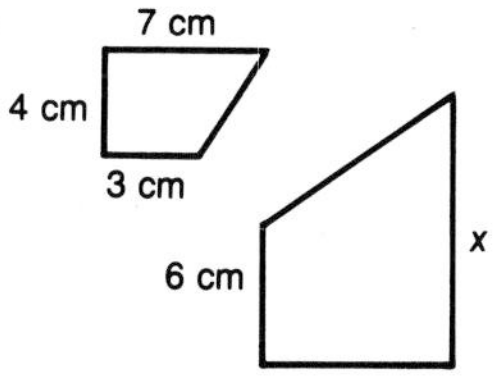

12. 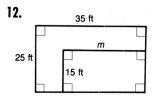 35 ft m 25 ft 15 ft

Applications

	Height of Pole	Shadow of Pole	Shadow of Tree
13.	3 m	8 m	42 m
14.	4 ft	5 ft	60 ft
15.	36 cm	60 cm	40 m
16.	32 in.	4 ft 2 in.	75 ft

Use the chart at the left to find the height of each tree given the height of the pole and the length of the shadows.

17. The shadow of a 4-foot pole is 6 feet long at the same time that the shadow of a tower is 52.5 feet long. How tall is the tower?

18. Sara is 5 feet tall. Her shadow is 9 feet long and the shadow of a building is 36 feet long. How tall is the building?

19. Brian is carving a model of a friend's car. The car is 172 in. long and 66 in. high. If the model is to be 6 in. long, how high should it be?

Art Connection

20. Craig drew a design measuring 5 in. wide by 8 in. long. He wants to enlarge the design to make prints that will be $12\frac{1}{2}$ in. wide. Find the dimensions for the enlarged design.

Critical Thinking

21. Suzanne ran for 20 minutes the first day of her fitness program. Each day she increased her time by a ratio of 2:3; that is, she ran half again as much. How many minutes did she run the fourth day?

Cooperative Groups

22. Find the height of the flagpole at your school by creating similar triangles. Measure the height of one student and the shadows of both the student and the flagpole.

9-7 PLANNING A TRIP

Objective

Locate distances and driving times on a map.

Christy and Zac Taylor are planning to drive from Jacksonville, FL, to Denver, CO. They want to visit friends in Memphis and Dallas before going on to Denver. About how many miles will the Taylors drive from Jacksonville to Denver?

Christy and Zac use a map that shows approximate distances and driving times between cities.

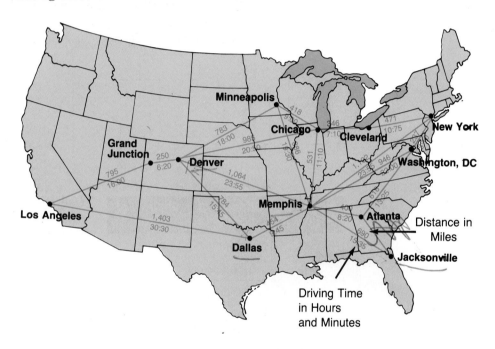

The map shows that the distance from Jacksonville to Memphis is 690 miles, from Memphis to Dallas is 454 miles, and from Dallas to Denver is 784 miles. Add to find the total mileage.

$$690 + 454 + 784 = 1,928$$

Christy and Zac will drive 1,928 miles from Jacksonville to Denver.

Understanding the Data

- On the map, what does the number above the line represent?
- On the map, what does the number below the line represent?
- What does the driving time 5:10 mean?

Guided Practice

Solve. Use the map above.

1. What is the distance from Chicago to Memphis?

2. What is the driving time from Washington, D.C., to Atlanta?

3. Between what two cities (no cities in between) is the driving time and distance the greatest?

Applications

Paper/ Pencil
Mental Math
Estimate
Calculator

Solve. Use the map on page 294.

4. How long will it take the Taylors to drive from Jacksonville to Denver via Memphis and Dallas?

5. Name a different route that the Taylor's can take to return home to Jacksonville.

6. Mrs. Salgado leaves Chicago at 7:20 A.M. and arrives in Minneapolis at 3:20 P.M. What is her average speed?

7. How long would it take you to travel from Los Angeles to Grand Junction at an average speed of 50 miles per hour?

Cooperative Groups

8. Use the map to plan a trip. Determine how many days the trip will take. If gasoline costs $1.39 a gallon, determine how much you will spend for gasoline. Remember to include your return trip.

Using Formulas

Solve. Use the formula d = rt.

9. A race car driver covers 1,043 kilometers in 3.5 hours. What is the average speed of the driver?

10. Min Toshio drove from Cleveland to New York in 10 hours 15 minutes at an average speed of 49 miles per hour. About how many miles did Min travel?

Calculator

On the map at the right, one inch represents 50 miles (1 in.:50 mi). Measure with a ruler the map distance between Glendale and Ashville. The map distance is 2.25 inches. What is the actual distance between the towns?

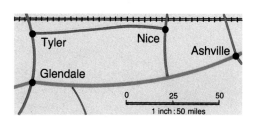

To find the actual distance, multiply 50 miles by the number of inches.

$$2.25 \boxed{\times} 50 \boxed{=} 112.5$$ The actual distance is 112.5 miles.

Use the map, the scale, and a ruler to find the actual distance between each city.

11. Nice to Ashville   **12.** Tyler to Nice   **13.** Glendale to Tyler

Mixed Review

Lesson 2-3

Subtract.

14.	**15.**	**16.**	**17.**
35	435	6,759	4,378
− 16	− 122	− 467	− 3,188

Lesson 6-2

Name the most reasonable unit for measuring each item. Use kilometer, meter, centimeter, or millimeter.

18. length of a car   **19.** the width of a shoelace

Lesson 6-8

20. Sandra bought milk for $2.19 a gallon at Convenient Mart. She later found milk for $1.20 for 64 ounces at Quick Stop 'n Shop. Which store had the better buy on milk?

9-8 RATIOS, PERCENTS, AND FRACTIONS

Objective

Write percents as fractions, and fractions as percents.

Blake hears the weather forecaster say there is an 80 percent chance of rain during the day. Should Blake take his umbrella with him? An 80 percent chance means that out of 100 days, it would probably rain during 80. Since the chance of rain is high, Blake should take his umbrella.

A ratio that compares a number to 100 may be written as a percent. **Percent** means *hundredths,* or *per 100.*

$$80 \text{ to } 100 = \frac{80}{100} = 80\%$$

The symbol % is the percent symbol. It means that a number is being compared to 100.

A percent can also be written as a fraction.

Method

1. ▶ Write the percent in the numerator and 100 in the denominator.
2. ▶ Write the fraction in simplest form.

Examples

Write each percent as a fraction in simplest form.

A 75%

1. ▶ $75\% = \frac{75}{100}$
2. ▶ $= \frac{\overset{3}{\cancel{75}}}{\underset{4}{\cancel{100}}}$
   $= \frac{3}{4}$

B 9.5%

1. ▶ $9.5\% = \frac{9.5}{100}$
2. ▶ $= \frac{9.5 \times 10}{100 \times 10}$
   $= \frac{95}{1,000}$
   $= \frac{19}{200}$

C $66\frac{2}{3}\%$

1. ▶ $66\frac{2}{3}\% = \frac{66}{100}$
2. ▶ $= 66\frac{2}{3} \div 100$
   $= \frac{\overset{2}{\cancel{200}}}{3} \times \frac{1}{\underset{1}{\cancel{100}}}$
   $= \frac{2}{3}$

A fraction can be written as a percent.

Method

1. ▶ Set up a proportion with the fraction equal to a ratio with a denominator of 100, $\frac{n}{100}$.
2. ▶ Use cross products to solve the proportion.

Examples

Write each fraction as a percent.

D $\frac{4}{5}$

1. ▶ $\frac{4}{5} = \frac{n}{100}$
2. ▶ $4 \times 100 = 5 \times n$
   $400 = 5 \times n$
   $\frac{400}{5} = \frac{5 \times n}{5}$
   $80 = n \quad \frac{4}{5} = 80\%$

$$\begin{array}{r} 80 \\ 5)\overline{400} \\ \underline{-40} \\ 0 \end{array}$$

E $\frac{3}{8}$

1. ▶ $\frac{3}{8} = \frac{n}{100}$
2. ▶ $3 \times 100 = 8 \times n$
   $300 = 8 \times n$
   $\frac{300}{8} = \frac{8 \times n}{8}$
   $37\frac{1}{2} = n \quad \frac{3}{8} = 37\frac{1}{2}\%$

$$\begin{array}{r} 37 \text{ R4} \\ 8)\overline{300} \\ \underline{-24} \\ 60 \\ \underline{-56} \\ 4 \end{array}$$

Guided Practice

Examples A-C

Write each percent as a fraction in simplest form.

1. 50%   **2.** 10%   **3.** 200%   **4.** 150%   **5.** $12\frac{1}{2}$%   **6.** 0.1%

Examples D, E

Write each fraction as a percent.

7. $\frac{9}{10}$ **8.** $\frac{1}{4}$ **9.** $\frac{5}{8}$ **10.** $\frac{17}{20}$ **11.** $\frac{1}{3}$ **12.** $\frac{1}{8}$

13.

Exercises

Write each percent as a fraction in simplest form.

13. 29%   **14.** 1%   **15.** 65%   **16.** 35%   **17.** 125%   **18.** 100%

19. 200%   **20.** 3%   **21.** 25%   **22.** $16\frac{2}{3}$%   **23.** 37.5%   **24.** 0.3%

f **UN with MATH**

Try fractions if you come home late from a date.
See page 309.

Write each fraction as a percent.

25. $\frac{21}{100}$ **26.** $\frac{1}{5}$ **27.** $\frac{1}{4}$ **28.** $\frac{2}{25}$ **29.** $\frac{41}{50}$ **30.** $\frac{19}{20}$

31. $\frac{1}{8}$ **32.** $\frac{3}{8}$ **33.** $\frac{1}{6}$ **34.** $\frac{5}{4}$ **35.** $\frac{4}{1}$ **36.** $1\frac{1}{2}$

Applications

Use the circle graph at the right.

37. What percent of 1989 car production in U.S. plants is represented by the graph?

38. What fraction of new cars were produced by General Motors or Ford?

39. What fraction of new cars were *not* produced by General Motors or Ford?

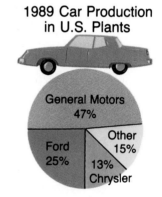

1989 Car Production in U.S. Plants

General Motors 47%

Other 15%

Ford 25%

13% Chrysler

40. Suppose you earned $24 in one week by doing odd jobs. If during the second week you increased your earnings by 150%, how much would you earn the second week?

41. Describe one topic in this chapter that you would liked to have spent more time on and explain why.

JOURNAL ENTRY

Mental Math

Some percents are used so often in computations that it will save time if you memorize their fractional equivalents.

Write each percent as a fraction. Study each column for a pattern.

42.	43.	44.	45.
a. 10%	**a.** 20%	**a.** 25%	**a.** $33\frac{1}{3}$%
b. 30%	**b.** 40%	**b.** 50%	**b.** $66\frac{2}{3}$%
c. 70%	**c.** 60%	**c.** 75%	**c.** 100%
d. 90%	**d.** 80%	**d.** 100%	
e. 100%	**e.** 100%		

9-9 PERCENTS AND DECIMALS

Objective

Write percents as decimals and decimals as percents.

Mrs. Ohnishi uses a calculator when computing her taxes. When she finds the percent of an amount, she changes the percent to an equivalent decimal.

Since percent means hundredths, a percent may be written as a decimal by dividing by 100.

United States Internal Revenue Service Building
← Visitors

Method

1▶ Divide the percent by 100. (Move the decimal point two places to the left.)

2▶ Omit the % symbol.

Examples

Write each percent as a decimal.

A 12.5% ▼1 ▼2

12.5% → 12.5% → 0.125

12.5% = 0.125

B 8% ▼1 ▼2

8% → 08.% → 0.08

8% = 0.08

To write a decimal as a percent, multiply by 100.

Method

1▶ Multiply the decimal by 100. (Move the decimal point two places to the right.)

2▶ Write the % symbol.

Examples

Write each decimal as a percent.

C 0.06 ▼1 ▼2

0.06 → 0.06 → 6%

0.06 = 6%

D 3.5 ▼1 ▼2

3.5 → 3.50 → 350%

3.5 = 350%

Guided Practice

Write each percent as a decimal.

Example A

1. 2.5%   2. 73.2%   3. 0.6%   4. 9.9%   5. $\frac{1}{4}$%

Example B

6. 75%   7. 7%   8. 100%   9. 9%   10. 50%

Write each decimal as a percent.

Example C

11. 0.12   12. 0.07   13. 0.015   14. 0.105   15. 0.9

Example D

16. 7.5   17. 1.0   18. 2.0   19. 6.15   20. 8.1

Write each percent as a decimal.

21. 12% **22.** 73% **23.** 30% **24.** 8%

25. 8.1% **26.** 200% **27.** 850% **28.** 125%

29. 38.2% **30.** $\frac{3}{10}$% **31.** $1\frac{1}{2}$% **32.** $2\frac{4}{10}$%

Write each decimal as a percent.

33. 0.68 **34.** 0.07 **35.** 0.7 **36.** 0.15

37. 0.01 **38.** 0.009 **39.** 0.056 **40.** 0.775

41. 1.38 **42.** 4.2 **43.** 3.0 **44.** 0.001

Applications

On pages 528–530, you can learn how automotive mechanics use mathematics in their jobs.

45. The Warriors won 7 out of 15 games. What percent of their games did they win?

46. The Tigers won about 55% of their games. If they played 18 games, how many did they win?

47. Jackie made 89 of 112 free throws. How many free throws did she miss?

48. Explain how to change $\frac{4}{5}$ to a decimal and then to a percent.

49. A hiker is at the campground. If there is an equal probability of choosing each path when a path divides, what percent of the time will the hiker end up in The Swamp?

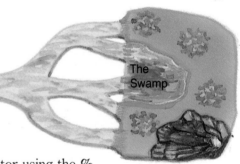

The Swamp

Calculator

You can enter a percent into a calculator using the % key to obtain its decimal equivalent.

Enter: ⬚ 6 ⬚ ⬚ % The calculator divides the
Display: 6 0.06 percent by 100.

Use a calculator to write each percent as a decimal.

50. 25% **51.** 5% **52.** 30% **53.** 100% **54.** 350% **55.** 0.5%

Mixed Review

Lesson 3-1

Multiply.

56. 23
    ×12

57. 320
    × 36

58. 435
    × 20

59. 701
    × 55

60. 1,002
    × 102

Lesson 7-1

Make a drawing of each figure.

61. ray XY **62.** line RS **63.** $\overline{DG}$

Lesson 8-3

64. Find the area of a triangle that has a base of 4.6 cm and has a height of 6.4 cm.

► Explore
► Plan
► Solve
► Examine

9-10 GUESS AND CHECK

Objective

Solve problems using guess and check.

Ms. Jenkins tells her math class that when a certain number is multiplied by itself, the product is 2,209. What is the number?

The guess-and-check strategy can be used to solve many problems. Here is an outline of how to use this strategy.
- Make a reasonable guess for a solution.
- Check the guess to see if it is correct.
- If it is not correct, decide how to improve the next guess.
- Make another guess and check it.
- Repeat these steps until the problem is solved.

► **Explore**

What is given?
- The product of a certain number multiplied by itself is 2,209.

What is asked?
- What is the number?

► **Plan**

Try multiples of ten to find a reasonable range for the number. Guess again using the ones digit as a clue.

► **Solve**

Think:
$$40 \times 40 = 1,600$$
$$50 \times 50 = 2,500$$

The number is between 40 and 50.

Try 45: $4 \: 5 \: \boxed{\times} \: 4 \: 5 \: \boxed{=} \: 2025$

The number is between 45 and 50.

Think: The product is 2,209, and we know $7 \times 7 = 49$.

Try a number that ends in a 7 like 47: $4 \: 7 \: \boxed{\times} \: 4 \: 7 \: \boxed{=} \: 2209$

The number is 47.

► **Examine**

Since $47 \times 47 = 2,209$, the number is 47.

Guided Practice

Solve. Use the guess-and-check strategy.

1. Carla buys film in rolls of 24 and 36 exposures. She buys 5 rolls and gets 132 exposures. How many rolls of each film did she buy?

2. The sum of three consecutive whole numbers is 54. What are the three numbers?

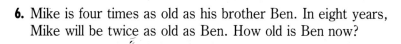

Solve. Use any strategy.

3. Maggie has eight coins. They are quarters, dimes, and nickels. Their total value is $1.10. How many of each coin does Maggie have?

4. The sum of three consecutive even numbers is 138. What are the three numbers?

5. When a certain number is multiplied by itself, the product is 1,521. What is the number?

6. Mike is four times as old as his brother Ben. In eight years, Mike will be twice as old as Ben. How old is Ben now?

7. The difference between two whole numbers is 15. Their product is 1,350. What are the numbers?

8. On a map, 3 inches represents 450 miles. How many miles are represented by $2\frac{1}{2}$ inches?

9. A basket filled with special candies costs $24. The basket and the candies can be purchased separately. The basket costs $10 less than the candies. How much does the basket cost?

10. Concert tickets cost $7.50 for adults and $3.50 for children. Aaron's family spent $48.00 for tickets. More adult tickets were bought than children's tickets. How many of each did they buy?

Critical Thinking

11. On a desk calendar like the one at the right, the dates must range from 01-31. What numbers are on the two cubes?

Mixed Review

Lesson 6-7

Complete.

12. 4 ft = _?_ in.   **13.** 72 in. = _?_ ft   **14.** $3\frac{1}{3}$ yd = _?_ in.

Lesson 7-3

Classify each angle as **right, acute,** *or* **obtuse.**

15.    **16.**    **17.**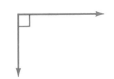

Lesson 8-4

18. The top of a circular plastic lid measures 8 cm from the center to the edge. What is the area of the lid?

REVIEW

Write the letter of the word or phrase that best matches each description.

1. a symbol used to stand for a quantity

2. a ratio that describes how likely it is that an event will occur

3. a comparison of two numbers

4. a ratio that compares a number to 100

5. a ratio that compares distances

6. a sentence that states that two ratios are equivalent

7. geometric figures with corresponding sides that are proportional

8. used to determine if two ratios form a proportion

a. cross products
b. equivalent
c. probability
d. percent
e. proportion
f. ratio
g. scale
h. similar
i. solution
j. variable

Lesson 9-1

Write each ratio as a fraction in simplest form.

9. 24 wins to 18 losses

10. 14 wins in 35 total games

11. 336 miles with 16 gallons

12. 16 passengers on 4 boats

Lesson 9-2

A die is rolled once. Write the probability of rolling each of the following.

13. a 0 or 6

14. a 2 or 4

15. a number greater than 4

Lesson 9-3

Determine if each pair of ratios forms a proportion.

16. 2 to 3, 3 to 4

17. 16 to 12, 48 to 36

18. 150:450, 13:39

Lesson 9-4

Solve each proportion.

19. $\frac{3}{8} = \frac{m}{40}$

20. $\frac{2}{5} = \frac{10}{b}$

21. $\frac{t}{7} = \frac{54}{63}$

22. $\frac{81}{n} = \frac{150}{100}$

23. A 16-ounce box of cereal costs $1.80. At that rate, how much should a 20-ounce box cost?

24. Leslie's car can go 95 miles on 5 gallons of gasoline. How many gallons of gasoline does Leslie need to drive 304 miles?

Lesson 9-5

Find the distance on a scale drawing for each actual distance. The scale is 1 in.:12 mi.

25. 36 mi   **26.** 54 mi   **27.** 96 mi   **28.** 3 mi

Find the missing length for each pair of similar figures.

29.

60 cm 100 cm

36 cm

n

30.

9 ft

m

9 ft

4 ft

6 ft

31.

25 m

40 m

x

15 m

20 m

12 m

32. The shadow of a tree is 32 meters long at the same time that the shadow of a meterstick is 1.6 meters long. How tall is the tree?

33. Anthony drives at an average speed of 52.5 miles per hour. After 4 hours 30 minutes, how many miles has he driven?

34. A race car driver drives 426 miles at an average speed of 142 miles per hour. How long does it take her to drive the 426 miles?

Copy and complete each table. Write fractions in simplest form.

	Percent	Fraction
35.	16%	
36.	92.3%	
37.		$\frac{9}{100}$
38.		$\frac{7}{20}$

	Percent	Decimal
39.	16%	
40.	3.1%	
41.		0.89
42.		2.35

	Percent	Fraction
43.	50%	
44.	$33\frac{1}{3}\%$	
45.		$\frac{1}{4}$
46.		$\frac{1}{5}$

Solve. Use the guess-and-check strategy.

47. The difference between two whole numbers is 7. Their sum is 99. What are the numbers?

48. Mr. Ling is four times as old as his daughter. In 20 years he will be twice as old as his daughter. What are their ages now?

TEST

Two pennies are tossed. Write the probability of each toss.

1. 2 heads   **2.** 1 head and 1 tail   **3.** no heads

Solve each proportion.

4. $\frac{5}{8} = \frac{n}{44}$ **5.** $\frac{3}{7} = \frac{48}{m}$ **6.** $\frac{72}{b} = \frac{87}{29}$

7. Greg's team ended the season with a 5:4 won-lost ratio. They lost 12 games. How many games did they win?

Find the distance on a scale drawing for each actual distance. The scale is 1 cm:20 km.

8. 160 km   **9.** 10 km   **10.** 65 km

Find the missing length for each pair of similar figures.

11.

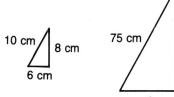

12.

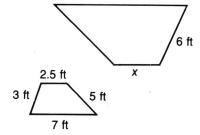

Solve.

13. The Dillers drove 1,102 miles from Memphis to New York. If it took them 24 hours, about what was their average speed?

Copy and complete each table. Write fractions in simplest form.

	Percent	Fraction
14.	32%	
15.	68.7%	
16.		$\frac{1}{5}$

	Percent	Decimal
17.		0.32
18.	2.8%	
19.		2.8

	Percent	Fraction
20.	$66\frac{2}{3}\%$	
21.		$\frac{3}{4}$
22.	$12\frac{1}{2}\%$	

23. Out of 50 students in the band, 16 are also in the chorus. What percent are also in the chorus?

Solve. Use the guess-and-check strategy.

24. Manuel has only nickels, dimes, and quarters. The six coins have a total value of $1.00. How many of each coin does Manuel have?

25. A whole number is multiplied by itself. When the whole number is subtracted from the product, the result is 182. What is the number?

▶ **BONUS:** Create a proportion in which a, b, and c are whole numbers, $a:b = b:c$, and $c = 16a$.

GASOLINE MILEAGE

Gasoline mileage is the number of miles a car travels per gallon of gasoline.

Each week Tamara computes her gas mileage (mpg). She starts each week with a full tank of gasoline. She records the odometer reading each time she fills her gas tank.

Start of week

1	8	3	9	8	.1

End of week

1	8	6	9	2	.1

Tamara subtracts the odometer readings to find the miles driven.

$$
\begin{array}{rl}
18{,}692.1 & \text{end of week} \\
-18{,}398.1 & \text{start of week} \\
\hline
294.0 & \text{miles driven}
\end{array}
$$

Tamara drove 294 miles.

The fill-up took 14.7 gallons. She uses this formula to find the gas mileage.

$$\text{mpg} = \frac{\text{miles driven}}{\text{number of gallons}}$$

$$= \frac{294}{14}$$

Tamara's car got 20 miles per gallon.

Find the miles per gallon for each car. Round decimal answers to the nearest tenth.

Car	Miles Driven	Number of Gallons
1. Dave's Honda	524.4	15.2
2. Kim's Ford	156.0	7.5

3. Julio drove a total of 320.6 miles and used 18.2 gallons of gasoline. What is his gas mileage?

4. How far can Jason travel on a tankful of gasoline if his car averages 18.5 mpg and the tank holds 12.5 gallons?

5. If Jon drives from New York to Cincinnati, a distance of 660 miles, and his car averages 25 mpg, to the nearest gallon how much gasoline does he need?

Free Response

Lesson 1-9

Find the value of each expression.

1. 25 × 4 ÷ 2

2. 3 × (8 − 6) × 4

3. 3 × [6 × (5 + 3)]

**Lessons 2-1
2-2
2-3
2-4**

Add or subtract.

4. 4,065 + 178

5. 246 + 93 + 10.7

6. 14,078 − 5,928

7. 85.22 − 3.17

8. 7.17 + 68.9

9. 6.03 − 1.945

**Lessons 3-1
3-4
3-7**

Multiply or divide.

10. 1,581 ÷ 23

11. 39,096 ÷ 362

12. 86 × 250

13. 783 × 207

14. 14.05 × 1,000

15. 0.7 ÷ 1,000

Lesson 5-10

16. Carol Kerns orders a truckload of topsoil. She uses $\frac{1}{2}$ in the garden, $\frac{1}{3}$ in the flower beds, and $\frac{1}{6}$ in the yard. How much topsoil is left?

Lesson 7-6

Name each polygon by the number of sides. Then state whether it is regular or not regular.

17.

18.

19.

20.

Lesson 8-1

21. Steve uses 50 feet of fence to enclose a circular area for a dog pen. To the nearest foot, what is the radius of the dog pen?

Lesson 8-4

Find the area of each circle whose radius is given. Use 3.14 for π. Round decimal answers to the nearest tenth.

22. 2 in.

23. 10 in.

Lesson 8-11

24. A soup can is 5.5 inches tall. The top and bottom of the can are circular and each have a 1-inch radius. To the nearest hundredth of a cubic inch, find the volume of the can.

Lesson 9-4

Solve each proportion.

25. $\frac{5}{6} = \frac{x}{18}$

26. $\frac{3}{4} = \frac{x}{100}$

27. $\frac{x}{7} = \frac{9}{100}$

28. $\frac{x}{1,000} = \frac{1}{8}$

29. $\frac{25}{x} = \frac{20}{100}$

Multiple Choice

Choose the letter of the correct answer for each item.

1. Name the digit in the thousandths place-value position in 97,842.013.
  a. 1   **c.** 7
  b. 3   **d.** 9

Lesson 1-1

2. Find the difference between 1,081 and 75.
  e. 906
  f. 1,006
  g. 1,014
  h. 1,016

Lesson 2-3

3. What is the product of 6.5 and 7.02?
  a. 42.471
  b. 45.63
  c. 424.71
  d. 456.3

Lesson 3-2

4. Replace the variable with a number to make a true statement. $\frac{3}{18} = \frac{n}{90}$
  e. 5   **g.** 75
  f. 15   **h.** *none of these*

Lesson 4-5

5. Change 13.5% to a fraction.
  a. 0.135
  b. $\frac{27}{200}$
  c. $\frac{27}{20}$
  d. $\frac{27}{2}$

Lesson 5-9

6. Which measurement is equivalent to 76 inches?
  e. 6 ft 4 in.   **g.** 2 yd
  f. 6 ft 6 in.   **h.** 2 yd 1 ft

Lesson 6-7

7. Classify the triangle shown below.
  a. isosceles, acute
  b. isosceles, obtuse
  c. equilateral, acute
  d. scalene, acute

Lesson 7-7

8. The formula for the volume of a pyramid is $V = \frac{1}{3}\ell wh$.
Find the volume.
  e. 18 cm³
  f. 600 cm³
  g. 720 cm³
  h. 1,800 cm³

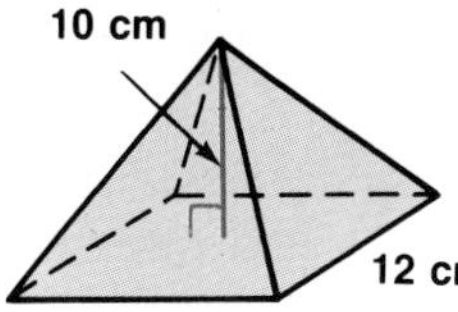

10 cm
12 cm
15 cm

Lesson 8-10

9. A bag contains 4 green marbles, 2 white marbles, and 3 black marbles. One marble is drawn. Find the probability that it is black or white.
  a. $\frac{6}{81}$ **c.** $\frac{3}{7}$
  b. $\frac{2}{7}$ **d.** $\frac{5}{9}$

Lesson 9-2

10. Albert is considering purchasing carpeting for his home. Which type of measurement will he need to be most concerned about?
  e. length of hallway
  f. total cubic feet
  g. total square feet
  h. width of living room

Lesson 8-5

fun with MATH

First schools Sumer (Babylon)	1800 BC	Reconnaissance kite China	AD 399
2500 BC	Stonehenge England	206 BC	Fabiola Roman physician

Why all the fuss about acid rain?

The burning of fossil fuels by industrial plants releases pollutants into the air. These pollutants often form acid compounds and travel vast distances before falling back to Earth as acid rain. Making a fuss may be the only way to get the polluters to listen and to get effective national and international laws passed to stop the production of the deadly pollutants.

MATH M·E·N·U

Mexican Cocoa

4 cups of milk
1 4-ounce milk chocolate candy bar
1 teaspoon ground cinnamon
1 teaspoon vanilla

Pour milk into a saucepan and heat at medium temperature. Add cut up chocolate and stir until it is melted. Remove from heat and add cinnamon and vanilla. Set aside. At serving time, return to heat and beat mixture with a whisk until frothy and hot. Pour into cups and serve.

COMICS

Honey is the only food that does not spoil. Honey, found in the tombs of Egyptian pharaohs, that was thousands of years old, was found safe to eat.

FUNKY WINKERBEAN

Multiplication sign (×) England		Photocopier USA	
AD 1631	First chocolate as food England	AD 1938	Blood plasma program USA

AD 1847 — (Multiplication sign)
AD 1940 — (Photocopier)

JOKE!

Q: Did you hear what happened to the plants in the math room?

A: They grew square roots!

How is it possible to sail faster than the wind? When wind fills the sail of a sailboat, the sail takes on a convex shape. As a sailboat sails at an angle into the wind, the sail splits the air flowing past. The air is squeezed and must speed up to get around the front of the sail. This reduces the air pressure outside and in front of the sail creating a suction force which pulls the boat along, in addition to its being pushed. It is because of the extra pulling force that a sailboat can actually sail faster than the wind.

RIDDLE

Q: A father told his daughter to be home from a date at a quarter of 12. She came home at 3. How did she explain the late hour of her arrival to her dad?

A: She reminded him that in math class she had learned that ¼ of 12 is 3.

QUESTION and ANSWER

How does a photocopier make copies?

A light, lens, and mirrors operate to project an image of the item to be copied onto a metal drum. The drum receives a negative electric charge that disappears wherever light-colored areas of the image strike the metal surface. The dark areas remain negatively charged. Positively-charged particles of dark toner powder are attracted to the negatively-charged dark areas. The image made up of dark toner is transferred to paper and sealed by a heater. A warm copy of your original item emerges.

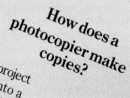

QUIZ TIME

The following are three views of the same cube. Name the side opposite A by carefully studying the three views. Then, name the side opposite B and the side opposite C.

10 APPLYING PERCENTS

SALE!!!   30% off everything

Do you check for advertised sales on items you buy? Do you watch for special savings on expensive items you would like to buy? Many businesses sell items at a certain percent off during seasonal sales and clearance sales. Businesses also use percents to express profits, losses, discounts, commissions, and interest.

Shirley Brobeck sells sweatshirts for Harper's Sweatshirts. In addition to her salary, she earns 2% of the retail price of any goods she sells. These earnings, based on sales, are called a commission. How much commission does Mrs. Brobeck earn for selling $189 worth of sweatshirts?

ACTIVITY: Percent

In this activity you will find the percent of a number. Suppose you surveyed your class to find the number of students who rented a video tape last week. Of the 100 students surveyed, 33 students said they rented a video tape last week. What percent of the students rented a video tape?

Materials: 10 × 10 grids, colored pencil (optional)

Cooperative Groups

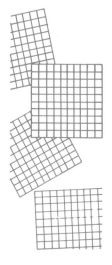

Work together in groups of three or four.

1. Shade 33 squares on a 10 × 10 grid. Write a fraction with a denominator of 100 that represents the shaded part of the grid.

2. Percents are ratios that compare numbers to 100. The word *percent* means per 100 or hundredths. The percent symbol (%) means that a number is being compared to 100. If 33 out of 100 students had rented a video last week, write the percent of students who had rented a video using the percent symbol (%).

3. Percents can be expressed as fractions without the percent symbol. Look at the fraction you wrote in Exercise 1. What is the fraction in simplest form? How would you write this fraction as a decimal?

4. Numbers can be expressed as a percent, as a fraction, as a decimal, and as a ratio. Write the number of students who rented a video tape in the last week in four ways. Each of the four ways represents the same percent.

5. Shade 46 squares on a 10 × 10 grid. Write the number of shaded squares as a fraction, a decimal, a ratio, and as a percent.

Extend the Concept

6. Decide with your group how to express 1 out of 100 and 0.5 out of 100 as a fraction, a decimal, and a percent.

7. Decide with your group how to express 2,000 out of 100 as a fraction, a decimal, and a percent.

8. Suppose you read an article that says the cost of oranges will be 240% of the usual price because of frost in Florida. With your group, write a sentence describing the increase in the cost of oranges.

You will study more about percent in this chapter.

10-1 FINDING THE PERCENT OF A NUMBER

Objective
Find the percent of a number.

Amanda has been shopping for a combination radio/tape player. She found the one she likes on sale at 40% off the regular price. If the radio usually sells for $90, how much would Amanda save if she buys the radio on sale? To answer the question, find 40% of $90.

RADIOS
40% OFF
regular price $90

To find the percent of a number, multiply the number by the percent. The equation below shows the relationship between the percent or rate r, the number or base b, and the percentage p.

$$\text{rate} \times \text{base} = \text{percentage}$$
$$r \times b = p$$

The percent of a number is called a percentage.

Method

1 ▶ Write the percent as a fraction or decimal.
2 ▶ Multiply the base by the decimal or fraction.

Examples

A How much will Amanda save?
Find 40% of $90.

1 ▶ 40% → 0.40

2 ▶ $0.40 \boxed{\times} 90 \boxed{=} 36$

40% of 90 = 36

Amanda will save $36.

B Find 105% of 72.

1 ▶ 105% → 1.05

2 ▶ $1.05 \boxed{\times} 72 \boxed{=} 75.60$

105% of 72 is 75.6.

It is easier to change some percents to fractions rather than decimals. Some convenient facts are given in the table at the right.

$25\% = \frac{1}{4}$	$33\frac{1}{3}\% = \frac{1}{3}$	$20\% = \frac{1}{5}$	$12\frac{1}{2}\% = \frac{1}{8}$
$50\% = \frac{1}{2}$	$66\frac{2}{3}\% = \frac{2}{3}$	$40\% = \frac{2}{5}$	$37\frac{1}{2}\% = \frac{3}{8}$
$75\% = \frac{3}{4}$		$60\% = \frac{3}{5}$	$62\frac{1}{2}\% = \frac{5}{8}$
		$80\% = \frac{4}{5}$	$87\frac{1}{2}\% = \frac{7}{8}$

Example C

Find $33\frac{1}{3}$% of 45.

1 ▶ $33\frac{1}{3}\% = \frac{1}{3}$

2 ▶ $\frac{1}{3} \times 45 = \frac{1}{\cancel{3}} \times \cancel{45}^{15}$

$= 15$

Guided Practice

Examples A, B

Find each percentage. Use a decimal for the percent.

1. 12% of 50   **2.** 18% of 70   **3.** 113% of 70   **4.** 3% of 45

Find each percentage. Use a fraction for the percent.

Example C

5. 10% of 50   **6.** $12\frac{1}{2}$% of 16   **7.** $66\frac{2}{3}$% of 27   **8.** 40% of 16

Find each percentage. Use a decimal for the percent.

9. 15% of 20   **10.** 12% of 50   **11.** 22% of 500   **12.** 36% of 300

13. 1% of 200   **14.** 5% of 40   **15.** 6% of 9   **16.** 8% of 7,250

Find each percentage. Use a fraction for the percent.

17. 25% of 40   **18.** 80% of 200   **19.** 50% of 86   **20.** 60% of 95

21. 80% of 915   **22.** $37\frac{1}{2}$% of 320   **23.** $87\frac{1}{2}$% of 8   **24.** 10% of 7.2

Find each percentage.

25. 75% of 16   **26.** 7% of 600   **27.** 240% of 25   **28.** 23% of 90

29. $37\frac{1}{2}$% of 120   **30.** 4% of 75   **31.** 15% of 200   **32.** $12\frac{1}{2}$% of 300

33. *True* or *False*: $37\frac{1}{2}$% of 24 is 8.

Talk Math

34. Describe the advantages of using a fraction instead of a decimal to find $33\frac{1}{3}$% of 45?

35. You know 50% $= \frac{1}{2}$ and $12\frac{1}{2}$% $= \frac{1}{8}$. How would you find the percents that are equivalent to $\frac{5}{8}$ and $\frac{3}{8}$?

Applications

36. A mountain pack usually sells for $48. The sale price is 80% of the usual price. What is the sale price?

| Paper/ Pencil |
| Mental Math |
| Estimate |
| Calculator |

37. Of the light bulbs produced at Manual Lighting, 2% are defective. If 2,500 light bulbs are produced each hour, how many defective bulbs are produced in an 8-hour shift?

38. Mrs. Roberts sells stock for her client. The total sale price of the stock is $25,000. If her commission is 3% of the sale price, how much commission does she receive?

Critical Thinking

39. Make 1,001 using only 7s and any combination of operations: $+, -, \times, \div$.

Mixed Review

Lesson 2-8

Find the perimeter.

40. square; side, 13.3 cm

41. rectangle; length, 4 in. width, 3.2 in.

Lesson 5-7

Divide.

42. $3\frac{1}{4} \div 2\frac{1}{2}$ **43.** $8\frac{2}{3} \div 2\frac{3}{4}$ **44.** $10\frac{3}{5} \div 3\frac{1}{3}$

Lesson 5-5

45. Andrew and Lamont can load 28 cartons a minute. If three friends help and all three work at the same rate, how long will it take all five friends to load 1,260 cartons?

10-2 FINDING WHAT PERCENT ONE NUMBER IS OF ANOTHER

Objective

Find the percent one number is of another.

Josh is buying a skateboard. The one he likes was originally priced for $120. The sale price is $90. $90 is what percent of the original amount?

In this problem you need to find the rate. To find the rate (the percent) when the base $120 and percentage $90 are known, use the equation $r \times b = p$ and solve for r.

Method

▶1 Write the equation $r \times b = p$.

▶2 Solve for r.

Example A

▶1 $r \times 120 = 90$

▶2 $\dfrac{r \times 120}{120} = \dfrac{90}{120}$ $b = 120$ $p = 90$

$r = $ 〔90〕〔÷〕〔120〕〔=〕〔0.75〕

or 75%

$90 is 75% of $120.

Example B

80 is what percent of 64?

▶1 $r \times 64 = 80$

▶2 $\dfrac{r \times 64}{64} = \dfrac{80}{64}$ $b = 64, p = 80$

$r = $ 〔80〕〔÷〕〔64〕〔=〕〔1.25〕 or 125%

80 is 125% of 64.

Guided Practice

Example A

Find each percent. Use an equation.

1. What percent of 90 is 72?

2. What percent of 80 is 72?

3. What percent of 200 is 30?

4. What percent of 16 is 24?

Example B

5. 15 is what percent of 40?

6. 14 is what percent of 28?

7. 40 is what percent of 25?

8. 4 is what percent of 5?

314   CHAPTER 10   APPLYING PERCENTS

Find each percent. Use an equation.

9. What percent of 48 is 32?

10. What percent of 200 is 60?

11. What percent of 35 is 175?

12. What percent of 36 is 300?

13. What percent of 80 is 2?

14. What percent of 75 is 3?

15. 180 is what percent of 120?

16. 16 is what percent of 12?

17. 150 is what percent of 200?

18. 32 is what percent of 128?

19. 12 is what percent of 16?

20. 15 is what percent of 75?

Applications

21. The mountain bike that Sue wants usually sells for $300. The sale price is $240. What percent of the original price is the sale price?

22. Troy missed 20 of the 125 points on a science test. What percent of the total points did Troy miss?

23. Hazel made 18 of her 24 free throw attempts in her last 4 games. What percent of her free throws did Hazel make?

24. Sam Strothers is a sales representative for Beacon Fashions, Inc. If he sells 115% of his first six months' quota (sales goal), Mr. Strothers will earn a three-day trip to Miami. If his quota is $18,500 and Mr. Strothers sold $22,200 worth of goods, what percent of his quota did he sell?

Suppose

25. Suppose 80 out of 200 people between the ages of 18 and 34 would find it exciting to visit another solar system. What percent of people between the ages of 18 and 34 would find it exciting to visit another solar system?

Collect Data

26. Survey your class. Find what percent would like to visit another solar system. Find what percent would like to visit a space station.

Estimation

At a restaurant, it is customary to tip the waiter or waitress. Many people leave 15% of the total bill as a tip. It is easy to calculate 15% of a number using the fact that 15% = 10% + 5%.

To estimate 15% of $24.80, take 10% of $25 and then add 5% of $25.

$$10\% \text{ equals } \frac{1}{10}, \text{ so } \frac{1}{10} \text{ of } \$25 \text{ equals } \$2.50.$$

$$5\% \text{ is half of } 10\%, \text{ so } \frac{1}{2} \text{ of } \$2.50 \text{ equals } \$1.25.$$

15% of $24.80 is about $2.50 + $1.25 or about $3.75.

Estimate a 15% tip for each amount.

27. $12

28. $28

29. $13.60

30. $9.20

10-3 FINDING A NUMBER WHEN A PERCENT OF IT IS KNOWN

Objective

Find a number when a percent of it is known.

After softball season, Stacy's coach told her that she got a hit 40% of the times she was at bat. If she had 48 hits, how many times was Stacy at bat? You need to find the number of which 48 is 40%.

To find a number when a percent of it is known, use $r \times b = p$ and solve for b.

Method

1 Write the percent as a decimal or fraction.

2 Write the equation using the variable b for the base.

3 Solve for b.

Example A

How many times was Stacy at bat? 40% of a number is 48. What is the number?

1 $40\% \rightarrow 0.40$ $r = 40\%, \quad p = 48$

2 $0.40 \times b = 48$ **3** $\dfrac{0.40 \times b}{0.40} = \dfrac{48}{0.40}$

$$48 \;\boxed{\div}\; .4 \;\boxed{=}\; 120$$

$$b = 120$$

40% of 120 is 48.

Stacy was at bat 120 times.

Examples

B $66\frac{2}{3}\%$ **of what number is 8?**

1 $66\frac{2}{3}\% \rightarrow \frac{2}{3}$

2 $\frac{2}{3} \times b = 8$

3 $\overset{1}{\underset{1}{\cancel{\frac{3}{2}}}} \times \overset{1}{\underset{1}{\cancel{\frac{2}{3}}}} \times b = \overset{}{\underset{1}{\cancel{\frac{3}{2}}}} \times \overset{4}{\cancel{8}}$

$$b = 12$$

$66\frac{2}{3}\%$ of 12 is 8.

C *11.7 is 200% of what number?*

1 $200\% \rightarrow 2.00$ or 2

2 $2 \times b = 11.7$

3 $\dfrac{2 \times b}{2} = \dfrac{11.7}{2}$

$$b = 5.85$$

11.7 is 200% of 5.85.

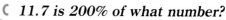

Guided Practice

Find each number. Use an equation.

Example A

1. 15% of the number is 3.

2. 125% of the number is 50.

3. 50% of what number is 12?

4. $33\frac{1}{3}$% of what number is 18?

5. 45 is 300% of what number?

6. 69.3 is 110% of what number?

Practice

Find each number. Use an equation.

7. 15% of the number is 6.

8. 27% of the number is 54.

9. 28% of the number is 14.

10. 7% of the number is 6.3

11. 25% of the number is 17.

12. 50% of the number is 29.

13. 30% of what number is 30?

14. 45% of what number is 18?

15. $33\frac{1}{3}$% of what number is 20?

16. 125% of what number is 15?

17. 200% of what number is 35?

18. 300% of what number is 9?

19. 10 is 20% of what number?

20. 6.2 is 40% of what number?

21. 4 is $12\frac{1}{2}$% of what number?

22. 18 is $66\frac{2}{3}$% of what number?

23. 150 is 200% of what number?

24. 8 is 100% of what number?

Application

25. In the election for freshman class officers, Joann received 52% of the votes. If Joann received 104 votes, how many votes were cast?

Using Data

Use the chart for Exercises 26 through 28.

26. Businesses owned and operated by women in 1987 accounted for 13.9% of business revenue in the United States. What was the total business revenue in the United States for firms owned by both men and women?

***f*UN with MATH**

Where and when was statistics established?
See page 373.

27. If women owned 30% of the number of United States businesses in 1987, how many businesses were owned by either men or both men and women in 1987?

28. How many more businesses were owned by women in 1987 than in 1982? This increase is what percent of the number of businesses owned by women in 1982?

Critical Thinking

29. Swapping two adjacent (side-by-side) markers constitutes one move in a game. What is the least number of moves required to change

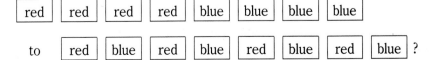

| red | red | red | red | blue | blue | blue | blue |

to

| red | blue | red | blue | red | blue | red | blue | ?

10-4 USING VENN DIAGRAMS

Objective

Solve problems using Venn diagrams.

The Blake High School *Gazette* reported the results of a survey of 46 students. Students were asked how many like to drive a car or a van. Jan wants to know how many like to drive neither type of vehicle.

Results of Car Survey	
Like cars	22
Like vans	28
Like both	7

▶ **Explore**

What is given?
● Seven students like both cars and vans.
● Twenty-two students like cars.
● Twenty-eight students like vans.
● The total number of students surveyed is 46.

What is asked?
● How many students like neither cars nor vans?

▶ **Plan**

Use a Venn diagram.

▶ **Solve**

The rectangle represents the total number of students surveyed, 46. Each circle represents a type of vehicle. The overlapping area of the two circles represents the number of students who like both. Write 7 in this area.

The number of students who like *only* vans is 28 − 7 or 21. Write 21 as shown.

The number of students who like *only* cars is 22 − 7 or 15. Write 15 as shown.

The number of students who like one type or both types of vehicle is found by adding 21 + 15 + 7 or 43.

The number of students who like neither type of vehicle is 46 − 43 or 3.

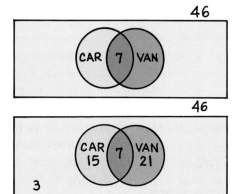

▶ **Examine**

Check your solution by adding all the possibilities from each part of the diagram.   15 + 21 + 7 + 3 = 46   The solution is correct.

Guided Practice

Solve. Use a Venn diagram.

1. Information Now, Inc. mailed 125 families a survey asking which types of movies they attended in the last six months. How many families did not return the survey?

Type of Movie	Attended
Comedy	79
Adventure	61
Cartoon	22
Comedy and Cartoon	6
Cartoon and Adventure	5
Comedy and Adventure	26
Comedy, Cartoon, and Adventure	4

Solve. Use any strategy.

2. Renley High School intramural participants were surveyed about the sports in which they participate. Thirty-five play baseball, 31 play basketball, 28 play soccer, 7 play baseball and basketball, 9 play basketball and soccer, 6 play baseball and soccer, and 5 play all three sports. How many students were surveyed?

3. Find the least common multiple of 2, 3, and 9.

4. Julie offers free samples of apple juice in Star's grocery store to encourage customers to buy the juice. She has 10 cases of juice to sell. If each case contains 12 bottles of juice and Julie sells 72 bottles, what percent of her stock does she sell?

5. Nick is a waiter at Cook's Family Restaurant. In 1 hour on Thursday night, he waited on 4 tables. The bills for the 4 tables came to $24.30, $26.80, $25.75, and $29.90. If Nick's customers left the customary 15% tip, *about* how much should Nick expect in tips for the 4 tables?

6. Jonelle delivers newspapers to 53 families 6 days a week. The Globe offers a flat rate of $21.50 per week or a rate of $0.07 a paper to paper carriers. Should Jonelle accept the flat rate or the per paper rate?

JOURNAL ENTRY

7. Describe the strategies you used to solve problems 2 through 6 above.

Critical Thinking

8. Count all the rectangles of any size in the figure.

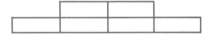

Mixed Review

Lesson 4-7

Replace each ● with <, >, or = to make a true statement.

9. $\frac{4}{5}$ ● $\frac{8}{10}$ **10.** $\frac{2}{3}$ ● $\frac{3}{4}$ **11.** $\frac{5}{6}$ ● $\frac{4}{5}$ **12.** $\frac{8}{9}$ ● $\frac{10}{12}$

Lesson 6-1

Complete.

13. 400 mm = $\underline{?}$ cm   **14.** 0.5 m = $\underline{?}$ cm   **15.** 3.9 cm = $\underline{?}$ mm

Lesson 7-5

16. If you were to choose an edge in a cube, at random, how many other edges would be parallel to the edge? How many edges in the same plane would be perpendicular to the edge?

10-5 ESTIMATING THE PERCENT OF A NUMBER

Objective

Estimate the percent of a number.

Mary McCoy bought a VCR so she could tape her favorite television show on Thursday night while she is at work. The sales tax on the VCR is 8%. If the VCR costs $299, about how much tax did Mrs. McCoy pay?

Percentages can be estimated when the exact amount is not necessary or when you wish to check the reasonableness of a calculated result.

Method

1 ► Round the percent to a familiar fractional equivalent.

2 ► Round any other factors to numbers that can be multiplied mentally.

3 ► Multiply.

Examples

A **About *how much tax did Mrs. McCoy pay? Estimate 8% of $299.***

1 ► 8% is a little less than 10%, which is $\frac{1}{10}$.

2 ► 299 rounds to 300, which is a multiple of 10.

3 ► $\frac{1}{10} \times 300 = 30$

Mrs. McCoy paid *about* $30 in sales tax.

B ***Estimate 40% of 86.***

1 ► 40% → $\frac{2}{5}$

2 ► Round 86 to 85 because 85 is a multiple of 5.

3 ► $\frac{2}{\overset{}{\underset{1}{5}}} \times \overset{17}{\cancel{85}} = 34$

40% of 86 is *about* 34.

Example C

Estimate 22% of $18.27.

1 ► 22% is a little more than 20%, which is $\frac{1}{5}$.

2 ► Round 18.27 to 20 because 20 is a multiple of 5.

3 ► $\frac{1}{5} \times 20 = 4$

22% of $18.27 is *about* $4.00.

Guided Practice

Estimate.

Example A

1. 9% of 50

2. 18% of 150

Example B

3. 10% of 19

4. 25% of 43

Example C

5. 77% of 315

6. 35% of $8.89

Estimate.

7. 11% of 108.9
8. 9% of 315
9. 63% of 120

10. 70% of 82
11. 15% of 58
12. 6% of 2,459

13. 82% of 27
14. 73% of 27
15. 22% of 27

16. 110% of 89
17. 121% of 70
18. $\frac{7}{8}$% of 200

19. $1\frac{1}{4}$% of 100
20. 0.5% of 800
21. 0.8% of 500

22. 0.1% of 90
23. 65% of 27
24. 1.2% of $340

25. Estimate 12% of $18 billion.

26. *True* or *false*: 19% of 26,000 is about 5,000.

Applications

27. Bill and Marie, and John and Sherri went to the Tiki Polynesian restaurant after the prom. The bill for dinner came to $49.56. *About* how much should they have left as a customary 15% tip?

28. Bev has started a savings account for college. She has $250 in the account. If Bev's money earns 6% annual interest, *about* how much will she receive in interest if she leaves the money in her account for 1 year?

29. Mack put $900 in a 12-month certificate of deposit that earns 9.5%. *About* how much interest will Mack receive after 1 year?

30. Lucy found a blouse on sale for $18.99 and slacks on sale for $31.90. The sales tax is 5%. Lucy has $60 in her purse. *About* how much sales tax would Lucy pay? Will Lucy be able to buy the blouse and slacks?

Make Up a
Problem

31. Make up a percent problem that has an estimated answer of *about* $25.

32. Make up a percent problem that has an estimated answer of *about* 15%.

Show Math

33. Draw a rectangle about 2 inches long and 1 inch wide. Separate the area of the rectangle into 4 parts as follows. One part should be about 49% of the area. Each of the other 3 parts should be about 17% of the area.

Estimation

One way to mentally estimate sales tax is as follows. Round the price to the nearest ten dollars. Then multiply the tax by the rounded price.

The sales tax is 6%. Estimate the total price, tax included, of a dress that costs $57.99.

A tax rate of 6% means 6¢ tax on every dollar.

$57.99 rounds to $60.   60 × 6¢ = 360¢ or $3.60

The total cost is *about* $58 + $3.60 or $61.60.

Estimate the sales tax and total cost for each price. The tax rate is 8%.

34. $18.99
35. $7.89
36. $51.99
37. $83.25
38. $238

SALES TAX

Jacob works part time for Henley's Pharmacy. For amounts under $11.00, Jacob uses a tax table to find the sales tax rather than computing the tax. If Joan's purchases totaled $9.68, what is the amount of the sales tax from the tax table shown below?

In most states, a sales tax is charged on purchases. The sales tax provides a way to raise money for states and cities. It is a percent of each sale.

SALES & USE TAX FOR STATE, COUNTY AND/OR
TRANSIT TAX—TOTAL 5 1/2% TAX LEVY
SALES 15¢ AND UNDER—NO TAX

Each Sale	Tax	Each Sale	Tax
.16 to .18	.01	5.47 to 5.64	.31
.19 to .36	.02	5.65 to 5.82	.32
.37 to .54	.03	5.83 to 6.00	.33
.55 to .72	.04	6.01 to 6.18	.34
.73 to .90	.05	6.19 to 6.36	.35
.91 to 1.09	.06	6.37 to 6.54	.36
1.10 to 1.27	.07	6.55 to 6.72	.37
1.28 to 1.46	.08	6.73 to 6.90	.38
1.47 to 1.64	.09	6.91 to 7.09	.39
1.65 to 1.82	.10	7.10 to 7.27	.40
1.83 to 2.00	.11	7.28 to 7.46	.41
2.01 to 2.18	.12	7.47 to 7.64	.42
2.19 to 2.36	.13	7.65 to 7.82	.43
2.37 to 2.54	.14	7.83 to 8.00	.44
2.55 to 2.72	.15	8.01 to 8.18	.45
2.73 to 2.90	.16	8.19 to 8.36	.46
2.91 to 3.09	.17	8.37 to 8.54	.47
3.10 to 3.27	.18	8.55 to 8.72	.48
3.28 to 3.46	.19	8.73 to 8.90	.49
3.47 to 3.64	.20	8.91 to 9.09	.50
3.65 to 3.82	.21	9.10 to 9.27	.51
3.83 to 4.00	.22	9.28 to 9.46	.52
4.01 to 4.18	.23	9.47 to 9.64	.53
4.19 to 4.36	.24	9.65 to 9.82	.54
4.37 to 4.54	.25	9.83 to 10.00	.55
4.55 to 4.72	.26	10.01 to 10.18	.56
4.73 to 4.90	.27	10.19 to 10.36	.57
4.91 to 5.09	.28	10.37 to 10.54	.58
5.10 to 5.27	.29	10.55 to 10.72	.59
5.28 to 5.46	.30	10.73 to 10.90	.60

Find the sales tax on a $9.68 purchase.

$9.68 is more than $9.55 but less than $9.72. The tax on $9.68 is $0.54.

For larger purchases, multiplication can be used to find the amount of tax.

Find the tax on a $180 purchase with a $5\frac{1}{2}\%$ sales tax.

1 8 0 $\times$ .0 5 5 $=$ 9.9   The tax is $9.90.

9.9 $+$ 1 8 0 $=$ 1 8 9.9   The total cost is $189.90.

Find the sales tax and total cost for each price. The tax rate is $5\frac{1}{2}\%$. Use the tax schedule or multiplication.

1. $9.95   2. $24.99   3. $114.95   4. $157.50

5. Jackie bought a blouse for $20, jeans for $32, socks for $4.95 and a sweater for $38. Find the total sales tax and the total cost.

SALES REPRESENTATIVE

Kelly Chapin sells paper goods to restaurants. She earns 3.5% **commission** on sales. This means she earns $0.035 on each $1.00 in sales. The sales for one week total $4,575. Estimate her commission.

To estimate a product, round each factor to its greatest place-value position. Then multiply.

$$\begin{array}{r} \$4,575 \\ \times\ 0.035 \end{array} \rightarrow \begin{array}{r} \$5,000 \\ \times\ 0.04 \\ \hline \$200.00 \end{array}$$

Kelly estimates that her commission earnings are *about* $200.

Compute the exact amount she earned.

$$4575 \ \boxed{\times} \ .035 \ \boxed{=} \ 160.125$$

Kelly earned $160.13 in commission.

Estimate. Then compute the exact amount.

1. $867.50
   × 0.09

2. 5,472
   × 0.055

3. $34,920
   × 0.085

4. Jan Hille sells used cars. She earns 4.5% commission on sales. Her sales for last week were $32,854. Estimate her commission. Then compute the exact amount.

5. Jill Kunkle sells jewelry for Kingman's Diamond Emporium. Jill earns 8% commission and her monthly sales total is $29,450. Estimate her commission. Then compute the exact amount.

6. Refer to Exercise 5. Estimate Ms. Kunkle's annual earnings assuming sales continue at the same rate.

7. If Alex Higgins' commission is 8%, how much will he receive for selling 25 boxes of computer disks at $32.25 per box?

8. Jose Clemente received a commission of $9,750 for selling a house for $150,000. What is his rate of commission?

9. Jason Baxter earns a salary of $400 plus 7% commission on sales *over* $4,500. If his sales are $6,218, what are his total earnings?

10-6 PERCENT OF CHANGE

Objective

Find the percent of increase or decrease in a quantity or price.

Steve works for the Green Thumb garden shop after school and on Saturdays. Last week his hourly wage was raised from $5.00 to $5.45. What is the percent of increase in Steve's hourly wage?

When the quantity or price of an item changes, the amount of increase or decrease from the original amount can be written as a percent. This percent is called the percent of change.

Method

1 ▶ Find the amount of increase or decrease.

2 ▶ Find what percent the amount of increase or decrease is of the original amount.

Example A

Find the percent of increase in Steve's hourly wage.

1 ▶ 5.45 ⊟ 5.00 ⊟ 0.45 ⊡ 5.00 ⊟ 0.09

amount of increase ⟶ **2**

Write the result as a percent. The percent of increase in Steve's hourly wage is 9%.

Example B

For the first month at college, Ann's phone bill was $48.50. Ann's phone bill for the second month was $38.80. Find the percent of decrease in Ann's bill.

Estimate: $38.80 is about $40 and $48.50 is about $50.

$50 − $40 = $10. $10 is $\frac{1}{5}$ of $50 or 20%.

1 ▶ 48.50 ⊟ 38.80 ⊟ 9.70 ⊡ 48.50 ⊟ 0.2

amount of decrease ⟶ **2**

Write the result as a percent. The percent of decrease in Ann's phone bill is 20%.

─ Guided Practice ─

Find the percent of increase or decrease. Round to the nearest percent.

Examples A, B

	Item	Original Price	New Price
1.	Shirt	$18	$20.25
2.	Tie	$12	$15
3.	Shoes	$36	$38.16
4.	Jeans	$39.50	$43.90

	City	Rainfall (in.) 1985	1987
5.	Buffalo	46.00	38.61
6.	Ashville	35.94	26.50
7.	Anchorage	15.51	14.32
8.	Norfolk	44.81	38.68

Practice

Find the percent of increase. Round to the nearest percent.

	Item	Original Price	New Price
9.	Milk	$1.50/gal	$1.79/gal
10.	Bread	$0.94/loaf	$1.09/loaf
11.	Yogurt	48¢/cup	59¢/cup

	Item	Original Price	New Price
12.	Grapes	80¢/pound	99¢/lb
13.	Candy	35¢/bar	45¢/bar
14.	Juice	69¢/bottle	79¢/bottle

Find the percent of decrease. Round to the nearest percent.

	Item	Original Price	New Price
15.	Calculator	$50	$15
16.	Radio	$65	$39
17.	Computer	$10,300	$2,884

	City	January Rainfall	April Rainfall
18.	New York	6.4 in.	5.44 in.
19.	Phoenix	0.44 in.	0.33 in.
20.	Chicago	3 in.	2.88 in.

Applications

21. Mrs. Hsu manages the coffee shop-deli at the Corporate Corners office building. In the last six months, sales of regular coffee have decreased from 160 cups to 85 cups per morning. Sales of decaffeinated coffee have increased from 48 cups to 72 cups per morning. What is the percent of decrease in sales of regular coffee?

22. A catalog ad from a 1909 issue advertised a disc talking machine (phonograph) for $14.90. A catalog ad from a 1990 issue advertised a compact disc player for $179.90. What is the percent of increase in price?

Research

23. In the library, from old newpapers or magazines at least thirty years old, research the price of a shirt, a pair of shoes, or a band instrument, such as a trumpet. Find the percent of increase in price.

Suppose

24. Suppose during the oil crisis in 1990, the price of a gallon of gasoline at your favorite gas station went up 34%. Then it dropped 10% after three months. If the price of a gallon of gasoline before the crisis was $1.19, what was the price of a gallon three months after the crisis began?

Mixed Review

Lesson 6-1

Name the larger unit.

25. 1 mm or 1 m

26. 1 liter or 1 kiloliter

Lesson 6-10

Find the elapsed time.

27. from 2:40 AM to 6:50 AM

28. from 10:45 AM to 10:40 PM

Lesson 8-1

29. Joni wants to decorate an empty cookie tin to hold odds and ends on her desk. If the can is 8 inches across, how long does the piece of fabric have to be to fit around the can?

10-7 Discount

Objective

Find the discount, the rate of discount, and the sale price.

Barb Bennett is moving into her own apartment. She wants to buy a microwave oven, a toaster, and a television. Kitchen Korner offers a microwave at a discount of 20%. If the regular price of the microwave is $349, find the discount and the sale price.

A **discount** is an amount subtracted from the regular price of an item. A **discount rate** is a percent of decrease in the regular price.

Method

▶ 1 Multiply the regular price by the rate of discount to find the discount.

▶ 2 Subtract the discount from the regular price to find the sale price.

Example A

What is the discount and the sale price of the microwave oven selling at a discount rate of 20% if the regular price is $349?

Estimate: 20% or $\frac{1}{5}$ of 350 is 70. The discount is about $70.

▶ 1 $349 \ominus 20 \boxed{\%} 69.8 \ominus 279.2$ or $279.20

discount ⟶   2

The discount is $69.80 and the sale price is $279.20.

If you know the regular price and the sale price, you can find the discount and the discount rate.

Method

▶ 1 Subtract the sale price from the regular price to find the discount.

▶ 2 Divide the discount by the regular price to find the discount rate.

Example B

A toaster is on sale for $23.95. Find the discount and the discount rate if the regular price is $32.00. Round the discount rate to the nearest percent.

▶ 1 $32. \ominus 23.95 \ominus 8.05 \div 32. \ominus 0.2515625$

discount ⟶   2

The discount is $8.05 and the discount rate is 25%.

Guided Practice

Example A

Find the discount and the sale price.

1. bus fare, $8.50, discount rate, 10%

2. picture, $55.60, discount rate, 25%

3. stereo, $900
discount rate, 35%

4. cassette tape, $6.50
discount rate, 50%

Find the discount and the discount rate. Round the discount rate to the nearest percent.

Example B

5. jacket, $56, sale price, $44.80

6. car, $12,650, sale price, $10,890

Practice

Find the discount and the sale price.

7. album, $9.80
discount rate, 10%

8. blouse, $22
discount rate, 20%

9. textbook, $28.75
discount rate, 5%

10. car, $9,200
discount rate, 7.5%

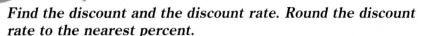

Find the discount and the discount rate. Round the discount rate to the nearest percent.

11. skis, $199
sale price, $150

12. boat, $2,780
sale price, $2,432.50

13. backpack, $20
sale price, $15.66

14. video recorder, $1,100
sale price, $792

Applications

15. Laluca's is having a 30%-off sale on all decorative living room accessories. What is the sale price of a vase that sells for $45?

16. Craig wants to buy a keyboard tray for his computer keyboard. He found one at Discount Office Warehouse for $28.49. Discount Office Warehouse advertises all merchandise always 30% off regular price. What is the regular price of the keyboard tray that Craig wants?

Suppose

17. Suppose Watkin's Department Store pays $48 for a denim jacket from the manufacturer. Watkin's added 25% to the cost, pricing the jacket at $60. Later Watkin's offered the jacket on sale at 25% off. Why is the sale price not equal to $48? What is the sale price?

Talk Math

18. If the discount on an item is 20%, what percent of the original price is the sale price?

Cooperative Groups

19. *In groups of three or four, copy and complete the table.*

Original Price	Discount Rate	Sale Price
$100	25%	▪
$100	50%	▪
$100	20%	▪
$100	30%	▪
$100	40%	▪

Discuss with your group the relationship between the discount rate and the sale price. Describe the relationship. Think of another way to find the sale price when you know the discount rate without finding the discount first. Explain your method.

Critical Thinking

20. Suppose February 10, 1965 was Wednesday. What day of the week was March 10, 1965?

10-8 INTEREST

Objective

Find simple interest.

Pat borrowed $1,200 from his mother to pay part of the tuition for his last year at Carver State. He will repay his mother in one year at an interest rate of $9\frac{1}{4}\%$. How much interest will Pat pay his mother?

If you borrow money from a bank or a person, you must pay **interest** on the amount of the loan. The amount borrowed is called the **principal**. Interest is a percent of the principal. Principal plus interest is the amount that must be repaid.

The interest you must pay depends on how much money you borrow, the yearly interest rate, and how long you borrow the money. The formula shows the relationship of interest to principal, rate, and time.

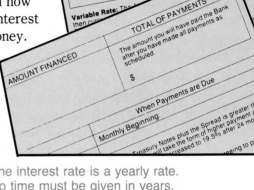

$$\text{Interest} = \text{principal} \times \text{rate} \times \text{time}$$
$$I \quad = \quad p \quad \times \quad r \quad \times \quad t$$

Method

1. Write interest rate (r) as a decimal.
2. Write the time (t) in years.
3. Use the formula $I = p \times r \times t$.

The interest rate is a yearly rate. So time must be given in years.

Examples

A *Find the interest charged on Pat's loan of $1,200 for 1 year at $9\frac{1}{4}\%$.*

1. $9\frac{1}{4}\% = 9.25\% \rightarrow 0.0925$
2. 1 year
3. $I = p \times r \times t$
   $= 1,200 \times 0.0925 \times 1$
   $= 111$

Pat will pay his mother $111 in interest.

B *Find the interest earned on a deposit of $200 for 20 months at $7\frac{3}{4}\%$. Round to the nearest cent.*

1. $7\frac{3}{4}\% = 7.75\% \rightarrow 0.0775$
2. 20 months $\rightarrow \frac{20}{12}$ or $\frac{5}{3}$ years
3. $I = p \times r \times t$
   $= 200 \times 0.0775 \times 5 \div 3$
   $= 25.833333$

The interest earned is $25.83.

Guided Practice

Find the interest owed on each loan. Then find the total amount to be repaid. Round to the nearest cent.

Example A

1. principal: $500
annual rate: 12%
time: 1 year

2. principal: $1,200
annual rate: 18%
time: 3 years

3. principal: $620
annual rate: $9\frac{3}{4}\%$
time: 6 months

Find the interest earned on each deposit.

4. principal: $230
annual rate: 6%
time: 1 year

5. principal: $80
annual rate: 8%
time: 18 months

6. principal: $500
annual rate: $8\frac{3}{4}$%
time: 9 months

Find the interest owed on each loan. Then find the total amount to be repaid.

Practice

7. principal: $90
annual rate: 12%
time: 2 years

8. principal: $2,500
annual rate: 14%
time: 36 months

9. principal: $2,500
annual rate: 12%
time: 48 months

10. principal: $2,500
annual rate: $14\frac{1}{4}$%
time: 2 years

11. principal: $205
annual rate: 18%
time: 6 months

12. principal: $205
rate: $1\frac{1}{2}$% per month
time: 6 months

Find the interest earned on each deposit.

13. principal: $120
annual rate: $9\frac{1}{2}$%
time: 3 years

14. principal: $10,500
annual rate: 10.25%
time: $1\frac{1}{2}$ years

15. principal: $6,700
annual rate: $9\frac{3}{4}$%
time: 10 months

16. principal: $1,000
annual rate: $9\frac{1}{4}$%
time: 10 years

17. principal: $150
annual rate: 8%
time: 18 months

18. principal: $750
annual rate: $9\frac{1}{2}$%
time: 10 months

Applications

19. Adel is saving for a car. Last month she deposited $350 in her savings account. If the account pays $5\frac{1}{2}$% interest per year, how much interest will Adel earn on $350 in 2 years?

20. Luther put $1,000 in a certificate of deposit for 24 months. The certificate pays 10.73% interest per year. How much interest will Luther earn in 24 months? What is the total amount Luther will have at the end of 24 months?

PORTFOLIO

21. Review the items in your portfolio. Make a table of contents of the items, noting why each item was chosen. Replace any items that are no longer appropriate.

You can find interest using mental math and fractional equivalents. For example, how much interest will a savings account deposit of $1,000 earn in 6 months if the interest rate is 10% per year?

Think: 10% is the same as $\frac{1}{10}$. 6 months is $\frac{1}{2}$ of a year.

$$I = 1,000 \times \frac{1}{10} \times \frac{1}{2} \rightarrow I = 100 \times \frac{1}{2} \text{ or } 50$$

The interest for 6 months on $1,000 at 10% per year is $50.

Find the interest earned using mental math.

22. p, $750
r, 10%
t, 8 months

23. p, $2,000
r, 5%
t, 2 years

24. p, $4,800
r, $12\frac{1}{2}$%
t, 6 months

10-9 COMPOUND INTEREST

Objective

Find compound interest.

Mark's savings account pays 6% annual interest compounded quarterly. At the end of 1 year, how much interest will Mark earn on $1,000?

Most savings accounts pay **compound interest.** Compound interest is computed at stated intervals. At the end of the first interval, the interest earned is added to the account. At the end of the next interval, interest is earned on the balance that includes the previous interest. Interest may be added annually, semiannually (2 times a year), quarterly (4 times a year), monthly, daily, or continuously.

Example

Find the savings total at the end of one year if $1,000 is deposited in a savings account at 6% annual interest compounded quarterly.

first quarter:
$I = p \times r \times t$
$I = $ `1 0 0 0` ⊠ `0.06` ⊡ `4` ⊟ `1 5`

$I = \$15.00$ Each quarter is $\frac{1}{4}$ of a year. To multiply by $\frac{1}{4}$ on the calculator, divide by 4.

The new balance or principal is `1 0 0 0` ⊞ `1 5` ⊟ `1 0 1 5` [STO]

or $1,015 after one quarter. Use the [STO] key to save the new balance after each computation.

After each of the next three quarters, interest is computed and added to the account.

second quarter: ⊠ `0.06` ⊡ `4` ⊞ [RCL] ⊟ `1 0 3 0.2 2 5` [STO]

new balance: $1,030.225   ↑ This recalls the last value put in memory; in this case, 1015.

third quarter ⊠ `0.06` ⊡ `4` ⊞ [RCL] ⊟ `1 0 4 5.6 7 8 4` [STO]

new balance: $1,045.68

fourth quarter: ⊠ `0.06` ⊡ `4` ⊞ [RCL] ⊟ `1 0 6 1.3 6 3 6`

new balance: $1,061.3636

The savings total at the end of one year is $1,061.36

Understanding the Data

- How much interest is earned at the end of the second quarter?
- What is the new balance at the end of the second quarter?
- How much interest is earned at the end of the fourth quarter?

Find the savings total for each account.

1. principal: $250
   annual rate: 6%
   compounded quarterly
   time: 1 year

2. principal: $500
   annual rate: 9%
   compounded annually
   time: 1 year

You can use the formula $S = P(1 + \frac{r}{n})^{n \times t}$ and the $\boxed{y^x}$ key on a calculator to compute powers. Find the savings total after 1 year for a deposit of $500 at an annual rate of 9% compounded quarterly.

$$S = 500 \times (1 + \frac{0.09}{4})^{4 \times 1}$$

S = savings total
P = beginning principal
r = annual rate
n = number of times compounded annually
t = number of years

Enter: $\boxed{1}\ \boxed{+}\ .\boxed{0}\boxed{9}\ \boxed{\div}\ \boxed{4}\ \boxed{=}$ The result of the operations inside the parentheses is 1.0225.

Enter: $\boxed{y^x}\ \boxed{4}\ \boxed{=}$ The value of $(1.0225)^4$ is 1.0930833.

Enter: $\boxed{\times}\ \boxed{5}\boxed{0}\boxed{0}\ \boxed{=}$ The product is 546.54166.

The savings total is $546.54.

Use the formula to find the savings total.

3. principal: $250
   annual rate: 6%
   compounded semiannually
   time: 1 year

4. principal: $500
   annual rate: 9%
   compounded quarterly
   time: 1 year

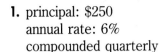

How does an electronic calculator add?
See page 373.

5. Tammara deposits $250 in her savings account that pays 8% interest compounded quarterly. After 6 months, how much will Tammara have in the account?

6. Khalid has $200 to start a savings account for a car. One account pays 6% annual interest compounded annually while another pays 5% interest compounded quarterly. What is the savings total Khalid can expect from each account after 1 year? Which account pays more interest?

7. Mrs. Schoby put $1,200 in a savings account for her daughter's college expenses. What savings total will she have in three years if the account earns 6.5% interest compounded semiannually?

Find the area of each triangle described below.

8. base, 12 cm; height, 10 cm

9. base, 3.2 mm; height, 8.1 mm

Determine if each pair of ratios forms a proportion.

10. 5 to 9, 6 to 10

11. 1 to 3, 4 to 12

12. 6 to 10, 12 to 20

13. A bag contains 2 red marbles, 3 green marbles, and 4 yellow marbles. What is the probability of choosing a red marble?

10-10 USING PERCENT

Objective
*Solve problems
using percents.*

As marketing manager for All-Sports Equipment, Bob Henderson keeps track of sales for each product. Sales are recorded for each month, quarter, half, and entire year. For the annual meeting, Bob needs to find the percent of increase in annual sales of headbands from 1989 to 1994.

Headband Sales per Year	
1989	$44,235
1990	46,889
1991	51,109
1992	53,153
1993	57,937
1994	63,831

▶ **Explore**

What is given?
● annual headband sales for 1989 through 1994

What is asked?
● What is the percent of increase in annual sales of headbands?

▶ **Plan**

Subtract the 1989 sales from the 1994 sales and find the percent of increase in sales.

▶ **Solve**

$63831 \boxed{-} 44235 \boxed{=}$
$19596 \boxed{÷} 44235 \boxed{=} 0.4429976$
↑—— amount of increase

The percent of increase in headband sales rounded to the nearest percent is 44%.

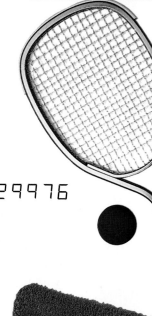

▶ **Examine**

The amount of increase in sales is about 20,000. 20,000 is a little less than half of 45,000 and 44% is a little less than half. The answer is reasonable.

Guided Practice

Solve. Use the chart. Round to the nearest percent.

1. The *amount* of increase from 1993 to 1994 is greater than the *amount* of increase from 1990 to 1991. Is the percent of increase from 1993 to 1994 *less than, greater than,* or *the same as* the percent of increase from 1990 to 1991?

2. If the estimated increase in sales for 1995 and 1996 is 8% each year, what is the amount of estimated sales for 1996 to the nearest dollar?

3. The research consultant Sally Martin estimated that sales will increase 12% from 1994 to 1995 if All-Sports introduces a new line of tri-colored headbands. What is the amount of estimated sales for 1995 if the estimated percent of increase is 12%?

4. A newspaper survey at Jason City High showed that 5 out of 6 students ride the bus to school. What percent of students ride the bus?

5. Don played in 60% of the football games this season and 50% of the games last season. How many games were there during this season if Don played in 6 games?

6. Dionne has $1,200 in her savings account for a car. If the account earns 6.5% compounded semiannually, how much will Dionne have in 3 years?

Using Formulas

7. The formula for the volume of a cylinder is $V = \pi r^2 h$. Mr. McCullogh is expanding his silo that measures 20 meters high and has a diameter of 15 meters. If he adds 4 meters of height to his silo, what is the percent of increase in the volume the silo can hold? Use 3.14 for π.

Using Data

f **UN with MATH**

When and where was democracy born?
See page 372.

Use the chart for Exercises 8-9.

8. There were 1,100 people surveyed. What percent of the people surveyed said they served on a jury because they feel it is a duty or a responsibility?

9. Of the 1,100 people surveyed, what percent served on a jury because they couldn't get out of it or because they wanted to serve?

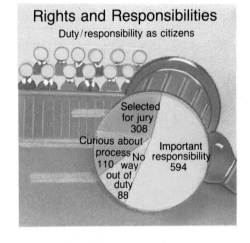

Rights and Responsibilities
Duty/responsibility as citizens

Selected for jury 308
Curious about process 110
No way out of duty 88
Important responsibility 594

Collect Data

10. Survey the members of your class and find the percent who would be willing to serve on a jury and the percent who would *not* be willing to serve on a jury.

Critical Thinking

11. What symbol comes next?

Mixed Review

Lesson 7-9

Classify each quadrilateral.

12.

13.

14.

Lesson 8-9

Find the volume of each rectangular prism.

15. length = 4 cm
width = 3 cm
height = 2 cm

16. length = 10 in.
width = 7 in.
height = 5 in.

17. length = 4.5 m
width = 3.2 m
height = 1.7 m

Lesson 9-5

18. On a map the scale is 1 inch:200 miles. How far is it between Dallas and Detroit if the distance on the scale drawing is 5.8 inches?

REVIEW

**Vocabulary /
Concepts**

*Use a word or number from the list at the
right to complete each sentence.*

1. To find the percent of a number, _?_ the
   number by the percent.
2. Salespersons may earn _?_ on each sale.
3. Written as a percent, $\frac{1}{5}$ is _?_ .
4. The _?_ is a percent of each purchase
   collected by the state or city.
5. The _?_ is a percent of decrease in price.
6. The cost of borrowing money is the _?_ .
7. _?_ is the amount of money borrowed.
8. Quarterly means _?_ a year.
9. _?_ interest is computed on the principal
   plus previous interest.

5%
20%
commission
compound
discount
discount rate
four times
interest
multiply
once
principal
sales tax
simple
twice

**Exercises /
Applications**

Lesson 10-1

Find each percentage. Use r × b = p.

10. 20% of 15
11. 16% of 80
12. 125% of 56
13. $33\frac{1}{3}$% of 1,890
14. $87\frac{1}{2}$% of 640
15. 6.2% of 115

Lesson 10-2

Find each percent. Use an equation.

16. What percent of 56 is 28?
17. What percent of 24 is 31.2?
18. What percent of 15 is 10?
19. 25 is what percent of 80?
20. 100 is what percent of 40?
21. 6 is what percent of 27?

Lesson 10-3

Find each number. Use an equation.

22. 29% of what number is 58?
23. 40% of what number is 8?
24. 150% of what number is 72?
25. 18 is 60% of what number?

Lesson 10-4

Solve. Use a Venn diagram.

26. At their 5-year class reunion, the class of '87 held a buffet dinner.
    If 43 people chose roast beef, 27 people chose chicken, 13 people
    chose both beef and chicken, and 6 chose neither, how many
    people attended the reunion?

Lesson 10-5

Estimate.

27. 34% of 210
28. 63% of 95
29. 15% of $3.50
30. $9\frac{3}{4}$% of 132

Find the percent of increase or decrease. Round decimal answers to the nearest percent.

	Item	1987	1989
31.	plane ticket	$320	$360
32.	house	$85,600	$93,200
33.	computer	$3,000	$2,848

	Name	1988 Salary	1990 Salary
34.	Snyder	$21,500	$20,000
35.	Jensen	$29,200	$31,500
36.	Holt	$42,000	$47,900

Find the discount, the sale price, or the discount rate.

37. coat, $120
discount rate, 22%
sale price, _?_

38. theater ticket, $24
discount, $2.40
discount rate, _?_

39. car, $9,400
discount rate, 7%
discount, _?_

Find the interest owed on each loan or the interest earned on each deposit.

40. deposit, $340
annual rate, 9%
time, 15 months

41. deposit, $1,200
annual rate, 9.75%
time, 2 years

42. loan, $500
annual rate, 12%
time, 3 months

43. loan, $200
annual rate, 10%
time, 8 months

44. Find the savings total for a deposit of $500 in an account that pays 6.5% compounded quarterly for two years.

45. The average American consumed 95.3 pounds of fresh vegetables in 1986 and 100.3 pounds in 1988. What was the percent of increase in consumption of fresh vegetables?

46. Mary Ann deposits $220 in a savings account that pays $9\frac{3}{4}\%$ annual interest. What will her savings total be at the end of 3 years?

47. Rita and Dan McDonald borrowed $3,000 at an interest rate of $9\frac{1}{4}\%$ compounded semiannually for 2 years. Find the total amount to be repaid.

48. In November, 1988, *TV Review* magazine reported that about 18% of U.S. households with televisions watched "Head of the Class" regularly. If 15,948,000 households watched "Head of the Class," *about* how many households with television are there in the U.S.?

TEST

Find each missing value. Use an equation.

1. 18% of 320 is what number?

2. 13 is $12\frac{1}{2}\%$ of what number?

3. Ninety-six students responded to a survey. Forty students said they like both rock and country, 55 students like rock, and 47 students like country. How many students like neither rock nor country? Use a Venn diagram.

Estimate.

4. 9.8% of $127

5. 22% of $80

6. 52% of $1,000

7. $5\frac{3}{4}\%$ of $200

Find the discount and the sale price.

8. album, $9.63; discount rate, 8%

9. ring, $350; discount rate, $12\frac{1}{2}\%$

Find the interest earned on each deposit.

10. principal: $90
    annual rate: 8%
    time: 8 months

11. principal: $375
    annual rate: $7\frac{3}{4}\%$
    time: 2 years

12. principal: $1,888
    annual rate: $7\frac{1}{4}\%$
    time: 15 months

Find the savings total for each account.

13. principal: $500
    6%, compounded
       semiannually
    time: 2 years

14. principal: $3,000
    5%, compounded
       quarterly
    time: 1 year

15. principal: $1,000
    $8\frac{1}{2}\%$, compounded
       semiannually
    time: 2 years

Solve.

16. Sung Lee bought a beach volleyball set for $45.00. If the original price was $62.50, what was the discount rate?

17. On a test with 200 questions, Sharon missed 32 questions. What percent of the questions did she answer correctly?

18. What is the sales tax and the total cost of a stereo that sells for $890? The tax rate is 5.5%.

19. As a handling fee, Phillips Corporation charges 2% of the total sale price for all stock transactions. What is the handling fee for a sale of $18,650?

20. Larry Bird made 3,356 free throws during his basketball career through the 1988-89 season. This is 89.6% of his free throw attempts. How many attempts did Larry Bird make?

▶ BONUS: In your own words, explain why an account that earns 6% compounded quarterly will earn more interest in one year than an account that earns 6% interest per year.

MAIL ORDERING

Glenda Steele ordered three stove-top popcorn poppers weighing 2 pounds 8 ounces each as gifts for her nephews and niece. She finds the weight of her total order to compute shipping and handling charges.

The customer sometimes pays shipping and handling charges. The rates for one company are shown below.

Delivery Weight	Local	Delivery Weight	Local	Delivery Weight	Local	Delivery Weight	Local
0– 0.5 lb	1.25	13.1–14 lb	2.51	27.1–28 lb	3.25	41.1–42 lb	4.00
0.6– 1 lb	1.50	14.1–15 lb	2.57	28.1–29 lb	3.31	42.1–43 lb	4.05
1.1– 2 lb	1.86	15.1–16 lb	2.62	29.1–30 lb	3.36	43.1–44 lb	4.10
2.1– 3 lb	1.98	16.1–17 lb	2.67	30.1–31 lb	3.41	44.1–45 lb	4.16
3.1– 4 lb	2.09	17.1–18 lb	2.72	31.1–32 lb	3.47	45.1–46 lb	4.21
4.1– 5 lb	2.19	18.1–19 lb	2.78	32.1–33 lb	3.52	46.1–47 lb	4.26
5.1– 6 lb	2.23	19.1–20 lb	2.83	33.1–34 lb	3.57	47.1–48 lb	4.31
6.1– 7 lb	2.26	20.1–21 lb	2.88	34.1–35 lb	3.63	48.1–49 lb	4.37
7.1– 8 lb	2.29	21.1–22 lb	2.94	35.1–36 lb	3.68	49.1–50 lb	4.42
8.1– 9 lb	2.32	22.1–23 lb	2.99	36.1–37 lb	3.73	Each additional lb	0.07
9.1–10 lb	2.35	23.1–24 lb	3.04	37.1–38 lb	3.78		
10.1–11 lb	2.39	24.1–25 lb	3.10	38.1–39 lb	3.84	*0.1 → means one tenth of a lb	
11.1–12 lb	2.42	25.1–26 lb	3.15	39.1–40 lb	3.89		
12.1–13 lb	2.46	26.1–27 lb	3.20	40.1–41 lb	3.94		

To find the total weight of her order, Mrs. Steele multiplies and adds the weights.

$$
\begin{array}{ll}
2 \text{ lb } 8 \text{ oz} & \text{3 popcorn poppers at} \\
\times \quad\quad 3 & \text{2 lb 8 oz each} \\
\hline
6 \text{ lb } \cancel{24} \text{ oz} & \text{Rename 24 oz} \\
+1 \text{ lb } 8 \text{ oz} & \text{as 1 lb 8 oz.} \\
\hline
7 \text{ lb } 8 \text{ oz} &
\end{array}
$$

The delivery weight for 3 popcorn poppers is 7 pounds 8 ounces. This is between 7.1 and 8 pounds. So the shipping and handling charge is $2.29.

Find the total weight for each order. Then find the shipping and handling charges.

1. 1 suitcase at 9 lb 8 oz and 2 flight bags at 4 lb 8 oz each

2. 3 necklaces at 4 oz each and 6 bracelets at 3.5 oz each

3. Jake placed an order for toys that weighed 6 lb 14 oz. The next week he ordered a set of dishes that weighed 12 lb 12 oz. How much would Jake save in shipping and handling charges if he ordered the toys and dishes at the same time?

Free Response

Lesson 1-1

Replace each ▓ with a number to make a true sentence.

1. 700 = ▓ tens
2. 200 = ▓ ones
3. 3,500 = ▓ hundreds
4. 0.018 = ▓ thousandths
5. 0.2 = ▓ hundredths
6. 0.7 = ▓ tenths

Lesson 3-1

Multiply.

7. 6,413 × 65
8. 2,117 × 39
9. 1,318 × 63
10. 8,185 × 225
11. 4,347 × 381

12. 121 × 7
13. 8 × 6,469
14. 444 × 16
15. 639 × 132
16. 671 × 669
17. 263 × 9,188

Lesson 4-9

Estimate.

18. $\frac{17}{20} + \frac{9}{12}$
19. $\frac{4}{5} - \frac{1}{10}$
20. $10\frac{8}{10} - 3\frac{1}{11}$

Lesson 5-1

Estimate.

21. $\frac{3}{7} \times 9\frac{1}{10}$
22. $4\frac{4}{5} \div 2\frac{5}{11}$
23. $10\frac{3}{5} \div 3\frac{1}{2}$

Lessons 6-2 6-4 6-5

Choose the most reasonable unit of measure.

24. length of a safety pin   a. millimeter   b. meter   c. kilometer
25. mass of a chair   a. milligram   b. gram   c. kilogram
26. capacity of a teaspoon   a. milliliter   b. liter   c. kiloliter

Lesson 7-8

27. A rectangle has a perimeter of 48 inches. The length of one side is 15 inches. What are the measures of the other three sides?

Lesson 8-4

28. The diameter of a circular go-cart racetrack is 250 meters. To the nearest meter, what is the area of the racetrack and the space in the center?

Lesson 9-4

Solve each proportion.

29. $\frac{2}{3} = \frac{x}{9}$
30. $\frac{5}{7} = \frac{y}{35}$
31. $\frac{1}{9} = \frac{n}{36}$

Lesson 9-8

Write each percent as a fraction in simpiest form.

32. 80%
33. 70%
34. 58%

Lesson 10-7

Find the discount and the sale price.

35. radio, $55
discount rate, 20%

36. compact disc, $14.95
discount rate, 15%

Multiple Choice

Choose the letter of the correct answer for each item.

1. What is the sum of one-fourth and one-eighth?

   a. $\frac{1}{12}$ **c.** $\frac{3}{8}$

   b. $\frac{1}{6}$ **d.** $\frac{3}{4}$

Lesson 4-10

2. Estimate the product of 0.93 and 17.2.

   e. 7

   f. 17

   g. 170

   h. 1,700

Lesson 3-2

3. Residents must pay 30% of the cost of putting curbs on their streets. How much will 50 feet of curbing at $3.25 per foot cost for a resident?

   a. $52.50   **c.** $49.25

   b. $48.75   **d.** $162.50

Lesson 10-10

4. What is the product of $7\frac{3}{14}$ and $10\frac{8}{9}$?

   e. $70\frac{4}{21}$

   f. $76\frac{23}{30}$

   g. $78\frac{5}{9}$

   h. $157\frac{1}{9}$

Lesson 5-3

5. Jamall earns $45 for every $1,000 worth of tires he sells. Which proportion can be used to find earnings on sales of $5,580?

   a. $\frac{45}{1,000} = \frac{5,580}{x}$

   b. $\frac{45}{5,580} = \frac{1,000}{x}$

   c. $\frac{1,000}{5,580} = \frac{x}{45}$

   d. $\frac{45}{1,000} = \frac{x}{5,580}$

Lesson 9-3

6. What is the area of a square with sides 3 inches long?

   e. 6 in²   **g.** 12 in²

   f. 9 in²   **h.** 27 in²

Lesson 8-2

7. Solve the equation $\frac{x}{36} = 4$.

   a. 8

   b. 9

   c. 124

   d. 144

Lesson 1-12

8. If there are 3 girls for every 8 boys, how many girls will there be if there are 1,000 boys?

   e. 37.5   **g.** $2,666.\overline{6}$

   f. 375   **h.** 3,000

Lesson 9-3

9. If Steven missed 14 out of 50 questions on an exam, what percent of the questions did he answer correctly?

   a. 14%

   b. 28%

   c. 72%

   d. 93%

Lesson 10-2

10. Debbie weighed 7 pounds at birth. She lost 4 ounces during the first week. How much did she weigh then?

   e. 3 pounds

   f. 6 pounds 4 ounces

   g. 6 pounds 8 ounces

   h. 6 pounds 12 ounces

Lesson 6-8

11 STATISTICS AND GRAPHS

Moving? Again?

American families now move an average of once every four years. Why do some families move more often than others? Have you moved more often than the average? Less often?

Melissa James is 18 years old. She and her family have moved six times since she was born. On the average, how often has the James family moved?

ACTIVITY: Averages

There are three types of averages. Do you know what they are called? In Chapter 3, you studied the most common average, the mean. You can find an average without having to compute. Use the following activity to find the average score on a test.

Materials: 9 index cards per student (or paper cut into 9 equal pieces)
pencil

1. On each index card, write a test score. You can write any scores you want. The scores must range from 0 to 100. Each person should have a set of nine cards.

2. Shuffle the cards.

Cooperative Groups

Work together in small groups.

3. Exchange your nine test-score cards with those of another member of your group.

4. Discuss with your group ways in which you could find an average test score without having to compute. Write a short paragraph about your ideas.

5. Use your index cards to arrange the scores in order from least to greatest. What is the middle score? The middle score is one kind of average for all the test scores.

6. Can you think of another way to find an average without having to compute? Discuss this with your group. (Hint: Do any of the scores in your set of cards appear more than once? If so, you have found another kind of average.)

Communicate Your Ideas

7. Discuss with your class the ideas from Problem 4.

8. Do you know what name is given to the average you found in Problem 5? in Problem 6?

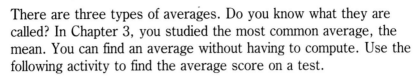

11-1 MEDIAN, MODE, AND RANGE

Objective

Find the median, mode, and range of a set of data.

For statistics class, Koji Ott needs to find the median, mode, and range for a set of data of his own choosing. To gather his data, he asked 11 classmates to estimate how many hours they talked on the telephone last week.

Number of Hours on the Telephone			
Sue	3	Lyn	7
Bob	6	Katy	1
Tim	2	John	4
Joe	6	Carole	8
Pete	0	Ida	6
Jill	3		

The **median** of a set of data is the middle number when the data are listed in order. The **mode** is the number that appears most often. The **range** is the difference between the greatest and the least number.

Method

1. List the numbers in order from least to greatest.
2. To find the median, locate the middle number or numbers. If there are two middle numbers, find their mean.
3. To find the mode, determine the number that appears most often. There can be more than one mode or no mode.
4. To find the range, subtract the least number from the greatest number.

Example A

Find the median, mode, and range for Koji's data.

1. List the numbers in order.   0,1,2,3,3,4,6,6,6,7,8

   5 numbers ↑ 5 numbers

2. The middle number is 4. So the median is 4.
3. The number 6 appears most often. So the mode is 6.
4. 8 − 0 = 8   The range is 8.

   The median number of hours that Koji's classmates talked on the telephone in one week is 4. The mode is 6 hours, and the range is 8 hours.

Example B

Find the median, mode, and range for the set of data.

343, 437, 265, 419, 356, 302

1. 265, 302, 343, 356, 419, 437
2. There are two middle numbers, 343 and 356.

   ⨝ 343 ⊕ 356 ⊘ 2 ⊜ 349.5   The median is 349.5.
3. Each number appears only once. So there is no mode.
4. ⨝ 437 ⊖ 265 ⊜ 172   The range is 172.

Guided Practice

Find the median, mode, and range for each set of data.

Example A   **1.** 6, 4, 5, 8, 2, 6, 9   **2.** 11, 17, 14, 15, 11   **3.** 75, 72, 71, 78, 75

Example B

4. 43, 28, 43, 23   **5.** 96, 51, 24, 55   **6.** 322, 325, 235, 232

Exercises

Practice

Find the median, mode, and range for each set of data.

7. 1, 5, 9, 1, 2, 5, 8, 2, 7, 2, 2   **8.** 3, 2, 4, 3, 6, 6, 6, 9, 1, 8, 4
9. 12, 98, 29, 12, 71, 37, 15   **10.** 53, 31, 54, 63, 84, 54, 30
11. 64, 25, 46, 50, 91, 86, 73, 91   **12.** 47, 49, 90, 6, 69, 89, 47, 66
13. 40, 42, 41, 43, 41, 40, 42, 43   **14.** 314, 179, 275, 341, 725
15. 3.4, 1.8, 2.6, 1.8, 2.3, 3.1   **16.** 0.6, 0.7, 0.7, 1.0, 0.9, 1.0, 1.2
17. 7.5, 8.3, 8.8, 8.3, 7.4, 8.35, 8.1, 7.6, 8.3, 7.59
18. 5, 7.4, 6.5, 5.25, 6, 4.5, 6, 6.25, 6.5, 4.6, 6.25, 8, 4.5, 6

Applications

Solve. Use the chart for Exercises 19–21.

19. Find the range for the high temperatures in each city for the year.

20. Find the median for high temperatures in each city for the year.

21. Find the mode for the high temperatures in each city for the year.

Average Daily High Temperatures (°F)			
	Juneau	**St. Louis**	**Honolulu**
Jan.	29.1	39.9	79.3
Feb.	33.9	44.2	79.2
Mar.	38.2	53.0	79.7
Apr.	46.5	67.0	81.4
May	55.4	76.0	83.6
June	62.0	84.9	85.6
July	63.6	88.4	86.8
Aug.	62.3	87.2	87.4
Sept.	56.1	80.1	87.4
Oct.	47.2	69.8	85.8
Nov.	37.3	54.1	83.2
Dec.	32.0	42.7	80.3

Suppose

22. Suppose Mark scored 81 and 78 on his first two science quizzes. What must his score be on the third quiz so that his mean score is 85?

Using Equations

23. If the range of a set of data is 102 and the least number is 256, what is the greatest number in the set? Write an equation and solve.

Make Up a Problem

24. Make up a set of data in which the mean does not equal either the mode or the median.

Estimation

The *clustering strategy* can be used to estimate the sum of a group of numbers. For example, the numbers in the chart *cluster* around 10,000. An estimate of the total number of births each week in the U.S. is 7 × 10,000 or 70,000.

Average U.S. Births per Day	
Sun.	8,532
Mon.	10,243
Tues.	10,730
Wed.	10,515
Thurs.	10,476
Fri.	10,514
Sat.	8,799

Use the clustering strategy to estimate each sum.

25. 82 + 77 + 82 + 76 + 79   **26.** 396 + 391 + 411 + 407 + 389
27. 6,637 + 5,952 + 5,848   **28.** 8.46 + 4.69 + 7.38 + 4.0

11-2 FREQUENCY TABLES

Objective

Make and interpret frequency tables.

Mrs. Simms lists the scores of her gymnastics team. She wants to organize the data in a way that makes it easy to study the scores.

A **frequency table** is a table for organizing a set of data.

Method

1 ▶ Make a table with three columns.

2 ▶ In the first column, list the items or numbers in the set of data.

3 ▶ In the second column, make a tally mark each time the item or number appears in the set of data.

4 ▶ In the third column, write the frequency, the number of times the item or number appears.

Example A

Make a frequency table for the scores of the gymnastics team given in Mrs. Simms' list. Then find the mean, median, and mode of the scores.

Score	Tally	Frequency
2					3
3				2
4									7
5										8

To find the mean, compute the total number of points scored and divide by the total number of scores.

The total number of points scored equals
$(2 \times 3) + (3 \times 2) + (4 \times 7) + (5 \times 8)$ or 80.

$$\text{mean} = \frac{80}{20} \text{ or } 4 \quad \begin{array}{l} \leftarrow \text{The total number of points scored is 80.} \\ \leftarrow \text{The number of scores is 20.} \end{array}$$

Since there are 20 scores, the median is the mean of the tenth and eleventh scores when listed in order.

$$\text{median} = \frac{4 + 4}{2} \text{ or } 4$$

The mode is 5. Why?

Guided Practice

Example A

Copy the table below and complete the frequency column.

1.

Height	Tally	Frequency
59 in.	||||| |||||	
62 in.	||||| ||||| |||	
65 in.	||||| ||||| ||||| ||	
68 in.	||||| |||||	
70 in.	||||| ||||	
72 in.	||||| ||||| |||||		

Exercises

Practice

Use the completed frequency table in Exercise 1 to find each of the following. Round to the nearest tenth.

2. mean   **3.** median   **4.** mode   **5.** range

Solve. Use the data at the right.

6. Make a frequency table for the set of data.

7. Find the mean.

8. Find the median.

9. Find the mode.

10. Find the range.

Number of miles run each day by members of the track team				
3	5	2	6	3
2	4	3	5	5
6	2	3	3	4
3	5	5	4	2

Collect Data

11. Collect data from your classmates about their favorite season or the kind of pet they own. Then organize the data into a frequency table.

Decision Making

12. The number of each size of track shoe sold last month by the Athlete's Locker is given in the table at the right. Which size should the store stock the most? the least?

Size	Tally	Frequency
9	ЖЖ II	12
$9\frac{1}{2}$	ЖЖЖ I	16
10	ЖЖЖЖЖ III	23
$10\frac{1}{2}$	ЖЖ I	11
11	ЖЖЖ	15
$11\frac{1}{2}$	ЖЖЖ IIII	19

Applications

13. Make a frequency table for the data: Friday Video rentals—comedy, drama, comedy, drama, adventure, comedy, comedy, adventure, drama, comedy, adventure, comedy, drama, drama, comedy, comedy, drama, comedy, comedy.

If the rental price of a comedy video is $2.00 and the rental price for all other videos is $1.50, what is the mean rental price for Friday's rentals?

Critical Thinking

14. Suppose you are responsible for the security of an ancient artifact being displayed in a rectangular room at the local museum. How would you position 10 guards so that there are the same number of guards along each wall?

Calculator

To find the total of the items in a frequency table, multiply before adding. When finding the mean, parentheses show that the sum is found before dividing by the total number of data values.

Score	Frequency
9.5	5
9	7
8.5	11
8	17

$$(\, 5 \, \times \, 9.5 \, + \, 7 \, \times \, 9 \, + \, 11 \, \times \, 8.5 \, + \, 17 \, \times \, 8 \,)$$

the sum of the scores $\div \, (\, 5 \, + \, 7 \, + \, 11 \, + \, 17 \,) \, = \, 8.5$

the sum of the frequencies

The mean is 8.5.

15. From Exercise 12, use a calculator to find the mean shoe size (to the nearest half-size) sold last month.

11-3 MISUSING STATISTICS

Objective
Interpret data.

The "average" salary of workers is of interest to both employers and employees. The chart at the right lists the income of each job classification at the JCH Compact Disc Company.

Job	Number of Employees	Salary
plant workers	20	$15,600
skilled workers	9	$20,600
supervisors	6	$25,000
managers	3	$32,500
vice-presidents	2	$56,500
president	1	$81,000

Understanding the Data

- What is the salary earned by a vice-president?
- Which job classification pays $25,000?
- How many employees are there at the JCH CD Company?

The "average" salary can be reported using the mean, median, or mode. The average chosen depends on the point of view a person takes.

Mean:

$$\frac{20(15,600) + 9(20,600) + 6(25,000) + 3(32,500) + 2(56,500) + 81,000}{41} = \$22,900$$

Median: The middle salary is $20,600 earned by a skilled worker.

Mode: The salary that appears most often is $15,600.

1. The company needs to hire more employees. Which "average" is likely to be used in advertisements? Why?

2. Which "average" *best* describes the salary for plant and skilled workers at the company? Why?

3. Which "average" *best* describes the salary for all employees at the company? Why?

4. Which "average" *best* describes the salary for supervisors and managers?

5. If a union representative negotiates for higher salaries, which job classifications are likely to be used to calculate the "averages"? Why?

6. Suppose the JCH CD Company adds a second shift of seven plant workers, three skilled workers, one supervisor, and one manager. Would you expect the mean, median, and mode, including the new employees, to increase or decrease? Why?

Talk Math

7. When a company president quotes one "average" and a union negotiator quotes another "average," they are using certain data for a specific reason. Describe a situation in which choosing the median would be better than choosing the mode.

Interpreting Data

The ages of the guests at a birthday party are given at the right.

19 25 88 23 24
26 20 23 22

8. Find the mean age and the median age of the guests.

9. If the youngest guest had been 13 and the oldest 45, how would the mean and median be affected?

The prices of several homes listed by a real estate agent are given at the right.

$87,500 $49,000
$200,000 $78,000
$62,500 $69,800

10. Find the mean price and the median price.

11. If the most expensive home costs $100,000 instead of $200,000, how would the median be affected? the mean?

Critical Thinking

12. Can you figure out what year the first professional football game was played if the year rounded to the nearest 10 is 1900 and the sum of its digits is 23?

Mixed Review

Lesson 1-11

Solve each equation. Check your solution.

13. $x - 4 = 18$ 14. $19 = 4 + y$

15. $0.23 + z = 4.21$ 16. $10 = b - 7$

Lesson 6-3

Complete.

17. $4.8 \text{ m} = \underline{?} \text{ km}$

18. $523 \text{ mm} = \underline{?} \text{ cm}$

19. $63 \text{ cm} = \underline{?} \text{ m}$

20. $3 \text{ km} = \underline{?} \text{ m}$

Lesson 7-7

21. Albert is cutting a piece of wood to make a pedestal for a lamp. The pedestal is to be in the shape of an equilateral triangle. One side is to have a length of 15 inches. What are the lengths of the other two sides?

11-4 BAR GRAPHS

Objective
Make bar graphs.

Mr. Delano is the owner of the Action Car Dealership. To increase sales, he uses a *bar graph* to show his sales representatives the number of cars a competitor sold that month.

A **bar graph** is used to compare quantities. The length of each bar represents a number. Use the steps below to make a bar graph.

Method

1 ▶ Label the graph with a title.

2 ▶ Draw and label the vertical axis and the horizontal axis.

3 ▶ Mark off equal spaces on one of the axes and label it with the scale that best represents the data.

Axes is the plural form of axis.

4 ▶ Draw bars to show the quantities. Label each bar.

Example A

Make a horizontal bar graph for the data in the table at the right.

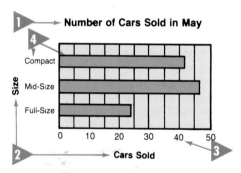

Number of Cars Sold in May	
Compact	42
Mid-size	47
Full-size	24

Example B

Make a vertical bar graph for the data in the table at the right.

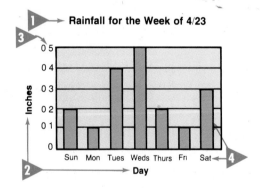

Rainfall for the Week of 4/23	
Sunday	0.2 inches
Monday	0.1 inches
Tuesday	0.4 inches
Wednesday	0.5 inches
Thursday	0.2 inches
Friday	0.1 inches
Saturday	0.3 inches

Guided Practice

Example A

Example B

Make the following bar graphs.

1. a vertical bar graph for the data in Example A

2. a horizontal bar graph for the data in Example B

Exercises

Practice

Where and when did the zero originate?
See page 372.

Make a vertical bar graph for each set of data.

3.

Highest Mountain on Each Continent (in feet above sea level)	
Asia	29,000
Europe	19,000
Africa	20,000
Australia	8,000
South America	23,000
Antarctica	17,000
North America	20,000

4.

Average High Temperature (°F)	
March	76°
April	81°
May	86°
June	90°
July	92°
August	93°
September	85°

Make a horizontal bar graph for each set of data.

5.

Runs Scored in 1 Week	
Reds	23
Braves	15
Padres	12
Dodgers	19
Astros	8
Giants	11

6.

Dollars Saved Each Week	
Week 1	$10
Week 2	$15
Week 3	$16
Week 4	$ 8
Week 5	$13

Applications

Solve. Use the bar graph.

7. Which cities have the same number of television stations?

8. Which city has the greatest number of television stations?

9. Which city has the least number of television stations?

10. How many television stations are in Dallas?

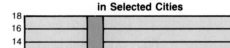

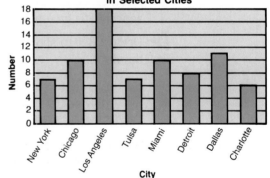

11. How many more stations are there in Chicago than in New York?

12. Amy bought a blouse for $30, a skirt for $36, a vest for $29, and a pair of shoes for $29. How much did Amy spend in all?

13. Dan ran five miles each day for five days and two miles each day for two days. How many miles did he run in all?

Collect Data

14. Survey your classmates to find their favorite rock star. Make a horizontal bar graph to illustrate the data.

15. Survey your classmates to find their favorite radio station. Make a vertical bar graph to illustrate the data.

11-5 LINE GRAPHS

Objective
Make line graphs.

Ms. Logan is a manager of the coat department for Mear's Department Store. Once a month she prepares a sales report for her supervisor. Ms. Logan uses a *line graph* as part of her report.

A **line graph** is used to show change and direction of change over a period of time. Use the steps listed below to make a line graph.

Method

1 ▶ Label the graph with a title.

2 ▶ Draw and label the vertical axis and the horizontal axis.   Explain the meanings of *vertical* and *horizontal.*

3 ▶ Mark off equal spaces on the vertical axis and label it with the scale that best represents the data.

4 ▶ Mark off equal spaces on the horizontal scale and label it with the appropriate time period.

5 ▶ Draw a dot to show each data point. Draw line segments to connect the dots.

Example A

Make a line graph for the data in the table below.

Monthly Coat Sales	
Jan.	$2,000
Feb.	$1,500
Mar.	$2,500
Apr.	$3,000
May	$3,500
June	$3,000

Guided Practice

Make a line graph for each set of data.

Example A

1.

Absences in Math Class	
Monday	3
Tuesday	1
Wednesday	4
Thursday	5
Friday	1

2.

Temperatures on Dec. 7	
10 A.M.	1°C
12 P.M.	3°C
2 P.M.	5°C
4 P.M.	6°C
6 P.M.	4°C

Practice

Make a line graph for each set of data.

3.

Estimated Population of the United States	
1820	10 million
1860	31 million
1900	76 million
1940	132 million
1980	227 million

4.

6 A.M. Barometer Readings	
Monday	29.5 inches
Tuesday	29.7 inches
Wednesday	30.1 inches
Thursday	30.2 inches
Friday	29.8 inches

5.

Growth of Saguaro Cactus (in feet)	
25 years	2
50 years	6
75 years	20
100 years	30
125 years	35

6.

Price for a Share of Ajax Stock	
Jan. 1	$23.00
Feb. 1	$21.00
Mar. 1	$22.50
Apr. 1	$24.25
May 1	$27.00

Applications

Solve. Use the line graph.

7. What was the height of the corn plant after 30 days?

8. About how many days after planting was the height 60 cm?

9. During which 10 days did the height increase the most?

10. During which 10 days did the height increase the least?

11. How much higher was the corn plant after 90 days than after 60 days?

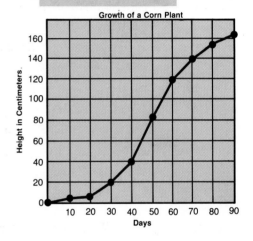

Critical Thinking

12. Choose a number. Multiply that number by 2. Then add 5 and multiply by 5. Subtract 25 from the result, and divide by 10. Try this with three different numbers. What can you say about the result each time? Explain.

Mixed Review

Lesson 3-8

Write in scientific notation.

13. 4,300,000   **14.** 421,000   **15.** 82,100   **16.** 942

Lesson 7-9

Name each shape.

17.   **18.**   **19.**

Lesson 8-6

20. Monica rides her bike 2 miles due east from her home. She then rides 1 mile south, then turns and rides 5 miles west, and turns again and rides 1 mile north. How far from home is she?

FLOWCHARTS

A **flowchart** can be used to plan a computer program. Each shape used in a flowchart has a special meaning. A *parallelogram* shows input or output. A *rectangle* shows assignment of variables. An *oval* shows the end of a program.

Make a flowchart for the following program.

```
10 READ M,T,W,R,F
20 DATA 10,6,8,9,7
30 LET SUM=M+T+W+R+F
40 LET MEAN=SUM/5
50 PRINT SUM
60 PRINT MEAN
70 END
```

DATA statements are usually not included in a flowchart. A word can be used as a variable.

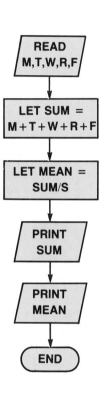

READ
M,T,W,R,F

↓

LET SUM =
M+T+W+R+F

↓

LET MEAN =
SUM/S

↓

PRINT
SUM

↓

PRINT
MEAN

↓

END

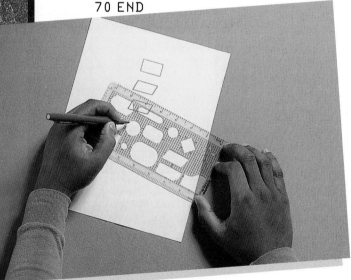

Make a flowchart for each program.

1.
```
10 LET Y=1
20 PRINT Y
30 PRINT Y↑3
40 LET Y=Y+1
50 PRINT Y
60 PRINT Y↑3
70 END
```

2.
```
10 READ D,E,F
20 PRINT D
30 PRINT E
40 PRINT F
50 PRINT D↑2+4*E+F
60 DATA 2,7,8
70 END
```

3. Draw a flowchart for a program to change any given number of feet to inches. Then write a program from the flowchart.

4. Draw a flowchart for a program to find the surface area of a rectangular prism. Then write a program from the flowchart.

INVENTORY SPECIALIST

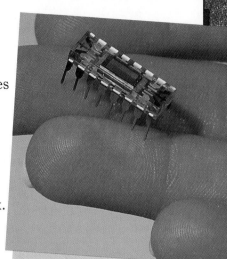

Kelly Woode is an inventory specialist for a computer firm. Each week she counts the items on hand and writes a report showing the number of items in stock and the number moving out of stock. She also makes sure that enough parts are in stock.

Kelly's company uses a certain kind of microchip. In the last six weeks, 35, 62, 44, 73, 36, and 68 microchips were used. Currently, there are 238 microchips in stock. Kelly approximates how long the supply will remain in stock.

1. First, she finds the mean number of microchips used per week.

$$(\; 35 \; + \; 62 \; + \; 44 \; + \; 73 \; + \; 36 \; + \; 68 \;) \; \div \; 6 \; = \; 53$$

2. Then she divides the number in stock by the mean, 53.

$$238 \; \div \; 53 \; = \; 4.490566$$

Thus, the supply should remain in stock about $4\frac{1}{2}$ weeks.

Approximate the number of weeks (to the nearest $\frac{1}{2}$ week)
each supply of parts will remain in stock.

Part	Number in Stock	Usage—Each of Last 6 Weeks					
1. circuit board	82	8	7	13	6	14	13
2. terminal	26	3	3	1	6	3	4
3. connector	112	10	2	6	5	0	14
4. port	19	3	1	4	0	5	2
5. switch	814	15	46	10	38 ·	74	51

6. Since the shipping time for circuit boards is six weeks, how many weeks after making this report should Kelly order circuit boards so that the supply will always be in stock?

7. If the firm doubles their production for the last week of the month, how many connectors would Kelly expect to be used during that week?

11-6   PICTOGRAPHS

Objective

Construct pictographs.

Mr. Bailey, a county extension agent, volunteered to give a speech on the decrease in the number of farms over the past few years. He used a *pictograph* to present the data visually and make it easier to remember.

A **pictograph** is used to compare data in a visually appealing way. Use the steps listed below to make a pictograph.

Method

1 Label the graph with a title. Write the scale on the vertical axis of the graph.

2 Choose a symbol and what it will represent. Write the definition of the symbol at the bottom of the graph.

3 Determine how many symbols will be used for each item by dividing.

4 Draw the symbols.

Example A

Make a pictograph for the data in the table at the right.

Number of Farms in the U.S.	
1965	3,300,000
1970	3,000,000
1975	2,500,000
1980	2,500,000
1985	2,300,000
1990	2,250,000

Number of Farms in the U.S.

1965	🏠 🏠 🏠 🏠 🏠 🏠 🏠
1970	🏠 🏠 🏠 🏠 🏠 🏠
1975	🏠 🏠 🏠 🏠 🏠
1980	🏠 🏠 🏠 🏠 🏠
1985	🏠 🏠 🏠 🏠 🏠
1990	🏠 🏠 🏠 🏠 🏠

2 🏠 = 500,000 farms

3 Divide 3,000,000 by 500,000. The result is 6. Draw 6 symbols next to 1970.

Guided Practice

Example A

Use the pictograph in Example A to answer each question.

1. How many farms does each symbol represent?

2. Which 5-year period had the greatest decrease in the number of farms?

3. Which 5-year period had no change in the number of farms?

4. How many more farms were there in 1965 than in 1990?

Make a pictograph for each set of data.

5.

Apple Production in 4 Orchards	
Adams	100,000
Lynd	275,000
Cooley	250,000
Smith	150,000

6.

Boxes of Cereal Sold	
Corn Flakes	550
Puffed Rice	250
Bran Flakes	300
Raisin Bran	400

Exercises

Practice

Make a pictograph for each set of data.

7.

Votes Received	
Johnson	800
Kuhn	850
Pitkin	1200
Van Dyke	600

8.

Number of Library Books Loaned	
Monday	60
Tuesday	40
Wednesday	45
Thursday	25
Friday	80

Mental Math

Look for a pattern. Divide mentally. Write only your answers.

9. $50 \div 5$ 10. $280 \div 40$ 11. $4,800 \div 60$ 12. $2,700 \div 9$

13. $63,000 \div 700$ 14. $250,000 \div 5,000$ 15. $7,200,000 \div 800$

16. $3,500,000 \div 500,000$ 17. $24,000,000 \div 30,000$

Applications

Solve. Use the pictograph.

18. *About* how many people lived in North Carolina in 1970?

19. Did the population of North Carolina increase or decrease from 1960 to 1970?

20. *About* how many more people lived in North Carolina in 1980 than in 1970?

21. *About* how many more people lived in North Carolina in 1960 than in 1950?

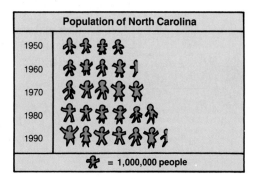

Population of North Carolina
1950
1960
1970
1980
1990
= 1,000,000 people

Collect Data

22. Survey your classmates to find their favorite sport. Make a pictograph to illustrate the data.

23. Survey the families in your neighborhood to find how many pets they have. Make a pictograph to illustrate the data.

Mixed Review

Lesson 5-6

Divide.

24. $\frac{8}{9} \div \frac{2}{3}$ 25. $\frac{3}{5} \div \frac{1}{10}$ 26. $\frac{6}{13} \div \frac{4}{7}$ 27. $4 \div \frac{3}{5}$

Lesson 8-7

Find the surface area of each rectangular prism.

28.
5 cm
8 cm

29. length = 14 in.
width = 8 in.
height = 2 in.

30. length = 9 mm
width = 4 mm
height = 3.5 mm

Lesson 7-10

31. At a busy intersection a stop light has a green arrow that allows 15 cars to turn left before changing. However, a semi-truck takes as much time as 3 cars in turning. How many cars can turn when there are 3 semi-trucks in line to turn?

11-7 CIRCLE GRAPHS

Objective

Construct circle graphs.

For social studies class, Mike made a *circle graph* to compare the number of persons in U.S. households for 1988.

A **circle graph** is used to compare parts of a whole. The whole amount is shown as a circle. Each part is shown as a percent of the whole. The percents should total 100%. Mike used the steps listed below to make a circle graph.

Method

1 ▶ Label the graph with a title.

2 ▶ Use a compass to draw a circle.

3 ▶ Multiply 360° by each percent to find the angle measure for each part. Round to the nearest degree.
There are 360° in a circle.

4 ▶ Use a protractor to draw the angles by placing the center of the protractor at the center of the circle.

5 ▶ Label each part of the circle.

Example A

Make a circle graph for the data in the table at the right.

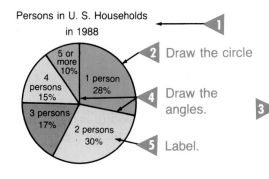

Persons in U. S. Households in 1988

Persons in U.S. Households in 1988	
1 person	28%
2 persons	30%
3 persons	17%
4 persons	15%
5 or more persons	10%

3 ▶ 28% of 360 = 0.28 × 360 ≈ 101°
30% of 360 = 0.30 × 360 ≈ 108°
17% of 360 = 0.17 × 360 ≈ 61°
15% of 360 = 0.15 × 360 ≈ 54°
10% of 360 = 0.10 × 360 ≈ 36°

Guided Practice

Example A

Use the circle graph in Example A to answer each question.

1. What percent of households have 4 persons?

2. The number of 4-person households is half of what size households?

3. What percent of households have 4 or more persons?

4. What percent of households have less than 3 persons?

Make a circle graph for the set of data.

5.

Earth's Surface	
Water	70%
Land	30%

Make a circle graph for each set of data.

6.

Water Use in the U.S.	
Agriculture	35%
Public Water	10%
Utilities	7%
Industry	48%

7.

Elements of Earth's Crust	
Oxygen	47%
Silicon	28%
Aluminum	8%
Iron	5%
Other	12%

8.

Chemical Composition of the Human Body	
Oxygen	65%
Carbon	18%
Hydrogen	10%
Nitrogen	3%
Other	4%

9.

Earth's Water	
Pacific Ocean	46%
Atlantic Ocean	23%
Indian Ocean	20%
Arctic Ocean	4%
Other	7%

10. Money for building homes comes from the following sources: savings and loans, 38.8%; commercial banks, 23.5%; life insurance, 18.8%; mutual savings banks, 16.3%; pension funds, 2.6%. Make a circle graph to illustrate these data.

Applications

Solve. Use the circle graph.

11. What percent of the sales is shirts?

12. Which item provided almost one-third of the sales?

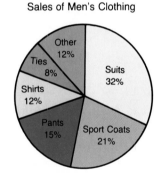

Sales of Men's Clothing

On pages 531–533, you can learn how biomedical equipment specialists use mathematics in their jobs.

13. The sales for suits is how many times greater than the sales for ties?

14. How many degrees is the angle labeled *sport coats?*

15. Which part has an angle of 115°?

16. Suppose the total sales for men's clothing is $2,500. How much was sales of pants?

17. Suppose Cindy Gannett budgets $240 per month, or 12% of her monthly salary, for food. What is her monthly salary?

Cooperative Groups

18. Make a frequency table of the hair color of your classmates. Then make a circle graph to illustrate the data. What conclusions can you make?

19. Make a circle graph to illustrate how you spend an average 24-hour day. Include the following categories: school, sleep, job, eating, leisure, and miscellaneous.

JOURNAL ENTRY

20. Name a situation where you would use a graph to prove a point to a committee. Tell what kind of graph you would use.

11-8 STEM-AND-LEAF PLOTS

Objective

Interpret and construct a stem-and-leaf plot.

Candice Brown checks the inventory at a video store. Each day she marks the number of tapes sold. The amounts sold each day during a three-week period are given at the right.

19	30	27	34	23
31	46	43	36	27
34	39	48	15	47
35	33	21	37	27	42

Numerical data can be organized in a **stem-and-leaf plot.** The greatest place value of the data is used for the stem. The next greatest place value is used for the leaves.

Method

1. Draw and label a plot with two columns.

2. In the first column, list the greatest place values of the data. Use each digit only once.

3. In the second column, record the next greatest place value of each number next to the correct stem.

4. Make a second stem-and-leaf plot to arrange the leaves in order from least to greatest.

Example A

Make a stem-and-leaf plot for the data above.

1.
2. The data range from 15 to 48. So the stems range from 1 to 4.

Stem	Leaf
1	9 5 3
2	7 3 7 1 7
3	0 5 1 6 3 9 4 4 7
4	6 3 8 7 2

4|6 represents the number 46.

4.

Stem	Leaf
1	5 9
2	1 3 7 7 7
3	0 1 3 4 4 5 6 7 9
4	2 3 6 7 8

Guided Practice

Example A

Use the stem-and-leaf plot in Example A to answer each question.

1. What is the least number of videos rented in one day?
2. What is the greatest number of videos rented in one day?
3. What does 2|1 represent?
4. How many days had 27 rentals?
5. What is the mode of the data?
6. What is the median of the data?
7. Each row of numbers represents an interval of 10. If the first line 1|5 9 represents the interval 10–19, in which interval do the days with the most rentals fall?

Practice

State the stems that you would use to plot each set of data.

8. 18, 36, 43, 25, 32, 4, 27

9. 12, 36, 24, 57, 10, 28, 39, 52

10. 158, 581, 182, 368, 404, 545

11. 6.4, 5.5, 7.6, 8.4, 5.1, 8.9, 7.2

Find the median and mode of the data in each stem-and-leaf-plot.

12.

Stem	Leaf
2	4 5
3	2 3 6
4	0 1 2 3 8 9

13.

Stem	Leaf
8	0 0 2 1
9	4 8 9 9
10	0 6 7 9

14.

Stem	Leaf
4	2 4 6
5	0 1 3 8 9
6	3 4 4 5 8
7	1 7

Applications

Use the science test scores at the right to complete Exercises 15–20.

15. Construct a stem-and-leaf plot of the data.

16. How many students took the test?

17. What is the lowest test score?

18. Find the range of the scores.

19. How many students had scores above 89?

20. In which interval did most students score?

100	62	85	72	99	87
87	93	77	86	96	79
100	86	94	68	75	90
88	99	87	73	66	89

21. Use the data at the right to construct a stem-and-leaf plot. With what age group was the concert the most popular?

Ages of People Attending a Concert

16	23	13	22	9	11	26	16	35	42
29	24	38	14	6	17	18	12	7	19
26	17	25	15	7	24	15	18	33	25

Critical Thinking

22. Carla ate 100 peanuts in five days. Each day she ate five more peanuts than she did the day before. How many peanuts did she eat the first day?

Mixed Review

Find the surface area of each cylinder described below. Use 3.14 for π. Round decimal answers to the nearest tenth.

Lesson 8-8

23.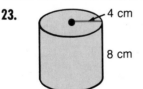

24.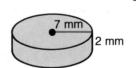

25.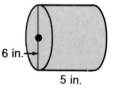

Lesson 9-1

Write each ratio as a fraction in lowest terms.

26. 6 teams for 7 coaches

27. 40 rooms for 1,000 students

Lesson 9-9

28. During the softball season, Carrie got on base 51 out of 85 times at bat. What percent of the times at bat did she get on base?

11-9 MEASURES OF VARIATION

Objective

Find quartiles and interquartile range.

Jamie Brooks is moving from San Francisco to Chicago. He compares the temperatures in both cities.

In some sets of data, the values are close together. In other sets, the values are far apart. The spread of values in a set of data is called the **variation.** The *range* is one such measure of variation. What are the ranges of the mean daily temperatures for San Francisco and Chicago?

Mean Daily Temperature (°F)		
	San Francisco	**Chicago**
January	49	21
February	52	26
March	53	36
April	55	49
May	58	59
June	61	69
July	62	73
August	63	72
September	64	65
October	61	54
November	55	40
December	49	28

Another measure of variation is the **interquartile range. Quartiles** are values that divide the ordered data set into four equal parts. The interquartile range is the difference between the upper and lower quartiles of the data.

Method

1. ▶ Find the median of the data.

2. ▶ Find the median of the upper half (upper quartile) and the median of the lower half (lower quartile) of the data.

3. ▶ To find the interquartile range, subtract the lower quartile value from the upper quartile value.

Example A

Find the interquartile range of the mean daily temperatures in San Francisco.

1. ▶ 49   49   52   53   55   55 ↑ 58   61   61   62   63   64

   The median of the data is $\frac{55 + 58}{2}$ or 56.5.

2. ▶ 49   49   52 ↑ 53   55   55 | 58   61   61 ↑ 62   63   64

         lower quartile   upper quartile

   The upper quartile is $\frac{61 + 62}{2}$ or 61.5.

   The lower quartile is $\frac{52 + 53}{2}$ or 52.5.

3. ▶ The interquartile range is 61.5 − 52.5 or 9. The middle half of the mean daily temperatures varies 9°F.

1. Find the median, upper quartile, lower quartile, and the interquartile range for the temperatures in Chicago.

2. Compare the interquartile ranges for San Francisco and Chicago. Does the temperature in Chicago vary more or less than the temperature in San Francisco?

Find the upper quartile, lower quartile, and interquartile range for each set of data.

3. 2, 6, 12, 16, 16, 18, 20, 22

4. 13, 2, 13, 2, 5, 4, 3, 10, 1, 2

5.

Stem	Leaf			
1	2			
2	1	3	5	
3	3	7	8	9
4	2	2		

4|2 represents 42

6.

Stem	Leaf			
8	2	5	6	8
9	0	2	4	
10	3	7		
11	9			

11|9 represents 119

7. What do the range and the interquartile range show about the San Francisco and Chicago temperatures that the mean does not?

8. The girls' basketball team has 15 wins. The points scored in each win are shown at the right. Make a stem-and-leaf plot for the data and find the interquartile range.

52	64	55
74	62	50
58	65	60
57	59	64
52	62	65

9. From the data in Exercise 8, would you conclude that the girls' basketball team is consistent or inconsistent in the number of points they score during a game?

The table below gives the milligrams of sodium in a slice for several kinds of pizza.

10. Find the range and the interquartile range of the milligrams of sodium in a slice for the brands of pizza in the table.

11. If the recommended daily allowance of sodium is 2,400 mg, how many slices of Candu's pizza could you eat without going over the recommended daily allowance?

Product	Sodium (mg)
Celentano Cheese Pizza	364
Chef Boyardee pizza mix	1,145
Candu's Cheese Pizza	697
Stouffer's French Bread Pizza	1,064
Totino's Microwave Pizza	1,436
Celeste Pizza	1,112
Croissant Pastry Pizza	953

12. Which brand could you eat the most slices of without going over the recommended daily allowance?

Evaluate each expression if n = 4.

13. $4n - 2$

14. $3(n + 6)$

15. $(n - 3) + 12$

11-10 BOX-AND-WHISKER PLOTS

Objective

Construct and interpret box-and-whisker plots.

Number of Hours of Satisfactory Performance	
38	44
22	46
57	60
58	53
44	33
42	48

Andi Greer tested twelve brands of batteries for the number of hours of satisfactory performance. She will display the data in a box-and-whisker plot for her article in *Wise Consumer*.

A **box-and-whisker plot** is a graph that shows the median, quartiles, and extremes (the least and greatest values) of a set of data.

Method

1. Draw a horizontal line and mark it with a number scale for the set of data.

2. Plot the median, the quartiles, and the extremes on the line.

3. Draw a *box* around the middle half of the data from the lower quartile to the upper quartile.

4. Indicate the median by drawing a vertical line through its point.

5. Draw the whiskers by connecting the lower extreme to the lower quartile and the upper quartile to the upper extreme with a line.

Example A

Draw a box-and-whisker plot to display the data from Andi's battery test results.

Guided Practice

Use this box-and-whisker plot to answer each question.

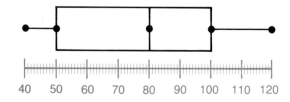

Example A

1. What is the range and interquartile range of the data?

2. The middle half of the data is between which two values?

3. What part of the data is greater than 100?

4. What part of the data is greater than 80?

5. What part of the data is less than 80?

Practice

Use the box-and-whisker plot in Example A to answer each question.

6. What information is easier to find on a box-and-whisker plot than on a stem-and-leaf plot?

7. What information is harder to find on a box-and-whisker plot than on a stem-and-leaf plot?

8. State the median, upper and lower quartiles, the range, and interquartile range of the data in Example A.

9. What fractional part of the data is included in each whisker? Is this true for every box-and-whisker plot? Why?

Applications

Compare the two box-and-whisker plots at the right. Then solve Exercises 10–12.

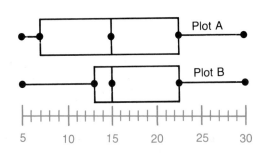

10. What is similar about the data in the two plots?

11. What is different about the data in the two plots?

Interpreting Data

12. If Plot A represents the number of hours of light from a test of ten Brand X light bulbs and Plot B represents the number of hours of light from a test of ten Brand Y light bulbs, which brand is the better buy? Why?

Collect Data

13. In a grocery store or consumer magazine, find the number of calories in a serving of potato chips for six different types or brands.
   a. Construct a box-and-whisker plot of the data.
   b. What is the range of the number of calories in a serving?
   c. Does the data indicate that different brands of potato chips are similar in the number of calories they have?

Critical Thinking

14. Rachel has 18 times more nickels than quarters. If Rachel's coins have a total value of $6.90, how many nickels does Rachel have?

Mixed Review

Lesson 6-9

Find the equivalent Celsius temperature to the nearest degree.

15. 82°F   **16.** 60°F   **17.** 43°F   **18.** 110°F

Lesson 8-10

Find the volume of each pyramid.

19.
6m   3m   4m

20.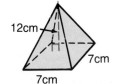
12cm   7cm   7cm

21. length = 10 in.
    width = 12 in.
    height = 11.4 in.

Lesson 9-4

22. Denise uses 3 gallons of gasoline for every 100 miles she drives her car. How much gasoline does she use if she drives 175 miles?

▶ Explore
▶ Plan
▶ Solve
▶ Examine

11-11 LOOK FOR A PATTERN

Objective

Solve verbal problems by looking for a pattern.

Debbie begins a physical fitness program. Debbie's goal is to do 100 sit-ups a day. On the first day of the program, she does 20 sit-ups. Every fifth day of the program, she increases the number of sit-ups by 10. After how many days will she reach her goal?

You can use a list, a table, or a drawing to find a pattern.

▶ **Explore**

What is given?
● Debbie begins with 20 sit-ups.
● Every fifth day the number increases by 10.

What is asked?
● On what day does Debbie do 100 sit-ups?

▶ **Plan**

First make a list showing the information you are given. Then look for a pattern and apply it to 100 sit-ups.

▶ **Solve**

Day	Increase	Number of Sit-Ups
1	0	20
5	0 + 10 = 10	20 + 10 = 30
10	10 + 10 = 20	20 + 20 = 40
15	20 + 10 = 30	20 + 30 = 50

Notice that the increase in sit-ups is twice the day number. Subtract 20 from 100 to find the increase. Then divide by 2 to find the day. Debbie will do 100 sit-ups on day $(100 - 20) \div 2$ or day 40.

▶ **Examine**

You can check the solution by extending the list to day 40. The solution is correct.

Guided Practice

1. Gena plants five strawberry plants in her garden. The number of plants triples every year. How many plants will Gena have in her garden in six years?

2. A volleyball team has 6 players. Suppose each player does a high five with every other player. How many high fives take place?

Problem Solving

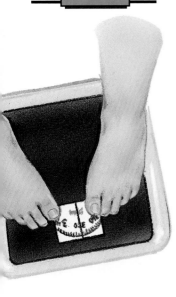

f UN with MATH

How about a recipe for Campfire Doughboys?
See page 373.

3. The pages in a book are numbered starting with 1. To number all the pages, the printer uses a total of 381 digits. How many pages are in the book?

4. Pat mails a recipe to five friends. Each of the five friends mails the recipe to five more friends, and so on. What is the total number of recipes in the sixth mailing?

5. Mrs. Burns buys numerals to put on the door of each apartment in a 99-unit apartment building. The apartments are numbered 1 through 99. How many of each digit (0, 1, 2, 3, 4, 5, 6, 7, 8, and 9) should Mrs. Burns buy?

6. Jessie wants to save $25 for a gift. She begins by saving 10¢ the first week. Each week she saves twice as much as the week before. In how many weeks will she have saved at least $25?

7. Mr. Mason weighed 170 pounds on his 40th birthday. Then he began gaining about 2 pounds every year. At this rate, how much will he weigh when he is 50 years old?

8. In a single-elimination softball tournament, teams are eliminated when they lose. If 8 teams are involved, how many games must be played to determine a tournament winner?

9.

Date	Oct.–May	June–Sept.
3 years ago	$35	$40
2 years ago	$38	$44
1 year ago	$41	$48
this year	$44	?

Motel rates change with the time of year. What do you expect the June–Sept. rates to be for this year if the same pattern continues?

Critical Thinking

10. The map shows the streets between the Tabor house and the Lee house. They are all one-way streets. How many different ways are there to go from the Tabor house to the Lee house?

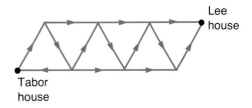

Mixed Review

Lesson 9-8

Write each fraction as a percent.

11. $\frac{15}{30}$ **12.** $\frac{9}{36}$ **13.** $\frac{17}{20}$ **14.** $\frac{4}{20}$ **15.** $\frac{3}{9}$

Lesson 10-1

Find the percent of each number.

16. 5% of 50   **17.** 20% of 340   **18.** 45% of 540

19. Select some of your work from this chapter that shows how you used a calculator or computer.

**Vocabulary/
Concepts**

*Write the letter of the term that best matches each
description.*

1. the difference between the greatest and
the least number

2. used to show change and direction of
change over a period of time

3. used to compare parts of a whole

4. the average of a set of data

5. the middle number of a set of data when
the data is listed in order

6. a way of organizing numbers or items in a
set of data

7. used to compare data in a visually
appealing way

8. the number that appears
most often in a set of data

9. used to show the median,
quartiles, and extremes of
a set of data

10. the median of the upper
half of a set of data

11. the difference between the
upper quartile and the lower
quartile in a set of data

a. bar graph
b. box-and-whisker plot
c. circle graph
d. frequency table
e. interquartile range
f. line graph
g. mean
h. median
i. mode
j. pictograph
k. range
l. stem-and-leaf plot
m. upper quartile

**Exercises/
Applications**

Lesson 11-1

Find the median, mode, and range for each set of data.

12. 8, 4, 8, 9, 6, 7, 4, 2, 5, 4, 8, 9, 8

13. 14, 12, 19, 14, 20, 14, 13, 15, 16,
16, 14, 12

Lesson 11-2

Solve. Use the data at the right.

14. Make a frequency table for the set of data.

15. Find the range.

16. Find the median.

17. Find the mode.

Scores on a History Quiz				
100	95	75	85	90
75	75	65	90	100
85	75	60	90	80
70	95	90	70	75

Lesson 11-3

The ages of the workers at a fast food
restaurant are given at the right.

16 16 17 15 40
18 19 39 63

18. Find the mean age and the median age of the workers.

19. Which average age should the restaurant use if it wants to attract
college age students?

Make the type of graph indicated for each set of data.

20. bar graph

High Temperatures in Tulsa, Oklahoma	
Sun.	94°F
Mon.	91°F
Tues.	87°F
Wed.	89°F
Thurs.	91°F
Fri.	95°F
Sat.	97°F

21. line graph

Average Monthly Rainfall (in inches)	
Mar.	1
Apr.	5
May	3
June	2
July	3
Aug.	1
Sept.	4
Oct.	3

22. pictograph

Favorite Radio Station	
WXNY	5
WNCI	12
WQFM	8
WRFD	9
WBBY	2

23. stem-and-leaf plot

The Ages of Each U.S. President on His First Inauguration						
57	61	57	57	58	57	61
65	52	56	46	54	49	50
55	51	54	51	60	62	43
54	68	51	49	64	50	48
47	55	55	54	42	51	56
55	56	61	52	69	64	46

24. Construct a box-and-whisker plot for the data in Exercise 23.

25. What percent of the members bike?

26. In which activity did over one-fourth of the members participate?

27. Suppose there are 200 members. How many participants are there in hiking?

28. How many degrees is the angle labeled swimming?

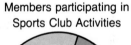
Members participating in Sports Club Activities

Find the upper quartile, lower quartile, and interquartile range for each set of data.

29. 4, 7, 8, 8, 10, 10, 12

30. 16, 22, 15, 18, 15, 22, 20, 16

Solve. Use look for a pattern.

31. Toby mails a math puzzle to three friends. Each of the three friends mails the puzzle to three more friends, and so on. What is the total number of puzzles in the fifth mailing?

TEST

Solve. Use the data at the right.

1. Make a frequency table for the set of data.

2. Find the range.

3. Find the median.

4. Find the mode.

Wages of Workers at Fouse Co.			
$14,500	$19,400	$13,000	$14,500
$11,300	$28,600	$14,500	$11,300
$13,000	$14,500	$16,000	$19,400
$13,000	$11,300	$13,000	$14,500
$16,000	$14,500	$16,000	$13,000

The prices of several used cars are listed at the right.

$2,500 $3,500 $4,000
$3,500 $5,000 $10,000

5. Find the mean price and the median price.

6. If the most expensive used car costs $20,000 instead of $10,000, how would the median be affected? the mean?

Make the type of graph indicated for each set of data.

7. bar graph

Height of Skyscrapers (in feet)	
Chrysler Building	1,046
John Hancock Center	1,127
Empire State Building	1,250
World Trade Center	1,350
Sears Tower	1,454

8. line graph

Average Monthly Temperatures in Oklahoma City, Oklahoma (°F)			
Jan.	35.9	July	82.1
Feb.	40.8	Aug.	81.1
March	49.1	Sept.	73.3
April	60.2	Oct.	62.3
May	68.4	Nov.	48.8
June	77.0	Dec.	39.9

9. stem-and-leaf plot

Height in Inches of Track Team Members					
61	60	67	62	59	72
68	63	72	78	60	65
72	55	66	70	59	77

10. circle graph

Monthly Budget	
Food	20%
Clothes	10%
Rent	40%
Other	15%
Transportation	15%

Use the box-and-whisker plot to answer each question.

11. What is the range?

12. What is the interquartile range?

13. What is the median?

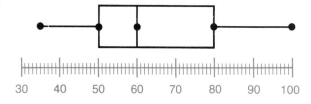

14. A business club has 8 members. Suppose each member shakes hands with every other member. How many handshakes take place?

▶ **BONUS:** Write two questions that can only be answered by using the box-and-whisker plot used in Exercises 11–13.

MISLEADING GRAPHS

Both graphs below show monthly sales for a small business. Notice that both graphs show the same data. However, the graphs look different because of the scales used along the vertical axes.

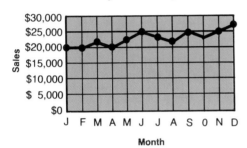

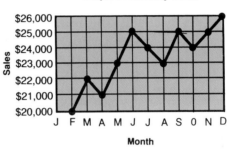

Graph B is *misleading*. It seems to show a very large increase in sales because the scale does not begin at zero.

The graphs below show the results of a survey on favorite restaurants.

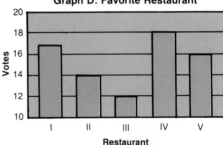

1. Do graphs C and D display the same data?
2. Find the number of votes for each restaurant.
3. In graph D, the bar for restaurant II is twice as long as the bar for restaurant III. Does this mean there were twice as many votes?
4. What causes the difference in voting to appear greater in graph D?
5. Why would restaurant III prefer graph C?
6. Why would restaurant IV prefer graph D?
7. Which graph best represents the result? Explain.

Free Response

Lessons 1-11
1-12

Solve each equation.

1. $23 + y = 130$

2. $x - 28 = 10$

3. $4 = \frac{x}{20}$

4. $18 = 3y$

5. $100 - 52 = y$

6. $8x = 7$

Lessons 2-1
2-3

Add or subtract.

7. $6,354 - 617$

8. $7,292 + 82$

9. $1,979 + 876$

10. $54,685 - 2,548$

11. $1,983 - 1,765$

12. $62 + 3.587 + 467$

Lesson 4-7

Replace each ● *with* $<$, $>$, *or* $=$ *to make a true sentence.*

13. $\frac{3}{7}$ ● $\frac{4}{7}$

14. $\frac{1}{2}$ ● $\frac{4}{8}$

15. $\frac{2}{3}$ ● $\frac{7}{9}$

16. $\frac{7}{15}$ ● $\frac{5}{12}$

17. $\frac{7}{8}$ ● $\frac{8}{9}$

Lesson 5-7

18. Four pizzas cut into eight pieces each were eaten by nine friends. If everybody had equal amounts, how much did each person eat?

Lessons 6-7
6-8

Complete.

19. 1 ft = ▓ in.

20. 48 in. = ▓ ft

21. 1 mi = ▓ ft

22. 1 gal = ▓ qt

23. 100 oz = ▓ qt ▓ oz

24. 10,000 lb = ▓ T

Lesson 7-1

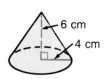

Use words and symbols to name each figure in as many ways possible.

25.
A   B

26.
S   T

27.
M   N

28.
Q

Lesson 8-12

Find the volume of each cone. Use 3.14 for π. *Round decimal answers to the nearest tenth.*

29.

6 cm
4 cm

30.

12 m
5 m

31.

15 in.
8 in.

Lesson 9-8

32. Ben's Bakery gives an extra donut with each dozen you buy. If you buy 80 donuts to take to a breakfast meeting, how many donuts will you get free?

Lesson 11-1

33. Jacob's quiz scores are 18, 10, 25, 13, and 19. Find the mean, median, and range of his quiz scores.

Multiple Choice

Choose the letter of the correct answer for each item.

1. Find the value of $4 \times 5^2 - 3 \times 2$.
a. 91
b. 94
c. 176
d. 394

Lesson 1-9

2. What is the sum of 40,592 and 3,287?
e. 37,305
f. 43,879
g. 43,889
h. 73,462

Lesson 2-1

3. Find the product when 6 is multiplied by 2,220.
a. 1,320
b. 1,332
c. 12,320
d. 13,320

Lesson 3-1

4. Find the lowest common multiple (LCM) of 24, 42, and 48.
e. 6
f. 96
g. 336
h. 48,384

Lesson 4-3

5. Janet needs $2\frac{7}{8}$ yards of fabric. She buys $3\frac{1}{4}$ yards. How much extra fabric did she buy?
a. $\frac{1}{4}$ yard
b. $\frac{5}{8}$ yard
c. $1\frac{3}{8}$ yards
d. *none of these*

Lesson 6-7

6. Change $0.12\frac{1}{2}$ to a fraction.
e. $\frac{1}{80}$
f. $\frac{1}{8}$
g. $\frac{1}{4}$
h. $\frac{25}{2}$

Lesson 5-9

7. A salt solution has 3 cups salt for every 8 cups water. What is the ratio of salt to water?
a. 3:8
b. 8:3
c. 3:11
d. 8:11

Lesson 9-1

8. In a class election, Brady received 60% of the 30 votes cast. How many more votes did Brady receive than the other candidate?
e. 8
f. 6
g. 10
h. 9

Lesson 10-1

9. 88 is what percent of 110?
a. 8%
b. 80%
c. 88%
d. 125%

Lesson 10-2

10. Ed deposits $150 in an account that pays 7% simple annual interest. How much interest will he earn in 2 years?
e. $7
f. $14
g. $21
h. $210

Lesson 10-8

fun with MATH

Noah's Ark Turkey	510 BC	The rotating fan China		AD 876
	2368 ? BC	Birth of Democracy Athens, Greece	200 BC	Zero, the symbol India

How far away was that lightning?

The sound of thunder travels at about 1,100 feet per second. It will take about five seconds for it to travel one mile. The next time you see lightning, count the number of seconds it takes for the thunder to arrive. Divide by five and you will have an estimate of how far away the lightning was.

MATH M·E·N·U

Campfire Doughboys

Add enough water slowly to 1 package of biscuit mix to make a soft, gooey dough. Wrap a small amount of dough around the end of a green stick. Hold the stick over the campfire, turning until the dough turns brown and appears to be done. Take the doughboy away from the fire and off the stick. Put butter and jelly on it.

COMICS

When "bowling at nine pins" was made illegal in Connecticut during the 18th century, bowlers added a tenth pin. Thus we have ten-pin bowling.

FOR BETTER OR WORSE

| Statistics England | AD 1946 | Symmetry in the universe? New York, USA | AD 1983 |

| AD 1662 | First electronic computer Pennsylvania, USA | AD 1956 | Genetic engineering Pennsylvania, USA |

JOKE!

Q: What do you call a person who counts on her fingers?

A: A digital computer

RIDDLE

Q: Use one word to describe what is shown here.

A: Hy pot en use

How does an electronic calculator add 4 plus 5? Pressing the three keys, 4, +, and 5, on your calculator closes a set of contacts for each key on the circuit board beneath the keyboard. This sends signals to the calculator's processing unit to store the codes for each key in memory. Pressing the equals key tells the processing unit to take the three codes from memory and to perform the addition operation. The answer is decoded and sent to the display as 9.

TEASER

Place four coins on your desk, all with tails up. A move consists of turning three coins over at a time. How many such moves do you need to arrange for all four heads to be up?

Ⓣ Ⓣ Ⓣ Ⓣ

Once you have solved this teaser you may like to investigate another teaser with five coins all tails up. In this teaser, a move consists of turning over any four coins.

QUIZ TIME

Try these number patterns to start you thinking.
1. Choose any single-digit number and multiply it by 9. The multiply the product by 12345679 on your calculator. Are you surprised at the answer? Try more numbers and see what happens. Then explain your answers.
2. What number would you substitute for 12345679 if you first multiplied your single-digit number by 7 instead of 9?

12 INTEGERS

-10

-5

0

52 (H)

10

45

32

(L)

70

How Cold Is It?

About how many days each year does the temperature drop below freezing in your community? Water freezes at 0°C. How would you write a temperature below zero? You can write *1°C below zero* as −1°C. How would you write 2°C below zero? 10° below zero? What are some reasons you need to know the temperature?

At 4:00 P.M. the temperature was 5°C. By 8:00 P.M. the temperature had fallen 7°C. What was the temperature at 8:00 P.M.?

ACTIVITY: Adding Integers

The temperature *6° below zero* can be written as −6°
and the temperature *6° above zero* can be written as +6°.
−6 is read as *negative 6* and +6 is read as *positive 6*.

Usually, the plus signs on numbers like +1 and +2 are omitted.

Numbers such as −6 and +6 are called integers. The set of
integers is listed below.

. . . , −6, −5, −4, −3, −2, −1, 0, +1, +2, +3, +4, +5, +6, . . .

In the activities that follow, positive and negative counters will
represent positive and negative integers. When the counters
are combined, every positive counter cancels exactly one
negative counter; that is, a negative paired with a positive
results in zero. The models below will help you understand how
to add integers.

Materials: counters

Cooperative Groups

Work with a partner. Study these models for each addition problem.

a.

$$3 + 2 = 5$$

b.

$$-3 + (-2) = -5$$

c.

$$3 + (-2) = 1$$

d.

$$-3 + 2 = -1$$

Communicate Your Ideas

Use counters to model each problem. Then write the sum.

1. 4 + 3   **2.** −5 + (−4)   **3.** −6 + 2   **4.** 3 + (−8)

5. −2 + (−2)   **6.** −8 + 4   **7.** −7 + 7   **8.** 1 + (−4)

Playing a Game

9. The "Game of 6" can be played with 2 to 5 players. Make a deck
of 52 cards by writing each integer from −6 through 6 on four
differently colored sets of thirteen cards. The dealer deals two
cards to each player. Each player mentally finds the sum of the
numbers on the cards. If the sum is close to 6, the player may
pass. If the sum is not close to 6, the player may receive up to
three more cards, one at a time, to try to reach that goal. The
winner is the player who has a sum of 6 or closest to 6.

INTEGERS AND THE NUMBER LINE

Objective

Identify, order, and compare integers.

Fran Cantu was a contestant on a TV quiz show. Her final dollar score was $1,050. Her opponent's final dollar score was −$1,050. Fran was declared the winner. In algebra, pairs of numbers like 1,050 and −1,050 are called **opposites**.

Every number has an opposite. On the number line a number and its opposite are the same distance from zero. −3 and 3 are opposites.
The opposite of 0 is 0.

Zero is neither positive nor negative.

The whole numbers and their opposites are called **integers**.

The number of units a number is from zero on the number line is called its **absolute value**. The absolute value of −3 is 3. It is written $|-3| = 3$. All absolute values are positive or zero.

You can compare two integers using the number line.

Method

1 ▶ Draw or imagine a number line.

2 ▶ The greater number is farther to the right on a horizontal number line. On a vertical number line, the greater number is above the lesser number.

Examples

A *Compare −4 and 3.*

2 ▶ −4 is to the left of 3.
So −4 < 3.

B *Compare −2 and −7.*

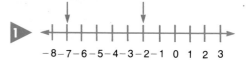

2 ▶ −2 > −7

C *Compare 0 and −3.*

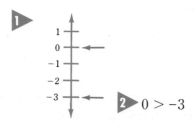

2 ▶ 0 > −3

D *Order −6, 6, −4, 4, 0 from least to greatest.*

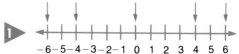

2 ▶ The order from least to greatest is −6, −4, 0, 4, 6.

Examples A, B, C

Replace each ● with <, >, or = to make a true sentence.

1. 3 ● 7   **2.** 5 ● 0   **3.** −2 ● 2   **4.** −5 ● 0

5. −4 ● −6   **6.** −3 ● 2   **7.** −1 ● −3   **8.** −14 ● 2

Example D

Order from least to greatest.

9. 4, −3, 0, −9, 5   **10.** 6, −6, 2, −8, 0

11. −1, 0, 11, −7, 6   **12.** −5, 0, 5, 3, −3

Exercises

Practice

Replace each ● with <, >, or = to make a true sentence.

13. −1 ● 0   **14.** 0 ● −4   **15.** −3 ● −4   **16.** −7 ● −1

17. 5 ● −2   **18.** 3 ● 8   **19.** −2 ● 3   **20.** 3 ● 4

21. −10 ● −15   **22.** −1,000 ● −1   **23.** $-\frac{1}{2}$ ● $-\frac{1}{4}$ **24.** −2 ● −2.1

25. |−4| ● |12| **26.** |−7| ● |15| **27.** |24| ● |−17| **28.** |2| ● |−8|

29. |0| ● |−9| **30.** |10| ● |−10| **31.** |15| ● |−4| **32.** |−3| ● |5|

Order from least to greatest.

33. 8, 3, −4, 9, 0   **34.** −7, 5, 0, −2, 2

35. −2, 0, −4, −1, 1   **36.** 0, −11, −20, 10, −15

37. 3.5, −3, 0, −3.5, 3   **38.** $-\frac{1}{2}, -\frac{1}{4}, \frac{1}{4}, \frac{1}{2}, 0$

Number Sense

39. If +10 indicates 10 seconds after takeoff, what integer indicates 10 seconds before takeoff? What integer represents the takeoff time?

Applications

Solve. Use the table at the right for Problems 40–42.

40. What is the lowest recorded temperature in North America?

41. Where is the world's lowest recorded temperature?

42. Order the lowest recorded temperatures of the countries and continents from 1 to 8, with 1 being the lowest.

43. The average daily low temperature for January in Great Falls, Montana, is about −5°F. In Chicago, it is about 2°F. Which city has the colder average?

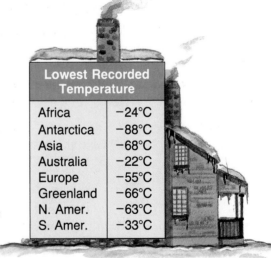

Lowest Recorded Temperature	
Africa	−24°C
Antarctica	−88°C
Asia	−68°C
Australia	−22°C
Europe	−55°C
Greenland	−66°C
N. Amer.	−63°C
S. Amer.	−33°C

Critical Thinking

44. At a party, everyone shook hands with everyone else exactly once. There were 36 handshakes. Find the number of people at the party.

Research

45. Look through magazines, newspapers, and journals to make a list of the different ways integers are used in our lives.

12-2 ADDING INTEGERS

Objective
Add integers.

In the spring after a period of heavy rain, the water level in Smyth Reservoir was 3 feet above its normal level of 0. In the summer, after a drought, the water level was 7 feet below the level of the heavy rains. What was the water level after the drought?

The number line can be used to show addition of integers. Move up when adding a positive number. Move down when adding a negative number.

Example A

Find 3 + (− 7) using the number line.

Start at 3. Then move down 7 units. The sum is −4.

3 + (−7) = −4

The water level in Smyth Reservoir is −4 or 4 feet below normal after the drought.

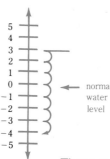

These rules also apply to positive and negative decimals and fractions.

Method

▷1 To add integers with the *same sign*, add their absolute values.

▷2 Give the result the same sign as the integers.

Examples

B 5 + 23

Both integers are positive.

▷1 |5| + |23| = 5 + 23
= 28

▷2 Give the result a positive sign.
5 + 23 = 28

C − 5 + (− 23)

Both integers are negative.

▷1 |−5| + |−23| = 5 + 23
= 28

▷2 Give the result a negative sign.
−5 + (−23) = −28

Method

▷1 To add integers with *different signs*, subtract their absolute values.

▷2 Give the result the same sign as the integer with the greater absolute value.

Examples

D 2 + (− 7)

One integer is positive and one is negative.

▷1 |−7| − |2| = 7 − 2
= 5

▷2 Since |−7| > |2|, the result is negative.
2 + (−7) = −5

E 13 + (− 6)

One integer is positive and one is negative.

▷1 |13| − |−6| = 13 − 6
= 7

▷2 Since |13| > |−6|, the result is positive.
13 + (−6) = 7

Example A

Write an addition sentence for each number line.

1.

2.

Add. Show your answer on a number line.

Examples B, C

3. $5 + 3$ **4.** $-4 + (-3)$ **5.** $-2 + (-4)$

Examples D, E

6. $2 + (-4)$ **7.** $8 + (-6)$ **8.** $9 + (-10)$

Practice

Add. Use a number line if necessary.

9. $-8 + 7$ **10.** $12 + 13$ **11.** $-8 + (-7)$ **12.** $-7 + (-6)$

13. $8 + (-8)$ **14.** $-2 + 2$ **15.** $-2 + (-2)$ **16.** $-8 + (-8)$

17. $24 + (-13)$ **18.** $-17 + 12$ **19.** $-14 + (-19)$ **20.** $-9 + (-11)$

21. $-3 + 16$ **22.** $-28 + 50$ **23.** $13 + (-51)$ **24.** $-74 + 21$

Number Sense

25. $-4.5 + 3.7$ **26.** $-5.2 + (-6)$ **27.** $-17 + (-31)$ **28.** $-\frac{4}{3} + \frac{1}{2}$

Write Math

29. Write a rule for adding two integers with different signs.

Applications

30. In a football game, the Denver Broncos lost 6 yards on one play and then lost 3 yards on the next. What was the net gain?

31. The temperature at dawn was $-12°C$. By noon the temperature was $18°$ higher. What was the temperature at noon?

32. A submarine at 1,150 feet below sea level descends an additional 1,250 feet. How far below sea level is the submarine now?

fUN with MATH

What is unusual about the day and the year on Venus?
See page 422.

33. At night the average surface temperature on the planet Saturn is $-150°C$. During the day the temperature rises $27°C$. What is the temperature on the planet's surface during the day?

34. A weather balloon rises 300 feet from the ground, drops 125 feet, and then rises 450 feet. Write an equation. Then find the height of the balloon.

Critical Thinking

35. Find the value of each letter.

$$\begin{array}{r} ODD \\ + ODD \\ \hline EVEN \end{array}$$

12-3 SUBTRACTING INTEGERS

Objective
Subtract integers.

Miami is 12 feet above sea level. New Orleans is 5 feet below sea level. To find the difference in their elevations subtract $12 - (-5)$.

Remember that every subtraction expression can be written as an addition expression. So, $12 - (-5)$ can be written as $12 + 5$.

$12 + 5 = 17 \quad 12 - (-5) = |12| + |-5|$
$= 12 + 5$
$= 17$

Adding an integer gives the same result as subtracting its opposite.

The difference in elevation is 17 feet.

The integers have different signs. Add their absolute values. Since $|12| > |-5|$, the sum is positive.

What number represents sea level?

Method

1. Write the subtraction expression as an addition expression.
2. Add.   These rules also apply to positive and negative decimals and fractions.

Examples

A $12 - 8$
To subtract 8, add -8
1. $12 - 8 = 12 + (-8)$
2. $= 4$
$12 - 8 = 4$

B $-13 - (-4)$
To subtract -4, add 4.
1. $-13 - (-4) = -13 + 4$
2. $= -9$
$-13 - (-4) = -9$

C $-18 - 5$
1. $-18 - 5 = -18 + (-5)$
2. $= -23$
The difference is -23.
Check. $-23 + 5 \stackrel{?}{=} -18$
$-18 = -18 \checkmark$

D $9 - 14$
1. $9 - 14 = 9 + (-14)$
2. $= -5$
So, $9 - 14 = -5$.
Check. $-5 + 14 \stackrel{?}{=} 9$
$9 = 9 \checkmark$

Guided Practice

Subtract.

Example A
1. $12 - (-5)$
2. $2 - (-2)$
3. $14 - (-6)$
4. $-3 - (-7)$

Example B
5. $-9 - (-4)$
6. $-6 - (-5)$
7. $-12 - (+7)$
8. $-5 - (-6)$

Example C
9. $-12 - 3$
10. $-8 - 8$
11. $-9 - 3$
12. $-5 - 3$

Example D
13. $15 - 15$
14. $18 - 28$
15. $19 - 11$
16. $9 - 10$

Subtract.

17. $-5 - 4$ **18.** $-6 - (-8)$ **19.** $7 - 13$ **20.** $-4 - 9$

21. $8 - 14$ **22.** $9 - (-2)$ **23.** $-8 - 5$ **24.** $11 - 7$

25. $15 - 19$ **26.** $-1 - 15$ **27.** $-3 - (-7)$ **28.** $2 - 9$

29. $-17 - (-17)$ **30.** $-23 - (-23)$ **31.** $-6 - 6$ **32.** $0 - 11$

Number Sense

33. $-57.3 - 27$ **34.** $-5.3 - 8$ **35.** $4.2 - 6.7$ **36.** $-\frac{1}{2} - \frac{3}{4}$

37. Find the sum of 8 and -14.

38. Find the difference of $59 - 74$.

Applications

Paper/ Pencil
Mental Math
Estimate
Calculator

39. At 8:30 P.M. the temperature was 5°C. What was the temperature at midnight if the temperature had dropped 12°?

40. On Monday the price of a share of stock is quoted as $28\frac{3}{8}$ points. It falls $\frac{1}{2}$ point on Tuesday and $\frac{1}{8}$ point on Wednesday. What is the new price?

41. In California, Mt. Whitney is the highest point and Death Valley is the lowest point. Their elevations are 14,494 feet and -282 feet respectively. Find the difference in their elevations.

42. Fred and Bob are both running backs for their high school football team. Fred ran for a total gain of 67 yards while Bob ran for a total of -14 yards. What is the difference in their yards run?

JOURNAL
ENTRY

43. Tell how you would explain adding and subtracting integers to a friend.

Using Expressions

Find the value of each expression if n = −4 and p = −2.

44. $n - 2$ **45.** $n - (-2)$ **46.** $p - 0$ **47.** $0 - p$

Calculator

Use the steps below to enter -14 on a calculator. Note that the change-sign key is pressed *after* the absolute value of the integer is entered.

Enter 14.   Then, press the change-sign key, [+/-]. -14

To add $-14 + 23$, enter: 14 [+/-] [+] 23 [=] 9

To subtract $-14 - 23$, enter: 14 [+/-] [−] 23 [=] -37

Use a calculator to find each sum or difference.

48. $4 + (-17)$ **49.** $-8 + 2$ **50.** $-21 + (-16)$ **51.** $-44 + 37$

52. $3 - 10$ **53.** $6 - (-22)$ **54.** $-57 - 60$ **55.** $-67 - (-96)$

12-4 WINDCHILL FACTOR

Objective

Use a windchill chart to determine equivalent temperatures.

Howie heard the weather forecaster on the radio say the low temperature in Chicago on January 11 was −10°F, but the wind made it feel like it was −45°F.

A thermometer measures the temperature of the air. However, if the wind is blowing, the temperature may feel much colder than the thermometer reading. The windchill factor depends on the actual temperature and the speed of the wind.

The chart below can be used to predict the equivalent temperature taking the windchill factor into account.

In Chicago on January 11, the low temperature was −10°F. The wind speed was 15 mph. The windchill factor made the temperature feel equivalent to −45°F with no wind.

Windchill Chart

Wind speed in mph	Actual temperature (°Fahrenheit)								
	50	40	30	20	10	0	−10	−20	−30
	Equivalent temperature (°Fahrenheit)								
0	50	40	30	20	10	0	−10	−20	−30
5	48	37	27	16	6	−5	−15	−26	−36
10	40	28	16	4	−9	−21	−33	−46	−58
15	36	22	9	−5	−18	−36	−45	−58	−72
20	32	18	4	−10	−25	−39	−53	−67	−82
25	30	16	0	−15	−29	−44	−59	−74	−88
30	28	13	−2	−18	−33	−48	−63	−79	−94

Understanding the Data

● Explain in your own words what windchill factor is.

● Explain what a temperature of −10°F and an equivalent temperature of −45°F with the windchill factor means.

● If the actual temperature is −10°F and the equivalent temperature is −45°F with the windchill factor, what is the wind speed?

Guided Practice

Solve. Use the windchill chart above.

1. On February 10, the actual temperature in New York was 0°F. The wind speed was 20 mph. What was the equivalent temperature with the windchill factor?

2. In Pittsburgh on December 20, the wind speed was 30 mph. The actual temperature was 10°F. What was the equivalent temperature with the windchill factor?

Exercises

Practice

Applications

Interpreting Data

Mixed Review

Lesson 2-2

Lesson 7-6

Use the windchill chart on page 382 to find each equivalent temperature.

3. 40°F, 5 mph   **4.** 10°F, 15 mph   **5.** 0°F, 30 mph

6. −10°F, 10 mph   **7.** −20°F, 25 mph   **8.** 20°F, 20 mph

9. 20°F, 0 mph   **10.** −30°F, 30 mph   **11.** −20°F, 30 mph

What is the difference between each actual temperature and the equivalent temperature when the wind speed is 20 mph?

12. 50°F   **13.** 30°F   **14.** 10°F   **15.** 0°F   **16.** −30°F

17. In Detroit on December 15, the actual temperature was 20°F. The equivalent temperature with the windchill factor was −15°F. What is the wind speed?

18. In Buffalo on February 3, the wind speed was 25 mph. The temperature with the windchill factor was −44°F. What was the actual temperature?

19. Suppose the actual temperature is −30°F and the wind speed is 20 mph. What is the difference between the actual temperature and the equivalent temperature?

Solve. Use the graph.

20. Did the government have a surplus in the budget? If so, in which year?

21. What is the difference in the budget surplus or deficit between 1960 and 1990?

22. How much greater is the budget surplus or deficit in 1990 than in 1940?

23. Does the budget tend to have a surplus or a deficit?

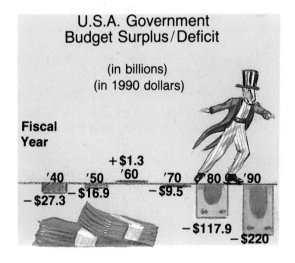

U.S.A. Government
Budget Surplus/Deficit

(in billions)
(in 1990 dollars)

Fiscal
Year

+$1.3

'40   '50   '60   '70   '80   '90

−$27.3   −$16.9   −$9.5

−$117.9
−$220

24. In which year was the deficit the least? the most?

Add.

25.   52.3
       +14.1

26.   91.47
       +23.59

27.   143.05
       + 93.61

28.   45.003
       +184.297

Draw an example of each polygon.

29. regular octagon   **30.** not regular triangle

CATERING MANAGER

Matthew Banks is the catering manager at a hotel. He organizes banquets for wedding receptions, parties, conventions, and business meetings.

Part of the menu available for business meetings is shown below.

Afternoon Break
Freshly Brewed Coffee, Regular or Decaffeinated, Tea, Assorted Chilled Soft Drinks, Mineral Water
$2.95 per person

Health Break
Assorted Fruit, Muffins, Chilled Assorted Fruit Juices, and Mineral Water
$4.95 per person

Harvest Break
Assorted Donuts and Hot Spiced Cider
$3.50 per person

To add a little variety to your mid-meeting breaks, may we suggest one or more of the following.

Assorted Fruit Yogurt	$1.95 per person
Whole Fresh Fruit	$2.25 per piece
Assorted Cookies	$3.95 per dozen
Ice Cream	$1.50 per serving

The Jackson Corporation is holding a business meeting for thirty people.

How much will the Afternoon Break and five dozen Assorted Cookies cost?

Afternoon Break: $2.95 per person × 30 people = $88.50

Assorted Cookies: $3.95 per dozen × 5 dozen = $19.75

Total: $88.50 + 19.75 = $108.25

The total cost is $108.25.

Find the cost of each order. Use the menu above.

1. Harvest Break and Assorted Fruit Yogurt, 40 people

2. Afternoon Break, 45 people, and 30 pieces Whole Fresh Fruit

3. Health Break, 60 people, and 30 servings Ice Cream

4. Mrs. Valdez is having an afternoon tea for 25 people. She orders the Health Break, Assorted Fruit Yogurt, and four dozen Assorted Cookies. Find the cost of Mrs. Valdez's order.

5. You are in charge of refreshments for the Class Party. $700 is budgeted for refreshments for 115 people. Plan a menu and calculate the cost.

INSULATION COSTS

Jeff Bayley wants to put insulation in his home. Insulation helps prevent heating and cooling loss. Thus, less energy is used and more money is saved. On the photo red indicates heat loss.

Suppose insulation costs \$990. The average monthly fuel savings are \$33. Mr. Bayley wants to find the number of years it will take for the fuel costs saved to pay for the insulation.

$$\begin{array}{rl} \$\ 33 & \text{monthly savings} \\ \times\ 12 & \text{months in a year} \\ \hline \$396 & \text{yearly savings} \end{array}$$

$$\text{yearly savings} \quad 396\overline{)\$990.0} = 2.5 \text{ years} \quad \text{cost of insulation}$$

It will take 2.5 years for the fuel savings to pay for the insulation.

Find the number of years it will take for the fuel cost savings to pay for the insulation. The charge for insulation and the average fuel savings per month are given. Express answers to the nearest tenth of a year.

1. \$787.50; \$18.75 **2.** \$2,112; \$42 **3.** \$1,987; \$32.50

4. \$918.55; \$21.20 **5.** \$648.85; \$19.20 **6.** \$2,018; \$47.75

7. The Alfred's monthly fuel bill is \$175. They estimate that insulation will cost \$885. What must the average monthly fuel cost savings be to pay for the insulation in 2 years?

8. If insulation costs \$1,500 and the fuel costs saved will pay for the insulation in 4 years, what is the average monthly fuel savings?

9. An insulation company advertises that their insulation will result in average monthly fuel savings of 25%. The Carter's average monthly fuel bill is \$185. The insulation costs \$1,110. How many years will it take for the fuel cost savings to pay for the insulation?

10. An insulation company advertises that their insulation will result in average monthly fuel savings of \$42.50. The company also claims that fuel costs saved will pay for the insulation in 2.5 years. How much does their insulation cost?

12-5 MULTIPLYING INTEGERS

Objective

Multiply integers.

Thick masses of ice that move slowly on land are called glaciers. Glaciers will melt and appear to move backwards when the temperature rises. The movement of the glacier can range from a few centimeters to a meter a day. Suppose a glacier melts 6 centimeters a day. What was its position 3 days ago? (The solution is in Example A.)

Look at some multiples of 3.

The product of two positive integers is positive. $\rightarrow$

$$\begin{cases} 2 \times 3 = 6 \\ 1 \times 3 = 3 \\ 0 \times 3 = 0 \end{cases}$$

Describe the pattern you see.

One factor is positive and one is negative. The product is negative. $\rightarrow$

$$\begin{cases} -1 \times 3 = -3 \\ -2 \times 3 = -6 \end{cases}$$

$0 - 3 = -3$

Look at some multiples of -3.

$$2 \times -3 = -6$$
$$1 \times -3 = -3$$
$$0 \times -3 = 0$$

Describe the pattern you see.

Both factors are negative. The product is positive. $\rightarrow$

$$\begin{cases} -1 \times -3 = 3 \\ -2 \times -3 = 6 \end{cases}$$

$0 + 3 = 3$

These rules also apply to positive and negative decimals and fractions.

Method

1 ▶ Find the product of the absolute values.

2 ▶ The product of two integers with the same sign is positive.
The product of two integers with different signs is negative.

Examples

A $-3 \times (-6)$

1 ▶ $|-3| \times |-6| = 18$

2 ▶ The factors have the same sign. The product is positive.

$-3 \times (-6) = 18$

Three days ago the glacier was 18 centimeters forward of where it is now.

B 5×14

1 ▶ $|5| \times |14| = 70$

2 ▶ The factors have the same sign. The product is positive.

$5 \times 14 = 70$

C -6×3

1 ▶ $|-6| \times |3| = 18$

2 ▶ The factors have different signs. The product is negative.

$-6 \times 3 = -18$

D $0 \times (-38)$

$|0| \times |-38| = 0$

Any number multiplied by zero is zero.

Multiply.

1. $-5 \times (-3)$ **2.** $-3 \times (-12)$ **3.** $-1 \times (-9)$ **4.** $-45 \times (-2)$

5. 7×5 **6.** 21×4 **7.** 8×9 **8.** 6×5

9. $4 \times (-3)$ **10.** -5×6 **11.** -1×32 **12.** -98×1

13. $7 \times (-11)$ **14.** $54 \times (-10)$ **15.** 0×15 **16.** -8×0

Multiply.

17. 5×8 **18.** $9 \times (-2)$ **19.** -3×2 **20.** $-5 \times (-4)$

21. $6 \times (-7)$ **22.** $0 \times (-5)$ **23.** -10×7 **24.** 9×21

25. $9 \times (-6)$ **26.** 5×15 **27.** $-8 \times (-7)$ **28.** $-1 \times (-34)$

29. -3×123 **30.** $-14 \times (-4)$ **31.** $35 \times (-6)$ **32.** -100×0

33. $6 \times (-2.1)$ **34.** $-0.3 \times (-0.3)$ **35.** $\frac{2}{3} \times (-\frac{6}{7})$ **36.** $-1\frac{2}{5} \times (-2\frac{2}{9})$

37. What is the product of -18 and 6?

38. If an odd number of negative factors are multiplied, is the product positive or negative?

39. On Cape Cod several cliffs recede 3.3 feet each year. How much do the cliffs recede in 5 years?

40. At 5:00 P.M. the temperature was 0°C. For the next 6 hours it dropped 2°C each hour. What was the temperature at 11:00 P.M.?

41. The record low temperature in Ohio is -39°C. The record high temperature is 45°C. What is the difference in the temperatures?

Find the value of each expression if x = − 8.

42. $4x$ **43.** $0x$ **44.** $-3x$ **45.** $-6x$

46. Suppose one person starts a story and tells it to four other people within 20 minutes. Each person then tells four other people within 20 minutes and so on. How long will it take for one million people to hear the story?

Simplify each fraction.

47. $\frac{12}{18}$ **48.** $\frac{9}{21}$ **49.** $\frac{6}{24}$

Find the missing length for each pair of similar figures.

50.   **51.**   **52.**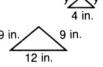

53. A store donated money to a charity based upon a percentage of total sales for a given day. If the total sales were above $5,000, the store would donate 2%. If the sales were above $7,500, the store would donate 3% of total sales. How much money was donated if the total sales for the day were $8,258?

12-6 DIVIDING INTEGERS

Objective
Divide integers.

Joe Johnson is a halfback for the community college he attends. In the latest game, he carried the ball the last two plays and his net gain was −12 yards. What was Joe's average net gain? (The problem is solved in Example D.)

| Every multiplication sentence has two related division sentences. | $3 \times 4 = 12$ has the related sentences:
$12 \div 4 = 3$ and
$12 \div 3 = 4$ | $5 \times (-2) = -10$ has the related sentences:
$-10 \div (-2) = 5$ and
$-10 \div 5 = -2$ |

You can use your knowledge about multiplying integers when you divide integers.

These rules also apply to positive and negative decimals and fractions.

Method

1 ▶ Divide the absolute values.

2 ▶ The quotient of two integers with the same sign is positive.
The quotient of two integers with different signs is negative.

Examples

A $45 \div 15$
1 ▶ $|45| \div |15| = 3$
2 ▶ The divisor and dividend have the same sign. So the quotient is positive.
$45 \div 15 = 3$
Check: $3 \times 15 \overset{?}{=} 45$
$45 = 45 \checkmark$

B $-16 \div (-2)$
1 ▶ $|-16| \div |-2| = 8$
2 ▶ The divisor and dividend have the same sign. So the quotient is positive.
$-16 \div (-2) = 8$
Check: $8 \times (-2) \overset{?}{=} -16$
$-16 = -16 \checkmark$

C $45 \div (-15)$
1 ▶ $|45| \div |-15| = 3$
2 ▶ The divisor and dividend have different signs. The quotient is negative.
$45 \div (-15) = -3$
Check: $-3 \times (-15) \overset{?}{=} 45$
$45 = 45 \checkmark$

D $-12 \div 2$
1 ▶ $|-12| \div |2| = 6$
2 ▶ The divisor and dividend have different signs. The quotient is negative.
$-12 \div 2 = -6$
Joe's average net gain was −6 yards.

Guided Practice

Divide.

Example A
1. $14 \div 2$
2. $25 \div 5$
3. $15 \div 60$
4. $5 \div 25$

Example B
5. $-18 \div (-3)$
6. $-24 \div (-4)$
7. $-2 \div (-4)$
8. $-30 \div (-4)$

Example C
9. $36 \div (-3)$
10. $45 \div (-5)$
11. $3 \div (-6)$
12. $0 \div (-7)$

Example D
13. $-56 \div 4$
14. $-81 \div 9$
15. $-1 \div 10$
16. $0 \div 5$

Practice

Divide.

17. $-12 \div (-3)$ **18.** $24 \div 4$ **19.** $18 \div (-2)$ **20.** $-27 \div (-3)$

21. $15 \div (-3)$ **22.** $-48 \div 6$ **23.** $-12 \div 6$ **24.** $0 \div 2$

25. $0 \div (-4)$ **26.** $36 \div 4$ **27.** $-42 \div 7$ **28.** $-133 \div 7$

29. $-56 \div (-4)$ **30.** $-200 \div 40$ **31.** $8 \div (-1)$ **32.** $45 \div (-9)$

33. $37 \div (-10)$ **34.** $-9 \div (-72)$ **35.** $4 \div (-4)$ **36.** $-4 \div 8$

37. $-75 \div (-7)$ **38.** $12 \div (-0.5)$ **39.** $-4.9 \div 1.4$ **40.** $\frac{1}{2} \div (-\frac{3}{4})$

41. Find the quotient of 165 and -3.

Number Sense

42. *True* or *false.* It is possible to have a positive integer as the quotient when dividing a negative integer by a positive integer.

Replace each ● with >, <, or = to make a true sentence.

43. $4 - 28 \div (-7) ● 4 - 4$ **44.** $-8 + 3 \times (-2) ● -8 \times 2$

45. $36 - 4 + 8 \div 4 ● -6 \times (-7)$ **46.** $-31 + 6 \times -3 ● -7 \times 7$

Using Equations

Solve each equation.

47. $a + 7 = 13$ **48.** $n \div (-3) = 4$ **49.** $6 \times b = -18$

50. $-24 \div g = -3$ **51.** $12 + t = -11$ **52.** $h - (-4) = 7$

Applications

53. During a 5-day period, a stock price had a change of -2. What was the average change per day?

***f*UN with MATH**

Try our recipe for Pineapple Chicken. See page 422.

54. From noon until 6:00 P.M., the temperature change was $-12°C$. Find the average change per hour.

55. The first recorded Olympic Games were held in 776 B.C. According to legend, Romulus founded Rome in 753 B.C. Which event occurred first?

Make Up a Problem

56. Make up a problem that matches the number sentence in Exercise 27.

Calculator

What is the quotient, q, when 6 is divided by 0?

Every division sentence has a related multiplication sentence. $(14 \div 2 = 7; 7 \times 2 = 14)$ So $6 \div 0 = q$ has the related sentence $q \times 0 = 6$. Since there is no number that can replace q to make a true statement, we say the quotient $6 \div 0$ is *undefined.* Your calculator will give an ERROR (E) message if you try to divide by 0.

Use a calculator to find each quotient.

57. $-8 \div 0$ **58.** $\frac{0}{3}$ **59.** $\frac{3}{0}$ **60.** $\frac{18}{-6 + 6}$ **61.** $0\overline{)12}$

	Explore
	Plan
	Solve
	Examine

12-7 WORK BACKWARDS

Objective

Solve verbal problems by working backwards.

A store is having a clearance sale to make room for new inventory. The store owner reduces the price of a stereo system by 50%. When the stereo still does not sell, that price is reduced another 25%. The stereo finally sells for half of the last price, or $150. What was the original price of the stereo system?

▶ **Explore**

What is given?
- A stereo sells for $150 after the original price is reduced by 50%, by 25%, and finally for half of the last price.

What is asked?
- What was the original price of the stereo?

▶ **Plan**

When you know the final price you can often work backwards step by step to find the original price.

▶ **Solve**

The store sells the stereo system for $150, one-half of the previous price. So its previous price was $150 × 2 or $300. This price had been reduced 25%, $300 is therefore 75% of the previous price. Before this it was reduced 50% from the original price. Use proportions to find the prices.

$$\frac{75}{100} = \frac{300}{s}$$
$$75 \times s = 100 \times 300$$
$$75s = 30,000$$
$$s = 400$$

$$\frac{50}{100} = \frac{400}{p}$$
$$50 \times p = 100 \times 400$$
$$50p = 40,000$$
$$p = 800$$

end result

$\frac{1}{2}$ $\frac{1}{2}$

$150 $150

$300

a reduction of 25% from

$400

a reduction of 50% from

$800
original price

The original price was $800.

▶ **Examine**

Start with $800. Fifty percent of $800 is $400. Since twenty-five percent of $400 is $100, the next price is $300. One-half of $300 is $150, so the solution is correct.

Guided Practice

Solve. Work backwards.

1. Marla picks some apples. She gives half to her mother and half of the remaining apples to her friend Andy. Andy receives 7 apples. How many apples did Marla pick?

Solve. Use any strategy.

2. Rico tries to sell his bicycle at a garage sale. He reduces the price by half. It still does not sell, so he reduces that price by 20%. He has to reduce the last price by 25% before it sells for $18. What was the original price?

3. A certain kind of microbe doubles its population every 12 hours. After 3 days there are 640 microbes. How many microbes were there at the beginning of the first day?

4. Forty percent of the cars Mr. Tibbs sells are compacts. Twenty percent of the compact cars he sells are hatchbacks. If Mr. Tibbs sells 34 hatchbacks, how many cars has he sold in all?

5. Sandy mailed letters and postcards for a total of $1.82. If it costs 29¢ to mail each letter and 19¢ to mail each postcard, how many of each did she mail?

6. Local library fines for overdue books are 12¢ for the first day, 6¢ for each of the next two days, and 4¢ each day thereafter. If Gail paid a fine of 80¢, how many days was her book overdue?

7. Four painters can paint 4 rooms in 4 hours. How many rooms of the same size can 8 painters paint in 8 hours?

8. Half of the students in a class are girls. One-third of the girls have blond hair. If five girls have blond hair, how many students are in the class?

9. Ms. James is making eggrolls for a big party. She needs a pound of cornstarch to seal the wrappers. It comes in 3-oz boxes for 36¢ and 6-oz boxes for 67¢. How many of each size should she purchase to get the best buy?

Critical Thinking

10. A magic square is a number square in which the numbers in all rows, columns, and diagonals have the same sum. This sum is called the magic sum. Find the missing numbers if the magic sum is −34.

-16	-2	-3	
-5	-11		
-9		-6	
	-14		-1

11. Name your favorite lesson from this chapter. Then write a paragraph explaining why it was your favorite.

Mixed Review

Lesson 6-5

Complete.

12. 7 L = $\underline{\ ?\ }$ mL   **13.** 148.1 mL = $\underline{\ ?\ }$ L   **14.** 91.2 L = $\underline{\ ?\ }$ kL

Lesson 11-11

15. Monica's soccer team enters a double-elimination tournament with three other teams. That is, a team is eliminated from winning the tournament only after two losses. Monica's team loses their first game, yet they win the tournament. How many games did they have to play?

REVIEW

Choose the word or phrase from the list at the right that best completes each sentence. You may use a word more than once.

Vocabulary/ Concepts

1. To subtract an integer, add its ⟨?⟩ .

2. On a number line, the distance that a number is from zero is its ⟨?⟩ .

3. The ⟨?⟩ and their opposites make up the set of ⟨?⟩ .

4. The product of two negative numbers is ⟨?⟩ .

5. The quotient of a negative number and a positive number is ⟨?⟩ .

6. To add integers with different signs, ⟨?⟩ their absolute values. Then give the same sign as the integer with the ⟨?⟩ absolute value.

7. Subtracting an integer gives the same result as ⟨?⟩ its opposite.

8. The product of a negative integer and a positive integer is ⟨?⟩ .

absolute value
adding
divide
greater
integers
lesser
multiply
negative
opposite
positive
subtract
whole numbers

Exercises/ Applications

Lesson 12-1

Replace each ● with <, >, or = to make a true sentence.

9. 0 ● −2
10. −3 ● −5
11. −7 ● 7
12. 2 ● −4
13. |−5| ● 5
14. |−3| ● 3
15. |−11| ● |7|
16. |−9| ● |9|

Order from least to greatest.

17. −7, 7, 9, −9, 0
18. 0, 100, 101, −100, −101

Lesson 12-2

Add.

19. 5 + (−8)
20. −3 + (−17)
21. −5 + 5
22. −68 + 33
23. −8 + 12
24. 0 + (−7)
25. −9 + (−13)
26. −14 + 0

Lesson 12-3

Subtract.

27. 9 − 16
28. 12 − 4
29. 14 − (−3)
30. 0 − 8
31. −9 − 0
32. 16 − (−16)
33. −5 − (+8)
34. −14 − (−8)

35. Water in a reservoir was 6 feet below normal. Water was pumped into the reservoir until the water level was 5 feet above normal. How many feet, in all, did the water rise or fall?

Solve. Use the windchill chart on page 382.

36. In Baltimore on January 24, the wind speed was 10 mph. The actual temperature was −10°F. What was the equivalent temperature with the windchill factor?

37. Suppose the actual temperature is 0°F and the wind speed is 25 mph. What is the difference between the actual temperature and the equivalent temperature?

Multiply.

38. −9 × 5 **39.** −3 × (−18) **40.** 7 × 8 **41.** −11 × (−9)
42. 14 × (−7) **43.** +7 × (−12) **44.** −16 × (8) **45.** −7 × (−12)

Divide.

46. 63 ÷ 7 **47.** −36 ÷ 4 **48.** 39 ÷ (−13) **49.** −15 ÷ 12
50. −5 ÷ 10 **51.** −48 ÷ (−6) **52.** −63 ÷ (−9) **53.** 0 ÷ (−4)

Solve. Use any strategy.

54. Joan spends 10 minutes eating a snack when she gets home from school. Then she studies math, science, and English for 25 minutes each. When she finishes, it is 5:30 P.M. At what time did she get home from school?

55. Pick a number, triple it, add 8, and divide by 5. If the result is 7, what was the original number?

56. The water level of a tank fell at the rate of $1\frac{1}{2}$ feet per hour for 6 hours. What is the water level now if the water level had been 38 feet?

57. The temperature rose for 8 hours at a rate of 2° per hour. What was the original temperature if the final temperature was −11°F?

TEST

Replace each ● with <, >, or = to make a true sentence.

1. 2 ● 8   **2.** 3 ● −5   **3.** −3 ● −11

4. −9 ● −5   **5.** −8 ● 9   **6** |−4| ● |2|

Add or subtract.

7. −3 + (−7)   **8.** −8 + (−1)   **9.** 14 + (−16)   **10.** 19 + (−13)

11. 20 − (−5)   **12.** −15 − 15   **13.** −6 − 4   **14.** 0 − (−3)

15. −8 − (−8)   **16.** 13 + (−14)   **17.** 2 − 3   **18.** −3 + 8

Solve. Use the chart.

19. In Terre Haute on December 28, the low temperature was 20°F. The wind speed was 10 mph. With the windchill factor, what was the equivalent temperature?

Windchill Chart

Wind speed in mph	Actual temperature (°Fahrenheit)						
	50	40	30	20	10	0	−10
	Equivalent temperature (°Fahrenheit)						
0	50	40	30	20	10	0	−10
5	48	37	27	16	6	−5	−15
10	40	28	16	4	−9	−21	−33

Multiply or divide.

20. 72 ÷ (−8)   **21.** −15 ÷ 3   **22.** 42 ÷ (−3)   **23.** −88 ÷ (−4)

24. −18 × 3   **25.** −5 × (−3)   **26.** 80 × 5   **27.** 0 × (−12)

28. −14 ÷ 2   **29.** 20 ÷ (−5)   **30.** −2 × 5   **31.** 5 × (−7)

Solve. Use working backwards.

32. Kenny entered a triathlon in which he biked twice as far as he ran. He ran five times as far as he swam. He biked 12 miles. How far did he swim?

33. Jolie picked some apples. She gave half to her mother and half of the remaining apples to her friend Tracey. Tracey received 9 apples. How many apples did Jolie pick?

▶ BONUS: If $n < 0$, is $n^2 < 0$? Explain.

CITY AND STATE INCOME TAXES

Many city and state governments levy their own income tax. Ned Davis lives in Westerville which levies both a city and state income tax. His employer is required to deduct these taxes from his wages.

The city rate is 1%.
Find 1% of $299.55.

$299.55	weekly wage
× 0.01	rate (1% = 0.01)
$2.9955	city tax

To the nearest cent, the weekly city tax deduction is $3.00.

The state rate is 2.5%.
Find 2.5% of $299.55.

$299.55	weekly wage
× 0.025	rate (2.5% = 0.025)
$7.48875	state tax

To the nearest cent, the weekly state tax deduction is $7.49.

Find the city tax deduction for each wage to the nearest cent. Use a 1% rate.

1. $237.35   **2.** $179.29   **3.** $322.56   **4.** $262.50

Find the state tax deduction for each wage to the nearest cent. Use a 2.5% rate.

5. $179.29   **6.** $256   **7.** $314.56   **8.** $278.50

9. Ms. Kosar pays 1.5% city tax and 2.75% state tax. Her annual salary is $36,000. How much is deducted from her monthly salary for city and state taxes?

10. Mr. Walters makes $650 a week. His employer deducts 1.25% for city tax and 2% for state tax. How much more state tax does he pay than city tax?

Free Response

Lesson 1-12

Solve each equation. Check your solution.

1. $5n = 25$ **2.** $\frac{t}{4} = 9$ **3.** $8y = 72$

Lesson 2-3

Add.

4. $93 + 89$ **5.** $232 + 295$ **6.** $1{,}751 + 156$

7. $1{,}638 + 4{,}534$ **8.** $83{,}346 + 1{,}714$ **9.** $54{,}648 + 77{,}059$

Lesson 4-9

Estimate.

10. $2\frac{1}{2} + 19\frac{1}{8}$ **11.** $27\frac{3}{4} + 4\frac{1}{8}$ **12.** $30 - 7\frac{9}{11}$ **13.** $10\frac{1}{8} - 6\frac{2}{3}$

Lesson 5-2

Multiply.

14. $\frac{1}{4} \times \frac{1}{4}$ **15.** $\frac{1}{2} \times \frac{1}{4}$ **16.** $\frac{3}{8} \times \frac{5}{6}$ **17.** $\frac{1}{6} \times \frac{1}{2}$

Lesson 8-2

18. Rudy's yard is in the shape of a parallelogram. How much area does his yard cover if the base length is 45 feet and the height is 120 feet?

Lesson 9-5

Find the distance on a scale drawing for each actual distance. The scale is 1 in.: 75 ft.

19. 300 ft   **20.** 50 ft

Lesson 10-7

21. Andy buys hockey equipment on sale for 30% off the original price. He buys skates that regularly sell for $85 and he buys a stick that had an original price of $58. How much did he pay for the two items?

Lesson 10-8

Find the simple interest earned on each deposit.

22. principal: $220
annual rate: 8%
time: 1 year

23. principal: $220
annual rate: 8%
time: 4 years

Lesson 11-1

Find the mean, median, and mode for each set of data.

24. 75, 90, 82, 79, 81, 82

25. 2, 6, 7, 9, 5, 6, 7, 9, 9, 4

26. −4, 7, 4, −7, 2, 8, 4

Lesson 12-2

27. At 6 A.M. the temperature was −4°F. At noon the temperature was 28°F. What was the change in temperature?

Multiple Choice

Choose the letter of the correct answer for each item.

1. What is the standard form for *five hundred and two hundredths?*
  a. 0.502
  b. 5.02
  c. 500.02
  d. 500.2

Lesson 1-1

2. Find the quotient when 13.62 is divided by 15.
  e. 0.908
  f. 0.98
  g. 1.101
  h. 1.113

Lesson 3-5

3. What is the difference when $1\frac{1}{2}$ is subtracted from $2\frac{1}{11}$?
  a. $\frac{13}{22}$ c. $1\frac{1}{9}$
  b. $1\frac{1}{11}$ d. $1\frac{13}{22}$

Lesson 4-13

4. What is the result when $\frac{5}{6}$ is divided by $\frac{1}{2}$?
  e. $\frac{2.5}{6}$
  f. $\frac{5}{12}$
  g. $1\frac{2}{5}$
  h. $1\frac{2}{3}$

Lesson 5-6

5. Sue works from 8:00 A.M. to 4:30 P.M. with 45 minutes for lunch each day. How many hours does she work in 10 days?
  a. 40 hours   c. 80 hours
  b. $77\frac{1}{2}$ hours   d. 85 hours

Lesson 6-10

6. Lines formed by the sides of a square tile floor form what kind of lines?
  e. parallel and perpendicular
  f. parallel and skew
  g. perpendicular and skew
  h. parallel only

Lesson 7-1

7. A certain brand of wheat crackers contains 1,060 mg of sodium for every 100 g of crackers. How much sodium is in 40 g of crackers?
  a. 265 mg   c. 265 g
  b. 424 mg   d. 424 g

Lesson 9-4

8. Gasoline costs $1.40 per gallon. The price drops 25%. What is the new price per gallon?
  e. 90¢   g. $1.45
  f. $1.00   h. $1.05

Lesson 10-6

9. By how much did monthly sales increase between April and June?
  a. $4,000
  b. $5,000
  c. $15,000
  d. $20,000

Monthly Sales

Lesson 11-5

10. Kaitlynn drives 20 miles east from her home. She then turns around and drives 42 miles west and then returns and drives 85 miles east. How far is she from her home?
  e. 63 miles   g. 107 miles
  f. 23 miles   h. 85 miles

Lesson 12-1

Chapter

13 EXTENDING ALGEBRA

▶ Decisions, Decisions!!!

Think about how many decisions you make in one day, decisions like "which books do I need from my locker for the morning classes, should I go to the library before track practice, and can I afford that cassette tape?" These are examples of everyday problems that you solve by making decisions. You use equations to solve problems, often without realizing it. Can you think of any problems that you solved with the help of an equation?

Jane deposits $250 into her checking account. Her balance is now $537. What was her balance before she made the deposit?

ACTIVITY: Equations

In this activity you will translate verbal sentences to mathematical sentences. You will also discover an "equational" magic trick to amaze your family and friends.

Cooperative Groups

Work in groups of three or four.

1. With your group, discuss how these two sentences are alike and how they are different.

   > Art threw the football.
   > The football was thrown by Art.

2. Just as different sentences may have the same meaning, different equations may have the same meaning or *solution*. Look at the equation $2x + 6 = 10$. Another equation that has the same solution is $2x = 4$.
   a. What operation, addition, subtraction, multiplication, or division, is used to get $2x = 4$ from the first equation?
   b. What operation is used to get $x = 2$?

3. With your group, write another equation that has the same solution as each given equation.
   a. $2t + 3 = 12$ b. $6y - 3 = 27$ c. $\frac{d}{5} + 1 = 3$

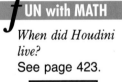

f UN with MATH

When did Houdini live?
See page 423.

4. Designate one member of your group as "Houdini" (the master magician). Houdini reads these instructions. "Choose any number. Add 3 to it. Multiply the sum by 100. Subtract 300 from the product. Divide the difference by 10. Write the result. I can tell you the original number." Houdini mentally divides the result by 10 and tells the person his or her original number.

5. The "secret" is that Houdini used an equation that has the same solution to create his magic. Look at the instructions.
   Choose a number. $\rightarrow n$
   Add 3 to it. $\rightarrow n + 3$
   Multiply by 100. $\rightarrow 100\ (n + 3)$ or $100n + 300$
   Subtract 300. $\rightarrow 100n$
   Divide by 10. $\rightarrow 10n$

   Note that under these specific conditions, the result will always be 10 times the original number. It's easy to divide mentally by 10 and give the original number as if by magic.

6. With your group, create your own magic instructions. Test it on your group. See if other groups can figure out your equivalent equations.

13-1 SOLVING EQUATIONS USING ADDITION OR SUBTRACTION

Objective

Solve equations involving integers and addition or subtraction.

Dawn was playing a game. In one turn she lost 17 points. Her accumulated score after losing 17 points was −9. What was her score before losing 17 points?

Solving an equation means finding the replacement for the variable that results in a true sentence. To solve an equation, get the variable by itself on one side of the equals sign. If a number has been added to the variable, subtract. If a number has been subtracted from the variable, add.

Method

▶1 Add or subtract the same number on each side of the equation to get the variable by itself.

▶2 Check the solution.

Example A

Find Dawn's score before losing 17 points. Let x = Dawn's score before losing 17 points.

$$x - 17 = -9 \qquad \text{17 is subtracted from } x.$$

▶1 $x - 17 + 17 = -9 + 17 \qquad$ Add 17 to each side.

$$x = 8$$

▶2 Check: $x - 17 \overset{?}{=} -9 \qquad$ In the original equation,
$$8 - 17 \overset{?}{=} -9 \qquad \text{replace } x \text{ with 8.}$$
$$-9 = -9 \checkmark \qquad \text{The solution is 8.}$$

Dawn's score before losing 17 points was 8.

Example B

Solve −210 + x = 100.

$$-210 + x = 100 \qquad \text{−210 is added to } x.$$

▶1 $-210 - (-210) + x = 100 - (-210) \qquad$ Subtract −210 from each side.

$$x = 310$$

▶2 Check: $-210 + x \overset{?}{=} 100$
$$-210 + 310 \overset{?}{=} 100 \qquad \text{Replace } x \text{ with 310.}$$
$$100 = 100 \checkmark \qquad \text{The solution is 310.}$$

Guided Practice

Solve each equation. Check your solution.

Example A

1. $x - 7 = 2$
2. $-1 = x - 7$
3. $y - 3 = -8$
4. $t - 25 = -5$
5. $-8 = p - 8$
6. $40 = y - 117$

Example B

7. $x + 7 = 4$
8. $7 + t = -2$
9. $4 = p + 8$
10. $1 + y = -7$
11. $x + 0 = -35$
12. $0 = y + (-29)$

Solve each equation. Check your solution.

13. $y + 7 = 5$
14. $-8 + p = -12$
15. $-15 + a = 19$

16. $y - 6 = -15$
17. $-17 = 14 + x$
18. $-8 = y + 6$

19. $y + (-2) = 5$
20. $b - (-7) = -3$
21. $x + (-4) = -7$

22. $-1 + y = -1$
23. $-2 = a - 17$
24. $x + (-20) = 6$

25. $283 = x + (-100)$
26. $-157 + x = 1$
27. $a = -32 - (-32)$

28. $x + 3.2 = 1.5$
29. $-1.5 + y = -2.3$
30. $p - 1 = -\frac{1}{3}$

Applications

31. Alma made a $300 deposit to her checking account. After making the deposit, her balance was $287. What was her balance before she made the deposit?

f UN with MATH
When was the first love song written?
See page 422.

32. The temperature in October at the base of Mt. Washington was 31°F. The temperature at the weather station on the top of the mountain was -7°F. What is the difference in temperature from the base to the top of the mountain?

JOURNAL ENTRY

33. Explain how equations are solved using addition or subtraction. Give an example of each.

34. Write the change in the average mortgage interest rate from June 28 to July 5, from July 5 to July 12, from July 12 to July 19.

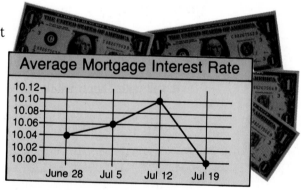
Average Mortgage Interest Rate

35. Find the change from June 28 to July 19. Compare this change to the sum of the changes from Exercise 34.

Mixed Review

Lesson 3-2

Multiply.

36. $\begin{array}{r} 4.5 \\ \times 1.8 \\ \hline \end{array}$
37. $\begin{array}{r} 16.3 \\ \times\ 4.7 \\ \hline \end{array}$
38. $\begin{array}{r} 19.5 \\ \times 21.6 \\ \hline \end{array}$
39. $\begin{array}{r} 0.082 \\ \times 4.1 \\ \hline \end{array}$

Lesson 8-2

Find the area of each parallelogram described below.

40. base, 12 m
height, 2.1 m

41. base, 9.3 cm
height, 6.5 cm

42. base, 18.1 mm
height, 23.5 mm

Lesson 10-3

43. In last year's basketball season, Trevor made 85% of his free throws. How many total free throws did he shoot if he made 34?

13-2 SOLVING EQUATIONS USING MULTIPLICATION OR DIVISION

Objective

Solve equations involving integers and multiplication or division.

Dan hopes to lose 12 pounds. If he loses 2 pounds a week, how many weeks will it take for him to lose 12 pounds? This problem can be expressed by the equation $-2x = -12$.

To solve an equation in which the variable is multiplied by a number, divide each side by that number. If the variable is divided by a number, multiply each side by that number.

Method

1 Multiply or divide each side of the equation by the same nonzero number to get the variable by itself.

2 Check the solution.

Example A

Find how many weeks it will take Dan to lose 12 pounds.

$-2x = -12$ x is multiplied by -2.

1 $\dfrac{-2x}{-2} = \dfrac{-12}{-2}$ Divide each side by -2.

$x = 6$

2 Check: $-2x = -12$ In the original equation, replace x with 6.

$-2 \times 6 \stackrel{?}{=} -12$

$-12 = -12$ ✓ The solution is 6.

It will take Dan 6 weeks to lose 12 pounds.

Example B

Solve $\dfrac{y}{4} = -48$.

$\dfrac{y}{4} = -48$ y is divided by 4.

1 $\dfrac{y}{4} \times 4 = -48 \times 4$ Multiply each side by 4.

$y = -192$

2 Check: $\dfrac{y}{4} = -48$ Replace the variable with -192.

$\dfrac{-192}{4} \stackrel{?}{=} -48$

$-48 = -48$ ✓ The solution is -192.

Guided Practice

Solve each equation. Check your solution.

Example A

1. $2x = 5$ **2.** $-60 = 12y$ **3.** $6 = 0.4x$ **4.** $\dfrac{3y}{5} = -45$

Example B

5. $\dfrac{y}{10} = -5$ **6.** $\dfrac{p}{4} = 10$ **7.** $y \div 8 = -24$ **8.** $x \div 5 = 40$

Solve each equation. Check your solution.

9. $-7q = 5$ **10.** $\frac{x}{-2} = 13$ **11.** $4x = -16$ **12.** $10 = -7x$

13. $\frac{1}{2}a = -2$ **14.** $5 = -35x$ **15.** $-36 = \frac{1}{4}x$ **16.** $15 = -\frac{2}{3}p$

17. $0.1x = 0.83$ **18.** $3t = \frac{1}{2}$ **19.** $\frac{1}{3}y = -1$ **20.** $13y = 0$

21. $5y = -4$ **22.** $y \div 9 = -3$ **23.** $-4 = -16x$ **24.** $-18p = -9$

25. $\frac{x}{6} = -7$ **26.** $-4 = \frac{t}{2}$ **27.** $\frac{2}{3}y = \frac{4}{9}$ **28.** $-4x = 0.8$

29. $-\frac{3}{2}x = -1$ **30.** $-0.1p = 10$ **31.** $-2 = -10y$ **32.** $-\frac{4}{5}x = -3\frac{1}{5}$

33. During the state playoffs, the Wildcats lost 18 yards in 3 plays. What was their average loss per play?

34. Troy, Deanna, and Jenny stopped selling T-shirts because their business was losing money. They shared the $141 loss equally. How much did each lose?

35. The product of two numbers is -20. Their sum is 1. What are the numbers?

36. Find out the average low temperature for each month of the year where you live and make a table to show the temperatures. Determine the amount of change from month to month.

Study these examples.

- The square of a negative number is positive.

$(-3)^2 = -3 \times (-3) = 9$

- The cube of a negative number is negative.

$(-3)^3 = \underbrace{-3 \times (-3)}_{9} \times (-3) = -27$

- The fourth power of a negative number is positive.

$(-3)^4 = \underbrace{-3 \times (-3) \times (-3)}_{-27} \times (-3) = 81$

From these examples, we can conclude that:

a. An even power of a negative number is positive.

b. An odd power of a negative number is negative.

Find each product.

37. $-2 \times (-2) \times (-2)$ **38.** $-4 \times (-4)$

39. $-1 \times (-1) \times (-1) \times (-1) \times (-1)$ **40.** $-2 \times (-3) \times (-1)$

41. $(-3)^5$ **42.** $(-7)^2$ **43.** $(-2)^4$

13-3 SOLVING TWO-STEP EQUATIONS

Objective

Solve equations involving integers and two operations.

Mrs. O'Hara ordered four pizzas for Jenny's after-prom party. She used a coupon for $3 off one pizza. The total bill after subtracting the value of the coupon was $39. What was the regular price of one pizza? This problem can be solved using the equation $4n - 3 = 39$.

There is often more than one operation in an equation. To solve a two-step equation, you must apply the *standard order of operations* in reverse.

Method

1 ▶ Add or subtract the appropriate number to each side of the equation to get all terms involving the variable on one side of the equation. Next, multiply or divide each side of the equation by the appropriate nonzero number.

2 ▶ Check the solution.

Example A

Find the regular price of one pizza.

$$4n - 3 = 39$$

1 ▶ $$4n - 3 + 3 = 39 + 3$$

$$4n = 42$$

$$\frac{4n}{4} = \frac{42}{4}$$

$$n = 10.5$$

First, n is multiplied by 4. Then 3 is subtracted from the product. Add 3 to each side.

Divide each side by 4.

The regular price of one pizza is $10.50.

2 ▶ **Check:** $4n - 3 = 39$

$$4(10.5) - 3 \overset{?}{=} 39$$

$$42 - 3 \overset{?}{=} 39$$

$$39 = 39 \quad \checkmark$$

In the original equation, replace n with 10.5.

Any $1.00 Off Any Lasagna Dinner
Please mention coupon when ordering. N...
other coupons or promotions.

$3.00 Off Any Large Pizza (14")
Please mention coupon when ordering. Not valid with other coupons or promotions.

NEW ITEM—Dragsters "Our Buffalo Wings" 8 For $1.45
Please mention coupon when ordering N...
other coupons or promotions.

Example B

Solve $-2(n - 3) = 8$.

$$-2(n - 3) = 8$$

1 ▶ $$\frac{-2(n - 3)}{-2} = \frac{8}{-2}$$

$$n - 3 = -4$$

$$n - 3 + 3 = -4 + 3$$

$$n = -1$$

First, 3 is subtracted from n. Then the difference is multiplied by -2. Divide each side by -2.

Add 3 to each side.

2 ▶ **Check:** $-2(n - 3) = 8$

$$-2(-1 - 3) \overset{?}{=} 8$$

$$-2(-4) \overset{?}{=} 8$$

$$8 = 8 \quad \checkmark$$

In the original equation, replace n with -1.

The solution is correct.

Guided Practice

Solve each equation. Check your solution.

Example A

1. $2x + 3 = -5$ **2.** $3y - 4 = 14$ **3.** $7y + (-2) = -51$

4. $4(n + 2) = -20$ **5.** $-5(x - 3) = 40$ **6.** $-15 = 2(x + 3)$

Solve each equation. Check your solution.

7. $49 = 7t + 14$ **8.** $2 - 3a = 8$ **9.** $8x + 5 = -45$

10. $\frac{x}{5} + 1 = -1$ **11.** $5(x + 1) = -15$ **12.** $6(x - 0.2) = 3$

13. $-2y - 1 = 9$ **14.** $-11 = 3d - 2$ **15.** $\frac{5}{9}(F - 32) = 100$

16. $0 = -3x + (-15)$ **17.** $\frac{2}{3}(y - 4) = 12$ **18.** $500 = 10(x + 30)$

19. $284 + -35y = 4$ **20.** $2y + 7 = -42$ **21.** $-6 = \frac{4 - y}{2}$

Paper/Pencil
Mental Math
Estimate
Calculator

22. Henry answered all six questions correctly on the history quiz. But he lost two points on each question because he did not answer using complete sentences. Henry's grade on the quiz is 48. If each question is worth the same number of points, how much is each question worth?

23. Leaman's is having a 30%-off everything sale. Chang buys a shirt at a sale price of $21. What is the original price of the shirt?

24. Mr. Thomas divides a case of floppy discs equally among the six members of the Computer Club. Kathy already has three discs of her own. With the new discs from Mr. Thomas, Kathy has seven discs. How many discs were there in the case?

25. Make up an equation that will require two steps to solve. Exchange with a classmate. Solve your classmate's equation. Return the equation and check your classmate's solution.

26. In your own words, describe how you would solve $\frac{2}{3}(x + 1) = 4$ using the least number of steps.

Divide.

27. $\frac{4}{5} \div \frac{1}{4}$ **28.** $\frac{3}{8} \div \frac{4}{7}$ **29.** $\frac{1}{5} \div \frac{2}{15}$ **30.** $6 \div \frac{3}{8}$

List the stems that you would use to plot the data on a stem-and-leaf plot.

31. 23, 25, 29, 31, 26, 43, 15   **32.** 6, 25, 19, 16, 46, 9, 33, 18

33. The sum of two whole numbers is 54. Their product is 680. What are the numbers?

FINANCE CHARGES

Many people often buy items on credit by borrowing money from a bank or by using a credit card. Most credit card companies charge a fee on the monthly balance of an account. This fee is a **finance charge.**

In April, Meagan Williams bought a television set for $525. She charged the bill to her credit card account. Ms. Williams' credit card charges a monthly finance charge of 1.54%. The first payment was due in May. Ms. Williams paid the minimum amount due, $20. What will the balance, including interest, be on her next statement if she does not add any more charges?

$$525 \ominus 20 \boxminus 505 \boxtimes .0154 \boxminus$$
$$7.777 \boxplus 505 \boxminus 512.777$$

The balance on the next statement will be $512.78.

If you do not need to know the amount of the finance charge, you can multiply by 1.0154.

$$525 \ominus 20 \boxminus 505 \boxtimes 1.0154 \boxminus 512.777$$

Multiplying 505 by 1.0154 is the same as multiplying by 0.0154 and adding 505.

Copy and complete the table. The finance charge is 1.5% a month.

	Previous Balance	Monthly Payment	Finance Charge	New Balance
1.	$125	$20	▨	▨
2.	$505	$50	▨	▨
3.	$752	$152	▨	▨
4.	$421	$40	▨	▨

5. Mrs. West has a balance of $680 on her credit card account. The monthly finance charge is 1.12%. One month she makes a payment of $75. The next month she makes a payment of $150. What is her new balance?

6. Refer to Exercise 5. Suppose Mrs. West pays $100 each month. How long will it take her to pay off the bill? How much will she pay in finance charges?

REGISTERED NURSE

Jon Farrell is a registered nurse (RN). He takes temperatures, blood pressures, gives injections, and administers medication. Mr. Farrell supervises licensed practical nurses (LPNs) and is under the direction of doctors. Sometimes he does special duty nursing in a private home, a burn unit, or a nursing home.

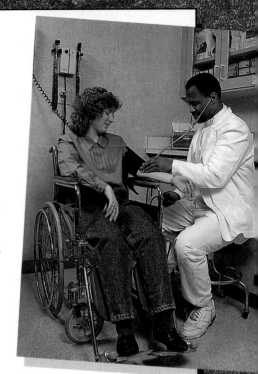

A formula for determining normal blood pressure *(B.P.)* is $110 + (\frac{1}{2} \times A)$ where *A* represents age.

Compute normal blood pressure for an 18-year-old.

$$B.P. = 110 + (\tfrac{1}{2} \times A)$$

$$= 110 + (\tfrac{1}{2} \times 18) \quad \text{Substitute 18 for } A.$$

$$= 110 + 9 \quad \text{Multiply } \tfrac{1}{2} \text{ times 18.}$$

$$= 119 \quad \text{Add.}$$

Normal blood pressure for an 18-year-old is 119.

Calculate a normal blood pressure reading for the following ages.

1. your age   **2.** 57   **3.** 26   **4.** 65

5. Mr. Smith has a normal blood pressure for his age. How old is he if his blood pressure is 135?

Type	Percent
A positive	38%
O positive	36%
B positive	8%
O negative	6%
A negative	6%
AB positive	3.5%
B negative	2%
AB negative	0.5%

The chart on the left shows different blood types and the percent of the population that has the blood type.

Find the approximate number of people out of 3,000 who have a blood type of O positive.

$$3000 \; \boxed{\times} \; .36 \; \boxed{=} \; 1080$$

Out of 3,000 people, about 1,080 have a blood type that is O positive.

Use the chart and find the approximate number of people out of a group of 1,500 who have each blood type.

6. AB negative   **7.** A positive   **8.** AB positive   **9.** O negative

10. Out of a group of 10,000 people, about how many have a blood type that is A positive or A negative?

13-4 THE COORDINATE PLANE

Objective

Name and graph points on the coordinate plane by using ordered pairs.

Axes is the plural form of axis.

Numbers are graphed on the number line. Ordered pairs of numbers, such as (2, 4), (−1, 5), (−3, −8), are graphed on the **coordinate plane.** The coordinate plane has two perpendicular number lines. Each number line is called an **axis.** The axes intersect at the **origin** labeled *O*. Every point on the coordinate plane has two coordinates. One corresponds to the horizontal axis and the other to the vertical axis. The coordinates are written as an **ordered pair.**

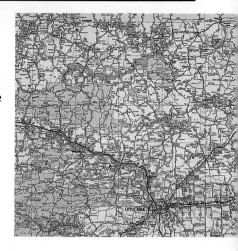

Method

1. ▶ The first coordinate in the ordered pair corresponds to a number on the horizontal axis. Numbers to the right of the origin are positive and to the left of the origin are negative.

2. ▶ The second coordinate in the ordered pair corresponds to a number on the vertical axis. Numbers above the origin are positive, numbers below the origin are negative.

3. ▶ Write the numbers that correspond to the location of the point.

Example A

Find the ordered pair for each point labeled on the coordinate plane.

1. ▶ Point *A* is 3 units to the right of the origin on the horizontal axis.

2. ▶ Point *A* is 4 units above the origin on the vertical axis.

3. ▶ The ordered pair is (3, 4).

1. ▶ Point *B* is 5 units to the left of the origin on the horizontal axis.

2. ▶ Point *B* is 2 units above the origin on the vertical axis.

3. ▶ The ordered pair is (−5, 2).

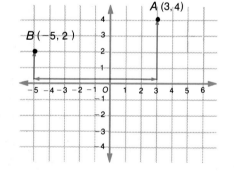

Example B

Graph the ordered pairs.

Point *D* (5, −3)   Start at the origin.

⟶ (5, −3) ⟵

1. ▶ Locate 5 on the horizontal axis.
2. ▶ Then move down 3 units and draw a dot.

Point *E* (−2, 0)   Start at the origin.

⟶ (−2, 0) ⟵

1. ▶ Locate −2 on the horizontal axis.
2. ▶ (Move 0 units.) Draw a dot.

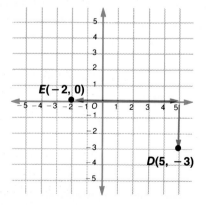

Find the ordered pair for each point labeled on the coordinate plane.

1. G **2.** H **3.** I
4. J **5.** K **6.** L

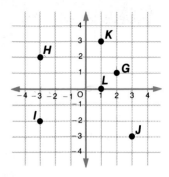

Draw a coordinate plane. Graph each ordered pair. Label each point with the given letter.

7. M (2, 3) **8.** N (–4, 5) **9.** O (0, 0)

10. P (4, –3) **11.** Q (–2, –5) **12.** R (4, 0)

13. S (0, 4) **14.** T (0, –4) **15.** U (–4, 0)

Find the ordered pair for each point labeled on the coordinate plane.

16. U **17.** V **18.** A

19. B **20.** E **21.** D

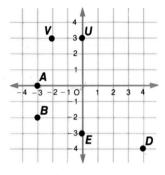

Draw a coordinate plane. Graph each ordered pair. Label each point with the given letter.

22. H (0, –1) **23.** I (2, 5) **24.** J (7, 0) **25.** K (–3, 6)

26. L (5, –3) **27.** M (–7, 3) **28.** N (3, –4) **29.** P (–2, –4)

30. Q (5, –1) **31.** R (–3, –3) **32.** S (4.5, 2.5) **33.** T (–4.5, 5)

34. Graph the following ordered pairs and connect the points in order: (2,1), (4,3), (6, 1), (6, –2), and (2, –2). What geometric figure is formed?

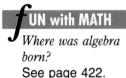

UN with MATH

Where was algebra born?
See page 422.

35. The ordered pairs for three vertices of a square are (2, 1), (–3, 1), and (–3, –4). What is the ordered pair for the other vertex?

36. The ordered pairs for three vertices of a parallelogram are (4, 2), (–2, 2), and (–4, –2). What is an ordered pair for another vertex?

37. Examine a map in which locations are identified by a letter and a number. Describe how locating a point on the map is similar to locating a point on the coordinate plane.

38. A line in the coordinate plane passes from the upper left, through the origin, and on to the lower right. What do the ordered pairs for all points except (0,0) along this line have in common?

13-5 GRAPHING EQUATIONS

Objective

Graph linear equations.

Jay wants to know what pairs of numbers have a sum of 4. The equation $x + y = 4$ describes this situation. There are two variables, x and y.

You have worked with equations that have one variable and one solution. An equation with two or more variables has more than one solution. You can use the coordinate plane to graph the solutions for an equation with two variables. Each solution can be written as an ordered pair.

Method

1 Find several ordered pairs that satisfy the equation.

2 Graph the ordered pairs.

3 Draw a line through the points. All of the ordered pairs for points on the line are solutions to the equation.

Example A

Graph the solution to x + y = 4. Rewrite the equation as y = 4 − x so it is easier to find values of y.

1 Make a table of possible values for x. Find the corresponding values for y.

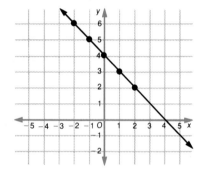

x	4 − x	y	(x, y)
−2	4 − (−2)	6	(−2, 6)
−1	4 − (−1)	5	(−1, 5)
0	4 − 0	4	(0, 4)
1	4 − 1	3	(1, 3)

2 Graph the ordered pairs.

3 Use a ruler to draw a line through the points.

Solutions include (3, 1), (4, 0) and (5, −1).

The coordinates of any ordered pair for points on the line $y = 4 - x$ have a sum of 4. All of these ordered pairs are solutions.

Example B

Graph the solution to $y = \frac{1}{2}x + 1$.

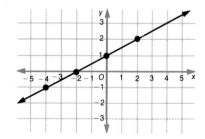

1

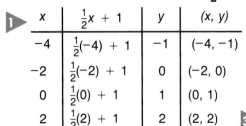

x	$\frac{1}{2}x + 1$	y	(x, y)
−4	$\frac{1}{2}(-4) + 1$	−1	(−4, −1)
−2	$\frac{1}{2}(-2) + 1$	0	(−2, 0)
0	$\frac{1}{2}(0) + 1$	1	(0, 1)
2	$\frac{1}{2}(2) + 1$	2	(2, 2)

2 Graph the ordered pairs.

3 All of the points on the line are solutions to the equation $y = \frac{1}{2}x + 1$.

Graph the solutions to each equation.

1. $x - 5 = y$ **2.** $x + 3 = y$ **3.** $\frac{x}{2} = y$ **4.** $\frac{1}{3}x = y$

Graph the solutions to each equation.

Practice

5. $x - 3 = y$ **6.** $2x = y$ **7.** $\frac{2}{3}x = y$

8. $x + 1 = y$ **9.** $1.5x = y$ **10.** $x + 2.5 = y$

11. $x - 1.5 = y$ **12.** $x = y$ **13.** $x + y = 5$

14. $x + y = -3$ **15.** $x - y = 4$ **16.** $x - y = 0$

17. Draw a graph to show all possible pairs of numbers whose sum is 13.

Applications

18. An oscilloscope is an electronic instrument that is used to show any kind of vibration, such as sound waves. The screen of an oscilloscope is marked with grids like the coordinate plane. The shape of a vibration is shown on the screen and can be measured using the grid. The amplitude (loudness) of a sound wave is the vertical distance from the rest position (the horizontal axis) to the highest point. The wave shown in blue has an amplitude of 20. What is the amplitude of the wave shown in red?

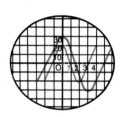

19. In Tuesday's *Daily News*, the financial page used a graph to show the increase in price per share of Americomp stock. If the price per share continues to increase at the rate shown on the chart, in which month will the price per share be $40?

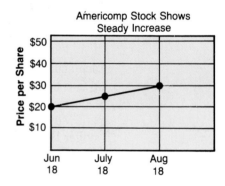

Problem Solving

20. Beth wants to make a rectangular garden with an area of 900 square feet. What dimensions should she use in order to use as little fence as possible around the garden?

21. A 50-pound bag of plant food contains 16 pounds of phosphoric acid. How many pounds of phosphoric acid are in a 25-pound bag of the same plant food?

22. Simon's age is 3 more than twice his brother's age. If Simon is 11, how old is his brother?

Critical Thinking

23. Graph the two ordered pairs (5, 2) and (10, 7) on a coordinate plane. Draw a straight line through the two points. Make a table of 5 ordered pairs that lie on the line. Use the values in the table to find the equation that represents the line.

13-6 READING A GRID MAP

Objective

Locate places on a grid map.

Marie is visiting her cousin Sharyn in California. They are going to a baseball game at Dodger Stadium. Marie and Sharyn locate Dodger Stadium on a grid map to check their directions before driving to the stadium.

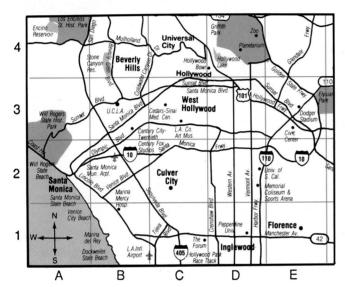

Index

Cedars-Sinai Medical Center, C-3
Dodger Stadium, E-3
The Forum, C-1
Griffith Park, D-4
Griffith Planetarium, D-4
Hollywood Park Race Track, D-1
LA International Airport, B-1
Los Angeles County Art Museum, C-3
Marina del Rey, B-1
Marina Mercy Hospital, B-1
Pepperdine University, D-1
University of California LA (UCLA), B-3
University of Southern Calif. (USC), D-2

Each square on a grid map can be identified by a letter-number pair. To find Dodger Stadium, check the index for the location. Dodger Stadium is in square E-3. Find the letter E along the bottom of the map. Move up to the square labeled 3 at the left of the map. Locate Dodger Stadium within square E-3.

Understanding the Data

● Where do you find the letters on the grid map?
● Where do you find the numbers on the grid map?
● What do you find in the index?
● What does a letter-number pair indicate?

Check for Understanding

Use the map and the index above.

1. What park is located in A-4?
2. What highway is near University of Southern California?
3. What airport is located in B-2?
4. Within what square do you find Marina del Rey?

Give the location of each as a letter-number pair.

5. Stone Canyon Reservoir
6. Hollywood Bowl
7. Twentieth Century Fox Studios
8. Florence

Practice

Use the map and index on page 412.

9. What zoo is located in D-4?

10. What highway is near Pepperdine University?

11. What airport is located in B-1?

12. What medical center is located in C-3?

Give the location of each with a letter-number pair.

13. Santa Monica State Beach

14. LA County Art Museum

15. Univ. of Southern California

16. Griffith Planetarium

Applications

f**UN with MATH**

Are there women heart surgeons?
See page 423.

17. Give directions from LA International Airport to The Forum.

18. Give directions from UCLA to LA International Airport.

19. Scott drove from the University of Southern California toward Dodger Stadium on Highway 110. After driving past what road did Scott begin to watch for Dodger Stadium?

20. If you are traveling from E-1 to A-4, are you traveling northeast or northwest?

21. Find a grid map of your area in the library or city hall. Make a list of four places and their letter-number location.

Mixed Review

Lesson 7-2

Trace each angle and extend the sides. Then find the measure of each angle.

22. **23.**  **24.**

Lesson 12-2

Add. Use a number line if necessary.

25. $-6 + 4$ **26.** $-10 + (-8)$ **27.** $18 + (-24)$

Lesson 10-4

28. At Ricardo's Pizza Shop one night, they sold 32 pizzas with pepperoni, 24 with mushrooms, and 21 with sausage. Of these, 7 had pepperoni and mushroom on one pizza, 3 had sausage and mushrooms, 6 had pepperoni and sausage and 4 had all three toppings. How many pizzas were sold?

13-7 WRITING AN EQUATION

Objective

Solve problems using equations.

Adam and Belinda compare projects for the science fair. Adam raised six hamsters on a special diet he created. The number of fish Belinda raised divided by six is two less than the number of hamsters Adam raised. How many fish did Belinda raise?

 Explore

What is given?
- Adam has six hamsters.
- The number of fish Belinda raised divided by six is two less than the number of hamsters.

What is asked?
- How many fish did Belinda raise?

 Plan

Let f equal the number of fish. Translate the problem into an equation using the variable f.

▶ **Solve**

The number of fish divided by six	is	two less than	the number of hamsters
$f \div 6$	$=$	-2	$+6$

$$\frac{f}{6} = -2 + 6 \qquad f \div 6 \text{ can be expressed as } \tfrac{f}{6}.$$

$$\frac{f}{6} = 4 \qquad -2 + 6 = 4$$

$$6 \times \frac{f}{6} = 6 \times 4 \qquad \text{Multiply each side by 6.}$$

$$f = 24$$

Belinda raised 24 fish.

▶ **Examine**

Read the problem again to see if the answer makes sense. Is 24 divided by 6 equal to 2 less than 6? Yes, the solution is correct.

 Guided Practice

Solve. Write an equation.

1. Susan has a total of 81 compact discs and tapes. If she has twice as many tapes as compact discs, how many of each does she have? Hint: Let x equal compact discs.

2. Glenn sells his stereo for $212. The sale price is $10.50 more than half the amount he originally paid for it. How much did he pay for his stereo?

Paper/ Pencil
Mental Math
Estimate
Calculator

Solve. Use any strategy.

3. The student council is selling school pins and mums for homecoming. They sell three times as many pins as mums. If they sell 117 pins, how many mums did they sell?

4. In triangle *ABC*, $\angle C$ measures 63°. Find the measures of the other two angles if the measure of $\angle B$ is twice the measure of $\angle A$. Remember, the sum of the measures of the angles in a triangle is 180°.

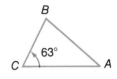

5. Pet Party pet shop has only parakeets and kittens located in the front corner of the store. If the animals in this corner have 18 heads and 52 feet, how many parakeets and kittens are there?

PORTFOLIO

6. Write a mathematical autobiography that describes your mathematical growth over the past year. Use correct punctuation, spelling, and good grammar.

7. Nicole read about Hartley's 25%-off sale. Hartley's has a blouse and a pair of slacks that Nicole wants. If the original price of the blouse is $28 and the original price of the slacks is $30, *about* how much money should Nicole take to the store?

Critical Thinking

***f*UN with MATH**

Try solving our story of a theft on the beach.
See page 423.

8. Four members of the track team finished the cross-country race at state competition. Don placed third. Lou finished six places behind Don. Keith finished three places in front of Lou, and George finished two places behind Keith. In which places did Lou, Keith, and George finish?

Mixed Review

Lesson 11-10

Use the box-and-whisker plot to answer each question.

9. The middle half of the data is between which two values?

10. What part of the data is greater than 50?

10 20 30 40 50 60 70 80

Lesson 12-3

Subtract.

11. $15 - (-6)$ **12.** $-21 - (-16)$ **13.** $-23 - 7$

Lesson 3-2

14. A mountain climber can descend 50 feet every minute by rappelling. How far can the climber descend in 8.5 minutes?

REVIEW

Vocabulary/
Concepts

Choose the word or phrase from the list at the right that best completes each sentence.

1. To solve an equation with one variable, add, subtract, multiply, or divide the __?__ on each side of the equation.

2. To check a solution, replace the __?__ with the possible solution.

3. A solution to an equation with two variables can be written as a(n) __?__ .

4. A solution to an equation with two variables can be graphed on a(n) __?__ .

5. The first coordinate in an ordered pair represents a value along the __?__ axis.

6. An equation in two variables has __?__ solution.

7. On a coordinate plane, the point that has coordinates $(0,0)$ is called the __?__ .

8. The graph of the ordered pair $(0,5)$ lies on the __?__ axis.

coordinate
   plane
horizontal
more than one
negative
one
opposite
ordered pair
origin
positive
same number
variable
vertical

Exercises/
Applications

Lesson 13-1

Solve each equation. Check your solution.

9. $y + (-7) = 8$

10. $-11 = x + 9$

11. $m - 9 = 10$

12. $-18 = y - 2$

13. $x - 30 = -5$

14. $y + 12 = -12$

Lesson 13-2

Solve each equation. Check your solution.

15. $4p = -8$

16. $-3t = 18$

17. $-104 = 8r$

18. $42 = \frac{y}{-3}$

19. $\frac{x}{8} = -24$

20. $\frac{p}{-2} = 14$

Lesson 13-3

Solve each equation. Check your solution.

21. $5m + 2 = -18$

22. $2y - 7 = 8$

23. $-23 = 6x + (-5)$

24. $3(n + 4) = -12$

25. $5(x - 2) = -18$

26. $8.75 = -2.5\,(7 + y)$

Lesson 13-4

Find the ordered pair for each point labeled on the coordinate plane.

27. A **28.** B **29.** C

Draw a coordinate plane. Then graph each ordered pair. Label each point with the given letter.

30. I (1, 4) **31.** J (1, −4) **32.** K (−3, 5)

33. M (2, −3) **34.** P (4, 1) **35.** Q (0, 6)

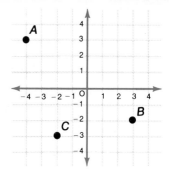

Lesson 13-5

Graph the solutions to each equation.

36. $12 - x = y$ **37.** $x + 2 = y$ **38.** $x - 7 = y$

39. $3x = y$ **40.** $\frac{1}{2}x = y$ **41.** $x + y = 5$

Lesson 13-6

Use the map to answer these questions.

42. What university is located in E-4?

43. What road runs from C-6 to E-5?

Give the location of each as a letter-number pair.

44. Georgetown U. **45.** McLean

46. Chesterbrook **47.** Cabin John

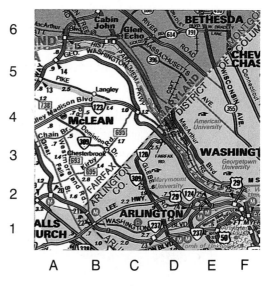

Lesson 13-7

Solve. Write an equation.

48. Mr. York bought 200 shares of stock at $31 per share. Then the price of each share went up $2.50. What was the total value of the 200 shares after the price rise?

49. From the surface, a diver descended to 30 meters below sea level where she obtained a water sample. She then rose 12 meters and collected another water sample. How far below the surface was she when she collected the second sample?

50. Frank and David buy plain sweatshirts from a wholesale dealer. They print their school's name and team's name on the sweatshirts and sell them for $7 more than the wholesale price. They sold 40 sweatshirts and collected $640. What is the wholesale price for each?

51. Mrs. Evearitt has two jobs. The first pays $50 more per week than the second. Her total earnings are $320 per week. How much does she earn from each job?

TEST

Solve each equation. Check your solution.

1. $m - 6 = 4$ **2.** $-8 + x = 2$ **3.** $8x = -32$ **4.** $-36 = y - 4$

Find the ordered pair for each point labeled on the coordinate plane.

5. P **6.** Q **7.** R

8. S **9.** T **10.** U

11. V **12.** W **13.** Z

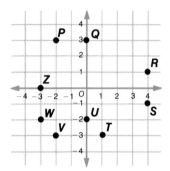

Draw a coordinate plane. Then graph each ordered pair. Label each point with the given letter.

14. $A\ (5,3)$ **15.** $B\ (-2, -4)$ **16.** $C\ (1, -6)$ **17.** $D\ (-2, 3)$

Graph the solutions to each equation.

18. $3 + x = y$ **19.** $\frac{1}{2}x + 4 = y$ **20.** $3x = y$ **21.** $\frac{1}{3}x = y$

22. Assume the letters of your grid map are on the horizontal edge and the numbers are on the vertical edge. Explain how to locate area C-5.

Solve. Write an equation.

23. Bill Raymond skied the slalom course in 138 seconds. This was 15 seconds faster than his father. What was his father's time?

24. Janet gave each of her three sisters an equal number of pictures from her collection of popular actors and actresses. Karrie already had 2 pictures. She has 8 pictures after getting some from Janet. How many pictures did Janet give away in all?

25. Doug weighs 180 pounds. He loses 2 pounds each week for 5 weeks, and then gains 1 pound each week for 3 weeks. How much does he weigh after the losses and gains?

▶ BONUS: Find two integers that have a sum of 6 and a difference of −20.

LATITUDE AND LONGITUDE

Most globes and maps of Earth show grid lines, called latitude lines and longitude lines. Because Earth is shaped like a sphere, these grid lines are actually circles or parts of circles. Latitude and longitude are measured in degrees because circles may be divided into degrees.

LATITUDE

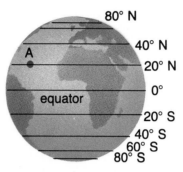

LONGITUDE

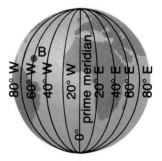

Any point on Earth is a certain number of degrees north or south of the equator. For example, point A is located at 20°N latitude.

Any point on Earth is within 180° east or west of the prime meridian. For example, point B is located at 40°W longitude.

Use the map at the right to find the latitude and longitude of each location to the nearest degree.

1. Providence, Rhode Island
2. Portsmouth, New Hampshire
3. Portland, Maine
4. Which city is farthest west? What is its latitude and longitude?
5. Which ship is farthest north? Which ship is farthest south? What is the approximate difference between their latitudes?

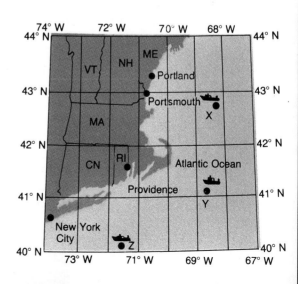

Free Response

Lesson 1-7
Estimate.

1. 318×7 **2.** 43×112 **3.** $285 \div 72$ **4.** $4,128 \div 59$

Lesson 2-1
5. A car dealer sold 1,774 cars one year and 1,967 cars the next year. How many were sold in the two years?

Lesson 4-10
Add.

6. $\frac{1}{5} + \frac{3}{5}$ **7.** $\frac{3}{7} + \frac{5}{7}$ **8.** $\frac{1}{6} + \frac{2}{3}$ **9.** $\frac{3}{4} + \frac{7}{8}$

Lesson 6-8
Complete.

10. $4 \text{ gal} = \underline{\overset{?}{}} \text{ qt}$ **11.** $2 \text{ qt} = \underline{\overset{?}{}} \text{ oz}$ **12.** $80 \text{ oz} = \underline{\overset{?}{}} \text{ lb}$

Lessons 8-9
8-10
Find the volume of each solid.

13.
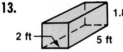
1.8 ft
2 ft
5 ft

14.

4 m
4 m
4 m

15.
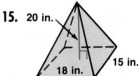
20 in.
18 in.
15 in.

Lesson 10-7
16. Janis buys 3 sweatshirts that normally sell for $24 each. Two of the shirts are on sale for 25% off and one shirt is discounted for 35% off regular price. How much does she pay for the 3 sweatshirts?

Lessons 11-2
11-4
Use the data at the right.

17. Make a frequency-table for the data.
18. Find the mean, median, mode, and range.
19. Make a bar graph for the data.

Ring Sizes of Senior Girls					
5	6	6	7	7	$6\frac{1}{2}$
7	5	$7\frac{1}{2}$	$6\frac{1}{2}$	$7\frac{1}{2}$	6
$7\frac{1}{2}$	$6\frac{1}{2}$	6	7	6	7

Lessons 12-2
12-3
Add or subtract.

20. $30 - 45$ **21.** $10 + (-17)$ **22.** $28 + (-57)$ **23.** $-71 + 100$

Lessons 12-5
12-6
Multiply or divide.

24. $3 \times (-8)$ **25.** 2×14 **26.** $20 \div (-2)$ **27.** $-45 \div 3$

Lesson 13-5
Graph the solutions to each equation.

28. $x + y = 0$ **29.** $3x + 2 = y$

Multiple Choice

Choose the letter of the correct answer for each item.

1. Write 6^3 in standard form.
   a. 18   **c.** 216
   b. 36   **d.** 729

Lesson 1-2

2. What is the quotient when 45 is divided by 100?
   e. 0.0045
   f. 0.45
   g. 0.045
   h. 4,500

Lesson 3-7

3. Divide $6\frac{1}{4}$ by $2\frac{1}{8}$. What is the result?
   a. $\frac{17}{50}$
   b. $2\frac{16}{17}$
   c. $3\frac{1}{2}$
   d. $13\frac{9}{32}$

Lesson 5-7

4. The formula for changing Fahrenheit degrees to Celsius degrees is $C = \frac{5}{9}(F - 32)$. Find the Celsius reading for 68°F.
   e. 30°C   **g.** 20°C
   f. 5.7°C   **h.** *none of these*

Lesson 6-9

5. The surface area of a solid is the sum of the areas of each surface. What is the surface area of the solid?
   a. 6 ft²
   b. $6\frac{1}{2}$ ft²
   c. $11\frac{1}{2}$ ft²
   d. 23 ft²

Lesson 8-7

6. 32 is 50% of what number?
   e. 16   **g.** 160
   f. 64   **h.** 1,600

Lesson 10-2

7. The frequency table lists the ages of sophomores at East High. How many sophomores are at East High?
   a. 48
   b. 162
   c. 489
   d. 537

Age	Frequency
15	112
16	324
17	53

Lesson 11-2

8. Which equation is graphed at the right?
   e. $x = y$
   f. $x = 2y$
   g. $x + 2 = y$
   h. $x - 2 = y$

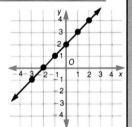

Lesson 13-5

9. Rachael spent half of her money for dinner and half of what remained for earrings. She had $4 left. How much did she spend for dinner?
   a. $20   **c.** $8
   b. $16   **d.** $4

Lesson 12-7

10. Suppose you were to read the numbers $-27, 17, -17, 27,$ and -16 on a number line. In which order would they appear from left to right?
   e. $-16, -17, -27, 17, 27$
   f. $-27, -17, -16, 17, 27$
   g. $-27, -16, -17, 17, 27$
   h. *none of the above*

Lesson 12-1

		Father of Algebra	
Human-made fire	2000 BC	Baghdad, Iraq	AD 1150
35,000–10,000 BC	First Love Song Sumer (Babylon)	AD 820	Rocketry China

Why recycle?

Recycling saves energy and other natural resources. You can earn extra cash by turning in recyclable products to a collection center. In 1986, 13.3 billion aluminum beverage cans were recycled. Reynolds Metals paid nearly $93 million to recyclers for 305 million pounds of aluminum, enough to make nearly 8 billion cans. A ton of recycled paper can save 15 to 17 trees. When glass is first made, it must be heated to 1,470° C; however, recycled glass melts at only 760° C. Imagine the enormous energy savings.

MATH M·E·N·U

Pineapple Chicken

4   chicken breasts (fillets)
1   can cream of chicken soup
1   can crushed pineapple (8¾-oz can)
½   teaspoon curry powder
1   tablespoon soy sauce
salt and pepper

Preheat oven to 375° F. Place chicken in a greased casserole dish. Mix other ingredients, add salt and pepper to taste, and pour over chicken. Cover dish and bake for 1 hr. 15 min. Serve with rice and a green salad. Serves 4.

COMICS

A day on the planet Venus is longer than its year. One rotation of Venus on its axis takes 243 Earth days while one revolution around the sun only takes 225 Earth days.

CALVIN & HOBBES

How does a robot work?

A robot is a machine that can be programmed to do different tasks. To program a robot to pick something up and put it down somewhere else,

the programmer moves each of the robot's joints by pressing buttons on a *teach unit* attached to the robot's arm. At the same time, the programmer engages a record button. This enters the program into the robot's computer memory.

The robot can carry out the entire task on its own from memory.

RIDDLE

Q: Why is simplifying a fraction like powdering your nose?

A: It improves the appearance without changing the value.

Find the indicated average for each letter.

ACROSS
- **A.** Mean: 5, 6, 9, 11, 2, 11, 40
- **B.** Median: 36, 37, 41, 43, 43
- **D.** Mode: 84, 10, 71, 10, 10, 69
- **E.** Mean: 421; 98; 602; 935; 1,084
- **G.** Mode: 256, 286, 95, 12, 286, 76

DOWN
- **A.** Mode: 105, 116, 78, 87, 116, 125
- **B.** Mean: 36, 37, 41, 43, 43
- **C.** Median: 4; 2,067; 583;5,300; 2,419
- **D.** Median: 45, 18, 67, 9, 3, 18
- **F.** Mode: 22, 23, 32, 28, 23, 33
- **G.** Mean: 28, 61, 0, 35, 19, 14, 28, 7

TEASER

QUIZ TIME

It's Saturday evening. There's been a theft. On Monday, the victim, Leonardo Fuentes, left his wallet in a towel on the beach and went in swimming. "That was at 3 P.M.," he said. "I came out at 3:30 and everything was gone." A local surfer called Wave is suspected, so I talk with her. "I hang out on the beach," she said. "But I didn't take that guy's stuff. The high tide must have washed it away. At 3:30 the tide was far up the beach." The lifeguard tells me that he couldn't remember what time high tide occurred on Monday. "Today, it was at 4:30," he said. "And it's gotten about 25 minutes later each day during this week."

Was Wave telling the truth? Why, or why not?

JOKE!

Q: What will happen to the inch worm when we go metric?

A: He'll become a centipede.

Chapter

14 PROBABILITY

You Win!

Many games and sports depend on skill. Others depend on chance. Have you ever won anything in a drawing or raffle at a fair or a carnival? Most drawings offer each player an equal opportunity to win. Such games are referred to as fair games. Some games do not offer an equal chance to every player. These games are unfair. There are ways to find the chances that a particular event will happen.

The Enfield Community Center is raffling off a car. Each ticket costs $10. Only 1,000 tickets will be sold. If you buy 1 ticket, what is the probability that you will win? Is this raffle a fair game? Explain.

ACTIVITY: Probability and Games

In this activity you will explore probability and its uses in games involving draws and chance. You will also discover what makes a game *fair* or *unfair*. Many people enjoy playing games that depend on chance. Knowing whether a game is fair or unfair will help you decide if you wish to take a chance.

Materials: coins, dice, spinners, paper and pencil

Cooperative Groups

Work in groups of three or four.

1. What outcomes are possible if you toss a coin and roll a die. Draw a tree diagram to show the possible outcomes. How many different outcomes are there? What is the probability of tossing a head and rolling a 5?

2. With your group play this game 50 times. Roll two dice. Add the numbers that face up on the dice. Player 1 wins if the sum of the numbers is odd. Player 2 wins if the sum of the numbers is even. One member of the group records the outcomes. How many wins did each player have out of 50?

3. Make a table with a list of the possible outcomes for the game in Exercise 2. How many different possible outcomes are there? How many sums will be even? How many will be odd? Decide with your group if the game is fair or unfair.

4. In a sentence, describe what makes a game fair or unfair.

Varying the Game

5. Make a list of the outcomes for this game—Roll two dice. Multiply the two numbers on the dice. Player 1 wins if the product is odd. Player 2 wins if the product is even. Decide with your group if the game is fair. Explain your reasons.

Extending the Idea

6. With your group create a game for two players using a spinner and a coin. Make the game fair. Then play the game and record the results. Were there an even number of wins by each player?

Communicate Your Ideas

7. Discuss with your group why a fair game does not always assure that each player will have the same number of wins in a specific number of games.

14-1 FINDING THE NUMBER OF OUTCOMES

Objective

Find the number of possible outcomes using tree diagrams or multiplication.

Melissa King is a finalist on the Perils game show. The show host will toss a nickel and a dime for the bonus. If Melissa calls the outcome correctly, she'll win a $500 bonus. Melissa wants to know the number of possible outcomes.

When more than one coin is tossed, you can use a tree diagram to determine the number of ways each outcome can occur.

A list of all the possible outcomes is a **sample space.** Any specific type of outcome is an **event.**

Example A

You can draw a tree diagram to show the possible outcomes when a nickel and a dime are tossed at the same time.

Nickel   Dime   Outcomes

```
        H ——————— H H
  H  <
        T ——————— H T

        H ——————— T H
  T  <
        T ——————— T T
```

In this tree diagram, the top branch shows the outcome of a head and a head. This is symbolized HH.

HT means that the nickel came up heads and the dime came up tails. TH means that the nickel came up tails and the dime came up heads.

The tree diagram shows four possible outcomes.

Example B

You can also use multiplication to find the number of possible outcomes when a nickel and a dime are tossed at the same time.

nickel		dime		nickel and dime
2	×	2	=	4
possible		possible		possible
outcomes		outcomes		outcomes

Guided Practice

Draw a tree diagram to show the possible outcomes.

Example A

1. tossing the coin and spinning the spinner

2. spinning each spinner once

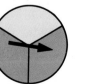

Example B

Use multiplication to find the number of possible outcomes.

3. rolling a red die and a blue die

4. tossing a coin and rolling a die

Practice

Use multiplication to find the number of possible outcomes. Draw a tree diagram to show the possible outcomes.

5. tossing a penny, a nickel, and a dime

6. spinning each spinner once

7. tossing a quarter, rolling a die, and tossing a penny

8. a choice of a desk or a wall phone in white, black, green, or tan

Applications

9. Mrs. Jergen's history quiz has four multiple choice questions. Each question has four answer choices. How many possible sets of answers are there for the quiz?

10. Four people are to sit in four chairs in a row. How many ways can the four people be seated? (Hint: once a person is seated, he or she is no longer a possible choice for the other chairs.)

11. A jar of marbles contains 490 blue marbles, 500 red marbles, and 10 green marbles. If Karen draws a green marble while she is blindfolded, she wins 100 free gallons of gasoline from the Corner-Stop station. What is the probability that Karen will draw a green marble?

Cooperative Groups

12. In groups of three or four, make two spinners with regions of different sizes. One must be one-half red, one-fourth green, and one-fourth yellow. The other must be two-thirds blue and one-third orange. Make a list of the possible outcomes for each spinner. With your group, explain which outcomes on each spinner should occur most often.

Mixed Review

Lesson 2-4

Subtract.

13. 18.4
        − 12.6

14. 25.7
        − 14.6

15. 38.2
        − 29.7

16. 183.01
        − 45.98

17. 44.878
        − 32.909

Lesson 9-2

A die is rolled once. Write the probability of each roll.

18. a 2 or a 5

19. an even number

20. a 1 or a 6

Lesson 12-7

21. Joel currently has 2,880 classic rock singles. Joel began his collection 7 years ago and has doubled the number in his collection each year. How many singles did Joel collect during the first year?

14-2 PROBABILITY AND PERCENTS

Objective

Express the probability of an event as a percent.

At the summer carnival in Shillington, Dr. Gomez has a chance at the $1,000 prize by spinning a spinner. The spinner is divided into ten equal-sized parts. Three parts are blue, three are red, two are yellow, and two are black. What is the probability Dr. Gomez will win if he chooses blue?

Remember that the probability of an event is a ratio.

$$\text{probability} = \frac{\text{number of ways event can occur}}{\text{number of ways all possible outcomes can occur}}$$

Probabilities are often written as percents.

Method

1. Find the probability as a ratio.

2. Express the probability as a percent.

Example A

Find the probability that Dr. Gomez will win if he chooses blue. Express the probability as a percent.

1. The probability of the spinner landing on blue is $\frac{3}{10}$.

$\frac{3}{10} = \frac{n}{100}$ Remember that percent means *per hundred*.

$3 \times 100 = 10 \times n$ Find the cross products.

$\frac{300}{10} = \frac{10 \times n}{10}$ Solve for *n*.

$30 = n$

The probability that Dr. Gomez will win if he chooses blue is 30%.

Example B

There are three blue pens, five red pens, and two black pens in a drawer. One pen is chosen without looking. Find the probability that a green pen will be chosen.

1. The probability of choosing a green pen is $\frac{0}{10}$ or 0.

2. A probability of 0 can be written as 0%.

There is a 0% probability that a green pen will be chosen.

Guided Practice

Use the spinner above. Find the probability that each event will occur. Express as a percent.

Examples A, B

1. landing on red   **2.** landing on yellow   **3.** landing on black

There are two quarters, two dimes, and one penny in a jar. One coin is chosen. Find the probability of choosing each event. Express as a percent.

4. a quarter   **5.** a penny   **6.** a dime or a quarter

7. a half-dollar   **8.** a penny or a nickel   **9.** a penny or a dime

Exercises

There are three red balls and two blue balls in a box. One ball is chosen. Find the probability that each event will occur. Express as a percent.

Practice

10. a red ball   **11.** a blue ball   **12.** a green ball

13. *not* a blue ball   **14.** *not* a red ball   **15.** *not* a green ball

16. a red or a blue ball   **17.** a yellow or a green ball

A red die and a blue die are rolled together. See the list of possible outcomes on page 283. Find the probability of rolling each event. Round to the nearest whole percent.

18. a red 6 and a blue 1   **19.** a blue 4 and a red 3

20. a red 4 and a blue 3   **21.** a 3 and a 2

22. a 2 on both dice   **23.** both numbers the same

24. both numbers odd   **25.** both numbers less than 3

Applications

26. The Long High School marching band sold 2,000 raffle tickets. Gerri bought 10 tickets. What is the probability she will win the raffle?

27. Special stickers have been attached to 1,000 cassette tapes in Mister Music's stock. Ten out of every 250 stickers win a free cassette. If all 1,000 tapes are sold and you buy one of them, what is the probability that you will win a free cassette?

28. The weather report states that there is a 40% chance of rain today. What is the probability that it will not rain today?

Write Math

29. Write a word problem that uses this tree diagram to solve the problem. Then solve the problem.

Critical Thinking

30. Suppose @ represents a certain phrase. If 12 @ 3 and 4, 8 @ 4 and 8, 6 @ 2 and 3, and 15 @ 3 and 5, what does @ represent?

14-3   MULTIPLYING PROBABILITIES

Objective

Find the probability of independent and dependent events.

Gary won third prize in New Mall's Grand Opening. He will draw a dime, not replace it, and draw another dime from the pot o' gold containing six very valuable dimes. One was minted in 1920, two in 1924, and three in 1925. What is the probability Gary will draw a 1925 dime both times?

You can multiply two probabilities to find the probability that one event *and* another event will occur. The key word is "and."

If one event is affected by the occurrence of another event, the events are **dependent**. Two events are **independent** if the occurrence of one event does not affect whether the other event occurs.

Method

1. ▶ Find the probability of the first event.
2. ▶ Find the probability of the second event.
3. ▶ Multiply the probabilities.

Example A

Suppose Gary chooses a dime, does not replace it, and draws another dime. What is the probability that he chooses a 1925 dime both times?

1. ▶ $P(1925) = \frac{3}{6}$ or $\frac{1}{2}$ These events are dependent. Since the first dime is not replaced, there are only five dimes left on the second draw. Also, we are assuming that a 1925 dime is chosen the first time. So there are only two 1925 dimes left.

2. ▶ $P(1925) = \frac{2}{5}$

3. ▶ $P(1925 \text{ both times}) = \frac{1}{2} \times \frac{2}{5} = \frac{1}{5}$

The probability that Gary will draw a 1925 dime both times is $\frac{1}{5}$.

Example B

Suppose Gary draws a dime, replaces it and draws another dime. What is the probability that he will draw a 1924 dime both times?

1. ▶ $P(1924) = \frac{2}{6}$ or $\frac{1}{3}$ These events are independent since the first dime is replaced.

2. ▶ $P(1924) = \frac{2}{6}$ or $\frac{1}{3}$

3. ▶ $P(1924 \text{ both times}) = \frac{1}{3} \times \frac{1}{3} = \frac{1}{9}$

Guided Practice

Find the probability of each event. Use the pot o' gold full of dimes. Each dime drawn is not replaced.

Example A

1. $P(\text{a 1924 dime both times})$
2. $P(\text{a 1920 dime, then a 1925 dime})$
3. $P(\text{a 1920 dime both times})$
4. $P(\text{a 1925 dime, then a 1924 dime})$

Find the probability of each event. Use the pot o' gold full of dimes on page 430. Each dime drawn is replaced.

5. P(a 1925 dime both times)

6. P(a 1920 dime both times)

7. P(a 1925, then a 1924 dime)

8. P(a 1920, then a 1924 dime)

A bag contains two white, three green, and five yellow marbles. Suppose you choose a marble and then without replacing it, choose another marble. Find the probability of choosing each event.

9. P(white, then green)

10. P(yellow, then white)

11. P(green, then blue)

12. P(white, then yellow)

13. P(white both times)

14. P(green both times)

15. P(yellow both times)

16. P(same color both times)

Suppose you choose a marble from the bag for Exercises 9–16, replace it, and then choose another marble. Find the probability of choosing each event.

17. P(white, then green)

18. P(yellow, then white)

19. P(green, then blue)

20. P(white, then yellow)

21. P(white both times)

22. P(green both times)

23. P(yellow both times)

24. P(same color both times)

Applications

25. In Mr. Morris's department, a jar contains twenty-five index cards with one name on each card. One card is drawn each month and returned to the jar. What is the probability that Mr. Morris's name will be drawn two months in a row?

26. Six members of the swim team qualified for the drawing for free swimsuits at Sport Haus. The coach put their names on a slip of paper and dropped them in a box. Two team members will receive a prize. The students' names are Linda, Lisa, David, Marianne, Shirley, and Keith. What is the probability that Lisa and David will win? Once a name is drawn, it is not replaced.

27. Use two nickels. Mark an X on each side of one. On the other mark an X on one side and a Y on the other. What is the probability of getting an XX on a toss? an XY?

28. Refer to Exercise 27. If you were playing with a partner and you get a point for every XX and your partner gets a point for every XY, would it be a fair game? Explain your answer.

Make Up a
Problem

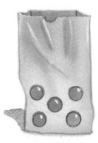

29. Look at the picture. Make up a problem using the information in the picture. Then solve the problem.

14-4 ADDING PROBABILITIES

Objective

Find the probability of one event or another.

The spinner shown below is divided into six equal regions. Every region is shaded alternately blue or red. You want to know the probability that you will spin an even number or a 5.

You can add two probabilities to find the probability that one event *or* another event will occur. The key word is "or."

Method

1 ▶ Find the probability of the first event.

2 ▶ Find the probability of the second event.

3 ▶ Add the probabilities. If the two events can occur at the same time, subtract the probability of both events occurring.

Example A

Find the probability of spinning an even number or a 5.

1 ▶ $P(\text{even number}) = P(2, 4, \text{ or } 6)$ or $\frac{3}{6}$

2 ▶ $P(5) = \frac{1}{6}$

3 ▶ $P(\text{even number or } 5) = \frac{3}{6} + \frac{1}{6}$

$= \frac{4}{6}$ or $\frac{2}{3}$

The probability of spinning an even number or 5 is $\frac{2}{3}$.

Example B

Find the probability of spinning blue or a number greater than 2.

1 ▶ $P(\text{blue}) = P(1, 3, \text{ or } 5)$ or $\frac{3}{6}$

2 ▶ $P(\text{number greater than 2}) = P(3, 4, 5, \text{ or } 6)$ or $\frac{4}{6}$

3 ▶ Add. $\frac{3}{6} + \frac{4}{6} = \frac{7}{6}$

> Remember that the probability of an event will be from 0 to 1, inclusive.

Since some of the blue numbers are greater than 2, some outcomes, namely spinning a 3 or a 5, have been counted twice. Subtract the probability of these events (blue 3 and blue 5) occurring.

$$\frac{7}{6} - \frac{2}{6} = \frac{5}{6}$$

The probability of blue or a number greater than 2 is $\frac{5}{6}$.

Guided Practice

Find each probability. Use the spinner shown above.

Example A

1. $P(2 \text{ or } 3)$ **2.** $P(1, 2, \text{ or } 3)$ **3.** $P(\text{an odd number or } 2)$

4. $P(3 \text{ or } 5)$ **5.** $P(\text{red or } 1)$ **6.** $P(\text{an even number or blue})$

Example B

Find each probability. Use the spinner shown above.

7. $P(\text{blue or a number less than 4})$ **8.** $P(\text{red or a number less than 3})$

The names of the students in Ms. Sparver's class who have qualified for the survival course are given below. Each name is written on a slip of paper and placed in a box. A slip of paper is drawn from the box. Find each probability.

CHARLES DOTTIE DREW SUKI RACHEL CATHY CORY DON

9. P(a name beginning with C or D)

10. P(a name ending with S or Y)

11. P(a 4-letter name beginning with C)

12. P(a 4-letter name beginning with S or D)

13. P(a name containing 4 letters or beginning with S)

14. P(a name containing 6 letters or beginning with R)

Find each probability. Use the spinner at the right.

15. P(green or blue)

16. P(green or red)

17. P(blue or a number greater than 5)

18. P(red or an even number)

19. P(a prime number or green)

20. P(a prime number or an odd number)

Applications

21. Sharon and Jeff are playing Travelog. Sharon rolls a die. She needs to roll a 3 or a 5 to win. What is the probability Sharon will roll a 3 or a 5?

Suppose

22. Suppose you make a spinner like the one used in Exercises 15–20, except that you place the four wedges of each color next to each other. Does this arrangement change the probability of spinning any one of the colors from the original probability? Explain.

23. Refer to Exercise 22. Does the new arrangement of colors affect the probability of spinning a blue or a number greater than 5? Explain.

Cooperative Groups

24. In small groups, copy the chart from Exercises 21–28 on page 283. Then find each probability.

a. P(doubles or a sum of 7)

b. P(a sum of 8 or sum of 9)

c. Now set up a table like the one shown at the right. Roll two dice 36 times and record your results. Find the probability of 24a and 24b using the results from your table. Compare your experimental probability with the probability you found in 24a and 24b.

Roll #	Result
1	?
2	?
⋮	⋮

SIMULATIONS

Computers are often used to simulate real-life situations. Sometimes pilots use computer simulations to practice maneuvers without endangering life and equipment. Scientists and engineers use simulations to test materials under extreme temperature and weight conditions.

The BASIC program below simulates an experiment in probability. Type it into a computer. To stop the output, push CTRL-C or BREAK.

```
100 FOR N=1 TO 10
200 IF RND (1) < 0.5 THEN GO TO 500
300 PRINT "-";
400 GOTO 600
500 PRINT "H";
600 NEXT N
700 PRINT
800 GOTO 100
RUN
H-HHHHHH-H
----H---HH
-H--H-----
-H--HHH-HH
H-HH--HHHH
HH--H-----
H-H-H--H-H
```

The command RND(1) tells the computer to choose a number between 0 and 1 at random.

Your output may look similar to this output.

Each row shows ten flips of a coin. The computer prints "H" for heads and "-" for tails.

Use the program and output shown above. Solve.

1. If you flip a coin ten times, is it possible to get nine heads in a row? How often would it happen?

2. If you flip several tails in a row, will the next few tosses be heads to make up for many tails?

3. Each row in the output shown above shows the outcomes for ten flips of a coin. In the first row, the experimental probability of tossing a tail is $\frac{1}{5}$ and a head is $\frac{4}{5}$. Will the experimental probability ever match the theoretical probability?

CAR RENTAL AGENCY MANAGER

Ramon Perez manages a car rental agency. One of his job responsibilities is to make sure that each car is in proper running condition. Another is to compute the rental charges for customers.

The charge for a mid-size car is $29 a day and 28¢ per mile. Tax is an additional 5.5%. If a customer keeps a car for three days and drives it 286 miles, what is the rental charge?

$29 \boxed{\times} 3 \boxed{+} 286 \boxed{\times} .28 \boxed{=} 167.08$

The rental charge for the mid-size car is $167.08.

To determine the total including tax, multiply $167.08 by 1.055. (100% + 5.5% = 105.5% or 1.055)

167.08 is already showing in the calculator.

$\boxed{\times} 1.055 \boxed{=} 176.2694$

The total cost including tax is $176.27

In Exercises 1–3, the number of miles and days are given. Use $29 a day and 22¢ a mile to find the rental charge. Assume there is no tax.

1. 113 miles, 2 days   **2.** 1,297 miles, 4 days   **3.** 853 miles, 3 days

4. Mrs. Gandi rented a compact car for 5 days at $19.95 a day and 19¢ a mile. When she rented the car, the mileage was 44,264.1. When she returned the car, the mileage was 45,179.5. What is the rental charge? Assume there is no tax.

5. Tom Bellisimo rented a full-size car for 2 weeks at $131.25 a week and 23¢ for every mile over 250. If Tom drove 1,987.3 miles and the tax is 5.5%, what was the total cost of renting the car? What was the rental charge?

6. Jennifer Norris needs to rent a car for a 5-day business trip. The trip is 1,248.5 miles. Action Rental charges $31 a day and 17.5¢ a mile. Fulton Rental charges $29.95 a day and 20¢ a mile. Which rental agency offers the better buy for this trip?

14-5 MARKETING RESEARCH

Objective

Find probabilities from market samples.

Joanie Mock is a marketing research assistant. She collects data on the interests and tastes of people who use products that her company makes and sells. From the data she collects, her company decides which products to make, how many to make, how to package the products, and how to advertise the products.

Ms. Mock's company, National Interior Design, wants to know which colors and patterns of wallpaper are liked best. Of course, every person cannot be questioned. A smaller, representative group called a **sample** is surveyed. A sample is representative when it has the same characteristics as the larger group. The results of Ms. Mock's sample are shown below.

Favorite Color	Number of People
Black	11
Blue	41
Green	25
Red	34
Yellow	39

Favorite Pattern	Number of People
Florals	46
Small prints	24
Solid	18
Stripes	62

Understanding the Data

- What is the total number of people surveyed?
- What color is liked by the most people?
- What pattern is liked by the most people?
- What color and pattern is liked by the least number of people?

Check for Understanding

Use the data from the tables above.

1. How many people don't prefer green?
2. How many people like small prints or stripes?
3. How many people don't prefer stripes?
4. What is the probability of a person liking green?
5. Last year, Ms. Mock's sample showed the following: florals, 28; small prints, 44; solids, 23; and stripes, 55. Based on the change in the sample, should Ms. Mock recommend adding a new stripe or a new floral design?

Find the probability of a person liking each of the following colors best. Use the survey on page 436.

Practice

6. blue

7. not yellow

8. neither red nor blue

Find the probability of a person liking each pattern best.

9. florals

10. solid or florals

11. not stripes

Assuming that the choices of color and pattern are independent, find the probability of a person liking each of the following color and pattern combinations.

12. black and florals

13. red and stripes

14. blue and solid

15. small prints and *not* yellow

16. blue and *not* stripes

Applications

17. Brad surveyed 50 grocery store customers. Tastee fishsticks were favored by 26 of the 50 customers. What is the probability that a customer favors Tastee?

18. The table below shows the results of surveying 40 people about their favorite brand of shampoo and conditioner.

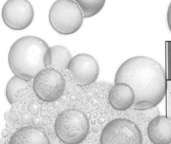

Favorite Shampoo	Number of People	Favorite Conditioner	Number of People
Jazz	21	Free	18
Fresh	10	Silken Mist	15
Potpourri	9	Gloss-on	7

a. What is the probability that a person likes Fresh and Free? Assume that the choices are independent.

b. Would the results of this survey affect the brand of shampoo the store should sell?

JOURNAL ENTRY

19. Explain how probability can be used in marketing research.

Critical Thinking

20. What is one half of 2^{60}?

Mixed Review

Lesson 10-5

Estimate.

21. 16% of 80

22. 25% of 121

23. 62% of 364

Lesson 13-3

24. Claudia bought 4 turtlenecks. She paid $12 cash and wrote a $20 check for the remaining amount. How much was each turtleneck?

14-6 ODDS

Objective

Find the odds for an event.

The student committee for the Read-a-thon fund raiser consists of three freshmen, four sophomores, a junior, and four seniors. Each student's name is placed in a hat and one name is chosen to be the chairperson of the committee. What are the odds that a freshman will be chosen?

The ratio of the number of ways an event can occur to the number of ways the event cannot occur is called the **odds** for the event.

$$\text{odds} = \frac{\text{number of ways an event can occur}}{\text{number of ways the event cannot occur}}$$

Method

1 ▶ Find the number of ways the event can occur.

2 ▶ Find the number of ways the event cannot occur.

3 ▶ Write the ratio of 1 ▶ to 2 ▶

Example A

Find the odds that a freshman will be chosen.

1 ▶ There are 3 ways a freshman could be chosen.

2 ▶ There are 9 ways a freshman could not be chosen.

3 ▶ The odds are $\frac{3}{9}$ or 3 to 9.

The odds are 3 to 9 that a freshman will be chosen.

Example B

Find the odds that a freshman or a sophomore will be chosen.

1 ▶ There are 7 students who are freshmen or sophomores.

2 ▶ There are 5 students who are not freshmen or sophomores.

3 ▶ The odds are $\frac{7}{5}$ or 7 to 5 that a freshman or a sophomore will be chosen.

Guided Practice

A die is rolled. Find the odds for each roll.

1. a 1   **2.** a 4   **3.** a 3   **4.** a 7

5. an odd number   **6.** an even number

7. a prime number   **8.** a multiple of 3

9. a number less than 3   **10.** a multiple of 2

11. a 2 or a 6   **12.** a 3 or a 5   **13.** a 1, 2, 4, or 6

14. a 3 or an even number   **15.** a 1 or a 7

16. a 6 or an odd number   **17.** a 1 or a number greater than 4

18. a 6 or a number less than 4   **19.** a 1 or a prime number

Exercises

Practice

Two dice are rolled. Find the odds for each roll.

20. a sum of 7   **21.** a sum of 10   **22.** a sum of 2

23. *not* a sum of 6   **24.** *not* a sum of 3   **25.** *not* a sum of 4

26. a sum of 8 with a 6 on one die

27. an odd sum or a sum less than 7

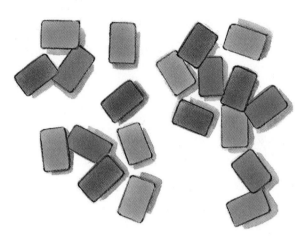

Suppose you choose a card from those shown at the left. Find the odds for each event. Then find the probability.

28. a blue card

29. a purple card

30. an orange card

31. *not* a purple card

32. *not* a blue card

33. a purple card or a blue card

34. *neither* a purple *nor* a blue card

Applications

35. Two coins are tossed. Find the odds for tossing at least one tail.

36. Jason's coach told him the odds that he will qualify for the state finals are 4 to 1. What is the probability that Jason will qualify?

37. For the gumball machine in Parker's Drugstore, the odds of getting a prize with the gum are 5 to 31. What is the probability of getting a prize?

Collect Data

38. Read an article about some sporting events that gives the odds of winning. Why is sports a common application of odds?

Mental Math

Subtracting from 100 can be done mentally as shown below.

Find 100 − 24.   **Find 100 − 53.**

$100 - 24 = \underbrace{100 - 20} - 4$ $100 - 53 = \underbrace{100 - 50} - 3$

$ = 80 - 4$ Think. $ = 50 - 3$ Think.

$ = 76$ $ = 47$

Find each difference.

39. 100 − 42   **40.** 100 − 65   **41.** 100 − 33

42. 100 − 79   **43.** 100 − 54   **44.** 100 − 22

14-7 USING SAMPLES TO PREDICT

Objective

Solve verbal problems using samples to make predictions.

Johnson's Department Store conducts a survey of 75 shoppers in the mall and asks if they have a Johnson's credit card. Thirty of the 75 shoppers say that they have one. If about 3,200 people shop in the mall each day, predict how many of them probably have a Johnson's credit card.

Predictions about a large group of people based on the choices of a sample population are *estimates*. The best estimates are made when the sample population is representative of the larger group. For example, if you stand at the door of Johnson's Department Store to conduct the survey described above, the sample results would probably not be representative.

▶ **Explore**

What is given?
● There are about 3,200 shoppers each day.
● In the sample of 75 shoppers, 30 have a credit card.

What is asked?
● How many shoppers in the mall probably have a Johnson's credit card?

▶ **Plan**

Use the sample of 75 shoppers. Let n represent the total number of shoppers who probably have a Johnson's credit card. The ratio of n to 3,200 should be equivalent to the ratio of 30 to 75. Set up a proportion and solve for n.

▶ **Solve**

$$\frac{n}{3,200} = \frac{30}{75}$$

$$3200 \boxed{\times} 30 \boxed{\div} 75 \boxed{=} 1280$$

The number of shoppers that probably have a credit card is *about* 1,280.

▶ **Examine**

30 out of 75 or less than half of the shoppers in the sample say they have a Johnson's credit card. Since half of 3,200 is 1,600, it seems reasonable to predict that about 1,280 shoppers have a Johnson's credit card.

Guided Practice

1. A sample of 100 seniors at Madison High School showed that 72 expect to continue their education after graduation. If there are 700 students in the class, predict the number of students who will continue their education.

2. A cashier in a supermarket noticed that 3 out of her first 20 customers had at least three coupons. Based on this sample, how many customers out of 100 had *less* than three coupons?

Paper/ Pencil
Mental Math
Estimate
Calculator

Sharp Cable Company has 4,800 customers. Use the results of a customer preference survey shown below for Exercises 3–6.

Which type of TV program do you prefer?	
News	5
Sports	18
Drama	7
Comedy	10

3. What is the probability that a person prefers comedy programs?

4. Predict how many of the 4,800 cable customers prefer comedy programs.

5. What is the probability that a person prefers news programs?

6. Predict how many of the 4,800 customers prefer news programs.

7. In a survey of 200 residents of Westville, 80 use Shine-on toothpaste. What is the probability that a Westville resident uses Shine-on toothpaste?

8. Review the items in your portfolio. Make a table of contents of the items, noting why each item was chosen. Your portfolio should include a variety of assignments and reflect your progress over the year. Replace any items that are no longer appropriate.

9. The choices of a large group may not agree with a prediction. Write two reasons why the actual choices may not agree with a prediction.

Critical Thinking

10. Arrange eight congruent equilateral triangles to form a parallelogram.

Mixed Review

Find the area of each circle whose radius is given. Use 3.14 for π. Round decimal answers to the nearest tenth.

Lesson 8-4

11. 7 cm   **12.** 8 in.   **13.** 24 mm   **14.** 3.5 cm

Lesson 11-2

Solve. Use data at the right.

15. Find the mean.

16. Find the median.

17. Find the mode.

Golf Team Scores				
49	42	38	38	43
38	36	50	37	

**Vocabulary/
Concepts**

*Choose a word from the list at the right
that best completes each sentence.*

1. The probability of an event is a _?_ that describes
how likely it is that the event will occur.

2. Two events are _?_ if the occurrence of one
event does not affect the occurrence of the other event.

3. The ratio of the number of ways an event can occur to
the number of ways the event cannot occur is called
the _?_ for the event.

4. An _?_ is a possible result.

5. A _?_ diagram can be used to determine the
possible outcomes.

6. An _?_ is a specific outcome or type of outcome.

7. Two events are _?_ if the occurrence of one of
the events affects the occurrence of the other event.

dependent
equally likely
event
independent
odds
outcome
random
ratio
replaced
tree

**Exercises/
Applications**

Lesson 14-1

*Use multiplication to find the
number of possible outcomes.
Then draw a tree diagram to
show the possible outcomes.*

8. a choice of a Probe or a Beretta in
blue, black, or white

9. a choice of a green, brown, or
black skirt with a choice of a
white, tan, or yellow blouse

Lesson 14-2

*A box contains three red chips, five black chips, and two
blue chips. One chip is chosen. Find the probability that
each event will occur. Express as a percent.*

10. a red chip
11. a black chip
12. a blue chip

13. *not* a red chip
14. *not* a black chip
15. *not* a blue chip

16. a white chip
17. *not* a white chip
18. a red or a blue chip

Lesson 14-3

*A box contains six red, eight green, and two blue marbles.
Find the probability of each event.*

19. draw a red marble, replace it, draw
a blue marble

20. draw a green marble, replace it,
draw a red marble

21. draw a green marble, do not
replace it, draw a blue marble

22. draw a blue marble, do not replace
it, draw a red marble

Lesson 14-4

The following words are written on slips of paper and placed in a box.
PEPPER   PARSNIP   CARROT   CABBAGE   CUCUMBER
A slip of paper is chosen from the box. Find each probability.

23. P(a word beginning with P or C)   **24.** P(a word ending in R or P)

25. P(a 6-letter or an 8-letter word)   **26.** P(a word beginning with B or D)

Lesson 14-5

Know-it-All Marketing surveyed 150 people for Better Bakers about the flavors they like best. Find the probability of a person liking each flavor or combination of flavors.

Favorite Flavor	Number of People
Vanilla	45
Chocolate	46
Cherry	32
Peach	20
Raspberry	7

27. vanilla   **28.** chocolate

29. cherry or peach   **30.** chocolate or raspberry

Lesson 14-6

A box contains six red, eight green, and two blue marbles. Find the odds for choosing each event.

31. *not* a green marble   **32.** a red or green marble

Lesson 14-7

33. In a sample of 50 car owners, 35 said they prefer to have regular maintenance performed by the dealer from whom they bought the car. Predict how many car owners out of 1,000 will prefer to have the dealer perform regular maintenance.

Paper/ Pencil
Mental Math
Estimate
Calculator

34. On the average Miss Chung, a sales representative for a publisher, sells a set of encyclopedias to 3 out of every 25 households she visits. If she visits 200 households, how many sets should she expect to sell?

35. Mr. Crady said the odds that the University of Kansas would win the NCAA basketball tournament are 1 to 2. What is the probability that the University of Kansas will win?

36. Andy Hayes and Becky Girard went to Paradise Cafe for lunch. The luncheon special has an entrée and 3 choices for a side dish. There are also some dessert choices. If there is a total of 18 different meals, how many dessert choices are there?

TEST

A die is rolled once and the spinner shown at the right is spun once.

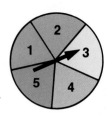

1. Use multiplication to find the number of possible outcomes.
2. Draw a tree diagram to show the possible outcomes.
3. If you roll three dice instead of one and spin the spinner once, how many different outcomes are possible?

The letters in the word PAPER are written on slips of paper and placed in a box. Find the probability of each draw.

4. draw an A, replace it, draw an R
5. draw an E, replace it, draw a P
6. draw a P, do not replace it, draw an A
7. draw a vowel both times without replacing the first one

Use the letters placed in the box from Exercises 4–7 above. One slip is drawn from the box. Find the probability that each event will occur. Express as a percent.

8. a P
9. not an R
10. a P or a vowel
11. an S
12. a P or an E
13. an R or a consonant

14. A survey of 200 registered voters in Fairfield showed 110 voters favor Jenkins for mayor. Predict how many will vote for Jenkins if 26,780 registered voters cast their votes.

A jar contains four white buttons, two red buttons, and three brown buttons. Find the odds for choosing each of the following.

15. a brown button
16. a white button
17. a red button
18. *not* a white button
19. a white or a brown button
20. *neither* a white *nor* a brown button

21. The odds that the Vikings will win are 2 to 3. Find the probability of the Vikings' winning.

22. A nickel, a dime, and a quarter are tossed. Find the probability of obtaining all tails.

23. In a survey of 300 voters in Huntsville, 120 favor passing the school levy. What is the probability that a voter favors passing the school levy?

24. A poll of 500 households in Eastown shows that 245 households watch the local news at 6:00 P.M. Predict how many of the 12,700 households in Eastown watch the local news at 6:00 P.M.

25. A box contains four balls numbered 1 through 4. The balls are drawn from the box one at a time and not replaced. In how many possible orders can they be drawn?

 BONUS: Write a situation represented by the tree diagram.

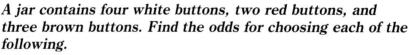

MARKUP

Tyson Bodey owns a sporting goods store. He sells most of the items in his store with at least a 50% markup. The amount Mr. Bodey pays for an item in his stock is the cost of an item. The **markup** is the amount added to the cost to obtain the selling price. The markup is usually a percent of the cost.

When Mr. Bodey sells an item, the markup is income for the store. This income is used to pay the expenses of operating the store, such as rent, salaries, insurance, and so on. A markup is necessary for a business to succeed.

Mr. Bodey pays \$16.50 for a volleyball and the markup is 50% of the cost. Find the selling price.

The selling price is found by adding the amount of the markup to the cost.

 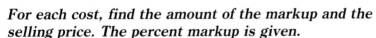

The selling price of the volleyball is \$24.75.

For each cost, find the amount of the markup and the selling price. The percent markup is given.

	Cost	% of Markup	Markup	Selling Price
1.	\$25	25%	▨	▨
2.	\$9.30	40%	▨	▨
3.	\$3.95	100%	▨	▨

4. The markup on a television is $\frac{1}{3}$ of the cost. If the cost of the television is \$345, find the markup and the selling price.

5. The selling price of a sweatshirt is \$21.75. If the cost is \$15, what is the markup? What is the percent of markup?

6. If the selling price of a football is \$27.70 and the markup is 55% of the cost, what is the cost of the football? Round to the nearest cent.

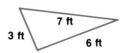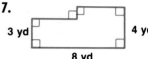

Free Response

Lesson 1-4

Round each number to the underlined place-value position.

1. 5,0<u>0</u>9 **2.** 54,<u>6</u>21 **3.** 30,<u>0</u>82 **4.** 9,<u>9</u>09

Lesson 2-8

Find the perimeter of each figure.

5.

2 m
3.7 m

6.

7 ft
3 ft
6 ft

7.

3 yd
4 yd
8 yd

Lesson 3-8

Write in standard form.

8. 4.6×10^5 **9.** 7.01×10^3 **10.** 9×10^4 **11.** 2.83×10^2

Lesson 5-8

Change each fraction to a decimal. Use bar notation to show a repeating decimal.

12. $\frac{1}{5}$ **13.** $\frac{2}{3}$ **14.** $\frac{5}{8}$ **15.** $\frac{1}{6}$ **16.** $\frac{8}{9}$

Lesson 7-10

17. The number 13 is to 169 as 18 is to what number?

Lesson 9-5

18. Darnell is 6 ft tall and his shadow is 15 ft long. Mark is standing next to him, and his shadow is 13 ft long. How tall is Mark? Round to the nearest inch.

Lesson 11-8

Find the median and mode of the data in each stem-and-leaf-plot.

19.

stem	leaf
1	2 3
2	4 6 6 7

20.

stem	leaf
2	7 8 8 9
3	1 2 3 5

Lessons 13-1 13-2

Solve each equation.

21. $x + 3 = -6$ **22.** $4x = -20$

23. $-5p = 10$ **24.** $-3 = y + 1$

25. $\frac{x}{7} = -2$ **26.** $\frac{y}{20} = -\frac{3}{4}$

Lesson 14-3

A drum contains forty balls, numbered 1 through 40. All of the even-numbered balls are red. The odd-numbered balls are blue. One ball is chosen. Find each probability.

27. P (red) **28.** P (a multiple of 7)

29. P (*neither* 7 *nor* 11) **30.** P (*not* a 13)

Multiple Choice

Choose the letter of the correct answer for each item.

1. Find the value of $3 \times 4^2 - 2$.

  a. 10   **c.** 46

  b. 42   **d.** 142

Lesson 1-9

2. Mr. Kasten bought a 13.5-pound turkey for $7.83. What was the price per pound?

  e. $0.17

  f. $0.58

  g. $0.60

  h. $1.73

Lesson 3-5

3. Twins are born weighing $5\frac{1}{2}$ pounds and $6\frac{3}{8}$ pounds. What is their total weight?

  a. $11\frac{2}{5}$ pounds

  b. $11\frac{7}{16}$ pounds

  c. $11\frac{1}{2}$ pounds

  d. $11\frac{7}{8}$ pounds

Lesson 4-11

4. Which measurement is equivalent to 750 mL?

  e. 0.75 L   **g.** 7.5 L

  f. 0.75 kL   **h.** 7.5 kL

Lesson 6-5

5. How many grams of vitamin granules should Jerry mix with his horse's food if the correct amount is 250 mg for every 50 pounds of the horse's weight? The horse weighs 1,500 pounds.

  a. 7.5 g   **c.** 7,500 g

  b. 6 g   **d.** 6,000 g

Lesson 9-3

6. Solve the equation $x + 12 = -4$.

  e. 8   **g.** -8

  f. -16 **h.** 20

Lesson 13-1

7. Use the bar graph. How many runs were scored by the Braves?

  a. 3

  b. 12

  c. 15

  d. 21

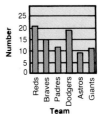

Runs Scored in 1 Week

Lesson 11-4

8. A $10.50 hat is on sale for 20% off. What is the sale price?

  e. $8.40

  f. $10.30

  g. $10.80

  h. $12.60

Lesson 10-7

9. Jim makes 5 out of every 8 free throws. What is the probability that he will make a free throw?

  a. 60%   **c.** 56%

  b. 37.5%   **d.** 62.5%

Lesson 14-2

10. If you buy 3 gallons of paint for $12.95 a gallon and then get a fourth gallon free, what is the percent of discount on 4 gallons of paint?

  e. 18%   **g.** 35%

  f. 25%   **h.** 50%

Lesson 10-7

EXTRA PRACTICE

Lesson 1-1   *Replace each ▓ with a number or a word to make a true sentence.*

1. 40 = ▓ tens

2. 600 = ▓ hundreds

3. 7,000 = ▓ ones

4. 9,000 = ▓ hundreds

5. 800 = ▓ tens

6. 80 = ▓ tens

7. 40,000 = ▓ thousands

8. 2,000 = ▓ ones

9. 500 = ▓ hundreds

10. 700,000 = ▓ ten thousands

11. 60,000 = ▓ ten thousands

12. 0.15 = ▓ hundredths

13. 0.2705 = ▓ ten-thousandths

14. 0.064 = ▓ thousandths

15. 3.47 = ▓ and ▓ hundredths

16. 8.0234 = ▓ and ▓ ten-thousandths

17. 56.071 = ▓ and ▓ thousandths

18. 0.13 = 13 ▓

19. 0.045 = 45 ▓

20. 2.053 = 2 ▓ 53 ▓

21. 9.0042 = 9 ▓ 42 ▓

22. 57.003 = 57 ▓ 3 ▓

Lesson 1-2   *Find the number named.*

1. 5^2

2. 7^2

3. 2^4

4. 10^3

5. 12^2

6. 2^5

7. 9^2

8. 3^3

9. 15^2

10. 10^5

11. 5^3

12. 13^1

Write using exponents.

13. $8 \times 8 \times 8$

14. 3×3

15. $2 \times 2 \times 2 \times 2 \times 2$

16. 20

17. 9×9

18. $6 \times 6 \times 6$

19. 11 squared

20. 2 cubed

Lesson 1-3   *Replace each ● with <, >, or = to make a true sentence.*

1. 0.36 ● 0.63

2. 1.74 ● 1.7

3. 4.03 ● 4.003

4. 0.06 ● 0.066

5. 10.5 ● 10.05

6. 3.0 ● 3

7. 5.632 ● 5.623

8. 0.423 ● 0.5

9. 2.020 ● 2.202

10. 0.93 ● 0.9

11. 0.205 ● 0.025

12. 0.46 ● 0.49

13. 13.100 ● 13.1

14. 0.062 ● 0.62

15. 9.99 ● 9.099

16. 0.030 ● 0.03

17. 20 ● 19

18. 79 ● 97

19. 101 ● 98

20. 3,654 ● 3,645

21. 5,274 ● 5,638

22. 4,951 ● 5,307

23. 8,674 ● 8,659

24. 65,870 ● 65,834

25. 36,924 ● 36,851

26. 59,430 ● 59,438

27. 72,619 ● 73,835

28. 94,367 ● 92,538

29. 274,910 ● 274,605

Lesson 1-4   *Round each number to the underlined place-value position.*

1. 1̲8   2. 2̲3   3. 3̲3   4. 5̲6   5. 9̲8   6. 1̲35

7. 8̲34   8. 9̲23   9. 1,2̲01   10. 3,1̲25   11. 6,3̲46   12. 9,9̲82

13. 110,7̲23   14. 14,6̲79   15. 25,6̲20   16. 33,4̲40   17. 38,4̲71   18. 9̲2,187

19. 0.7̲28   20. 1.4̲75   21. 3.2̲81   22. 5.81̲2   23. 6̲.777   24. 0.99̲5

25. 0.07̲9   26. 0̲.821   27. 16.5̲81   28. 21.6̲31   29. 19.7̲41   30. 9.6̲10

Write the name for each number in words.

31. 485   32. 10.123   33. 1.53   34. 673.2   35. 4.03

36. 19,002   37. 9.0427   38. 58.71   39. 62.501   40. 1.001

Lesson 1-6   *Estimate.*

1.	2.	3.	4.	5.	6.
36 + 42	56 + 89	62 − 24	91 − 75	300 + 238	539 + 251

7.	8.	9.	10.	11.	12.
14.5 4.2 + 5.3	5.63 8.5 + 2.149	1.63 7.4 + 0.219	0.92 0.6 + 0.475	$5.26 − 3.95	17.3 − 5.86

13. 5,406 + 788   14. 4,511 − 653   15. 8,783 + 9,963

16. 19,023 + 4,449   17. 23,861 − 7,139   18. 4,232 − 1,776

19. 63,792 − 43,065   20. 25,565 + 19,536   21. 9,639 − 183

22. 14.9 − 9.4   23. 63.3 − 18.46   24. 0.473 − 0.29

25. $4.19 + $3.90 + $1.75   26. 93.9 − 62   27. $40 − $6.80

Lesson 1-7   *Estimate.*

1.	2.	3.	4.	5.	6.
25 × 18	92 × 125	86 × 325	123 × 84	631 × 29	856 × 51

7.	8.	9.	10.	11.	12.
383 × 121	456 × 702	1,321 × 412	3,502 × 1,619	8,829 × 7,312	9,900 × 331

13. 63 × 5   14. 93 × 46   15. 68 × 582   16. 986 × 31   17. 5,463 × 56

18. 74 × 8,561   19. 463 × 538   20. 965 × 349   21. 3,496 × 580   22. 249 × 5,892

23. 636 ÷ 19   24. 456 ÷ 26   25. 682 ÷ 65   26. 935 ÷ 42   27. 858 ÷ 88

28. 2,349 ÷ 23   29. 5,465 ÷ 38   30. 1,980 ÷ 82   31. 7,562 ÷ 64   32. 9,642 ÷ 55

33. 833 ÷ 15   34. 9,621 ÷ 120   35. 8,121 ÷ 61   36. 6,423 ÷ 56   37. 8,327 ÷ 72

38. 439 ÷ 63   39. 8,444 ÷ 131   40. 9,731 ÷ 82   41. 7,176 ÷ 43   42. 8,133 ÷ 99

Lesson 1-9   *Find the value of each expression.*

1. $12 + 3 - 6$ 2. $25 + 2 - 3$ 3. $19 - 6 + 7$ 4. $24 - 19 + 8$

5. $4 \times 3 \div 2$ 6. $18 + 4 \div 2$ 7. $23 - 7 \times 3$ 8. $3 \times 12 + 4$

9. $4 \div 1 + 10$ 10. $17 + 7 \times 4$ 11. $19 \times 2 + 7$ 12. $16 \times 4 \div 8$

13. $16 + 2 - 4$ 14. $5 \times 2 + 3$ 15. $15 \div 3 - 2$ 16. $3 \times 2 \div 3$

17. $5 \times 6 \div 10 + 1$ 18. $12 + 6 \div 3 - 5$ 19. $7 - 2 \times 8 \div 4$ 20. $27 \div 9 \times 2 + 6$

21. $36 \div 9 \times 2 \div 4$ 22. $6 \times 7 \div 3 \div 2$ 23. $5^2 - 6 \times 2$ 24. $17 - 2^3 + 5$

25. $27 \div 3^2 \times 2$ 26. $4 + 4^2 \times 2 - 8$ 27. $12 \div 3 - 2^2 + 6$ 28. $3^3 \times 2 - 5 \times 3$

29. $(10 \times 2) + (7 \times 3)$ 30. $(14 \div 7) - (12 \div 6)$ 31. $3 \times [4 \times (7 + 1)]$

32. $8 \times [7 + (5 - 1) \div 2]$ 33. $10 \times [(4 + 1) \times 6 \div 2]$ 34. $6 \times [7 \times (4 \times 4) + 2]$

Lesson 1-10   *Write an expression for each phrase.*

1. 12 more than a number

2. 3 less than a number

3. a number divided by 4

4. a number increased by 7

5. a number decreased by 12

6. 8 times a number

7. the product of a number and 2

8. a number plus 13

9. a number subtracted from 21

10. 15 divided by a number

11. 28 multiplied by m

12. the sum of a number and 33

13. 54 divided by n

14. 18 increased by y

15. q decreased by 20

16. n times 41

17. the product of 17 and b

18. c subtracted from 36

Lesson 1-11   *Solve each equation.*

1. $q - 7 = 7$ 2. $g - 3 = 10$ 3. $b + 7 = 12$ 4. $a + 3 = 15$

5. $r - 3 = 4$ 6. $t + 3 = 21$ 7. $s + 10 = 23$ 8. $7 + a = 10$

9. $5 + a = 8$ 10. $14 + m = 24$ 11. $9 + n = 13$ 12. $13 + v = 31$

13. $s - 0.4 = 6$ 14. $x - 1.3 = 12$ 15. $y + 3.4 = 18$ 16. $z + 0.34 = 3.1$

17. $0.013 + h = 4.0$ 18. $63 + f = 71$ 19. $7 + g = 91$ 20. $19 + j = 29$

21. $z - 12.1 = 14$ 22. $w - 0.1 = 0.32$ 23. $v - 18 = 13.7$ 24. $r - 12.2 = 1.3$

25. $s + 1.3 = 18$ 26. $t + 3.43 = 7.4$ 27. $x + 7.4 = 23.5$ 28. $y - 7.1 = 6.2$

29. $p + 3.1 = 18$ 30. $q - 2.17 = 21$ 31. $g - 0.12 = 7.1$ 32. $h + 15 = 18.4$

33. $j - 3 = 7.4$ 34. $k - 6.23 = 8$ 35. $m + 6 = 10.01$ 36. $n - 0.05 = 23$

Lesson 1-12   *Solve each equation.*

1. $4x = 36$ **2.** $3y = 39$ **3.** $4z = 16$ **4.** $9w = 54$

5. $2m = 18$ **6.** $42 = 6n$ **7.** $72 = 8k$ **8.** $20r = 20$

9. $420 = 5s$ **10.** $325 = 25t$ **11.** $14 = 2p$ **12.** $18q = 36$

13. $40 = 10a$ **14.** $100 = 20b$ **15.** $416 = 4c$ **16.** $45 = 9d$

17. $2g = 0.6$ **18.** $3h = 0.12$ **19.** $5k = 0.35$ **20.** $12x = 144$

21. $\frac{x}{6} = 6$ **22.** $\frac{z}{7} = 8$ **23.** $\frac{c}{10} = 8$ **24.** $\frac{x}{2} = 4$

25. $\frac{m}{7} = 5$ **26.** $\frac{n}{3} = 6$ **27.** $\frac{p}{4} = 4$ **28.** $\frac{r}{5} = 5$

29. $\frac{s}{9} = 8$ **30.** $\frac{t}{5} = 6$ **31.** $\frac{w}{7} = 8$ **32.** $\frac{c}{8} = 2$

Lesson 2-1   *Add.*

1. $\begin{array}{r} 38 \\ + 21 \end{array}$ **2.** $\begin{array}{r} 836 \\ + 74 \end{array}$ **3.** $\begin{array}{r} 64 \\ + 718 \end{array}$ **4.** $\begin{array}{r} 439 \\ + 199 \end{array}$ **5.** $\begin{array}{r} 834 \\ + 689 \end{array}$ **6.** $\begin{array}{r} 593 \\ + 152 \end{array}$

7. $\begin{array}{r} 4{,}762 \\ + 5{,}851 \end{array}$ **8.** $\begin{array}{r} 5{,}967 \\ + 533 \end{array}$ **9.** $\begin{array}{r} 72{,}752 \\ + 15{,}601 \end{array}$ **10.** $\begin{array}{r} 6{,}120 \\ + 19{,}932 \end{array}$ **11.** $\begin{array}{r} 83{,}496 \\ + 815 \end{array}$

12. $26 + 85$ **13.** $65 + 305$ **14.** $438 + 226$

15. $336 + 986$ **16.** $2{,}382 + 4{,}609$ **17.** $6{,}549 + 385$

18. $83{,}402 + 74{,}379$ **19.** $2{,}367 + 40{,}493$ **20.** $30{,}260 + 979$

21. $4{,}301 + 7{,}812$ **22.** $3{,}111 + 4{,}222$ **23.** $18{,}123 + 850$

24. $42{,}017 + 287$ **25.** $51{,}007 + 987$ **26.** $25{,}435 + 720$

Lesson 2-2   *Add.*

1. $\begin{array}{r} 0.46 \\ 0.72 \\ + 0.81 \end{array}$ **2.** $\begin{array}{r} 13.7 \\ 2.6 \\ + 4.9 \end{array}$ **3.** $\begin{array}{r} 48.3 \\ 0.91 \\ + 9.85 \end{array}$ **4.** $\begin{array}{r} \$ 4.68 \\ 23.99 \\ + 17.10 \end{array}$ **5.** $\begin{array}{r} 15 \\ 6.02 \\ + 3.8 \end{array}$ **6.** $\begin{array}{r} 8.9 \\ 12 \\ + 0.38 \end{array}$

7. $5.61 + 0.09$ **8.** $0.38 + 2.46 + 0.19$ **9.** $5.9 + 0.45 + 1.13$

10. $\$14.70 + \$3.65 + \$2.40$ **11.** $17.9 + 18$ **12.** $4.075 + 3.6 + 0.08$

13. $3.06 + 0.17 + 0.097$ **14.** $38.786 + 14.5$ **15.** $7.92 + 0.792 + 9.2$

16. $0.46 + 0.5 + 5$ **17.** $14 + 7.41 + 0.6$ **18.** $0.446 + 44 + 6.44$

19. $0.51 + 1.3 + 6$ **20.** $18 + 9.32 + 2.1$ **21.** $6.7 + 3.2 + 5.05$

22. $8.5 + 3.1 + 0.01$ **23.** $19.2 + 7.36 + 4.2$ **24.** $18 + 0.7 + 16.5$

Lesson 2-3   *Subtract.*

| 1. | 38
 − 25 | 2. | 436
 − 178 | 3. | 600
 − 274 | 4. | 381
 − 56 | 5. | 515
 − 288 | 6. | 642
 − 375 |

| 7. | 4,963
 − 575 | 8. | 8,902
 − 4,266 | 9. | 60,202
 − 9,786 | 10. | 36,436
 − 8,718 | 11. | 76,814
 − 27,925 |

12. 57 − 38   13. 706 − 88   14. 531 − 64

15. 413 − 326   16. 633 − 208   17. 8,920 − 465

18. 3,636 − 1,597 19. 7,437 − 5,548 20. 46,462 − 4,483

21. 5,877 − 4,186 22. 12,145 − 387   23. 19,123 − 6,421

24. 23,411 − 3,418 25. 43,761 − 21,421 26. 56,776 − 32,411

27. 83,976 − 82,135 28. 95,163 − 4,832 29. 123,411 − 95,109

Lesson 2-4   *Subtract.*

| 1. | 0.89
 − 0.65 | 2. | 7.3
 − 4.9 | 3. | 0.836
 − 0.75 | 4. | 6.3
 − 4.28 | 5. | 7
 − 2.36 |

6. 4.91 − 2.32   7. 17.83 − 0.24   8. 0.763 − 0.49   9. 2.63 − 1.8

10. 0.563 − 0.08   11. 2.87 − 1.965   12. 5.19 − 0.238   13. 0.8 − 0.526

14. $30 − $14.16   15. 73 − 0.45   16. 63.6 − 48.48   17. 155 − 6.78

18. 24.96 − 8.088   19. $12 − $2.39   20. 15.9 − 0.999   21. $138 − $19.88

22. 19 − 3.076   23. 42.6 − 18.7   24. 23.61 − 14.77

25. 82.063 − 52.13 26. 92.6 − 26.071 27. 52.67 − 18.66

28. 26.007 − 18.3   29. 135.68 − 23.77 30. 25.68 − 2.007

31. 77.077 − 25.1   32. 147.002 − 96.01 33. 45.79 − 2.3

Lesson 2-6   *Write the next three numbers in each sequence.*

1. 14, 21, 28, ? , ? , ?   2. 36, 42, 48, ? , ? , ?

3. 2, 6, 10, ? , ? , ?   4. 18, 16, 14, ? , ? , ?

5. 3, 7, 11, ? , ? , ?   6. 25, 20, 15, ? , ? , ?

7. 80, 70, 60, ? , ? , ?   8. 100, 85, 70, ? , ? , ?

9. 25, 45, 65, ? , ? , ?   10. 6, 12.5, 19, ? , ? , ?

11. 36, 34, 32, ? , ? , ?   12. 3, 7.2, 11.4, ? , ? , ?

13. 7.2, 8.3, 9.4, ? , ? , ?   14. 18.6, 18.5, 18.4, ? , ? , ?

15. 93, 193, 293, ? , ? , ?   16. 8, 128, 248, ? , ? , ?

Lesson 2-8   *Find the perimeter of each figure.*

1.
7 in. / 4 in.

2.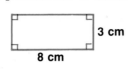
8 cm / 3 cm

3.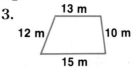
13 m / 12 m / 10 m / 15 m

4.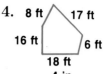
8 ft / 17 ft / 16 ft / 6 ft / 18 ft

5.
11 in. / 13 in. / 12 in.

6.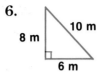
10 m / 8 m / 6 m

7.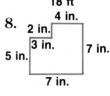
16 cm

8.
4 in. / 2 in. / 3 in. / 7 in. / 5 in. / 7 in.

9.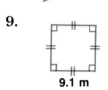
9.1 m

10.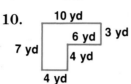
10 yd / 6 yd / 3 yd / 7 yd / 4 yd / 4 yd

11.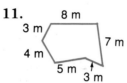
8 m / 3 m / 7 m / 4 m / 5 m / 3 m

12.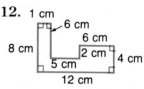
1 cm / 6 cm / 6 cm / 8 cm / 2 cm / 4 cm / 5 cm / 12 cm

Lesson 3-1   *Multiply.*

1. 5,338 × 9

2. 82 × 14

3. 496 × 60

4. 735 × 37

5. 5,302 × 53

6. 3,791 × 68

7. 7,482 × 78

8. 643 × 205

9. 491 × 372

10. 1,445 × 470

11. 2,339 × 608

12. 8,516 × 926

13. 6 × 36

14. 834 × 9

15. 7 × 4,566

16. 23 × 82

17. 28 × 416

18. 556 × 28

19. 65 × 790

20. 4,562 × 13

21. 87 × 1,762

22. 39 × 43,625

23. 409 × 600

24. 314 × 233

25. 4,820 × 701

26. 1,526 × 370

27. 8,614 × 129

28. 5,621 × 683

29. 3,183 × 425

30. 4,606 × 523

31. 8,660 × 4,320

32. 9,002 × 6,230

33. 8,463 × 987

34. 7,630 × 5,005

Lesson 3-2   *Multiply.*

1. 9.4 × 8

2. 14 × 0.3

3. 4.56 × 7

4. 7.3 × 11

5. $3.57 × 23

6. 0.236 × 8

7. 0.5 × 0.8

8. 8.76 × 3.9

9. 43.68 × 2.07

10. 0.004 × 5.6

11. 4.8 × 32

12. 8.31 × 58

13. 46 × 2.4

14. 5 × 6.125

15. 4.304 × 16

16. $9.14 × 48

17. 346 × 0.67

18. 28 × $0.86

19. 0.4 × 0.23

20. 4.59 × 0.3

21. 65.1 × 0.38

22. 4.13 × 2.09

23. 214.2 × 0.8

24. 9.76 × 1.12

25. 0.07 × 3.28

26. 14.25 × 0.06

27. 83.3 × 4.5

28. 633.45 × 16.2

29. 18.073 × 14.2

30. 73.45 × 18.73

31. 23.63 × 43.85

32. 96.081 × 19

Lesson 3-4   *Divide.*

1. $5\overline{)345}$ 2. $8\overline{)416}$ 3. $9\overline{)489}$ 4. $4\overline{)2,354}$ 5. $8\overline{)10,036}$ 6. $14\overline{)126}$

7. $35\overline{)276}$ 8. $42\overline{)2,730}$ 9. $58\overline{)4,284}$ 10. $92\overline{)8,470}$ 11. $180\overline{)4,500}$ 12. $287\overline{)3,784}$

13. $370 \div 5$ 14. $215 \div 6$ 15. $5,285 \div 9$ 16. $2,040 \div 8$

17. $780 \div 60$ 18. $870 \div 36$ 19. $567 \div 96$ 20. $4,089 \div 47$

21. $2,075 \div 28$ 22. $5,846 \div 16$ 23. $18,900 \div 54$ 24. $12,834 \div 22$

25. $3,200 \div 200$ 26. $7,626 \div 305$ 27. $8,502 \div 654$ 28. $31,611 \div 123$

29. $4,620 \div 400$ 30. $8,320 \div 35$ 31. $8,615 \div 56$ 32. $4,323 \div 173$

33. $7,315 \div 26$ 34. $6,426 \div 12$ 35. $4,777 \div 77$ 36. $5,325 \div 70$

37. $1,008 \div 58$ 38. $7,998 \div 2,600$ 39. $14,691 \div 36$ 40. $8,430 \div 121$

Lesson 3-5   *Divide.*

1. $6\overline{)1.26}$ 2. $8\overline{)23.2}$ 3. $6\overline{)89.22}$ 4. $15\overline{)54.75}$

5. $13\overline{)128.31}$ 6. $9\overline{)2.583}$ 7. $47\overline{)11.28}$ 8. $26\overline{)32.5}$

9. $0.5\overline{)18.45}$ 10. $0.08\overline{)5.2}$ 11. $2.6\overline{)0.65}$ 12. $1.3\overline{)12.831}$

13. $0.87\overline{)5.133}$ 14. $2.54\overline{)24.13}$ 15. $3.7\overline{)35.89}$ 16. $14.5\overline{)142.1}$

17. $7.2 \div 8$ 18. $0.036 \div 9$ 19. $1.75 \div 7$ 20. $167.5 \div 25$

21. $37.1 \div 14$ 22. $5.88 \div 0.4$ 23. $3.7 \div 0.5$ 24. $6.72 \div 2.4$

25. $41.4 \div 18$ 26. $9.87 \div 0.3$ 27. $8.45 \div 2.5$ 28. $7.8 \div 2$

29. $90.88 \div 14.2$ 30. $33.6 \div 8.4$ 31. $25.389 \div 4.03$ 32. $85.92 \div 4.8$

33. $63.18 \div 16.2$ 34. $18.49 \div 4.3$ 35. $9.06 \div 0.003$ 36. $1.02 \div 0.3$

Lesson 3-7   *Multiply or divide.*

1. 14.7×10 2. 3.36×10 3. 8.404×100 4. 7.9×100

5. 0.49×100 6. 0.643×100 7. $7.5 \times 1,000$ 8. 260×100

9. 46.9×10^2 10. 18×10^3 11. $5.95 \times 1,000$ 12. 0.37×10

13. $0.63 \div 10$ 14. $14.6 \div 100$ 15. $3.68 \div 10$ 16. $17.9 \div 100$

17. $23.1 \div 1,000$ 18. $436.6 \div 100$ 19. $42.9 \div 1,000$ 20. $0.07 \div 10$

21. $4.04 \div 10^2$ 22. $80.9 \div 10^1$ 23. $7.73 \div 10^3$ 24. $56,707 \div 100$

25. 18.6×10 26. 25.6×100 27. $0.06 \times 1,000$ 28. 83.4×10

29. 56.8×10^3 30. 6.07×10^4 31. $47.6 \div 100$ 32. $963 \div 1,000$

33. $7.6 \div 10^2$ 34. $9.6 \div 10^3$ 35. $4.63 \div 10^3$ 36. $73.3 \div 10^2$

Lesson 3-8

Write each in scientific notation.

1. 350   **2.** 628   **3.** 1,423   **4.** 800   **5.** 1,600

6. 3,450   **7.** 9,220   **8.** 7,100   **9.** 19,000   **10.** 25,500

11. 63,350   **12.** 123,300   **13.** 401,000   **14.** 898,000   **15.** 923,000

16. 4,500,000   **17.** 18,000,000   **18.** 967,000   **19.** 42,010,000   **20.** 8,000,000

Write in standard form.

21. 5.23×10^2 **22.** 6.8×10^3 **23.** 8.4×10^6 **24.** 9.1×10^5

25. 7.04×10^3 **26.** 8.001×10^7 **27.** 4.35×10^4 **28.** 8.2×10^3

29. 4×10^6 **30.** 6.07×10^2 **31.** 8.9×10^8 **32.** 6.7×10^4

33. 3.01×10^5 **34.** 8.102×10^4 **35.** 7.65×10^6 **36.** 8.3×10^6

Lesson 3-9

Find the common ratio and write the next three terms in each sequence.

1. 2, 6, 18, . . .   **2.** 3, 9, 27, . . .

3. 100, 50, 25, . . .   **4.** 180, 144, 115.2, . . .

5. 5, 25, 125, . . .   **6.** 1, 6, 36, . . .

7. 900, 270, 81, . . .   **8.** 1,600, 800, 400, . . .

9. 90, 63, 44.1, . . .   **10.** 14, 28, 56, . . .

11. 5, 15, 45, . . .   **12.** 625, 125, 25, . . .

13. 96, 48, 24, . . .   **14.** 500, 375, 281.25, . . .

15. 1,530, 459, 137.7, . . .   **16.** 1,100, 990, 891, . . .

17. 2, 8, 32, . . .   **18.** 5, 10, 20, . . .

Lesson 3-10

Solve for d, r, or t using the formula d = rt.

1. $d = 250, r = 50, t = \underline{\ ?\ }$ **2.** $d = 300, r = 60, t = \underline{\ ?\ }$

3. $d = 1{,}500, r = 60, t = \underline{\ ?\ }$ **4.** $d = 2{,}000, r = 50, t = \underline{\ ?\ }$

5. $d = 35, r = 35, t = \underline{\ ?\ }$ **6.** $d = 1{,}800, r = 90, t = \underline{\ ?\ }$

7. $d = 300, r = 2.5, t = \underline{\ ?\ }$ **8.** $d = 500, t = 20, r = \underline{\ ?\ }$

9. $d = 3{,}000, r = 40, t = \underline{\ ?\ }$ **10.** $d = 600, t = 15, r = \underline{\ ?\ }$

11. $r = 55, t = 3, d = \underline{\ ?\ }$ **12.** $r = 65, t = 4, d = \underline{\ ?\ }$

Solve for r, y, or n using the formula r = y ÷ n.

13. $y = 112, n = 16, r = \underline{\ ?\ }$ **14.** $y = 231, n = 21, r = \underline{\ ?\ }$

15. $y = 150, n = 20, r = \underline{\ ?\ }$ **16.** $y = 315, n = 30, r = \underline{\ ?\ }$

Lesson 3-11   *Find the area of each rectangle.*

1. 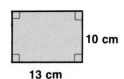 10 cm, 13 cm

2. 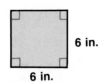 6 in., 6 in.

3. 15 ft, 8 ft

4. 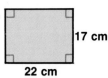 17 cm, 22 cm

5. 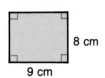 8 cm, 9 cm

6. 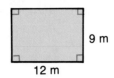 9 m, 12 m

7. 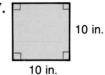 10 in., 10 in.

8. 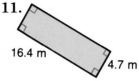 14 mm, 3 mm

9. 8.6 m, 4.5 m

10. 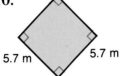 5.7 m, 5.7 m

11. 16.4 m, 4.7 m

12. 4.3 mm, 12.6 mm

Lesson 3-12   *Find the mean for each set of data. Round to the nearest tenth.*

1. 6, 3, 5, 6, 7, 9

2. 8, 10, 7, 3, 2, 5, 7

3. 43, 36, 72, 15, 44

4. 63, 65, 66, 60, 71

5. 12, 15, 38, 11, 43, 46

6. 55, 56, 36, 59, 64

7. 24, 32, 42, 14, 12, 13, 10

8. 25, 52, 45, 54, 39, 93

9. 204, 430, 680, 190

10. 563, 560, 565, 600

11. 2.6, 3.5, 8.2, 5.3

12. 5.7, 7.8, 8.6, 6.5

13. 20.6, 17.9, 18.5

14. $6.35, $7.86, $5.92, $4.46

15. 103.6, 101.4, 190.8

16. 10.48, 12.62, 17.17

17. 12, 13, 9, 6, 7, 15, 13, 12, 10, 8

18. 23, 36, 65, 58, 70, 60, 30

19. 4, 8, 46, 39, 78, 6, 16, 17, 9, 11

20. 14, 96, 102, 17, 28, 84, 58

21. 556, 672, 783, 491, 186, 781

22. 93, 781, 430, 80, 760, 490

23. $7.50, $6.90, $8.30, $7.70, $4.50

24. 5.8, 7.9, 6.3, 5.7, 7.6, 4.9

Lesson 4-1   *Find all the factors of each number.*

1. 4   **2.** 10   **3.** 15   **4.** 25   **5.** 30

6. 55   **7.** 42   **8.** 120   **9.** 60   **10.** 96

State whether each number is divisible by 2, 3, 5, 9, or 10.

11. 42   **12.** 55   **13.** 63   **14.** 72   **15.** 99

Lesson 4-2   *State whether each number is prime or composite.*

1. 7   2. 9   3. 15   4. 17   5. 23   6. 27

7. 33   8. 39   9. 41   10. 55   11. 27   12. 61

13. 62   14. 75   15. 77   16. 79   17. 84   18. 89

Write the prime factorization of each number.

19. 9   20. 12   21. 16   22. 24   23. 28   24. 31

25. 36   26. 38   27. 45   28. 48   29. 52   30. 56

31. 120   32. 135   33. 140   34. 144   35. 201   36. 203

37. 319   38. 420   39. 444   40. 600   41. 635   42. 725

Lesson 4-3   *Find the GCF and the LCM of each group of numbers.*

1. 6, 7   2. 3, 12   3. 5, 15   4. 10, 15   5. 8, 16

6. 9, 12   7. 15, 20   8. 20, 25   9. 18, 36   10. 14, 21

11. 16, 24   12. 30, 40   13. 35, 49   14. 27, 36   15. 19, 57

16. 24, 46   17. 56, 16   18. 36, 42   19. 28, 20   20. 21, 35

21. 3, 6, 12   22. 6, 8, 10   23. 10, 20, 30   24. 12, 16, 20

25. 4, 8, 20   26. 9, 12, 15   27. 2, 9, 15   28. 2, 8, 13

29. 4, 7, 9   30. 5, 10, 16   31. 8, 9, 11   32. 7, 9, 11

33. 8, 12, 20   34. 3, 5, 7   35. 6, 7, 10   36. 10, 12, 15

Lesson 4-5   *Replace each ▦ with a number so that the fractions are equivalent.*

1. $\frac{1}{2} = \frac{▦}{12}$ 2. $\frac{3}{4} = \frac{▦}{16}$ 3. $\frac{2}{3} = \frac{▦}{9}$ 4. $\frac{4}{5} = \frac{▦}{20}$

5. $\frac{9}{10} = \frac{▦}{100}$ 6. $\frac{8}{12} = \frac{▦}{3}$ 7. $\frac{12}{16} = \frac{▦}{4}$ 8. $\frac{8}{9} = \frac{▦}{72}$

9. $\frac{25}{30} = \frac{▦}{6}$ 10. $\frac{15}{20} = \frac{▦}{4}$ 11. $\frac{25}{75} = \frac{▦}{3}$ 12. $\frac{19}{38} = \frac{▦}{2}$

13. $\frac{6}{7} = \frac{▦}{42}$ 14. $\frac{36}{40} = \frac{▦}{10}$ 15. $\frac{75}{100} = \frac{▦}{4}$ 16. $\frac{8}{3} = \frac{▦}{9}$

17. $\frac{4}{3} = \frac{▦}{12}$ 18. $\frac{50}{25} = \frac{▦}{1}$ 19. $\frac{18}{6} = \frac{▦}{1}$ 20. $\frac{15}{3} = \frac{▦}{1}$

21. $\frac{7}{3} = \frac{▦}{21}$ 22. $\frac{75}{70} = \frac{▦}{14}$ 23. $\frac{9}{3} = \frac{▦}{9}$ 24. $\frac{24}{8} = \frac{▦}{1}$

25. $\frac{12}{16} = \frac{▦}{4}$ 26. $\frac{4}{1} = \frac{▦}{4}$ 27. $\frac{10}{4} = \frac{▦}{12}$ 28. $\frac{17}{3} = \frac{▦}{9}$

Lesson 4-6 *Simplify each fraction.*

1. $\frac{4}{6}$
2. $\frac{5}{10}$
3. $\frac{10}{15}$
4. $\frac{16}{24}$
5. $\frac{7}{14}$
6. $\frac{7}{21}$

7. $\frac{9}{16}$
8. $\frac{12}{20}$
9. $\frac{19}{38}$
10. $\frac{15}{25}$
11. $\frac{4}{10}$
12. $\frac{8}{12}$

13. $\frac{15}{30}$
14. $\frac{40}{60}$
15. $\frac{42}{63}$
16. $\frac{27}{36}$
17. $\frac{36}{42}$
18. $\frac{50}{75}$

19. $\frac{4}{8}$
20. $\frac{3}{15}$
21. $\frac{4}{20}$
22. $\frac{5}{30}$
23. $\frac{2}{20}$
24. $\frac{11}{44}$

25. $\frac{7}{28}$
26. $\frac{8}{64}$
27. $\frac{9}{99}$
28. $\frac{10}{10}$
29. $\frac{28}{56}$
30. $\frac{9}{21}$

31. $\frac{18}{24}$
32. $\frac{18}{54}$
33. $\frac{9}{63}$
34. $\frac{10}{80}$
35. $\frac{40}{90}$
36. $\frac{36}{72}$

37. $\frac{12}{48}$
38. $\frac{21}{28}$
39. $\frac{33}{33}$
40. $\frac{42}{77}$
41. $\frac{64}{96}$
42. $\frac{16}{60}$

43. $\frac{14}{30}$
44. $\frac{12}{64}$
45. $\frac{15}{63}$
46. $\frac{17}{34}$
47. $\frac{18}{22}$
48. $\frac{22}{55}$

49. $\frac{43}{51}$
50. $\frac{30}{100}$
51. $\frac{13}{39}$
52. $\frac{17}{27}$
53. $\frac{16}{24}$
54. $\frac{8}{28}$

Lesson 4-7 *Replace each ● with <, >, or = to make a true sentence.*

1. $\frac{2}{3}$ ● $\frac{1}{3}$
2. $\frac{3}{4}$ ● $\frac{1}{4}$
3. $\frac{5}{11}$ ● $\frac{9}{11}$
4. $\frac{1}{6}$ ● $\frac{5}{6}$
5. $\frac{5}{8}$ ● $\frac{3}{8}$

6. $\frac{1}{4}$ ● $\frac{3}{12}$
7. $\frac{3}{7}$ ● $\frac{6}{14}$
8. $\frac{1}{2}$ ● $\frac{3}{4}$
9. $\frac{3}{5}$ ● $\frac{7}{10}$
10. $\frac{2}{3}$ ● $\frac{6}{7}$

11. $\frac{5}{8}$ ● $\frac{4}{5}$
12. $\frac{2}{9}$ ● $\frac{1}{3}$
13. $\frac{4}{7}$ ● $\frac{4}{11}$
14. $\frac{3}{8}$ ● $\frac{3}{9}$
15. $\frac{5}{12}$ ● $\frac{7}{16}$

16. $\frac{3}{3}$ ● $\frac{10}{10}$
17. $\frac{1}{6}$ ● $\frac{1}{9}$
18. $\frac{5}{24}$ ● $\frac{8}{18}$
19. $\frac{3}{4}$ ● $\frac{5}{6}$
20. $\frac{4}{9}$ ● $\frac{12}{16}$

21. $\frac{8}{10}$ ● $\frac{9}{12}$
22. $\frac{3}{4}$ ● $\frac{4}{8}$
23. $\frac{5}{15}$ ● $\frac{3}{9}$
24. $\frac{7}{18}$ ● $\frac{6}{27}$
25. $\frac{10}{25}$ ● $\frac{3}{15}$

27. $\frac{7}{10}$ ● $\frac{6}{5}$
28. $\frac{8}{12}$ ● $\frac{6}{10}$
29. $\frac{4}{6}$ ● $\frac{3}{8}$
30. $\frac{9}{10}$ ● $\frac{11}{12}$
31. $\frac{4}{7}$ ● $\frac{8}{11}$

32. $\frac{12}{15}$ ● $\frac{4}{5}$
33. $\frac{9}{13}$ ● $\frac{8}{10}$
34. $\frac{7}{22}$ ● $\frac{8}{23}$
35. $\frac{5}{17}$ ● $\frac{6}{19}$
36. $\frac{10}{24}$ ● $\frac{8}{20}$

37. $\frac{6}{24}$ ● $\frac{8}{32}$
38. $\frac{7}{27}$ ● $\frac{8}{30}$
39. $\frac{11}{13}$ ● $\frac{12}{17}$
40. $\frac{9}{31}$ ● $\frac{6}{23}$
41. $\frac{7}{29}$ ● $\frac{9}{31}$

Lesson 4-8 *Change each number to a mixed number in simplest form.*

1. $\frac{7}{3}$
2. $\frac{5}{2}$
3. $\frac{8}{3}$
4. $\frac{9}{4}$
5. $\frac{10}{5}$
6. $\frac{8}{4}$

7. $\frac{6}{2}$
8. $\frac{7}{2}$
9. $\frac{18}{3}$
10. $\frac{5}{4}$
11. $\frac{6}{5}$
12. $\frac{21}{4}$

13. $\frac{23}{5}$
14. $\frac{18}{4}$
15. $\frac{35}{6}$
16. $\frac{25}{3}$
17. $\frac{26}{4}$
18. $\frac{19}{3}$

19. $\frac{20}{6}$
20. $\frac{25}{7}$
21. $\frac{28}{9}$
22. $\frac{29}{7}$
23. $\frac{18}{7}$
24. $\frac{33}{4}$

25. $\frac{85}{6}$
26. $\frac{81}{7}$
27. $\frac{65}{6}$
28. $\frac{62}{9}$
29. $\frac{45}{3}$
30. $\frac{42}{6}$

31. $\frac{95}{3}$
32. $\frac{72}{6}$
33. $\frac{65}{4}$
34. $\frac{55}{10}$
35. $\frac{19}{9}$
36. $\frac{46}{7}$

37. $\frac{53}{11}$
38. $\frac{58}{13}$
39. $\frac{67}{15}$
40. $\frac{72}{19}$
41. $\frac{96}{14}$
42. $\frac{25}{14}$

43. $\frac{83}{21}$
44. $\frac{76}{31}$
45. $\frac{96}{25}$
46. $\frac{25}{9}$
47. $\frac{76}{22}$
48. $\frac{81}{33}$

Lesson 4-9   *Estimate.*

1. $\frac{1}{2} + \frac{1}{3}$
2. $\frac{2}{5} + \frac{2}{3}$
3. $\frac{4}{5} + \frac{9}{13}$
4. $\frac{7}{10} + \frac{7}{8}$

5. $\frac{2}{3} + \frac{10}{11}$
6. $\frac{14}{15} + \frac{19}{20}$
7. $\frac{2}{5} + \frac{1}{8}$
8. $\frac{9}{10} + \frac{1}{6}$

9. $\frac{8}{9} - \frac{1}{2}$
10. $\frac{4}{5} - \frac{2}{3}$
11. $\frac{16}{17} - \frac{4}{7}$
12. $\frac{7}{16} - \frac{1}{12}$

13. $\frac{5}{8} + \frac{3}{4}$
14. $\frac{1}{6} + \frac{3}{8}$
15. $\frac{7}{9} - \frac{5}{12}$
16. $\frac{1}{4} + \frac{5}{8}$

17. $5\frac{1}{4} + 6\frac{2}{3}$
18. $3\frac{1}{4} + 7\frac{3}{7}$
19. $10\frac{1}{5} + 11\frac{1}{8}$
20. $14\frac{3}{5} + 7\frac{1}{3}$

21. $2\frac{7}{8} + 3\frac{9}{10}$
22. $8\frac{1}{5} + 4\frac{1}{9}$
23. $18\frac{2}{7} + 2\frac{4}{9}$
24. $12\frac{2}{3} + 14\frac{1}{8}$

25. $7\frac{1}{7} - 4\frac{2}{3}$
26. $16\frac{3}{5} - 14\frac{1}{4}$
27. $19\frac{2}{3} - 4\frac{1}{4}$
28. $22\frac{1}{8} - 15\frac{7}{8}$

29. $17 - \frac{5}{6}$
30. $9\frac{5}{7} - 2\frac{1}{3}$
31. $8\frac{9}{11} - 4\frac{2}{5}$
32. $6\frac{1}{5} - 2\frac{1}{3}$

Lesson 4-10   *Add.*

1. $\frac{5}{11} + \frac{9}{11}$
2. $\frac{1}{8} + \frac{5}{8}$
3. $\frac{7}{10} + \frac{7}{10}$
4. $\frac{5}{12} + \frac{9}{12}$
5. $\frac{1}{3} + \frac{1}{2}$

6. $\frac{2}{9} + \frac{1}{3}$
7. $\frac{1}{2} + \frac{3}{4}$
8. $\frac{1}{4} + \frac{3}{12}$
9. $\frac{3}{7} + \frac{6}{14}$
10. $\frac{2}{5} + \frac{2}{3}$

11. $\frac{1}{4} + \frac{3}{5}$
12. $\frac{4}{9} + \frac{1}{2}$
13. $\frac{5}{7} + \frac{4}{6}$
14. $\frac{3}{4} + \frac{1}{6}$
15. $\frac{5}{12} + \frac{5}{16}$

16. $\frac{3}{5} + \frac{3}{4}$
17. $\frac{2}{3} + \frac{1}{8}$
18. $\frac{9}{10} + \frac{1}{3}$
19. $\frac{8}{15} + \frac{2}{9}$
20. $\frac{5}{6} + \frac{7}{8}$

21. $\frac{6}{7} + \frac{6}{9}$
22. $\frac{3}{7} + \frac{3}{4}$
23. $\frac{5}{7} + \frac{5}{9}$
24. $\frac{7}{8} + \frac{5}{6}$
25. $\frac{3}{4} + \frac{4}{9}$

26. $\frac{3}{10} + \frac{24}{25}$
27. $\frac{6}{7} + \frac{1}{3}$
28. $\frac{7}{9} + \frac{4}{5}$
29. $\frac{2}{3} + \frac{7}{10}$
30. $\frac{1}{4} + \frac{5}{6}$

31. $\frac{7}{12} + \frac{11}{18}$
32. $\frac{9}{16} + \frac{13}{24}$
33. $\frac{8}{15} + \frac{2}{3}$
34. $\frac{5}{6} + \frac{13}{24}$
35. $\frac{5}{14} + \frac{11}{28}$

36. $\frac{11}{12} + \frac{7}{8}$
37. $\frac{3}{8} + \frac{1}{6}$
38. $\frac{1}{2} + \frac{5}{7}$
39. $\frac{4}{9} + \frac{1}{6}$
40. $\frac{5}{6} + \frac{7}{15}$

41. $\frac{4}{9} + \frac{1}{6}$
42. $\frac{5}{12} + \frac{7}{8}$
43. $\frac{4}{9} + \frac{5}{18}$
44. $\frac{1}{2} + \frac{7}{18}$
45. $\frac{5}{12} + \frac{3}{8}$

46. $\frac{9}{20} + \frac{2}{15}$
47. $\frac{5}{6} + \frac{4}{5}$
48. $\frac{2}{3} + \frac{2}{7}$
49. $\frac{7}{20} + \frac{4}{5}$
50. $\frac{4}{5} + \frac{17}{25}$

Lesson 4-11   *Add.*

1. $2\frac{1}{3} + 1\frac{1}{3}$
2. $5\frac{2}{7} + 2\frac{3}{7}$
3. $6\frac{3}{8} + 7\frac{1}{8}$
4. $1\frac{3}{4} + 2\frac{1}{4}$
5. $4\frac{6}{7} + 9\frac{6}{7}$

6. $5\frac{1}{2} + 3\frac{1}{4}$
7. $2\frac{2}{3} + 4\frac{1}{9}$
8. $7\frac{4}{5} + 9\frac{3}{10}$
9. $3\frac{3}{4} + 5\frac{5}{8}$
10. $3\frac{2}{5} + 7\frac{6}{15}$

11. $10\frac{2}{3} + 5\frac{6}{7}$
12. $17\frac{2}{9} + 12\frac{1}{3}$
13. $6\frac{5}{12} + 12\frac{5}{8}$
14. $7\frac{1}{4} + 15\frac{5}{6}$
15. $8\frac{2}{15} + 9\frac{5}{9}$

16. $6\frac{1}{8} + 4\frac{2}{3}$
17. $7 + 6\frac{4}{9}$
18. $8\frac{1}{12} + 12\frac{6}{11}$
19. $7\frac{2}{3} + 8\frac{1}{4}$
20. $9\frac{3}{5} + 4\frac{1}{4}$

21. $12\frac{3}{11} + 14\frac{3}{13}$
22. $21\frac{1}{3} + 15\frac{3}{8}$
23. $19\frac{1}{7} + 6\frac{1}{4}$
24. $9\frac{2}{5} + 8\frac{1}{3}$
25. $12\frac{1}{3} + 6\frac{1}{4}$

26. $21\frac{3}{8} + 17\frac{1}{5}$
27. $6\frac{2}{5} + 8\frac{1}{9}$
28. $13\frac{1}{2} + 14\frac{3}{8}$
29. $23\frac{5}{6} + 2\frac{1}{5}$
30. $16\frac{4}{7} + 12\frac{1}{8}$

Lesson 4-12 — *Subtract.*

1. $\frac{12}{13} - \frac{7}{13}$
2. $\frac{15}{18} - \frac{12}{18}$
3. $\frac{10}{14} - \frac{3}{14}$
4. $\frac{7}{20} - \frac{5}{20}$
5. $\frac{2}{3} - \frac{1}{2}$
6. $\frac{5}{9} - \frac{1}{3}$
7. $\frac{5}{8} - \frac{2}{5}$
8. $\frac{3}{4} - \frac{1}{2}$
9. $\frac{7}{8} - \frac{3}{16}$
10. $\frac{8}{9} - \frac{2}{6}$
11. $\frac{11}{12} - \frac{5}{18}$
12. $\frac{5}{6} - \frac{3}{14}$
13. $\frac{11}{15} - \frac{7}{25}$
14. $\frac{9}{12} - \frac{3}{18}$
15. $\frac{7}{9} - \frac{2}{15}$
16. $\frac{5}{8} - \frac{3}{5}$
17. $\frac{7}{9} - \frac{2}{3}$
18. $\frac{13}{16} - \frac{5}{8}$
19. $\frac{3}{4} - \frac{7}{12}$
20. $\frac{4}{5} - \frac{2}{7}$
21. $\frac{7}{8} - \frac{5}{6}$
22. $\frac{5}{7} - \frac{1}{4}$
23. $\frac{9}{10} - \frac{1}{2}$
24. $\frac{8}{9} - \frac{2}{3}$
25. $\frac{5}{6} - \frac{1}{3}$
26. $\frac{2}{3} - \frac{1}{6}$
27. $\frac{9}{16} - \frac{1}{2}$
28. $\frac{5}{8} - \frac{11}{20}$
29. $\frac{14}{15} - \frac{2}{9}$
30. $\frac{1}{4} - \frac{1}{6}$
31. $\frac{11}{12} - \frac{5}{6}$
32. $\frac{14}{15} - \frac{2}{3}$
33. $\frac{13}{16} - \frac{5}{8}$
34. $\frac{19}{20} - \frac{2}{5}$
35. $\frac{49}{100} - \frac{3}{25}$
36. $\frac{4}{5} - \frac{1}{6}$
37. $\frac{23}{25} - \frac{27}{50}$
38. $\frac{19}{25} - \frac{1}{2}$
39. $\frac{5}{6} - \frac{13}{16}$
40. $\frac{15}{64} - \frac{7}{32}$

Lesson 4-13 — *Subtract.*

1. $2\frac{7}{10} - 1\frac{4}{10}$
2. $8\frac{6}{7} - 2\frac{5}{7}$
3. $7\frac{5}{12} - 2\frac{3}{12}$
4. $6\frac{13}{14} - 5\frac{6}{14}$
5. $13\frac{7}{12} - 9\frac{1}{4}$
6. $9\frac{4}{7} - 3\frac{5}{14}$
7. $11\frac{2}{3} - 8\frac{11}{12}$
8. $15\frac{6}{9} - 13\frac{5}{12}$
9. $3\frac{4}{7} - 1\frac{2}{3}$
10. $7\frac{1}{8} - 4\frac{1}{3}$
11. $18\frac{1}{9} - 12\frac{2}{5}$
12. $12\frac{3}{10} - 8\frac{3}{4}$
13. $13\frac{1}{5} - 10$
14. $29\frac{5}{8} - 6$
15. $4 - 1\frac{2}{3}$
16. $16\frac{1}{4} - 12\frac{1}{5}$
17. $15\frac{2}{5} - 6\frac{1}{4}$
18. $23\frac{1}{2} - 2$
19. $18\frac{1}{5} - 6\frac{1}{4}$
20. $23\frac{2}{3} - 4\frac{1}{2}$
21. $5\frac{2}{3} - 3\frac{1}{2}$
22. $16\frac{1}{4} - 7\frac{1}{5}$
23. $43 - 5\frac{1}{5}$
24. $16\frac{3}{5} - 7\frac{1}{7}$
25. $6\frac{1}{2} - 5\frac{1}{4}$
26. $8\frac{3}{5} - 2\frac{1}{5}$
27. $21\frac{5}{8} - 3\frac{1}{4}$
28. $26\frac{2}{3} - 6\frac{1}{5}$
29. $8\frac{1}{5} - 4\frac{1}{4}$
30. $6\frac{3}{7} - 2\frac{2}{9}$
31. $14\frac{1}{6} - 3\frac{2}{3}$
32. $25\frac{4}{7} - 21$
33. $26 - 4\frac{1}{9}$
34. $17\frac{3}{9} - 4\frac{3}{5}$
35. $18\frac{3}{10} - 14\frac{1}{8}$
36. $26\frac{1}{4} - 3$
37. $19\frac{2}{3} - 3\frac{1}{4}$
38. $18\frac{1}{9} - 1\frac{3}{7}$
39. $6 - 4\frac{3}{5}$
40. $12\frac{2}{3} - 10$

Lesson 5-1 — *Estimate each product or quotient.*

1. $3\frac{1}{8} \times 4\frac{1}{4}$
2. $7\frac{1}{3} \times 3\frac{3}{4}$
3. $7\frac{1}{2} \times 8\frac{3}{5}$
4. $6\frac{1}{9} \times 4\frac{1}{2}$
5. $7\frac{2}{3} \times 1\frac{3}{4}$
6. $8\frac{3}{6} \times 5\frac{2}{3}$
7. $2\frac{3}{8} \times 4\frac{7}{8}$
8. $6\frac{5}{6} \times 5\frac{1}{4}$
9. $12 \times 5\frac{6}{7}$
10. $9\frac{2}{5} \times 2\frac{3}{7}$
11. $1\frac{3}{10} \times 4\frac{7}{9}$
12. $7\frac{6}{7} \times 8\frac{1}{5}$
13. $10\frac{1}{8} \times 5\frac{6}{7}$
14. $4\frac{1}{10} \times 8\frac{1}{9}$
15. $1\frac{3}{5} \times 7\frac{6}{9}$
16. $2\frac{7}{11} \times 3\frac{10}{11}$
17. $8\frac{2}{3} \times 8\frac{1}{5}$
18. $2\frac{6}{13} \times 1\frac{7}{10}$
19. $3\frac{1}{3} \times 3\frac{5}{6}$
20. $4\frac{2}{7} \times 4\frac{2}{5}$
21. $3\frac{1}{5} \div 3\frac{1}{9}$
22. $21\frac{2}{3} \div 2\frac{4}{9}$
23. $16\frac{1}{4} \div 4\frac{1}{5}$
24. $25\frac{1}{4} \div 4\frac{2}{3}$
25. $16\frac{1}{8} \div 3\frac{3}{4}$
26. $15\frac{1}{4} \div 4\frac{4}{5}$
27. $12\frac{1}{8} \div 6\frac{1}{9}$
28. $16\frac{2}{5} \div 8\frac{1}{4}$
29. $23\frac{2}{3} \div 3\frac{1}{5}$
30. $20\frac{2}{3} \div 7\frac{1}{4}$
31. $17\frac{1}{8} \div 5\frac{5}{6}$
32. $23\frac{5}{8} \div 11\frac{7}{9}$
33. $19\frac{5}{8} \div 3\frac{2}{3}$
34. $12\frac{1}{4} \div 11\frac{5}{6}$
35. $27\frac{5}{6} \div 6\frac{3}{4}$

Lesson 5-2   *Multiply.*

1. $\frac{2}{3} \times \frac{4}{5}$
2. $\frac{1}{6} \times \frac{2}{5}$
3. $\frac{4}{9} \times \frac{3}{7}$
4. $\frac{5}{12} \times \frac{6}{11}$
5. $\frac{7}{10} \times \frac{5}{14}$

6. $\frac{3}{8} \times \frac{8}{9}$
7. $\frac{3}{5} \times \frac{1}{12}$
8. $\frac{2}{5} \times \frac{5}{8}$
9. $\frac{7}{15} \times \frac{3}{21}$
10. $\frac{6}{10} \times \frac{2}{3}$

11. $\frac{5}{6} \times \frac{15}{16}$
12. $\frac{6}{14} \times \frac{12}{18}$
13. $\frac{2}{3} \times \frac{3}{13}$
14. $\frac{4}{9} \times \frac{1}{6}$
15. $\frac{1}{5} \times 4$

16. $\frac{3}{4} \times \frac{5}{6}$
17. $\frac{9}{10} \times \frac{3}{4}$
18. $\frac{8}{9} \times \frac{2}{3}$
19. $\frac{6}{7} \times \frac{4}{5}$
20. $\frac{3}{8} \times \frac{4}{5}$

21. $\frac{8}{11} \times \frac{11}{12}$
22. $\frac{5}{6} \times \frac{3}{5}$
23. $\frac{6}{7} \times \frac{7}{21}$
24. $\frac{8}{9} \times \frac{9}{10}$
25. $\frac{2}{3} \times \frac{5}{8}$

26. $\frac{2}{3} \times \frac{5}{7}$
27. $\frac{3}{4} \times \frac{5}{6}$
28. $\frac{1}{5} \times \frac{12}{13}$
29. $\frac{9}{10} \times \frac{1}{4}$
30. $\frac{1}{2} \times \frac{1}{2}$

31. $\frac{7}{11} \times \frac{12}{15}$
32. $\frac{7}{9} \times \frac{5}{7}$
33. $\frac{8}{13} \times \frac{2}{11}$
34. $\frac{4}{7} \times \frac{2}{9}$
35. $\frac{3}{11} \times \frac{7}{15}$

36. $\frac{4}{9} \times \frac{24}{25}$
37. $\frac{1}{9} \times \frac{6}{13}$
38. $\frac{4}{7} \times 6$
39. $\frac{7}{10} \times 5$
40. $\frac{4}{9} \times 6$

Lesson 5-3   *Multiply.*

1. $3 \times \frac{1}{9}$
2. $5 \times \frac{6}{7}$
3. $\frac{3}{5} \times 15$
4. $3\frac{1}{2} \times 4\frac{1}{3}$
5. $2\frac{2}{5} \times 1\frac{1}{5}$

6. $3\frac{5}{8} \times 4\frac{1}{2}$
7. $\frac{4}{5} \times 2\frac{3}{4}$
8. $6\frac{1}{8} \times 5\frac{1}{7}$
9. $2\frac{2}{3} \times 2\frac{1}{4}$
10. $1\frac{4}{5} \times \frac{3}{5}$

11. $6\frac{2}{3} \times 7\frac{3}{5}$
12. $3\frac{1}{2} \times 2\frac{4}{7}$
13. $10 \times 2\frac{2}{3}$
14. $8 \times 7\frac{1}{8}$
15. $3\frac{5}{6} \times 12$

16. $5\frac{1}{4} \times 6\frac{2}{3}$
17. $7\frac{1}{5} \times 3\frac{1}{4}$
18. $8\frac{3}{4} \times 2\frac{2}{5}$
19. $4\frac{1}{3} \times 2\frac{1}{7}$
20. $8\frac{1}{9} \times 2\frac{1}{4}$

21. $1\frac{3}{5} \times 6\frac{2}{5}$
22. $8\frac{2}{5} \times 3\frac{4}{7}$
23. $9\frac{1}{4} \times 3\frac{1}{3}$
24. $4\frac{3}{4} \times 2\frac{2}{3}$
25. $8\frac{3}{4} \times 3\frac{2}{7}$

26. $12\frac{1}{4} \times 1\frac{1}{7}$
27. $16\frac{1}{5} \times 2\frac{1}{3}$
28. $3\frac{1}{4} \times 4\frac{1}{6}$
29. $1\frac{9}{16} \times 4\frac{4}{5}$
30. $9 \times 7\frac{1}{4}$

31. $1\frac{2}{5} \times 6\frac{3}{4}$
32. $5\frac{3}{7} \times 14$
33. $4\frac{1}{8} \times 2\frac{2}{3}$
34. $9\frac{1}{4} \times 9$
35. $1\frac{1}{2} \times 1\frac{11}{14}$

36. $11\frac{1}{9} \times 2\frac{1}{8}$
37. $6\frac{1}{4} \times 2\frac{2}{3}$
38. $7\frac{2}{7} \times 8$
39. $3\frac{1}{5} \times 10$
40. $6\frac{3}{7} \times 8\frac{2}{5}$

Lesson 5-6   *Divide.*

1. $\frac{2}{3} \div \frac{1}{2}$
2. $\frac{3}{5} \div \frac{2}{5}$
3. $\frac{7}{10} \div \frac{3}{8}$
4. $\frac{5}{9} \div \frac{2}{3}$
5. $\frac{7}{12} \div \frac{7}{9}$

6. $4 \div \frac{2}{3}$
7. $8 \div \frac{4}{5}$
8. $9 \div \frac{5}{9}$
9. $\frac{2}{7} \div 2$
10. $\frac{4}{11} \div 4$

11. $\frac{1}{14} \div 7$
12. $\frac{2}{13} \div \frac{5}{26}$
13. $\frac{4}{7} \div \frac{6}{7}$
14. $\frac{7}{8} \div \frac{1}{3}$
15. $\frac{10}{11} \div \frac{4}{5}$

16. $15 \div \frac{3}{5}$
17. $\frac{9}{14} \div \frac{3}{4}$
18. $\frac{8}{9} \div \frac{5}{6}$
19. $\frac{4}{9} \div 36$
20. $49 \div \frac{13}{14}$

21. $\frac{3}{5} \div \frac{2}{3}$
22. $\frac{8}{9} \div \frac{4}{5}$
23. $\frac{3}{4} \div \frac{15}{16}$
24. $6 \div \frac{1}{5}$
25. $\frac{4}{3} \div \frac{1}{5}$

26. $\frac{11}{12} \div \frac{5}{6}$
27. $\frac{9}{10} \div \frac{5}{6}$
28. $\frac{4}{5} \div 8$
29. $\frac{9}{11} \div \frac{3}{11}$
30. $\frac{4}{7} \div \frac{9}{14}$

31. $\frac{15}{16} \div 10$
32. $\frac{4}{5} \div \frac{7}{10}$
33. $4 \div \frac{1}{3}$
34. $15 \div \frac{5}{7}$
35. $\frac{12}{13} \div \frac{11}{13}$

36. $\frac{7}{15} \div \frac{7}{9}$
37. $\frac{4}{9} \div \frac{8}{21}$
38. $\frac{5}{12} \div \frac{25}{36}$
39. $\frac{11}{12} \div 33$
40. $\frac{13}{20} \div \frac{39}{40}$

Lesson 5-7 *Divide.*

1. $\frac{3}{5} \div 1\frac{2}{3}$ 2. $2\frac{1}{2} \div 1\frac{1}{4}$ 3. $4\frac{3}{5} \div 4\frac{1}{5}$ 4. $3\frac{2}{9} \div \frac{3}{4}$ 5. $12 \div 2\frac{2}{5}$

6. $7 \div 4\frac{9}{10}$ 7. $5\frac{1}{9} \div 5$ 8. $1\frac{3}{7} \div 10$ 9. $1\frac{3}{4} \div 2\frac{3}{8}$ 10. $9\frac{1}{3} \div 5\frac{2}{5}$

11. $3\frac{3}{5} \div \frac{4}{5}$ 12. $8\frac{2}{5} \div 4\frac{1}{2}$ 13. $6\frac{1}{3} \div 2\frac{1}{2}$ 14. $5\frac{1}{4} \div 2\frac{1}{3}$ 15. $6\frac{1}{5} \div 7\frac{1}{2}$

16. $4\frac{1}{8} \div 3\frac{2}{3}$ 17. $6\frac{1}{4} \div 2\frac{1}{5}$ 18. $2\frac{5}{8} \div \frac{1}{2}$ 19. $4\frac{2}{5} \div 1\frac{1}{9}$ 20. $6\frac{3}{7} \div 2\frac{1}{2}$

21. $5\frac{2}{3} \div 4$ 22. $4\frac{1}{8} \div 1\frac{2}{3}$ 23. $12\frac{1}{4} \div 3\frac{1}{2}$ 24. $5\frac{1}{2} \div 3\frac{1}{4}$ 25. $7 \div 1\frac{1}{4}$

26. $5\frac{1}{2} \div 3\frac{2}{3}$ 27. $7\frac{1}{5} \div 2$ 28. $9\frac{3}{7} \div 2\frac{1}{5}$ 29. $4 \div 3\frac{1}{3}$ 30. $6\frac{2}{3} \div 10$

31. $12\frac{1}{5} \div 6\frac{2}{3}$ 32. $6\frac{2}{3} \div 5\frac{5}{6}$ 33. $12\frac{1}{4} \div 8$ 34. $7\frac{9}{16} \div 2\frac{3}{4}$ 35. $4\frac{3}{8} \div 5$

36. $1\frac{5}{6} \div 3\frac{2}{3}$ 37. $21 \div 5\frac{1}{4}$ 38. $18 \div 2\frac{1}{4}$ 39. $12 \div 3\frac{3}{5}$ 40. $16\frac{3}{5} \div 4$

Lesson 5-8 *Change each fraction to a decimal. Use bar notation to show a repeating decimal.*

1. $\frac{3}{4}$ 2. $\frac{5}{8}$ 3. $\frac{3}{25}$ 4. $\frac{1}{20}$ 5. $\frac{5}{11}$ 6. $\frac{11}{18}$

7. $\frac{1}{3}$ 8. $\frac{4}{9}$ 9. $\frac{37}{40}$ 10. $\frac{1}{15}$ 11. $\frac{7}{12}$ 12. $\frac{3}{16}$

13. $\frac{3}{50}$ 14. $\frac{14}{45}$ 15. $\frac{5}{12}$ 16. $\frac{1}{16}$ 17. $\frac{10}{33}$ 18. $\frac{4}{15}$

19. $\frac{1}{5}$ 20. $\frac{6}{11}$ 21. $\frac{4}{9}$ 22. $\frac{7}{10}$ 23. $\frac{5}{8}$ 24. $\frac{4}{25}$

25. $\frac{9}{16}$ 26. $\frac{17}{20}$ 27. $\frac{1}{8}$ 28. $\frac{13}{16}$ 29. $\frac{9}{20}$ 30. $\frac{8}{30}$

31. $\frac{5}{18}$ 32. $\frac{1}{6}$ 33. $\frac{2}{3}$ 34. $\frac{7}{8}$ 35. $\frac{9}{25}$ 36. $\frac{5}{9}$

37. $\frac{22}{45}$ 38. $\frac{12}{25}$ 39. $\frac{33}{50}$ 40. $\frac{16}{25}$ 41. $\frac{9}{11}$ 42. $\frac{2}{33}$

43. $\frac{11}{25}$ 44. $\frac{39}{40}$ 45. $\frac{15}{16}$ 46. $\frac{5}{12}$ 47. $\frac{17}{18}$ 48. $\frac{41}{45}$

Lesson 5-9 *Change each decimal to a fraction.*

1. 0.6 2. 0.9 3. 0.45 4. 0.08 5. 0.96

6. 0.39 7. 0.55 8. 0.36 9. 0.79 10. 0.404

11. 0.565 12. 0.083 13. 0.208 14. 0.566 15. 0.734

16. 0.005 17. 0.004 18. 0.061 19. 0.072 20. 0.009

21. 0.085 22. 0.601 23. 0.432 24. 0.088 25. 0.074

26. $0.85\frac{5}{7}$ 27. $0.83\frac{1}{3}$ 28. $0.66\frac{2}{3}$ 29. $0.55\frac{5}{9}$ 30. $0.77\frac{7}{9}$

31. $0.28\frac{4}{7}$ 32. $0.37\frac{1}{2}$ 33. $0.31\frac{1}{4}$ 34. $0.22\frac{2}{9}$ 35. $0.16\frac{2}{3}$

36. $0.14\frac{2}{7}$ 37. $0.85\frac{5}{7}$ 38. $0.33\frac{1}{3}$ 39. $0.09\frac{1}{11}$ 40. $0.11\frac{1}{9}$

41. 0.56 42. 0.63 43. 0.09 44. 0.02 45. 0.72

Lesson 6-1   *Complete.*

1. 1 liter = $\underline{?}$ deciliters
2. 1 centiliter = $\underline{?}$ milliliters
3. 1 gram = $\underline{?}$ milligrams
4. 1 dekagram = $\underline{?}$ grams
5. 1 meter = $\underline{?}$ millimeters
6. 1 hectometer = $\underline{?}$ meters
7. 1 kilogram = $\underline{?}$ grams
8. 1 millimeter = $\underline{?}$ decimeters

Name the larger unit.

9. 1 meter or 1 decimeter
10. 1 kilogram or 1 gram
11. 1 milligram or 1 gram
12. 1 dekaliter or 1 hectoliter
13. 1 kilometer or 1 decimeter
14. 1 gram or centigram
15. 1 centiliter or 1 deciliter
16. 1 decimeter or 1 dekameter

Lesson 6-2   *Choose the most reasonable measurement.*

1. length of scissors	20 cm	20 m	20 km
2. length of house	15 m	15 cm	15 km
3. height of street sign	3 km	3 m	3 cm
4. width of door	120 mm	120 km	120 cm
5. length of child's shoe	100 mm	100 m	100 cm
6. width of shoelace	5 mm	5 cm	5 m
7. height of a tree	18 m	18 cm	18 km

Use a metric ruler to measure each of the following segments. Give the measurement in centimeters and millimeters.

1. _____
2. _____
3. _____
4. _____
5. _____
6. _____
7. _____
8. _____

Lesson 6-3   *Complete.*

1. 1 kg = ▨ g
2. 1 mm = ▨ m
3. 1 cm = ▨ m
4. 400 mm = ▨ cm
5. 4 km = ▨ m
6. 660 cm = ▨ m
7. 0.3 km = ▨ m
8. 30 mm = ▨ cm
9. 84.5 m = ▨ km
10. 4.8 cm = ▨ mm
11. 14 m = ▨ mm
12. 31.8 mm = ▨ m
13. 36 km = ▨ m
14. 1,838 m = ▨ km
15. 415 m = ▨ cm
16. 43 mm = ▨ cm
17. 93 m = ▨ cm
18. 1.7 m = ▨ cm
19. 54 cm = ▨ m
20. 18 km = ▨ cm
21. 45 cm = ▨ mm

Lesson 6-4 *Complete.*

1. 4 kg = ■ g
2. 632 mg = ■ g
3. 4,497 g = ■ kg
4. 15 g = ■ mg
5. 30 g = ■ kg
6. 3 mg = ■ g
7. 7.82 g = ■ mg
8. 38.6 kg = ■ g
9. 8.9 g = ■ kg
10. 6 kg = ■ g
11. 9.5 mg = ■ g
12. 63.4 kg = ■ g
13. 12.6 g = ■ mg
14. 5.6 g = ■ kg
15. 0.5 kg = ■ g
16. 21 g = ■ mg
17. 61.2 mg = ■ g
18. 61 g = ■ mg
19. 1.02 kg = ■ g
20. 10.3 mg = ■ g
21. 23.5 g = ■ mg
22. 53 mg = ■ kg
23. 95 mg = ■ kg
24. 65 mg = ■ g
25. 0.51 kg = ■ mg
26. 0.63 kg = ■ g
27. 563 g = ■ kg
28. 96.4 mg = ■ g
29. 5,347 mg = ■ g
30. 732 mg = ■ g

Lesson 6-5 *Complete.*

1. 4 L = ■ kL
2. 8 kL = ■ L
3. 2.9 kL = ■ L
4. 12 L = ■ kL
5. 5 mL = ■ L
6. 13.5 L = ■ kL
7. 1.3 L = ■ kL
8. 6.1 mL = ■ L
9. 3.5 kL = ■ L
10. 3,330 L = ■ kL
11. 4,301 mL = ■ L
12. 16 L = ■ mL
13. 0.351 kL = ■ L
14. 16.35 kL = ■ L
15. 18.5 mL = ■ L
16. 5.6 mL = ■ L
17. 63 L = ■ mL
18. 853 kL = ■ L
19. 321 L = ■ kL
20. 6.53 kL = ■ L
21. 485 mL = ■ L
22. 0.538 kL = ■ mL
23. 0.721 L = ■ mL
24. 1,471 mL = ■ L
25. 3.5 L = ■ kL
26. 402 L = ■ kL
27. 6.5 kL = ■ mL

Lesson 6-7 *Complete.*

1. 7 ft = ■ in.
2. 3 yd = ■ ft
3. 5 yd = ■ in.
4. 5 mi = ■ yd
5. 48 in. = ■ ft
6. 31,680 ft = ■ mi
7. $\frac{1}{4}$ mi = ■ ft
8. 30 in. = ■ ft
9. $3\frac{1}{2}$ yd = ■ in.
10. 69 in. = ■ ft ■ in.
11. 9 ft = ■ in.
12. 24 in. = ■ ft
13. 12 ft = ■ yd
14. 36 ft = ■ yd
15. 36 in. = ■ yd
16. 5 yd = ■ ft
17. 6 yd = ■ in.
18. 2 yd = ■ in.
19. 2 mi = ■ ft
20. 1 mi = ■ in.
21. 6 ft = ■ in.
22. 72 in. = ■ ft
23. 27 ft = ■ yd
24. 15 ft = ■ yd

Lesson 6-8   *Complete.*

1. 4,000 lb = ■ T

2. 5 T = ■ lb

3. 2 lb = ■ oz

4. 12,000 lb = ■ T

5. $\frac{1}{4}$ lb = ■ oz

6. 6 lb 2 oz = ■ oz

7. 122 oz = ■ lb ■ oz

8. 24 fl oz = ■ c

9. 8 pt = ■ c

10. 10 pt = ■ qt

11. $2\frac{1}{4}$ c = ■ fl oz

12. $1\frac{1}{2}$ pt = ■ c

13. 4 gal = ■ qt

14. 4 qt = ■ fl oz

15. 12 pt = ■ c

16. 9 lb = ■ oz

17. 15 qt = ■ gal

18. 4 pt = ■ c

19. 2,000 lb = ■ T

20. 3 T = ■ lb

21. 6 lb = ■ oz

22. 2 gal = ■ fl oz

23. 20 pt = ■ qt

24. 18 qt = ■ pt

25. 3 gal = ■ pt

26. 24 pt = ■ gal

27. 20 lb = ■ oz

Lesson 6-9   *Find the equivalent Fahrenheit temperature to the nearest degree.*

1. 20°C

2. 15°C

3. 85°C

4. 90°C

5. 100°C

6. 45°C

7. 62°C

8. 19°C

9. 2°C

10. 5°C

11. 18°C

12. 22°C

13. 26°C

14. 30°C

15. 35°C

16. 27°C

17. 36°C

18. 46°C

19. 48°C

20. 52°C

Find the equivalent Celsius temperature to the nearest degree.

21. 4°F

22. 8°F

23. 20°F

24. 35°F

25. 38°F

26. 100°F

27. 98°F

28. 90°F

29. 15°F

30. 150°F

31. 180°F

32. 21°F

33. 32°F

34. 45°F

35. 63°F

36. 72°F

37. 78°F

38. 95°F

39. 200°F

40. 210°F

Lesson 6-10   *Complete.*

1. 8 min = ■ s

2. 120 h = ■ d

3. 4 h = ■ min

4. 480 min = ■ h

5. 600 s = ■ min

6. 30 min = ■ h

7. $\frac{3}{4}$ h = ■ min

8. 4 h 40 min = ■ min

9. 9 min 20 s = ■ s

10. 5 d = ■ h

11. 12 h = ■ min

12. 16 h = ■ min

13. 12 min = ■ s

14. 360 s = ■ h

15. 4 h = ■ s

16. 12 d = ■ h

17. 72 h = ■ d

18. 180 min = ■ h

19. 12 h = ■ d

20. 14 h = ■ min

21. $5\frac{1}{4}$ h = ■ min

22. 30 d = ■ h

23. 20 h = ■ min

24. 360 s = ■ min

25. 3,600 s = ■ h

26. 18 min = ■ s

27. 5.5 h = ■ min

Lesson 7-1   *Use words and symbols to name each figure.*

1.
M   N

2.
X   Y

3.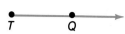
T   Q

4.
R   S   T

5.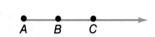
A   B   C

6.
X   Y   Z

Lesson 7-2   *Use symbols to name each angle in three ways.*

1.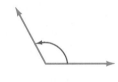

2.

3.

4.

Measure each angle.

5.

6.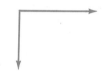

7.

8.

9.

10.

11.

12.

Lesson 7-3   *Classify each angle.*

1.

2.

3.

4.

5.

6.

7.

8.

9.

10.

11.

12.

Lesson 7-5

State whether each pair of lines is parallel, perpendicular, or skew. Use symbols to name all parallel and perpendicular lines.

1.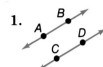

2.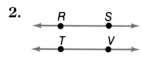

3.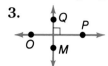

4.

5.

6.

7.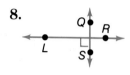

8.

Lesson 7-6

Name each polygon by the number of sides. Then state whether it is regular or not regular.

1.

2.

3.

4.

5.

6.

7.

8.

Lesson 7-7

Classify each triangle by its sides.

1.

2.

3.

4.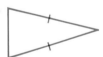

5.

6.

7.

8.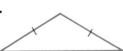

Classify each triangle by its angles.

9.

10.

11.

12.

13.

14.

15.

16.

Lesson 7-8   *Classify each quadrilateral.*

1.

2.

3.

4.

5.

6.

7

8.

9.

10.

11.

12.

Lesson 7-9   *Name each shape.*

1.

2.

3.

4.

5.

6.

7.

8.

Lesson 8-1   *The diameter or radius of a circle is given. Find the circumference to the nearest tenth. Use 3.14 for π.*

1. 14 mm, diameter

2. 18 cm, diameter

3. 24 in., radius

4. 42 m, diameter

5.
20 mm

6.
3.5 m

7.
6 m

8.
4 in.

9.
16 ft

10.
2.4 cm

11.
56 mm

12.
35 in

Lesson 8-2

Find the area of each parallelogram.

1. base, 6 cm
height, 13 cm

2. base, 14 in.
height, 23 in.

3. base, 12.5 in.
height, 6 in.

4. base, 3 mm
height, 2.5 mm

5.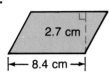

6.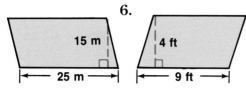

7.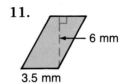

8.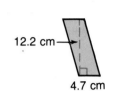

9.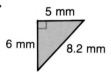

10.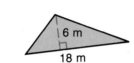

11.

12.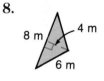

Lesson 8-3

Find the area of each triangle.

1. base, 6 ft
height, 3 ft

2. base, 4.2 in.
height, 6.8 in.

3. base, 13.2 in.
height, 16.2 in.

4. base, 9.1 m
height, 7.2 m

5.

6.

7.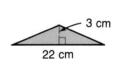

8.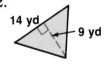

9.

10.

11.

12.

Lesson 8-4

Find the area of each circle whose radius or diameter is given. Use 3.14 for π. Round decimal answers to the nearest tenth.

1. radius, 4 m

2. diameter, 6 in.

3. radius, 12 in.

4. diameter, 16 m

5. diameter, 11 in.

6. radius, 5 in.

7. radius, 9 cm

8. diameter, 24 mm

9.

10.

11.

12.

13.

14.

15.

16.

Lesson 8-7   *Find the surface area for each rectangular prism.*

1. length = 2 in.
   width = 1 in.
   height = 10 in.

2. length = 18 m
   width = 7 m
   height = 14 m

3. length = 2.5 cm
   width = 1 cm
   height = 4.5 cm

4. length = 6 mm
   width = 4 mm
   height = 10 mm

5. length = 14 in.
   width = 7 in.
   height = 14 in.

6. length = 10 cm
   width = 10 cm
   height = 10 cm

7. length = 4.5 mm
   width = 3.6 mm
   height = 10.6 mm

8. length = 18 cm
   width = 12 cm
   height = 11 cm

9. length = 12.6 m
   width = 6.8 m
   height = 10.4 m

Lesson 8-8   *Find the surface area of each cylinder. Use 3.14 for π. Round decimal answers to nearest tenth.*

1.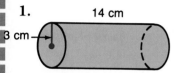
14 cm, 3 cm

2.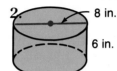
8 in., 6 in.

3.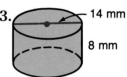
14 mm, 8 mm

4.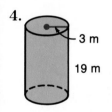
3 m, 19 m

5.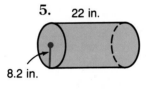
22 in., 8.2 in.

6.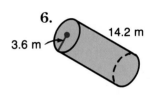
14.2 m, 3.6 m

7. radius = 4.2 cm
   height = 12.4 cm

8. radius = 5 in.
   height = 10 in.

9. radius = 6.3 in.
   height = 4.6 in.

Lesson 8-9   *Find the volume of each rectangular prism.*

1.
3 m, 3 m, 3 m

2.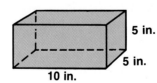
5 in., 5 in., 10 in.

3.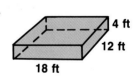
4 ft, 12 ft, 18 ft

4.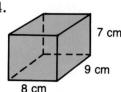
7 cm, 9 cm, 8 cm

5.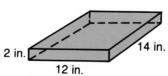
2 in., 14 in., 12 in.

6.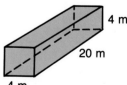
4 m, 20 m, 4 m

7. length = 4 mm
   width = 12 mm
   height = 1.5 mm

8. length = 16 cm
   width = 20 cm
   height = 20.4 cm

9. length = 8.5 m
   width = 2.1 m
   height = 7.6 m

Lesson 8-10

Find the volume of each pyramid.

1. $\ell = 4$ cm
   $w = 12$ cm
   $h = 6.3$ cm

2. $\ell = 8$ m
   $w = 10$ m
   $h = 2.3$ m

3. $\ell = 5$ in.
   $w = 8$ in.
   $h = 12$ in.

4.
   5 cm
   3 cm   4 cm

5.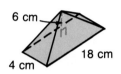
   60 m
   60 m
   60 m

6.
   24 ft
   12 ft   10 ft

7.
   3 ft
   4 ft
   2 ft

8.
   6 cm
   4 cm   18 cm

9.
   14 in.
   7 in.
   7 in.

Lesson 8-11

Find the volume of each cylinder. Use 3.14 for π. Round to the nearest tenth.

1. $r = 6$ cm
   $h = 12$ cm

2. $r = 4$ m
   $h = 12$ m

3. $r = 3.5$ mm
   $h = 4.2$ mm

4. $r = 9$ in.
   $h = 13$ in.

5. $r = 15$ mm
   $h = 20$ mm

6. $r = 8$ in.
   $h = 11$ in.

7.
   6 in.
   11 in.

8.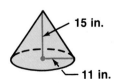
   8 m
   13 m

9.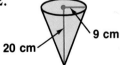
   7 cm
   30 cm

Lesson 8-12

Find the volume of each cone. Use 3.14 for π. Round decimal answers to nearest tenth.

1. $r = 6$ m
   $h = 10$ m

2. $r = 5$ in.
   $h = 13$ in.

3. $r = 1$ cm
   $h = 8.2$ cm

4. $r = 8$ mm
   $h = 4.5$ mm

5. $r = 4$ cm
   $h = 7.2$ cm

6. $r = 9$ in.
   $h = 14.2$ in.

7. $r = 3$ cm
   $h = 7$ cm

8. $r = 7$ m
   $h = 12.7$ m

9. $r = 2$ mm
   $h = 6.4$ mm

10. 12 yd
    7 yd

11. 15 in.
    11 in.

12. 20 cm
    9 cm

Lesson 9-1   *Write each ratio as a fraction in lowest terms.*

1. 14 wins to 10 losses

2. 12 boys to 20 girls

3. 30 pennies to 25 dimes

4. 20 tickets for $300

5. 5 in. of snow in 10 hours

6. 18 cars on 3 trucks

7. 285 students for 16 classrooms

8. 300 miles in 6 hours

9. 211 seniors in 2 schools

10. 58 stores in 3 shopping malls

11. 48 wins to 12 losses

12. 24 pairs of socks to 2 feet

13. 1,800 logs for 10 fireplaces

14. 12 teachers for 300 students

15. 2 tickets for $35

16. 8 lb of fruit for $3.20

Lesson 9-2   *A die is rolled once. Write the probability of rolling each.*

1. a 1 or 4

2. a 2 or 5

3. a 0

4. a 7 or 1

5. a 3 or 6

6. a 1 or 5

7. an even number

8. a number larger than 3

9. a number less than 2

10. *not* a 3

11. *not* a 2 or 4

12. an odd number

A box contains two yellow pencils, four red pencils, and three blue pencils. One pencil is chosen. Find the probability of each event.

13. P (red)

14. P (blue)

15. P (green)

16. P (yellow)

17. P (red or yellow)

18. P (blue or yellow)

19. P (red or blue)

20. P (*not* blue)

21. P (*not* red)

In a bag there are three quarters, two nickels, and four dimes. One coin is chosen. What is the probability of each?

22. P (quarter)

23. P (nickel)

24. P (dime)

25. P (penny)

26. P (quarter or dime)

27. P (dime or nickel)

28. P (*not* quarter)

29. P (*not* dime)

30. P (quarter, nickel, or dime)

Lesson 9-3   *Determine if each pair of ratios forms a proportion.*

1. 3 to 5, 5 to 10

2. 8 to 4, 6 to 3

3. 10 to 15, 5 to 3

4. 2 to 8, 1 to 4

5. 6 to 18, 3 to 9

6. 14 to 21, 12 to 18

7. 4 to 20, 5 to 25

8. 9 to 27, 1 to 3

9. 4 to 9, 5 to 10

10. 18 to 20, 27 to 30

11. 15 to 18, 5 to 6

12. 25 to 9, 42 to 8

13. 9 to 4, 9 to 5

14. 18 to 9, 20 to 10

15. 42 to 3, 28 to 2

16. 8 to 9, 16 to 18

17. 5 to 1, 7 to 2

18. 26 to 21, 21 to 26

Lesson 9-4

Solve each proportion.

1. $\frac{2}{3} = \frac{a}{12}$ 2. $\frac{7}{8} = \frac{c}{16}$ 3. $\frac{3}{7} = \frac{21}{d}$ 4. $\frac{2}{5} = \frac{18}{x}$ 5. $\frac{9}{10} = \frac{27}{m}$

6. $\frac{3}{5} = \frac{n}{21}$ 7. $\frac{5}{12} = \frac{b}{5}$ 8. $\frac{4}{36} = \frac{2}{y}$ 9. $\frac{3}{10} = \frac{z}{36}$ 10. $\frac{4}{5} = \frac{r}{100}$

11. $\frac{4}{5} = \frac{8}{b}$ 12. $\frac{12}{3} = \frac{x}{1}$ 13. $\frac{9}{2} = \frac{y}{6}$ 14. $\frac{14}{7} = \frac{7}{m}$ 15. $\frac{21}{3} = \frac{r}{15}$

16. $\frac{5}{4} = \frac{n}{100}$ 17. $\frac{3}{2} = \frac{y}{18}$ 18. $\frac{4}{7} = \frac{16}{r}$ 19. $\frac{8}{11} = \frac{y}{33}$ 20. $\frac{15}{17} = \frac{30}{w}$

21. $\frac{9}{2} = \frac{9}{n}$ 22. $\frac{3}{8} = \frac{m}{40}$ 23. $\frac{7}{11} = \frac{21}{y}$ 24. $\frac{15}{21} = \frac{5}{z}$ 25. $\frac{22}{25} = \frac{k}{10}$

26. $\frac{24}{48} = \frac{j}{50}$ 27. $\frac{9}{27} = \frac{t}{42}$ 28. $\frac{3}{19} = \frac{c}{38}$ 29. $\frac{40}{100} = \frac{2}{g}$ 30. $\frac{18}{45} = \frac{4}{f}$

31. $\frac{5}{9} = \frac{p}{10}$ 32. $\frac{7}{9} = \frac{q}{24}$ 33. $\frac{18}{7} = \frac{z}{2}$ 34. $\frac{14}{1} = \frac{7}{y}$ 35. $\frac{18}{19} = \frac{s}{38}$

36. $\frac{2}{3} = \frac{t}{4}$ 37. $\frac{9}{10} = \frac{r}{25}$ 38. $\frac{16}{19} = \frac{v}{48}$ 39. $\frac{7}{8} = \frac{a}{12}$ 40. $\frac{14}{3} = \frac{12}{m}$

Lesson 9-5

Find the distance on a scale drawing for each actual distance. The scale is 1 in.: 20 ft.

1. 40 ft   2. 80 ft   3. 120 ft   4. 90 ft   5. 60 ft

6. 130 ft   7. 10 ft   8. 5 ft   9. 15 ft   10. 220 ft

11. 350 ft   12. 400 ft   13. 420 ft   14. 170 ft   15. 200 ft

16. 85 ft   17. 75 ft   18. 55 ft   19. 125 ft   20. 95 ft

21. 2 ft   22. 18 ft   23. 24 ft   24. 16 ft   25. 12 ft

26. 82 ft   27. 66 ft   28. 112 ft   29. 225 ft   30. 164 ft

Lesson 9-6

Find the missing length for each pair of similar figures.

1.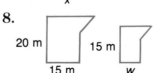

2.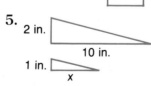

3.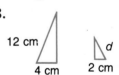

4.

5.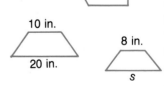

6.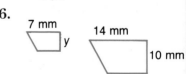

7.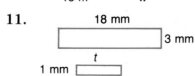

8.

9.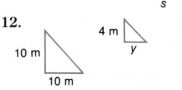

10.

11.

12.

Lesson 9-8

Write each percent as a fraction in simplest form.

1. 40% **2.** 10% **3.** 15% **4.** 23% **5.** 2%

6. 8% **7.** 4% **8.** 80% **9.** 400% **10.** 49%

11. 310% **12.** 550% **13.** 24.5% **14.** 46.25% **15.** 13.75%

16. 42% **17.** 4.5% **18.** 92.75% **19.** 16.5% **20.** 735%

Write each fraction as a percent.

21. $\frac{2}{100}$ **22.** $\frac{26}{100}$ **23.** $\frac{2}{5}$ **24.** $\frac{1}{2}$ **25.** $\frac{3}{25}$ **26.** $\frac{3}{10}$

27. $\frac{2}{50}$ **28.** $\frac{3}{20}$ **29.** $\frac{20}{25}$ **30.** $\frac{21}{50}$ **31.** $\frac{8}{10}$ **32.** $\frac{10}{25}$

33. $\frac{10}{16}$ **34.** $\frac{7}{8}$ **35.** $\frac{5}{12}$ **36.** $\frac{3}{8}$ **37.** $\frac{4}{6}$ **38.** $\frac{1}{3}$

39. $\frac{7}{10}$ **40.** $\frac{4}{20}$ **41.** $\frac{1}{4}$ **42.** $\frac{2}{3}$ **43.** $\frac{9}{20}$ **44.** $\frac{19}{25}$

Lesson 9-9

Write each percent as a decimal.

1. 0.2% **2.** 4.6% **3.** 7.8% **4.** 25.4% **5.** 16.8%

6. 19% **7.** 25% **8.** 14% **9.** 98% **10.** 72%

11. 145% **12.** 223% **13.** 104% **14.** 23.7% **15.** 0.08%

16. 0.45% **17.** 0.621% **18.** 2.56% **19.** 22.71% **20.** 14.06%

Write each decimal as a percent.

21. 0.35 **22.** 0.23 **23.** 0.06 **24.** 0.08 **25.** 0.9 **26.** 0.006

27. 0.066 **28.** 0.036 **29.** 0.132 **30.** 0.778 **31.** 0.48 **32.** 0.39

33. 4.83 **34.** 5.56 **35.** 2.34 **36.** 1.8 **37.** 2.6 **38.** 5.35

39. 7.65 **40.** 0.79 **41.** 14.23 **42.** 12.17 **43.** 6.21 **44.** 9.65

Lesson 10-1

Find each percentage.

1. 25% of 20 **2.** 10% of 90 **3.** 16% of 30 **4.** 39% of 40

5. 250% of 100 **6.** 6% of 86 **7.** 78% of 50 **8.** 3% of 46

9. $66\frac{2}{3}$% of 60 **10.** $12\frac{1}{2}$% of 160 **11.** 9% of 29 **12.** 18% of 350

13. 74% of 600 **14.** 89% of 47 **15.** 435% of 30 **16.** 156% of 78

17. 19% of 200 **18.** 48% of 15 **19.** 28% of 4 **20.** 77% of 100

21. 34% of 38 **22.** 5% of 420 **23.** 55% of 134 **24.** 68% of 68

25. 25% of 48 **26.** 39% of 126 **27.** 14% of 40 **28.** 93% of 63

29. 40% of 45 **30.** 18% of 90 **31.** 31% of 13 **32.** 206% of 65

33. 22% of 300 **34.** 42% of 150 **35.** 24% of 340 **36.** 90% of 140

Lesson 10-2   *Find each percent. Use an equation.*

1. What percent of 10 is 5?
2. What percent of 16 is 4?
3. What percent of 25 is 5?
4. What percent of 12 is 3?
5. What percent of 56 is 14?
6. What percent of 63 is 56.7?
7. What percent of 80 is 20.8?
8. What percent of 400 is 164?
9. What percent of 550 is 61.6?
10. What percent of 42 is 14?
11. What percent of 4 is 20?
12. What percent of 5 is 45?
13. 15 is what percent of 10?
14. 28 is what percent of 7?
15. 27 is what percent of 54?
16. 21 is what percent of 84?
17. 23.4 is what percent of 65?
18. 111 is what percent of 148?
19. 24 is what percent of 72?
20. 61.5 is what percent of 600?

Lesson 10-3   *Find each number. Use an equation.*

1. 14% of the number is 63. Find the number.
2. 75% of the number is 27. Find the number.
3. 63% of what number is 63?
4. 55% of what number is 33?
5. 20% of what number is 5?
6. 30% of what number is 27?
7. $66\frac{2}{3}$% of what number is 40?
8. $33\frac{1}{3}$% of what number is 15?
9. 500% of what number is 45?
10. 150% of what number is 54?
11. 39 is 5% of what number?
12. 30.8 is 35% of what number?
13. 29.7 is 55% of what number?
14. 72 is 24% of what number?
15. 108 is 18% of what number?
16. 3 is $37\frac{1}{2}$% of what number?
17. 9 is $33\frac{1}{2}$% of what number?
18. 57 is 300% of what number?
19. 300 is 150% of what number?
20. 125 is 500% of what number?

Lesson 10-5   *Estimate.*

1. 17% of 36
2. 86% of 24
3. 11% of 20
4. 27% of 48
5. 33% of 12
6. 48% of 76
7. 68% of 66
8. 63% of 40
9. 39% of 50
10. 89% of 200
11. 73% of 84
12. 85% of 72
13. 9% of 32
14. 24% of 84
15. 78% of 20
16. 65% of 85
17. 48% of $23.95
18. 98% of $5.50
19. 1.5% of 135
20. 125% of 100
21. 0.6% of 205

Lesson 10-6   *Find the percent of increase. Round to the nearest percent.*

Item	Original Price	New Price		Item	Original Price	New Price
1. soup	43¢/can	52¢/can	2.	apples	99¢/pound	$1.05/lb
3. butter	88¢/lb	$1.09/lb	4.	gum	35¢/pack	48¢/pack
5. cookies	$2.39/lb	$2.59/lb	6.	soda	99¢/liter	$1.19/liter

Find the percent of decrease. Round to the nearest percent.

Item	Original Price	New Price		Item	Original Price	New Price
7. phone	$35	$29	8.	toaster	$55	$46
9. radio	$28	$19	10.	tool box	$88	$72
11. TV	$550	$425	12.	shoes	$78	$44

Lesson 10-7   *Find the discount and sale price.*

1. piano, $4,220
   discount rate, 35%

2. sweater, $38
   discount rate, 25%

3. scissors, $14
   discount rate, 10%

4. compact disc, $15.95
   discount rate, 20%

5. book, $29
   discount rate, 40%

6. answering machine, $69
   discount rate, 15%

7. motorcycle, $3,540
   discount rate, 30%

8. pants, $45
   discount rate, 50%

9. tire, $65
   discount rate, 33%

10. VCR, $280
    discount rate, 25%

Lesson 10-8   *Find the interest owed on each loan. Then find the total amount to be repaid.*

1. principal, $300
   annual rate, 10%
   time, 3 years

2. principal, $4,000
   annual rate, 12.5%
   time, 4 years

3. principal, $3,200
   annual rate, 8%
   time, 5 years

4. principal, $10,200
   annual rate, 9.5%
   time, 6 years

5. principal, $20,000
   annual rate, 14%
   time, 20 years

6. principal, $6,300
   annual rate, 6.5%
   time, 24 months

Find the interest earned on each deposit.

7. principal, $500
   annual rate, 7%
   time, 2 years

8. principal, $2,500
   annual rate, 6.5%
   time, 3 years

Lesson 11-1   *Find the median, mode, and range for each set of data.*

1. 1, 2, 4, 6, 1, 2, 3, 1
2. 5, 4, 3, 8, 9, 7, 8, 8, 9
3. 38, 92, 92, 38, 46
4. 19, 17, 83, 82, 81, 80
5. 236, 49, 55, 237, 49
6. 45, 46, 45, 45, 46, 47
7. 296, 926, 692, 296
8. 173, 171, 172, 171, 173
9. 6.6, 7.9, 8.3, 4.5, 8.3
10. 9.4, 3.8, 9.4, 4.9, 3.8
11. 12.1, 12.1, 13.2, 10.8
12. 14.4, 14.8, 14.3, 14.8, 14.9
13. 20.46, 20.64, 20.66
14. 17.94; 18.86; 17.94
15. 15, 19, 91, 51, 51, 55, 55, 56, 55
16. 43, 34, 42, 45, 43, 43, 45, 42
17. 136, 163, 163, 136, 636, 136
18. 719, 983, 919, 917, 919, 719, 917
19. 5.8, 8.9, 8.8, 8.6, 8.8, 8.8, 8.9
20. 0.4, 0.8, 1.3, 0.9, 2.0, 0.6, 0.8, 4.0
21. 1.9, 8.0, 6.3, 3.6, 1.0, 5.7, 5.6
22. 4.95, 5.95, 1.39, 4.59, 4.59, 5.59

Lesson 11-2   *Solve. Use the data at the right.*

1. Make a frequency table for the set of data.
2. Find the mean.
3. Find the median.
4. Find the mode.
5. Find the range.

Shoe Sizes Sold at Athletic Shoe Store				
10	$10\frac{1}{2}$	13	$10\frac{1}{2}$	13
9	$10\frac{1}{2}$	$10\frac{1}{2}$	11	8
$6\frac{1}{2}$	11	11	11	$8\frac{1}{2}$
8	12	$9\frac{1}{2}$	12	11

Lesson 11-4   *Make a vertical bar graph for each set of data.*

1.

Average Snowfall	
Nov.	1.2 inches
Dec.	2.6 inches
Jan.	4.8 inches
Feb.	5.3 inches
Mar.	4.2 inches
April	1.6 inches

2.

Average Points Scored per Game	
Lakers	123
Pistons	127
76ers	119
Bulls	120
Cavs	114
Pacers	118

Make a horizontal bar graph for each set of data.

3.

Cars Sold per Month	
Jan.	52
Mar.	56
May	76
July	62
Sept.	95
Nov.	100

4.

Average Hours Worked per Day	
Monday	8
Tuesday	6
Wednesday	6
Thursday	7
Friday	7
Saturday	4

Lesson 11-5   *Make a line graph for each set of data.*

1.

Temperature on July 10	
8 AM	70° F
10 AM	75° F
12 noon	82° F
2 PM	86° F
4 PM	87° F
6 PM	87° F
8 PM	85° F

2.

Height of a Child	
2 yr	2'8"
4 yr	3'6"
6 yr	4'2"
8 yr	4'5"
10 yr	4'11"
12 yr	5'4"
14 yr	5'10"

3.

Number of Students at Central	
1950	850
1960	911
1970	948
1980	1,120
1990	1,450

4.

Inches of Rainfall	
1980	23 inches
1982	40 inches
1984	38 inches
1986	42 inches
1988	36 inches

Lesson 11-6   *Make a pictograph for each set of data.*

1.

Cans of Soda Sold	
Smith's	820
Carl's	510
Amy's	220
Green's	450

2.

Total Number of Coaching Wins	
Smith	421
Allen	310
Craig	120
Lewis	266

3.

Number of Refrigerators Sold	
Salesperson 1	26
Salesperson 2	32
Salesperson 3	18
Salesperson 4	21

4.

Number of Restaurants per City	
Calico	42
Benz	31
Smithfield	26
Longsville	27

Lesson 11-7   *Make a circle graph for each set of data.*

1.

Household Income	
primary job	82%
secondary job	9%
investments	5%
gifts	3%
other	1%

2.

Sporting Goods Sales	
shoes	44%
apparel	28%
equipment	22%
magazines	6%

3.

Energy Use in Home	
Heating/cooling	51%
Appliances	28%
Lights	15%
Other	6%

4.

Students in North High School	
White	30%
Black	28%
Hispanic	24%
Asian	18%

Lesson 11-8   *Find the median and mode of the data in each stem-and-leaf plot.*

1. Stem	Leaf
0	2 4 6
1	6 6 6 7 9
2	6 6 6 5 5
3	2 2 3 3
4	1 2 2

2. Stem	Leaf
4	5 5 6
5	1 2 3 5
6	8 8 8 9 9
7	2 2 2 3 5
8	1 1 1 2

3. Stem	Leaf
0	3 6 7 8
1	2 4 6 7
2	1 1 1 3
3	7 7 8 9

Use the weights of 24 elementary school students to complete.

4. Construct a stem-and-leaf plot of the data.

5. What is the smallest weight for a student?

6. Find the range of the weights.

7. How many students weighed more than 75?

8. What is the median weight?

86	53	80	71
76	58	44	74
78	61	58	76
55	65	68	78
62	72	82	56
92	73	90	95

Lesson 11-9   *Find the upper quartile, lower quartile, and interquartile range for each set of data.*

1. 4, 6, 10, 11, 21, 25, 29, 37

2. 2, 9, 13, 17, 20, 36, 38, 51

3. 17, 27, 31, 38, 53, 42, 47

4. 38, 44, 59, 73, 84, 92, 95

5. Stem	Leaf
1	0 1
2	0 3 4
3	1 1 2
4	2 5 6
5	8 9 9
6	0 1

6. Stem	Leaf
3	5 7 9
4	5 5 5
5	2 4 4
6	8
7	2 3
8	1

7. Stem	Leaf
4	0 1 2
5	8 8 9
6	2 3 7
7	4 6 7 7
8	1
9	5

Lesson 11-10   *Use the box and whisker plot to answer each question.*

1. What is the range and interquartile range of the data?

2. The middle half of the data is between which two values?

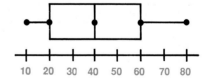

3. What part of the data is greater than 60?

4. What part of the data is greater than 40?

5. What part of the data is less than 40?

6. What numbers do 10 and 80 represent?

7. What is the median?

8. What are the upper and lower quartiles?

Lesson 12-1

Replace each ● with <, >, or = to make a true sentence.

1. −3 ● 0
2. −1 ● −2
3. −5 ● −4
4. 6 ● −7
5. 8 ● 10
6. −6 ● 6
7. −11 ● −20
8. −8 ● 2
9. −13 ● −12
10. 5 ● 2
11. 9 ● −8
12. 19 ● −19
13. |−2| ● |5|
14. |13| ● |−19|
15. |−6| ● |2|
16. |14| ● |−14|
17. |0| ● |−4|
18. |23| ● |−20|
19. |−75| ● |75|
20. |−32| ● |30|
21. −10 ● 18
22. −20 ● −38
23. −71 ● 72
24. −15 ● −35

Order from least to greatest.

25. −2, −8, 4, 10, −6, −12,
26. 19, −19, −21, 32, −14, 18
27. −5, −3, −4.5, −6.6, 1.8
28. 4.2, 5.6, −6.5, −6.6, −4.2
29. 18, 23, 95, −95, −18, −23, 2
30. 46, −48, −47, −52, −18, 12

Lesson 12-2

Add.

1. −4 + 8
2. 14 + 16
3. −7 + (−7)
4. −9 + (−6)
5. 5 + (−5)
6. −3 + (−3)
7. −18 + 11
8. −4 + 17
9. −13 + (−11)
10. −36 + 40
11. −23 + (−36)
12. −42 + 29
13. 18 + (−32)
14. −26 + 74
15. 42 + (−18)
16. −25 + 12
17. −33 + (−12)
18. 18 + (−63)
19. −38 + (−39)
20. −15 + (−10)
21. 25 + (−32)
22. 62 + (−95)
23. 82 + (−63)
24. 47 + 12
25. −96 + (−18)
26. −67 + (−14)
27. −91 + (−11)
28. 60 + (−42)
29. −81 + (−17)
30. −69 + (−32)
31. −100 + 98
32. 95 + (−5)
33. −120 + (−2)
34. 120 + (−2)
35. −120 + 2
36. −46 + (−3)

Lesson 12-3

Subtract.

1. 3 − 7
2. −5 − 4
3. −6 − 2
4. −4 − 15
5. 8 − 13
6. 6 − (−4)
7. 12 − 9
8. −2 − 23
9. 63 − 78
10. 0 − (−45)
11. −20 − 0
12. −5 − (−9)
13. −19 − (−19)
14. −8 − (−12)
15. −18 − (−26)
16. 26 − (−14)
17. 43 − (−18)
18. −18 − 23
19. −26 − 42
20. −61 − 21
21. −23 − (−42)
22. −60 − (−36)
23. −18 − (−6)
24. −43 − (−41)
25. 55 − 33
26. 58 − (−10)
27. 72 − (−19)
28. 84 − (−61)
29. −41 − 15
30. −81 − 21
31. −67 − 28
32. −51 − 47
33. −86 − (−61)
34. −4.5 − (−8.5)
35. −16.4 − (−6.5)
36. −24.6 − 8.5

Lesson 12-5

Multiply.

1. $5 \times (-2)$
2. $6 \times (-4)$
3. 4×21
4. -13×4

5. $-11 \times (-5)$
6. -6×45
7. $-9 \times (-38)$
8. -50×0

9. $64 \times (-10)$
10. -8×114
11. -3×14
12. -18×6

13. $4 \times (-20)$
14. $-4 \times (-16)$
15. $-12 \times (-12)$
16. -15×12

17. $16 \times (-5)$
18. -3×16
19. $18 \times (-10)$
20. $-5 \times (-32)$

21. $-16 \times (-12)$
22. $-80 \times (-5)$
23. 5×-12
24. $-13 \times (-3)$

25. 14×7
26. -29×10
27. $-11 \times (-11)$
28. $15 \times (-8)$

29. -9×15
30. $-7 \times (-21)$
31. $9 \times (-12)$
32. $-12 \times (-11)$

33. $(-8)(-8)(2)$
34. $5(-7)(-4)$
35. $6(3)(-2)$
36. $(-4)(-2)(-7)$

Lesson 12-6

Divide.

1. $4 \div (-2)$
2. $16 \div (-8)$
3. $-14 \div (-2)$
4. $32 \div 8$

5. $18 \div (-3)$
6. $0 \div (-1)$
7. $-42 \div 6$
8. $-63 \div (-9)$

9. $8 \div (-8)$
10. $-100 \div (-20)$
11. $15 \div (-1)$
12. $-45 \div (-10)$

13. $14 \div (-7)$
14. $-21 \div 3$
15. $25 \div (-5)$
16. $-50 \div 10$

17. $-42 \div (-7)$
18. $64 \div (-4)$
19. $-32 \div (-8)$
20. $63 \div (-7)$

21. $-81 \div (-9)$
22. $-100 \div (-20)$
23. $48 \div (-6)$
24. $-72 \div 3$

25. $-55 \div 11$
26. $28 \div (-7)$
27. $-40 \div 8$
28. $-144 \div (-4)$

29. $-36 \div (-6)$
30. $18 \div (-2)$
31. $99 \div (-9)$
32. $-56 \div (-8)$

33. $-121 \div (-11)$
34. $-81 \div 9$
35. $68 \div (-2)$
36. $169 \div (-13)$

Lesson 13-1

Solve each equation. Check your solution.

1. $x + 4 = 9$
2. $y - 3 = 15$
3. $-4 + b = 12$

4. $z - 10 = -8$
5. $-7 = x + 12$
6. $m + (-2) = 6$

7. $r - (-8) = 14$
8. $t - 13 = -3$
9. $-2 + n = -2$

10. $-19 = y - 19$
11. $a + 6 = -9$
12. $-14 + c = -12$

13. $m - 5 = 3$
14. $x - 2 = 14$
15. $r - 6 = 12$
16. $y - 9 = 15$

17. $s - 1.3 = 2.1$
18. $3.8 = t - 4.6$
19. $w - \frac{1}{4} = \frac{3}{4}$
20. $\frac{1}{3} = z - 2\frac{5}{6}$

21. $a + 7 = 10$
22. $k + 5 = 6$
23. $q + 7 = 20$
24. $h + 12 = 14$

25. $m + 2.6 = 9.8$
26. $4.3 + b = 5.8$
27. $3.1 = 2.5 + n$
28. $a + \frac{1}{3} = \frac{1}{2}$

29. $\frac{2}{5} + c = 1\frac{1}{10}$
30. $2 = q + \frac{7}{8}$
31. $1\frac{3}{4} = \frac{1}{4} + x$
32. $2\frac{5}{8} = 1\frac{3}{5} + y$

33. $\frac{3}{4} + m = -\frac{5}{8}$
34. $3 = t - 6\frac{1}{3}$
35. $\frac{4}{7} = -\frac{8}{9} + s$
36. $3\frac{1}{3} = -6\frac{1}{8} + r$

Lesson 13-2   Solve each equation.

1. $3m = -15$
2. $4x = 12$
3. $-324 = 30y$
4. $\frac{c}{-4} = 10$
5. $\frac{1}{2}w = -7$
6. $8 = \frac{y}{-4}$
7. $5q = -40$
8. $-6h = -5$
9. $r \div 7 = -8$
10. $\frac{2}{3}x = \frac{8}{15}$
11. $-12 = \frac{1}{5}y$
12. $-1 = -\frac{4}{5}m$
13. $\frac{1}{3}d = 15$
14. $\frac{1}{5}t = 25$
15. $\frac{-3}{10}f = 12$
16. $\frac{-3}{4}a = 6$
17. $0 = 6r$
18. $\frac{y}{12} = -6$
19. $\frac{-3}{8} = \frac{3}{8}k$
20. $\frac{3}{8} = \frac{1}{2}x$
21. $-1 = \frac{a}{21}$

Lesson 13-3   Solve each equation. Check your solution.

1. $2x + 4 = 14$
2. $5p - 10 = 0$
3. $5 - 6a = 41$
4. $\frac{x}{3} - 7 = 2$
5. $\frac{2}{3}y + 10 = 22$
6. $15 = \frac{1}{4}m - 6$
7. $3(r - 1) = 9$
8. $-18 = -6(q - 4)$
9. $0 = -4x + (-28)$
10. $3x - 1 = -5$
11. $-2x + 3 = 1$
12. $\frac{1}{2}a - 3 = -14$
13. $5 - \frac{3}{4}y = -7$
14. $\frac{1}{3}z + 5 = -3$
15. $-3(x - 5) = 12$
16. $\frac{1}{2}(q - 1) = 10$
17. $\frac{1}{2}x - 3 = 7$
18. $\frac{3}{4}t + 6 = 9$
19. $9a - 8 = 73$
20. $5 - 2y = 15$
21. $3(c - 4) = 3$

Lesson 13-4   Find the ordered pair for each point labeled on the coordinate plane.

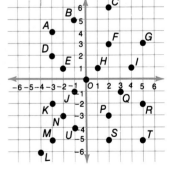

1. A
2. B
3. C
4. D
5. E
6. F
7. G
8. H
9. I
10. J
11. K
12. L
13. M
14. N
15. O
16. P
17. Q
18. R
19. S
20. T

Draw a coordinate plane. Graph each ordered pair. Label each point with the letter given.

1. A (4, 2)
2. B (2, 0)
3. C (-3, 0)
4. D (0, -3)
5. E (4, 1)
6. F (-4, 0)
7. G (-2, 1)
8. H (-2, -1)
9. I (-4, 3)
10. J (0, 5)
11. K (0, -2)
12. L (6, 5)
13. M (-6, 5)
14. N (6, -5)
15. O (-6, -3)
16. P (1, 3)
17. Q (-1, -3)
18. R (3, 4)
19. S (4, 5)
20. T (2, -2)
21. U (-3, -6)
22. V (-2, -4)
23. W (3, -1)
24. X (1, -4)
25. Y (0, 0)

Lesson 13-5 *Graph the solutions to each equation.*

1. $4x = y$ **2.** $x - 7 = y$ **3.** $5x - 2 = y$

4. $\frac{1}{2}x + 1 = y$ **5.** $3x + 2 = y$ **6.** $2x = y$

7. $4x + 2 = y$ **8.** $3x - 1 = y$ **9.** $\frac{1}{2}x = y$

10. $x + y = 6$ **11.** $x = y$ **12.** $x + 1 = y$

13. $\frac{1}{3}x = y$ **14.** $2 + x = y$ **15.** $3x = y$

16. $x - 2 = y$ **17.** $x - 5 = y$ **18.** $x - y = 0$

Lesson 14-1 *Draw a tree diagram to show the possible outcome.*

1. tossing the two coins.

2. tossing the three coins

Use multiplication to find the number of possible outcomes.

3. tossing a coin and rolling a die

4. tossing a quarter and dime and rolling a die

5. a choice of strawberry, vanilla or chocolate for a three scoop ice cream cone

6. a choice of a queen or king size bed with a firm, or super firm mattress

7. a choice of blue or gray pants with a choice of white, yellow, or blue stripe oxford shirt

8. a choice of roses or carnations in red, yellow, pink, or white

Lesson 14-2 *A drawer contains two red socks, four blue socks, and six black socks. One sock is chosen from the drawer. Find the chance of choosing each of the following.*

1. a black sock   **2.** a red sock   **3.** a blue sock

4. a brown sock   **5.** *not* a red sock   **6.** *not* a blue sock

7. *not* a black sock   **8.** a red or blue sock   **9.** a black or red sock

10. a green or blue sock   **11.** a blue or black sock   **12.** a blue or a white sock

There are three quarters, five dimes, and twelve pennies in a bag. One coin is chosen. Find the chance of choosing each of the following.

13. a quarter   **14.** a penny   **15.** a dime

16. a nickel   **17.** a quarter or a dime   **18.** a dime or a penny

19. a nickel or a dime   **20.** not a dime   **21.** not a penny

22. a quarter or a penny   **23.** not a quarter   **24.** not a nickel

Lesson 14-3

A bag contains four blue chips, five red chips, and three green chips. A chip is chosen from the bag. Find the probability of each event.

1. P(a blue chip)
2. P(a red chip)
3. P(a green chip)
4. P(a white chip)
5. P(a red or blue chip)
6. P(a red or green chip)
7. P(*not* a green chip)
8. P(*not* a blue chip)
9. P(*not* a red chip)
10. P(*not* a white chip)
11. P(*not* a blue or red chip)
12. P(*not* a green or blue chip)

Lesson 14-4

Use the same situation in Lesson 14-3 to find the probability of each event.

1. draw a red chip, replace it, draw a green chip
2. draw a blue chip, replace it, draw a blue chip
3. draw a green chip, replace it, draw a red chip
4. draw a green chip, replace it, draw a blue chip
5. draw a blue chip, do not replace it, draw a green chip
6. draw a red chip, do not replace it, draw a green chip

A box contains three red, five blue, and two purple marbles. Suppose you choose a marble, replace it, and choose another marble. Find the probability of each event.

7. P (red, then blue)
8. P (blue, then purple)
9. P (purple, then red)
10. P (blue, then red)
11. P (red both times)
12. P (blue both times)
13. P (purple both times)
14. P (same color both times)

Lesson 14-6

A box contains ten blue blocks, three yellow blocks, and five purple blocks. One block is chosen from the box. Find the odds for each event.

1. P (a blue block)
2. P (a yellow block)
3. P (a purple block)
4. P (a white block)
5. P (a green block)
6. P (*not* a blue block)
7. P (*not* a yellow block)
8. P (*not* a purple block)
9. P (a blue or purple block)
10. P (a yellow or blue block)
11. P (a green or blue block)
12. P (a green or white block)
13. P (a purple or yellow block)
14. P (a purple or blue block)
15. P (a purple or white block)
16. P (*not* a purple or blue block)
17. P (*not* a blue or yellow block)
18. P (*not* a green or blue block)
19. P (*not* a yellow or purple block)
20. P (*not* a blue or purple block)
21. P (*not* a green or yellow block)
22. P (*not* a red, yellow, or blue block)
23. P (*not* a yellow, blue, or black block)

GLOSSARY

absolute value (376)   The number of units a number is from zero on the number line.

acute angle (216)   An angle with measure between 0° and 90°.

acute triangle (226)   A triangle with all acute angles.

angle (214)   Two rays with a common endpoint.

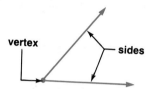

arc (211)   A part of a circle.

area (94)   The number of square units that cover a surface.

arithmetic sequence (54)   A sequence where the difference between consecutive numbers is the same.

axis (408)   Number line that is in the coordinate plane.

bar graph (348)   A graph which is used to compare quantities. The length of each bar represents a number.

bases (230)   The two polygon shaped parallel sides of a polyhedron.

BASIC (222)   Beginner's All-Purpose Symbolic Instruction Code. Language used for micro computers.

basic units (178)   Units used in the metric system to show length, mass, and capacity (gram, liter, meter).

bisect (219)   Separate into two congruent parts.

box-and-whisker plot (362)   A graph that shows the median, quartiles, and extremes of a set of data.

center (230)   A given point of a sphere from which all points on the sphere are an equal distance.

chance (428)   The probability of an event expressed as a percent.

circle (244)   A closed path of points in a plane, all the same distance from a fixed point called the center.

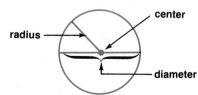

circle graph (356)   A graph used to compare parts of a whole. The whole amount is shown as a circle and each part is shown as a percent of the whole.

circumference (244)   The distance around a circle found by using formula $C = 2\pi r$.

commission (310)   Earnings based on a percent of sales.

common ratio (88)   The same number used in a geometric sequence to multiply or divide by to obtain consecutive numbers.

compass (211)   A tool used to draw circles and parts of circles.

complementary (217)   Two angles whose sum is 90°.

composite number (110)   A number that has more than two factors.

compound interest (330)   Interest computed at stated intervals and added to the principal at the end of the interval.

cone (230)   A three dimensional figure with a circular base and one vertex.

congruent triangles (248)   Triangles that have the same size and shape.

constructions (211)   Drawings that are made with only a compass and a straight edge.

coordinate plane (408)   A plane on which ordered pairs are graphed consisting of two perpendicular number lines that are the axes.

cross products (286)   A way to determine if two ratios form a proportion. If the cross products are equal then the ratio forms a proportion. In the proportion $\frac{2}{3} = \frac{8}{12}$, the cross products are 2×12 and 3×8.

cube (230)   A rectangular prism that has six square faces.

cup (194)   A customary unit of capacity equal to 8 fluid ounces.

cylinder (230)   A three dimensional figure with two parallel congruent circular bases.

data (96)   Information in the form of numbers.

decagon (224)   A polygon with 10 sides.

degree (214)   Common unit used in measuring angles.

denominator (116)   Tells the number of objects or equal-sized parts. The bottom number in a fraction.

dependent event (430)   An event that is affected by the occurrence of another event.

diameter (244)   A line segment through the center of a circle with endpoints on the circle. See *circle.*

difference (14)   The result of subtracting one number from another.

$$9 - 4 = 5$$
$$\uparrow \text{ difference}$$

discount (326)   An amount subtracted from the regular price of an item.

discount rate (326)   The percent of decrease in the regular price of an item.

dividend (76)   The original number into which the divisor is divided.

$$40 \div 5 = 8 \qquad 5\overline{)40}^{\,8}$$
$$\uparrow\!\!—\text{dividend}—\!\!\uparrow$$

divisor (76)   The number divided into the dividend.

$$40 \div 5 = 8$$
$$\uparrow$$
$$\text{divisor}$$

down payment (74)   A small portion of the total price to be paid at the time of purchase.

edges (230)   The intersection of faces of a polyhedron.

equation (26)   A mathematical sentence with an equals sign.

equilateral triangle (226)   A triangle with 3 congruent sides.

equivalent fractions (116)   Fractions that name the same number. $\frac{3}{4}$ and $\frac{6}{8}$ are equivalent fractions.

event (282)   A specific outcome or type of outcome.

expanded form (6)   A method of writing numbers using place value and addition.

$$739 = 700 + 30 + 9 \text{ or}$$
$$(7 \times 100) + (3 \times 10) + (9 \times 1) \text{ or}$$
$$(7 \times 10^2) + (3 \times 10^1) + (9 \times 10^0)$$

exponent (6)   The number of times the base is used as a factor. In 10^3, the exponent is 3.

faces (230)   The flat surfaces of a polyhedron.

factors (108)   Numbers that divide into another number so that the remainder is zero.

finance charge (74)   The difference between the credit price and the cash price of an item.

flowchart (352)   A diagram used to help plan steps in a computer program.

fluid ounce (194)   A customary unit of capacity.

foot (190)   A customary unit of length equal to 12 inches.

formula (90)   An equation that shows how certain quantities are related.

fraction (116)   A number used to name part of a whole or group.

frequency table (344)   A table for organizing a set of data that shows the number of times each item or number appears.

gallon (194)   A customary unit of capacity equal to 4 quarts.

geometric sequence (88)   A sequence of numbers in which succeeding terms are found by multiplying the preceeding term by the same number (common ratio.)

gram (178)   A basic unit of mass in the metric system.

greatest common factor (GCF) (112)   The greatest number that is a factor of two or more numbers. The greatest common factor of 24 and 30 is 6.

gross pay (114)   Total wage before deductions are made.

hexagon (224)   A polygon with six sides.

hypoteneuse (257)   The longest side of a right triangle, the side opposite the right angle. See *right triangle.*

improper fraction (124)   A fraction that has a numerator greater than or equal to the denominator.

inch (190)   A customary unit of length.

independent event (430)   An event that is *not* affected by the occurrence of another event.

integers (376)   The whole numbers and their opposites.

interest (328)   A percent of the principal that is an amount earned on a deposit or an amount owed on a loan.

interquartile range (360)   A measure of variation, it is the difference between the upper and lower quartiles.

inverse operations (26)   Pairs of operations that undo each other. Addition and subtraction are inverse operations. Multiplication and division are inverse operations.

isosceles triangle (226)   A triangle that has at least 2 congruent sides.

kilowatt hour (kWh) (51)   The unit used to measure electrical energy equal to 1 kilowatt of electricity used for 1 hour.

least common denominator (122)   The least common multiple of the denominators of fractions. The least common denominator of $\frac{4}{5}$ and $\frac{5}{6}$ is 30.

least common multiple (LCM) (112)   The least nonzero number that is a multiple of two or more numbers. The least common multiple of 6 and 8 is 24.

line (212)   All the points on a never-ending straight path. A representation of line AB $(\overleftrightarrow{AB})$ is shown below.

line graph (350)   A graph used to show change and direction of change over a period of time.

line segment (212)   Two endpoints and the straight path between them. A representation of line segment CD $(\overline{CD})$ is shown below.

liter (178)   A basic unit of capacity in the metric system.

mathematical expression (24)   Variables and/or numbers combined by symbols of operations.
$$x + 3 \quad a \times b \quad 4 - 2$$

mean (94)   An average of a set of data.

measure of variation (360)   The spread of values in a set of data.

median (342)   Middle number of the set of data when listed in order.

meter (178)   A basic unit of length in the metric system.

mile (190)   A customary unit of length equal to 5,280 feet or 1,760 yards.

mixed decimals (162)   Fractions that are equivalent to repeating decimals can be written in this form.
$$\frac{1}{3} = 0.33\frac{1}{3} \quad \frac{1}{6} = 0.16\frac{2}{3} \quad \frac{6}{7} = 0.85\frac{5}{7}$$

mixed number (124)   A number that indicates the sum of a whole number and a fraction.

mode (342)   The number that appears most often in a set of data.

multiple (112)   A number obtained by multiplying a given number by any whole number. A multiple of 12 is 24.

net pay (114)   Take home pay found after subtracting total and personal deductions from gross pay.

numerator (116)   Tells the number of objects or parts being considered. The top number in a fraction.

obtuse angle (216)   An angle with measure between 90° and 180°.

obtuse triangle (226)   A triangle with an obtuse angle.

octagon (224)   A polygon with eight sides.

odds (438)   A ratio of the number of ways an event can occur to the number of ways the event cannot occur.

opposites (376)   Two integers are opposites if their sum is 0. The integers 2 and -2 are opposites because $2 + (-2) = 0$.

ordered pair (408)   A pair of numbers where order is important. Ordered pairs may be graphed on a coordinate plane.

origin (408)   The point at which the axes intersect in the coordinate plane.

ounce (194)   A customary unit of weight.

outcome (282)   A possible result of a probability experiment. When a coin is tossed, one outcome is tails.

output (50)   The results of computer processing. Two output devices and the cathode-ray tube (CRT) and the printer.

parallel lines (220)   Lines that are in the same plane that do not intersect.

parallelogram (228)   A quadrilateral with two pairs of parallel sides.

pentagon (224)   A polygon with five sides.

percent (296)   A way of expressing hundredths using the percent symbol (%). Thus, 7% = $\frac{7}{100}$ or 0.07.

perimeter (P) (58)   The distance around a polygon.

perpendicular lines (220)   Two lines that intersect to form right angles.

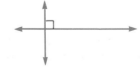

pi (π) (244)   The ratio of the circumference of a circle to the diameter of a circle. π is approximately equal to 3.14.

pictograph (354)   A graph used to compare data in a visually appealing way.

pint (194)   A customary unit of capacity equal to 2 cups.

place value (4)   A system for writing numbers. In this system the position of a digit determines its value.

plane (212)   All the points in a never-ending flat surface.

point (212)   An exact location in space.

polygons (224)   Closed plane figures formed by line segments.

polyhedron (230)   A solid figure with flat surfaces.

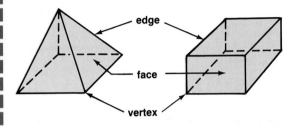

pound (194)   A customary unit of weight equal to 16 ounces.

powers (6)   Expressions written with exponents.
$$10^3 \quad 3^4 \quad 8^3$$

prime factorization (110)   A composite number that is expressed as the product of prime numbers. The prime factorization of 12 is $2 \times 2 \times 3$.

prime number (110)   A number that has exactly two factors, 1 and the number itself.

principal (328)   The amount of money borrowed or invested on which interest is based.

prism (230)   A polyhedron with two parallel congruent bases that are shaped like polygons.

probability (282)   A ratio that describes how likely it is that an event will occur.

product (16)   The result of multiplying two or more numbers.
$$3 \times 5 = 15$$

proper fraction (124)   A fraction that has a numerator less than the denominator.

proportion (284)   A mathematical sentence that states two ratios are equivalent.

protractor (214)   A device used in measuring angles.

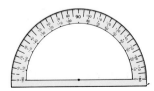

pyramid (230)   A polyhedron with a single base shaped like a polygon.

quadrilateral (224)   A polygon with four sides.

quart (194)   A customary unit of capacity equal to 2 pints.

quartiles (360)   Values that divide data into four equal parts.

quotient (76)   The result of dividing one number by another.

radius (244)   A line segment from the center of a circle to any point on the circle. See *circle.*

range (342)   The difference between the greatest and least number in a set of data.

ratio (280)   A comparison of two numbers by division. A comparison of 9 and 12 can be written as 9 to 12, 9:12, and $\frac{9}{12}$.

ray (212)   All the points in a never-ending straight path extending in only one direction. A representation of ray *DE* ($\overrightarrow{DE}$) is shown below.

reciprocals (156)   Two numbers whose product is 1. Since $\frac{5}{12} \times \frac{12}{5} = 1$, the reciprocal of $\frac{5}{12}$ is $\frac{12}{5}$.

rectangle (228)   A parallelogram with four right angles.

regular polygon (224)   A polygon in which all sides are congruent and all angles are congruent.

remainder (76)   The whole number left after one number is divided into another number.

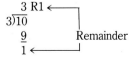

repeating decimal (162)   A decimal whose digits repeat in groups of one or more.
$$0.666\ldots \text{ or } 0.\overline{6}$$

rhombus (228)   A parallelogram with all sides congruent.

right angle (216)   An angle that measures 90°.

right triangle (226)   A triangle with one right angle.

hypotenuse

sample (436)   A smaller, representative group of a larger group surveyed when it is unreasonable to survey the large group.

sample space (426)   A list of all possible outcomes from an action, such as tossing a coin or rolling a die.

scale drawing (288)   A geometrically similar representation of something too large or too small to be conveniently drawn to actual size.

scalene triangle (226)   A triangle with no congruent sides.

scientific notation (86)   A way of expressing numbers as the product of a number that is at least 1, but less than 10 and a power of 10. In scientific notation 4,100 is written 4.1×10^3.

sequence (54)   A list of numbers that follows a certain pattern.

sides (214)   The two rays that form an angle.

similar figures (292)   Two geometric figures that have the same shape but may differ in size. Corresponding sides of similar figures are proportional.

simplest form (118)   The form of a fraction when the GCF of the numerator and denominator is 1.

skew lines (220)   Lines that do not intersect and are not parallel.

sphere (232)   A solid with all points the same distance from a given point.

square (228)   A parallelogram with all sides and angles congruent.

standard form (4)   A number written using only the digits and place value. The standard form of seven hundred thirty nine is 739.

stem-and-leaf plot (358)   A graph for organizing data in which each number is separated into two parts. One part forms the stem and one part forms the leaves. The data is displayed in two columns, the stem on the left and the leaves on the right.

straightedge (211)   Any object that can be used to draw a straight line. Important tool in geometry.

sum (14)   The result of adding two or more numbers.
$$5 + 3 = 8$$

surface area (258)   The sum of the areas of each surface of a three-dimensional figure.

symmetry (233)   A shape such that a line may be placed on a figure so that the shape on one side of the line matches the shape on the other side or that the parts are congruent.

terminating decimal (162)   A quotient in which the division ends or terminates with a remainder of zero.

ton (194)   A customary unit of weight equal to 2,000 pounds.

trapezoid (228)   A quadrilateral with only one pair of parallel sides.

triangle (224)   A polygon with three sides.

undefined quotient (389)   Since multiplication and division are inverse operations, division by zero results in a contradiction. If $5 \div 0 = n$, then there is *no* related statement $0 \times n = 5$ because there is no number that can replace n and result in a true statement. Therefore, we say that division by zero is undefined.

unit fraction (151)   A fraction that has a numerator of 1.

variable (24)   A symbol, usually a letter, used to represent a number.

vertex (214)   The common endpoint of two rays that form an angle.

volume (262)   The number of cubic units required to fill a space.

yard (190)   A customary unit of length equal to 3 feet or 36 inches.

INDEX

SELECTED ANSWERS

CHAPTER 1 APPLYING NUMBERS AND
VARIABLES

Pages 4-5 Lesson 1-1
1. tens **2.** thousands **3.** ten thousands **4.** tenths
5. hundredths **6.** thousandths **7.** hundreds **8.** ones
9. six hundred three **11.** two thousand, seven and eight
tenths **13.** 0.420 **15.** 0.040 **17.** 50 **19.** 8 **21.** 0
23. 2 **25.** 7 **27.** 1 **29.** one hundred and five
hundredths **31.** three hundred three and three
hundredths **33.** 3,000 **35.** 0.48 **37.** 0.025
39. 4, 1, 3, 5, 2 **41.** seventy-three billion, seventy
three thousand million **43.** 0.030

Pages 6-7 Lesson 1-2
1. $10 \times 10 \times 10$; 1,000 **2.** 5×5; 25 **3.** $2 \times 2 \times 2$
$\times 2$; 16 **4.** 12×12; 144 **5.** $20 \times 20 \times 20$; 8,000
6. 7^3 **7.** 25^2 **8.** 5^4 **9.** $(1 \times 10^2) + (5 \times 10^1) +$
(6×10^0) **10.** $(2 \times 10^3) + (9 \times 10^2) + (3 \times 10^1) +$
(3×10^0) **11.** $(7 \times 10^2) + (8 \times 10^0)$
12. $(4 \times 10^3) + (7 \times 10^1) + (6 \times 10^0)$
13. $(9 \times 10^4) + (6 \times 10^3)$ **15.** $3 \times 3 \times 3$; 27
17. 8^2 **19.** 3^4 **21.** 11^2 **23.** 475 **25.** $(3 \times 10^2) +$
$(4 \times 10^1) + (5 \times 10^0)$ **27.** $(3 \times 10^3) + (7 \times 10^2) +$
(5×10^0) **29.** 3,136 **31.** 662,596 **33.** 181,476
35. 56,715,961 **37.** 25 tiles **39.** 10^{-1} **40.** 3
41. 1 **42.** 9 **43.** 8 **44.** 500 **45.** 1,600 **46.** 50
47. $28,000

Page 9 Lesson 1-3
1. < **2.** = **3.** > **4.** < **5.** < **6.** > **7.** =
8. 5, 5.05, 5.105, 5.15, 5.51 **9.** 237, 273, 327, 372,
723, 732 **11.** = **13.** > **15.** < **17.** 0.128, 0.821,
1.28, 1.82 **19.** 1,002, 1,008, 1,035, 1,156
21. Ehrhardt, Komisova, Donkova **26.** tens
27. hundredths **28.** thousands **29.** hundreds
30. 4^4 **31.** 15^3 **32.** 5^3 **33.** $9

Page 11 Lesson 1-4
1. 4,000 **2.** 1,000 **3.** 1,000 **4.** 10,000 **5.** 45,000
6. 1 **7.** 101 **8.** 50 **9.** 0 **10.** 10 **11.** 0.78
12. 12.86 **13.** 0.10 **14.** $8.55 **15.** $0.40 **17.** 60
19. 410 **21.** 1,000 **23.** 6,730 **25.** 80,000
27. 93,000,000 **29.** $40.00 **31.** 0.0 **33.** 0.2700
35. 1.000 **37.** six thousand, thirty-five

39. twenty-five thousandths **41.** $35 **43.** 7 shares
45. 910 miles **47.** 240 miles **49.** exact
51. exact

Pages 12-13 Lesson 1-5
1. $3,056 **2.** $2,930 **3.** $2,944 **4.** $3,126 **5.** $2,972
6. $3,084 **7.** $3,182 **8.** $3,070 **9.** $3,427 **11.** $2,989
13. $19,549 **17.** $2,951 **19.** They pay the same
amount. **21.** > **22.** < **23.** > **24.** 20
25. 300 **26.** 50,000 **27.** 0.880 **28.** 115,124

Pages 14-15 Lesson 1-6
1. 500 **2.** 10,000 **3.** 4,600 **4.** 0.7 **5.** 11.4 **6.** $23
7. 148 **8.** 300 **9.** 6,200 **10.** 5.000 **11.** 0.1
12. $8.00 **13.** 20,000 **14.** $80 **15.** 900 **17.** 2,000
19. 100.00 **21.** 28.000 **23.** 2.00 **25.** 4,100
27. $3.70 **29.** 31 **31.** $2 **33.** no **35.** 7 minutes
37. no The estimate is about $50. Even with tax $67.56
is too much.

Pages 16-17 Lesson 1-7
1. 1,800 **2.** 180,000 **3.** 0 **4.** 600,000 **5.** $77 **6.** 16
7. 2 **8.** 8 **9.** 4 **10.** 60 **11.** 200 **13.** 20 **15.** 120
17. 21,000 **19.** 68 **21.** 7 **23.** 55 **25.** $5
27. 20,000 **29.** $600 **31.** 30 **33.** 33 **35.** Did not
multiply $4,000 \times 30$ correctly. Four zeros are
required. **37.** 23 mpg **39. a.** 60 **41. b.** 225 million

Pages 20-21 Lesson 1-8
1. John-1; Aaron-2; Lyn-3; Al-5; Diane-6
2. Dan-teacher; William-lawyer; Laurie-pediatrician
3. Paul-pepperoni and anchovies; David-sausage;
Ben-pepperoni and sausage **5.** 28 pieces **7.** Jack-rock;
Kevin-heavy metal; Jason-jazz **9.** 21st floor **11.** 3
12. 46 **13.** 9 **14.** 64 **15.** 1,200 **16.** 6 **17.** 5,200
18. 570 **19.** 200 miles

Pages 22-23 Lesson 1-9
1. 50 **2.** 18 **3.** 48.3 **4.** 35 **5.** 2.4 **6.** 0 **7.** 2
8. 8 **9.** 10 **10.** 54 **11.** 64 **12.** 16 **13.** 14
15. 12 **17.** 120 **19.** 20 **21.** 9 **23.** 80 **25.** 10
27. 60 **29.** $2 **31.** The tapes alone would weigh 360
pounds. **35.** 16 **37.** 20 **38.** 1.6 **39.** 110 **40.** $14

41. 48 **42.** 50 **43.** 6,400 **44.** Jenny-Ft. Myers, Betty-Daytona, Ashley-Miami

Page 25   Lesson 1-10
1. $7 + n$ **2.** $n - 5$ **3.** $16 \div n$, $\frac{16}{n}$ **4.** $8 \times n$, $8n$
5. 39 **6.** 28 **7.** 30 **8.** 1 **9.** $n + 8$ **11.** $3 - n$
13. $7n$ **15.** $n - 3$ **17.** 17 **19.** 88 **21.** 6 **23.** 1
25. 12 **27.** seven more than a number **29.** a number divided by 9 **31.** a number decreased by 26 **33.** 4; 11
35. 4; 17 **37.** 5.61 **39.** 22.4196

Page 27   Lesson 1-11
1. 5; 14 **2.** 20; 50 **3.** 0.6; 0.8 **4.** 9; 18 **5.** 3; 2
6. 4.7; 3.2 **7.** 40; 50 **8.** 4; 5 **9.** 16 **11.** 14 **13.** 2
15. 11 **17.** 18 **19.** 1 **21.** 0.3 **23.** 4 **25.** 16
27. 50 tapes **29.** 3 empty backpacks

Page 29   Lesson 1-12
1. 3; 4 **2.** 7; 3 **3.** 20; 4 **4.** 3; 0.2 **5.** 5; 35
6. 10; 300 **7.** 2; 0.6 **8.** 25; 250 **9.** 8 **11.** 0
13. 0.04 **15.** 48 **17.** 42 **19.** 0.8 **21.** 400
23. 400 **25.** 12 **27.** 400 **29.** $20
31. $400,900,000,000

Pages 30-31   Lesson 1-13
1. a. $632 + 346 = y$; 978 people **2. a.** $20 \times 5 = z$; 100 audio tapes **3.** $590 **5.** 700 calories **7.** 2,100 seats **9.** 26 weeks **11.** $11,125 **14.** 216 **15.** 64
16. 144 **17.** 6,561 **18.** $6 + n$ **19.** $11 - n$
20. $12n$ **21.** $23 + n$ **22.** n = Frank's age now; $n + 14 = 30$; $n = 16$

Pages 32-33   Chapter 1 Review
1. $(6 \times 10^2) + (4 \times 10^0)$ **3.** estimating **5.** 8.030
7. equation **9.** variable **11.** 360 **13.** 50
15. $(2 \times 10^1) + (4 \times 10^0)$ **17.** $(2 \times 10^3) + (6 \times 10^2) + (5 \times 10^1)$ **19.** $(7 \times 10^3) + (5 \times 10^0)$ **21.** 10.708, 10.78, 10.87, 10.871 **23.** 10,000 **25.** 10.7
27. $3,056 **29.** $2,876 **31.** 3,200 **33.** $6.00
35. 24,000 **37.** 100 **39.** 90 **41.** 24.6 **43.** 7
45. 11 **47.** 2 **49.** 81 **51.** $y = 5$ **53.** $k = 7$
55. $p = 70$ **57.** $m = 7$ **59.** $y = 8$ **61.** $65
63. yes **65.** $220

Pages 36-37   Cumulative Review/Test
Free Response: **1.** 40 **3.** 200 **5.** 7 **7.** 2
9. 6 **11.** 200 **13.** 81 **15.** 343 **17.** 1,000,000
19. $(4 \times 10^3) + (3 \times 10^2) + (8 \times 10^1) + (1 \times 10^0)$ **21.** $(3 \times 10^4) + (3 \times 10^3) + (3 \times 10^2) + (3 \times 10^1) + (2 \times 10^0)$ **23.** 43, 44, 45 **25.** Bill Huck
27. 10.500 **29.** 25.12 **31.** 0.1 **33.** 7.70
35. 6,700 **37.** 1 **39.** 0.7 **41.** 500 **43.** 7
Multiple Choice: **1.** d **3.** b **5.** c **7.** a **9.** b

CHAPTER 2   PATTERNS: ADDING AND SUBTRACTING

Pages 42-43   LESSON 2-1
1. 79 **2.** 38 **3.** 98 **4.** 128 **5.** 137 **6.** 919 **7.** 880
8. 673 **9.** 602 **10.** 33,264 **11.** 46,360 **12.** 76,886
13. 96 **15.** 83 **17.** 1,837 **19.** 5,100 **21.** 69,009
23. 1,102 **25.** 163,703 **27.** 19,084 **29.** 290 km
31. $455 **34.** 5 **35.** 28 **36.** 20 **37.** 8 **38.** 34
39. 9 **40.** 82 **41.** $2.30

Pages 44-45   Lesson 2-2
1. 0.97 **2.** 7.978 **3.** 30.1 **4.** 30.79 **5.** 57.91
6. 92.378 **7.** 22.03 **8.** 197.28 **9.** 12.49 **10.** 21.38
11. 947.79 **12.** 21,259.2 **13.** 0.41 **15.** 2.71
17. $2.00 **19.** $51.59 **21.** 8.565 **23.** $942.46
25. $18.13 **27.** 1.886 **29.** 43.79 seconds **31.** 5.4%
33. $t + 0.12$ **37.** less

Pages 46-47   Lesson 2-3
1. 25 **2.** 23 **3.** 39 **4.** 114 **5.** 108 **6.** 3,461
7. 4,705 **8.** 4,065 **9.** 3,409 **10.** 1,342 **11.** 32
13. 23 **15.** 418 **17.** 792 **19.** 10,102 **21.** 1,078
23. 964 **25.** 40,711 **27.** 1,182 feet **29.** $5.50
32. 3 **33.** 12 **34.** 108 **35.** 87 **36.** 184 **37.** 1,422
38. 33,608 **39.** 1,001

Pages 48-49   Lesson 2-4
1. 1.8 **2.** 8.18 **3.** 77.42 **4.** 0.337 **5.** 1.16 **6.** 2.95
7. 5.37 **8.** 213.75 **9.** 28.88 **10.** 2,210.5 **11.** 2.4
13. 0.722 **15.** 0.82 **17.** 2.668 **19.** 1.308 **21.** 0.11
23. 0.146 **25.** 534.327 **27.** 0.045 **29.** $341 million

Pages 52-53   Lesson 2-5
1. Numbers cannot be altered by someone else.
2. Words cannot be altered by someone else. **3.** and
5. Twenty-nine and $\frac{00}{100}$ **7.** One hundred eight and $\frac{97}{100}$
9. $1,209.25 **11.** $451.26 **13.** $0.10 \times n = s$
15. $0.25 **16.** $13z$ **17.** $x - 5$ **18.** $18 \div n$, $\frac{18}{n}$
19. $n + 14$ **20.** 10.98 **21.** 28.278 **22.** 108.975
23. 19.062 **24.** 1,234 miles

Pages 54-55   Lesson 2-6
1. 2; 14, 16, 18 **2.** 3.0; 12.6, 15.6, 18.6 **3.** 1; 5, 4, 3
4. 3; 40, 37, 34 **5.** 68, 60, 52 **7.** 44, 54, 64
9. 175, 150, 125 **11.** 19.6, 21.3, 23 **13.** no
15. yes; 21.7, 26.9, 32.1 **17.** 27, 45, 54
19. 42, 57.8, 89.4 **21.** 7.0, 5.6, 4.2 **23.** 3 weeks
25. 36, 42; 3, 4, 5 **27.** $12 + (n - 1) \times 6$ **29.** 250
31. 1,000

499

Pages 56-57   Lesson 2-7
1. $189.75; calculator   **2.** $9.00; estimate   **3.** $603;
mental computation   **5.** square, blue; circle, red; triangle,
green   **7.** $13; estimate   **9.** 36 chickens, 18 pigs
10. 109   **11.** 1,576   **12.** 12,753   **13.** 71,793   **14.** 9
15. 455   **16.** 881   **17.** 3,329   **18.** $319.32

Pages 58-59   Lessson 2-8
1. 95 cm   **2.** 27.1 m   **3.** 22 m   **4.** 140 mm   **5.** 48 cm
6. 88 ft   **7.** 56.4 cm   **9.** 24 ft   **11.** 56 ft
13. 218 inches   **15.** 114 plants   **17.** 350 ft   **19.** 312 m
21. Multiply the measure of one side by 6.

Pages 60-61   Lesson 2-9
1. $3.38; The cost of the gasoline is not needed.
2. missing the cost of the darkroom supplies   **3.** missing
the original amount of flour   **5.** $455   **7.** Pour 5 L into
the 5 L bucket. There are 2 L left. Fill the 2-L bucket
from the 5-L bucket. Add the 2 L to the 2 L in the
largest bucket.   **9.** 0.002   **10.** 200   **11.** 200   **12.** 1.8
13. $78.00, calculator or paper/pencil

Pages 62-63   Chapter 2 Review
1. e   **3.** g   **5.** f   **7.** 84   **9.** 1,436   **11.** 0.510
13. 18.4   **15.** 4,741   **17.** 4.842   **19.** 774
21. 1.14   **23.** 2.286   **25.** 42,479   **27.** 37,537
29. $363.51   **31.** $794.37   **33.** yes; 17.9, 20.1, 22.3
35. yes; 55.1, 45, 34.9   **37.** 360 miles, estimate
39. 34 inches   **41.** 28 m   **43.** 88 ft
45. missing initial weight   **47.** 33; do not need number
of pompoms

Pages 66-67   Cumulative Review/Test
Free Response: 1. 750   **3.** 600   **5.** 4,800   **7.** 390,000
9. 780   **11.** 5,500   **13.** 100,000   **15.** 14   **17.** 62
19. 252   **21.** 79   **23.** 4,523   **25.** 62,608   **27.** 9,556
29. 161   **31.** 903   **33.** 198   **35.** 131   **37.** 1,661
39. 25 CDs
Multiple Choice: 1. c   **3.** b   **5.** b   **7.** b   **9.** c

**CHAPTER 3   PATTERNS: MULTIPLYING
AND DIVIDING**

Pages 70-71   Lesson 3-1
1. 1,260   **2.** 930   **3.** 1,425   **4.** 2,336   **5.** 1,508
6. 7,810   **7.** 5,978   **8.** 33,640   **9.** 2,320   **10.** 22,260
11. 3,268   **13.** 960   **15.** 2,460   **17.** 31,122
19. 6,634   **21.** 197,756   **23.** 145,830   **25.** 392,424

27. 1,252,368   **29.** 448 pages   **31.** 15   **33.** 7
35. 132   **39.** 126   **41.** 180

Pages 72-73   Lesson 3-2
1. 55.2   **2.** 58.8   **3.** 60.8   **4.** 4.75   **5.** 27.72   **6.** 0.06
7. 0.094   **8.** 0.045   **9.** 0.0014   **10.** 0.0012   **11.** 3.5
13. 0.40   **15.** 1.173   **17.** 0.021   **19.** 29.88   **21.** 4.045
23. 0.0453   **25.** 0.3410   **27.** 0.4710   **29.** 15.6
31. $97.75   **33.** 15,000 seats; yes; $1\frac{1}{3}$ of 44,702 is
59,454, so the Kingdome has more than $1\frac{1}{3}$ the seating
35. 1.4   **37.** correct

Pages 74-75   Lesson 3-3
1. $442; $58   **2.** $2,301.64; 401.64   **3.** $7,112; $1,116
4. $3,738; $688   **5.** $1,594; $394   **7.** $24   **9.** $1,800
13. $9,455   **14.** 25   **15.** 28   **16.** 54   **17.** 12 cm
18. 32 in.   **19.** 24 mm   **20.** 780 miles

Page 77   Lesson 3-4
1. 27   **2.** 96   **3.** 163 R2   **4.** 228 R1   **5.** 925
6. 3 R2, 1 digit   **7.** 3 R3, 1 digit   **8.** 6 R28, 1 digit
9. 12 R29, 2 digits   **10.** 7 R4, 1 digit   **11.** 1 digit
13. 2 digits   **15.** 54   **17.** 64   **19.** 4 R20   **21.** 6 R1
23. 4 R18   **25.** 8 R150   **27.** 39 R48   **29.** 11 R51
31. approximately 52 weeks   **33.** 19%, 190; 12%, 120;
8%, 80; 23%, 230

Pages 78-79   Lesson 3-5
1. 65.4   **2.** 8,204   **3.** 4,500   **4.** 340,000   **5.** 49   **6.** 512
7. 62,900   **8.** 370   **9.** 0.00448   **10.** 9.41   **11.** 0.145
12. 3.412   **13.** 0.637   **14.** 0.0239   **15.** 3.9   **16.** 0.048
17. 1,750   **19.** 372   **21.** 178   **23.** 14,700   **25.** 230.4
27. 28,400   **29.** 0.093   **31.** 0.529   **33.** 0.2817
35. 0.0118   **37.** 0.00605   **39.** 5.07   **43.** 1,000 $10 bills
45. 0.345   **47.** 0.001   **49.** 4.71   **50.** 15.17
51. 80.996   **52.** 6.132   **53.** 35 books

Pages 80-81   Lesson 3-6
1. 55 blocks   **2.** 135   **3.** 64 cans   **5.** $46.50
7. 337 miles   **9.** 7 min 49.7 sec   **10.** # = 1, * = 0;
10.1 × 10 = 101.0   **11.** 0.84   **12.** 11.98   **13.** 0.49
14. 1.01   **15.** 70.24   **16.** 16.609   **17.** 4.581
18. 237.293   **19.** 10

Pages 82-83   Lesson 3-7
1. 5.1   **2.** 4.3   **3.** 18.9   **4.** 3.03   **5.** 0.013   **6.** 19
7. 2.8   **8.** 0.25   **9.** 24.3   **10.** 60   **11.** 1.12   **13.** 0.26
15. 20   **17.** 680   **19.** 2.6   **21.** 20.4   **23.** 40   **25.** 0.8
27. 30   **29.** 71   **31.** 0.7   **33.** 1.12   **35.** 2.0   **37.** 5.22
39. 0.4   **41.** 1.44 yards   **43.** 2.56   **45.** 1.75   **47.** 4.3

Pages 86-87   Lesson 3-8

1. 4.68×10^2 **2.** 1.4×10^1 **3.** 1.543×10^4
4. 2.66×10^5 **5.** 5×10^4 **6.** 1.4×10^2 **7.** 2.1×10^7
8. 3.5×10^6 **9.** 2.07×10^3 **10.** 5.4006×10^5
11. 575   **12.** 3,100   **13.** 271,000   **14.** 76,800
15. 38,000,000   **16.** 1,090,000   **17.** 4,002
18. 100,000   **19.** 3.5×10^5 **21.** 9.2372×10^4
23. 6.7×10^6 **25.** 1.5×10^1 **27.** 6×10^6
29. 4.17×10^{11} **31.** 431   **33.** 53,000   **35.** 217,000
37. 750,000   **39.** 6,380,000,000   **41.** 5,350,000
43. 25,730,000 miles; 2.573×10^7 miles
47. 1.42×10^8; 142,000,000   **49.** 1.6×10^8;
160,000,000   **51.** 1×10^{11}; 100,000,000,000

Page 89   Lesson 3-9

1. 2; 34.4, 68.8, 137.6   **2.** 10; 56,000, 560,000,
5,600,000   **3.** 5; 937.5 4,687.5 23,437.5   **4.** 0.1; 0.091,
0.0091, 0.00091   **5.** 0.4; 12.8, 5.12, 2.048
6. 0.5; 2.5, 1.25, 0.625   **7.** 4; 1,472, 5,888, 23,552
9. 5, 250, 1,250, 6,250   **13.** 1.2; 2.0736, 2.48832,
2.985984   **15.** 3; 54, 162, 486   **17.** 0.5; 8, 4, 2
19. 2,048 vibrations   **21.** sixth birthday   **23.** *about* 1
25. *about* 0.1   **27.** *about* 10   **29.** 77.8   **31.** 7,780
33. 12,309   **35.** 4.253   **37.** 0.04253

Pages 90-91   Lesson 3-10

1. 4.8 hr   **2.** 5 miles   **3.** 2.5 mph   **4.** 1,430 miles
5. 11 yards per carry   **6.** 294 yards   **7.** 9.5 yards per
carry   **8.** 25 carries   **9.** 4 mph   **11.** 1.5 hours
13. 30 carries   **15.** 6.6 yards per carry   **17.** 22 mpg
19. 24.1 mpg   **21.** 44 mpg   **23.** $s = 216 \div 8$, 27 mpg
25. 30 ohms   **27.** 12   **28.** 71.2   **29.** 5.47   **30.** 20
31. 70   **32.** 9   **33.** 10   **34.** $1.44

Pages 92-93   Lesson 3-11

1. 10 in²   **2.** 32 in²   **3.** 36 cm²   **4.** 154 cm²   **5.** 256 m²
6. 90 in²   **7.** 240 yd²   **8.** 1,600 cm²   **9.** 20 cm²
11. 360 ft²   **13.** 8,100 ft²   **15.** $126   **17.** 8 feet
21. 5.98×10^{15}; 5,980,000,000,000,000

Pages 94-95   Lesson 3-12

1. 6   **2.** 4   **3.** 6   **4.** 17   **5.** 38   **6.** 24   **7.** 59.7
8. 10.4   **9.** 119.7   **10.** 0.6   **11.** 6.1   **12.** 4.1   **13.** 115
15. 175   **17.** 1.2   **19.** 6   **21.** 646.5   **25.** $50,480
29. $22.99   **31.** 80   **33.** 6,000

Pages 96-97   Lesson 3-13

1. 1.1 m   **2.** Larry   **3.** $53.30   **5.** $1.00   **11.** $x \div 4$
13. $(t \div 3) + 10$ **14.** 128   **15.** 5,772   **16.** 14,513
17. 140,418   **18.** 15 R4   **19.** 9   **20.** 102 R41
21. 13 R330   **22.** $286

Pages 98-99   Chapter 3 Review

1. two, left   **3.** dividend   **5.** before   **7.** 9.52×10^9
9. mean   **11.** 522   **13.** 2,646   **15.** 22.47   **17.** 689.91
19. 1.392   **21.** 70 R4   **23.** 1.457   **25.** 6   **27.** 6,570
29. 0.0148   **31.** 0.625   **33.** 3.5   **35.** 6.8   **37.** 0.475
39. 0.5625   **41.** 7.99×10^4 **43.** 8.6×10^7 **45.** 490
47. 9,700   **49.** 21.6, 6.48, 1.944   **51.** 87.75, 131.625,
197.4375   **53.** $t = 3$ h   **55.** 160 m²   **57.** 5.5
59. $6.90   **61.** $1.83   **63.** 840,000 people

Pages 102-103   Cumulative Review/Test

Free Response: 1. $n + 7$ **3.** $6m$ **5.** $g + 32$ **7.** 21
9. 15   **11.** $\frac{1}{2}$ **13.** 1,200 miles   **15.** 3.05   **17.** 26.13
19. 6.09   **21.** $11.95   **23.** 252   **25.** 924   **27.** 19,296
29. $11   **31.** 19.482   **33.** 0.36   **35.** 50.076   **37.** 0.75
Multiple Choice: 1. d   **3.** d   **5.** b   **7.** d   **9.** b

**CHAPTER 4   FRACTIONS: ADDING AND
SUBTRACTING**

Page 109   Lesson 4-1

1. 1, 2, 3, 6   **2.** 1, 2, 5, 10   **3.** 1, 2, 3, 5, 6, 10, 15,
30   **4.** 1, 5, 11, 55   **5.** 1, 2, 3, 4, 5, 6, 8, 10, 12, 15,
20, 24, 30, 40, 60, 120   **6.** 2, 3, 9   **7.** 2   **8.** 2, 3
9. 3   **10.** 5   **11.** 1, 3, 9   **13.** 1, 2, 3, 6, 9, 18
15. 1, 3, 7, 21   **17.** 1, 2, 3, 4, 6, 9, 12, 18, 36
19. 1, 2, 3, 6, 7, 14, 21, 42   **21.** none   **23.** 5
25. 2, 3, 9   **27.** 3, 5, 9   **29.** 2, 3   **31.** 1,050 and
1,260   **33.** 2003   **35.** 19 fish   **37.** in packages of 1, 3,
9 or 81   **39.** 3

Pages 110-111   Lesson 4-2

1. P   **2.** P   **3.** C   **4.** C   **5.** P   **6.** C   **7.** $2^3 \times 3$
8. $2 \times 5 \times 7$ **9.** $2 \times 3 \times 17$ **10.** 11^2 **11.** $2^2 \times 41$
12. $5^2 \times 3^2$ **13.** P   **15.** P   **17.** P   **19.** P   **21.** P
23. C   **25.** $2^2 \times 5$ **27.** $2^2 \times 13$ **29.** $2^2 \times 7$
31. 5×31 **33.** $2^5 \times 3$ **35.** 2×5^4 **37.** 2, 3, 5, 7,
11, 13, 17, 19, 23, 29   **39.** 53   **41.** All even numbers
are divisible by 2.   **43.** 234; 156; 36; yes
45. The number is divisible by the product of any
combination of the relatively prime numbers.
46. 49   **47.** 751   **48.** 289   **49.** 3,687   **50.** 21,429
51. 40 R3   **52.** 100 R10   **53.** 140 R4   **54.** 15 R399
55. A-$0.22, B-$0.21, Store B

501

Page 113 Lesson 4-3

1. 3 **2.** 10 **3.** 1 **4.** 27 **5.** 6 **6.** 12 **7.** 84
8. 50 **9.** 420 **10.** 24 **11.** 2 **13.** 4 **15.** 8
17. 5 **19.** 32 **21.** 30 **23.** 60 **25.** 15 **27.** 48
29. 42 **31.** $76.88 **33.** $44.97 **35.** $2 \times 5 \times 5$
37. June **39.** He suggested that every even whole number greater than or equal to 4 is the sum of two primes.

Pages 114-115 Lesson 4-4

1. $225.03, $344.16 **3.** $141.83 **5.** $201.84
7. $2,224.04 **9.** $105.66 **11.** gross pay—total earnings; net pay—take-home pay **13.** compare it to previous statements, find the error, contact your employer or payroll department immediately
15. 410 **16.** 690 **17.** 130,312 **18.** 693,684
19. 40 mph **20.** 478.5 miles **21.** 3 hours **22.** 12 yd²

Pages 116-117 Lesson 4-5

1. 24 **2.** 7 **3.** 4 **4.** 6 **5.** 14 **6.** 1 **7.** 2 **8.** 1
9. 2 **10.** 1 **11.** 15 **13.** 24 **15.** 9 **17.** 1 **19.** 5
21. 9 **23.** 1 **25.** 18 **27.** 0 **29.** 18 **31.** $\frac{4}{6}$ **33.** $\frac{2}{8}$
35. $\frac{1}{2}$ **37.** two fractions equivalent to $\frac{6}{20}$
39. two fractions equivalent to $\frac{9}{20}$ **41.** 15 million barrels **43.** decreases

Pages 118-119 Lesson 4-6

1. $\frac{3}{4}$ **2.** $\frac{7}{8}$ **3.** $\frac{3}{4}$ **4.** $\frac{1}{2}$ **5.** $\frac{1}{3}$ **6.** $\frac{3}{4}$ **7.** $\frac{1}{4}$ **8.** $\frac{1}{2}$ **9.** $\frac{1}{2}$
10. $\frac{3}{5}$ **11.** $\frac{2}{5}$ **12.** $\frac{1}{3}$ **13.** $\frac{4}{5}$ **14.** $\frac{2}{5}$ **15.** $\frac{3}{5}$ **16.** $\frac{14}{19}$
17. $\frac{3}{4}$ **18.** $\frac{3}{4}$ **19.** $\frac{5}{6}$ **20.** $\frac{9}{20}$ **21.** $\frac{8}{21}$ **22.** $\frac{2}{3}$ **23.** $\frac{3}{8}$
24. $\frac{6}{31}$ **25.** $\frac{9}{16}$ **27.** $\frac{3}{8}$ **29.** $\frac{3}{4}$ **31.** $\frac{5}{9}$ **33.** $\frac{3}{8}$
35. $\frac{1}{3}$ **37.** $\frac{4}{5}$ **39.** $\frac{8}{9}$ **41.** $\frac{3}{5}$ **43.** $\frac{2}{3}$ **45.** $\frac{22}{25}$
47. $(5 + 5) \times 5$ **49.** $5 \div (5 + 5)$ **53.** 150
54. 600 **55.** 7,100 **56.** 4,000 **57.** 0.128
58. 13.1066 **59.** 602.58 **60.** 38.28319
61. 32 pencils

Pages 122-123 Lesson 4-7

1. > **2.** > **3.** < **4.** < **5.** > **6.** > **7.** > **8.** <
9. = **10.** < **11.** < **13.** > **15.** > **17.** <
19. = **21.** = **23.** > **25.** > **27.** < **29.** <
31. $\frac{1}{5}, \frac{1}{4}, \frac{1}{3}, \frac{1}{2}$ **33.** $\frac{1}{3}, \frac{3}{8}, \frac{5}{6}, \frac{7}{8}$ **35.** $\frac{1}{4}, \frac{5}{16}, \frac{1}{2}, \frac{11}{16}$
37. Africa **39.** about 5 pounds **41.** no **43.** no
45. yes **47.** yes **49.** no

Page 125 Lesson 4-8

1. $1\frac{2}{3}$ **2.** $1\frac{1}{6}$ **3.** $1\frac{1}{2}$ **4.** $5\frac{1}{4}$ **5.** $2\frac{2}{3}$ **6.** $2\frac{4}{5}$ **7.** 1 **8.** 3
9. 2 **10.** 4 **11.** 6 **12.** 9 **13.** $\frac{11}{8}$ **14.** $\frac{19}{12}$ **15.** $\frac{11}{3}$
16. $\frac{22}{5}$ **17.** $\frac{26}{7}$ **18.** $\frac{35}{3}$ **19.** $1\frac{1}{7}$ **21.** 3 **23.** $1\frac{2}{7}$
25. $2\frac{1}{5}$ **27.** $2\frac{2}{3}$ **29.** $8\frac{4}{9}$ **31.** $\frac{7}{5}$ **33.** $\frac{15}{8}$ **35.** $\frac{29}{10}$
37. one-half **39.** two and seven-eighths **41.** $\frac{3n + 1}{3}$
43. $1\frac{7}{10}$ acres **45.** 11 weeks **46.** 24 **47.** 23
48. 44.8, 89.6, 179.2 **49.** 27, 9, 3 **50.** $\frac{1}{3}$ year

Page 127 Lesson 4-9

1. 1 **2.** $\frac{1}{2}$ **3.** $\frac{1}{2}$ **4.** 0 **5.** 0 **6.** 1 **7.** 7 **8.** 18
9. 28 **10.** 32 **11.** 13 **12.** 20 **13.** 1 **14.** $1\frac{1}{2}$
15. 4 **16.** 12 **17.** $\frac{1}{2}$ **18.** 1 **19.** $2\frac{1}{2}$ **20.** 1 **21.** $\frac{1}{2}$
23. $\frac{1}{2}$ **25.** $\frac{1}{2}$ **27.** $12\frac{1}{2}$ **29.** 7 **31.** 13
35. $1\frac{7}{16}$-mile race **37.** $4 \times 4 \times 4 = 64$
38. $2 \times 2 \times 2 \times 2 \times 2 \times 2 = 64$ **39.** $5 \times 5 = 25$
40. $8 \times 8 = 64$ **41.** 200 **42.** 12 **43.** 150,000
44. 22 **45.** 2 times

Page 129 Lesson 4-10

1. $\frac{3}{4}$ **2.** $\frac{4}{5}$ **3.** $\frac{7}{8}$ **4.** $\frac{2}{3}$ **5.** $1\frac{1}{4}$ **6.** $1\frac{1}{4}$ **7.** $1\frac{1}{2}$ **8.** $1\frac{3}{5}$
9. $\frac{7}{8}$ **10.** $\frac{7}{24}$ **11.** $\frac{31}{35}$ **12.** $1\frac{4}{21}$ **13.** $\frac{4}{5}$ **15.** $\frac{2}{3}$ **17.** $1\frac{1}{4}$
19. $1\frac{1}{12}$ **21.** $1\frac{1}{5}$ **23.** $1\frac{17}{24}$ **25.** $1\frac{5}{6}$ **27.** $1\frac{4}{9}$ **29.** $1\frac{13}{36}$
31. $\frac{7}{12}$ can **33.** $1\frac{1}{8}$ **35.** $1\frac{5}{8}$ **37.** $\frac{1}{2}$ inch

Pages 130-131 Lesson 4-11

1. $11\frac{3}{5}$ **2.** $13\frac{8}{11}$ **3.** $15\frac{1}{2}$ **4.** $9\frac{3}{4}$ **5.** $15\frac{1}{2}$ **6.** $7\frac{7}{8}$ **7.** $8\frac{5}{6}$
8. $12\frac{2}{3}$ **9.** $9\frac{5}{7}$ **11.** $8\frac{1}{4}$ **13.** $16\frac{19}{24}$ **15.** $11\frac{3}{4}$ **17.** $10\frac{1}{4}$
19. $11\frac{1}{2}$ **21.** $14\frac{5}{6}$ **23.** $24\frac{1}{2}$ **25.** $8\frac{3}{4}$ **27.** $5\frac{1}{4}$ **29.** $25\frac{7}{8}$
31. $8\frac{3}{8}$ **33.** $13\frac{7}{8}$ **35.** $8\frac{3}{8}$ yards **37.** Since all operations are addition, the sum is the same with or without parentheses. **39.** 161.8 **40.** $99.02 **41.** 73.63
42. 76.812 **43.** 21, 27, 30 **44.** 1.3, 11.2, 14.5 **45.** $17; do not need bold type cost

Pages 132-133 Lesson 4-12

1. $\frac{1}{3}$ **2.** $\frac{1}{9}$ **3.** $\frac{4}{7}$ **4.** $\frac{4}{11}$ **5.** $\frac{1}{5}$ **6.** $\frac{1}{6}$ **7.** $\frac{5}{12}$ **8.** $\frac{23}{90}$

9. $\frac{1}{12}$ **10.** $\frac{1}{6}$ **11.** $\frac{1}{2}$ **13.** $\frac{7}{20}$ **15.** $\frac{2}{3}$ **17.** $\frac{1}{20}$ **19.** 0
21. $\frac{1}{24}$ **23.** $\frac{1}{2}$ **25.** $\frac{7}{18}$ **27.** $\frac{11}{24}$ **29.** $\frac{2}{3}$ **31.** $\frac{1}{2}$ cup
33. 14 students liked both. **35.** 19

Page 135   Lesson 4-13
1. $2\frac{2}{3}$ **2.** $1\frac{1}{2}$ **3.** $5\frac{1}{4}$ **4.** $3\frac{3}{5}$ **5.** $3\frac{1}{20}$ **6.** $4\frac{2}{15}$ **7.** $11\frac{1}{8}$
8. $6\frac{4}{9}$ **9.** $3\frac{5}{6}$ **10.** $6\frac{17}{20}$ **11.** $4\frac{7}{24}$ **12.** $4\frac{23}{24}$ **13.** $3\frac{1}{3}$
14. $4\frac{3}{4}$ **15.** $9\frac{3}{8}$ **16.** $13\frac{3}{7}$ **17.** $10\frac{3}{8}$ **19.** $3\frac{2}{3}$ **21.** $8\frac{1}{3}$
23. $1\frac{33}{40}$ **25.** $4\frac{5}{6}$ **27.** $2\frac{1}{2}$ **29.** $13\frac{5}{7}$ **31.** $1\frac{3}{8}$ **33.** $5\frac{1}{2}$
35. $10\frac{5}{8}$ **37.** $1\frac{19}{24}$ **39.** $\frac{22}{75}$ **41.** $\frac{5}{12}$ **43.** 1, 5, 3, 2, 4

Pages 136-137   Lesson 4-14
1. 16 sections **2.** 4 pictures **3.** 51 inches **5.** 9 days
7. 20 persons **11.** 19 ways **12.** 31.2
13. 41.5 **14.** 765.5 **15.** 3.573 **16.** 1,840 **17.** 96
18. 3,720 **19.** 628 **20.** 0.084 **21.** 0.624
22. 0.0159
23. 4.6 **24.** no, $\frac{1}{8}$

Pages 138-139   Chapter 4 Review
1. 30 **3.** $\frac{2}{3}$ **5.** GCF **7.** mixed number **9.** 1, 2, 3, 4,
6, 12 **11.** 1, 29 **13.** 1, 2, 3, 6, 7, 14, 21, 42 **15.** 3
17. 2, 3, 9 **19.** $2 \times 7 \times 7$ **21.** $2 \times 3 \times 3 \times 5$
23. $2 \times 2 \times 3 \times 13$ **25.** 16 **27.** 2 **29.** 28
31. 24 **33.** 30 **35.** \$5,701.80 **37.** 12 **39.** 2 **41.** 6
43. 21 **45.** 4 **47.** $\frac{3}{4}$ **49.** $\frac{1}{7}$ **51.** $\frac{19}{36}$ **53.** > **55.** <
57. $7\frac{1}{3}$ **59.** $6\frac{3}{4}$ **61.** $7\frac{4}{5}$ **63.** $\frac{37}{8}$ **65.** $\frac{20}{3}$ **67.** $\frac{71}{9}$
69. 1 **71.** $8\frac{1}{2}$ **73.** 6 **75.** $1\frac{1}{4}$ **77.** $1\frac{1}{20}$ **79.** $6\frac{2}{9}$
81. $9\frac{9}{11}$ **83.** $12\frac{11}{12}$ **85.** $\frac{3}{5}$ **87.** $\frac{11}{18}$ **89.** $2\frac{2}{5}$ **91.** $3\frac{5}{12}$
93. $17\frac{1}{2}$ **95.** Cut the cake into thirds horizontally and
then cut the cake one fourth of the way vertically.

Pages 142-143   Cumulative Review/Test
Free Response: 1. 600 **3.** 4,000 **5.** 850,600
7. 5,400 **9.** 7 **11.** 8,700 **13.** 180,000 **15.** 300,000
17. $483 + n = 625$; 142 **19.** 929 **21.** 468 **23.** 248
25. 171 **27.** 4,758 **29.** 3,886 **31.** 146 **33.** 9,496
35. 9,940 **37.** 12 weeks **39.** $3 \times 2 \times 2 \times 17$
41. 2×37 **43.** 70 **45.** 4 **47.** 6
Multiple Choice: 1. c **3.** b **5.** c **7.** b **9.** c

CHAPTER 5   FRACTIONS: MULTIPLYING AND DIVIDING

Pages 146-147   Lesson 5-1
1. 21 **2.** 2 **3.** 12 **4.** 9 **5.** 2 **6.** 5 **7.** 2 **8.** 4
9. less than **10.** greater than **11.** 15 **13.** 2 **15.** 7
17. 6 **19.** less than **21.** less than **23.** about \$9
25. about 4 weeks **27.** 5 **29.** 36 **31.** 6.94
32. 5.42 **33.** 21.62 **34.** 48,000 **35.** 6,230
36. 901,000 **37.** $1\frac{5}{12}$ cups

Pages 148-149   Lesson 5-2
1. $\frac{3}{16}$ **2.** $\frac{5}{9}$ **3.** $\frac{8}{15}$ **4.** $\frac{3}{10}$ **5.** $\frac{1}{3}$ **6.** $\frac{3}{7}$ **7.** $1\frac{5}{16}$ **8.** 15
9. $4\frac{3}{8}$ **10.** 6 **11.** $\frac{4}{63}$ **13.** $\frac{40}{63}$ **15.** $\frac{4}{9}$ **17.** $\frac{7}{9}$ **19.** $3\frac{3}{4}$
21. 10 **23.** $\frac{3}{8}$ **25.** $\frac{9}{40}$ **27.** true **29.** \$48.75
31. $1\frac{5}{12}$ **33.** $2\frac{2}{3}$ **35.** 65.2 **36.** 17.03 **37.** 4.05
38. 182 **39.** 360 **40.** 3,213 **41.** \$33

Pages 150-151   Lesson 5-3
1. $4\frac{1}{8}$ **2.** $8\frac{2}{3}$ **3.** $7\frac{7}{8}$ **4.** $18\frac{2}{3}$ **5.** 39 **6.** 130 **7.** $1\frac{19}{30}$
8. $3\frac{31}{63}$ **9.** $1\frac{7}{9}$ **10.** 30 **11.** $12\frac{2}{3}$ **12.** 8 **13.** $4\frac{23}{28}$
15. $3\frac{3}{5}$ **17.** 25 **19.** $4\frac{7}{12}$ **21.** 1 **23.** $51\frac{1}{4}$ **25.** 45
27. $15\frac{1}{5}$ **29.** $42\frac{1}{4}$ **31.** 5 hours **35.** yes **37.** 35
39. 40

Pages 152-153   Lesson 5-4
1. $\frac{11}{20}$ ton **2.** $\frac{1}{5}$ ton **3.** $\frac{1}{8}$ ton **4.** 1 ton **5.** $\frac{1}{4}$ ton
6. $\frac{1}{4}$ ton **7.** $\frac{1}{2}$ ton **8.** $\frac{1}{6}$ ton **9.** $\frac{1}{40}$ ton
11. $\frac{13}{20}$ of the total coal deposits **13.** Appalachian
Plateau **15.** $\frac{9}{50}$ of the total coal deposits
17. *about* \$65 **19.** *about* 30 million tons **20.** 4^3, 8^2
21. 3^4, 9^2 **22.** 1^2, 1^3 **23.** 10^4, 100^2 **24.** 1, 2, 4, 8,
16, 32 **25.** 1, 3, 5, 9, 15, 45 **26.** 1, 2, 3, 4, 6,
8, 12, 24 **27.** 1, 2, 3, 6, 9, 18, 27, 54
28. 700 students

Pages 154-155   Lesson 5-5
1. 48 lawns **2.** Since the last cut yields two pieces, only
15 cuts are necessary. **3.** $(337 + 4) \times 100 = 34,100$
feet. **5.** 2,678,400 seconds, calculator **7.** 6 socks
11. 22.3 **12.** 5.5 **13.** 0.8 **14.** 2×3^2 **15.** $2^2 \times 5$
16. $2^2 \times 7$ **17.** $2 \times 3 \times 7$ **18.** $2 \times 3 \times 17$
19. $7\frac{5}{6}$ hours

Page 159 Lesson 5-6

1. $\frac{9}{8}$ **2.** $\frac{3}{2}$ **3.** $\frac{20}{3}$ **4.** $\frac{7}{5}$ **5.** $\frac{1}{5}$ **6.** $\frac{1}{12}$ **7.** $5\frac{1}{3}$ **8.** 16

9. 18 **10.** 14 **11.** $2\frac{2}{3}$ **12.** $\frac{14}{25}$ **13.** $\frac{3}{40}$ **14.** $\frac{1}{12}$

15. $\frac{20}{19}$ **17.** $\frac{16}{11}$ **19.** $1\frac{7}{8}$ **21.** $\frac{5}{8}$ **23.** 16 **25.** 4

27. $\frac{1}{36}$ **29.** $\frac{1}{20}$ **31.** $1\frac{3}{4}$ **33.** no; $\frac{7}{8} \div \frac{1}{6} = 5\frac{1}{4}$

35. Six rectangles, each divided into thirds shows a quotient of 18 smaller rectangles.

37. $\frac{2}{3}$ **39.** $\frac{1}{4}$

Pages 160-161 Lesson 5-7

1. $1\frac{7}{18}$ **2.** $1\frac{37}{62}$ **3.** 6 **4.** $\frac{2}{11}$ **5.** $\frac{2}{29}$ **6.** $1\frac{47}{88}$ **7.** $\frac{9}{14}$

8. $1\frac{1}{4}$ **9.** $\frac{5}{6}$ **10.** $\frac{7}{8}$ **11.** $\frac{1}{12}$ **12.** 5 **13.** $1\frac{7}{8}$ **14.** $7\frac{1}{2}$

15. $\frac{8}{11}$ **16.** $1\frac{1}{6}$ **17.** $\frac{3}{8}$ **19.** $2\frac{1}{2}$ **21.** $6\frac{2}{3}$ **23.** $\frac{3}{10}$

25. $\frac{2}{5}$ **27.** $1\frac{1}{10}$ **29.** 2 **31.** $2\frac{8}{13}$ **33.** 5

35. 30 pieces **37.** 20 boards **41.** $18 **43.** $29 **45.** $34

Pages 162-163 Lesson 5-8

1. 0.4 **2.** 0.375 **3.** 0.75 **4.** 0.3125 **5.** $0.\overline{1}$ **6.** $0.\overline{15}$

7. $0.8\overline{3}$ **8.** $0.58\overline{3}$ **9.** $0.66\frac{2}{3}$ **10.** $0.53\frac{11}{13}$ **11.** $0.88\frac{8}{9}$

12. $0.54\frac{1}{6}$ **13.** 0.7 **15.** 0.55 **17.** $0.\overline{27}$ **19.** $0.\overline{6}$

21. $0.\overline{7}$ **23.** $0.6\overline{2}$ **25.** 0.1875 **27.** $0.91\overline{6}$

29. $0.63\frac{1}{3}$ **31.** $0.33\frac{1}{3}$ **33.** 0.008 **35.** 0.875 inch

37. Answers may vary. A good answer refers to prime factors other than 2 or 5. That is, denominators without prime factors 2 or 5 result in terminating decimals.
39. 89% = 0.89; 1,051 students

Pages 164-165 Lesson 5-9

1. $\frac{3}{500}$ **2.** $\frac{21}{250}$ **3.** $\frac{1}{8}$ **4.** $\frac{13}{20}$ **5.** $\frac{51}{125}$ **6.** $\frac{49}{50}$ **7.** $\frac{53}{100}$

8. $\frac{31}{100}$ **9.** $\frac{16}{25}$ **10.** $\frac{1}{20}$ **11.** $\frac{1}{7}$ **12.** $\frac{2}{3}$ **13.** $\frac{2}{9}$ **14.** $\frac{6}{7}$

15. $\frac{5}{7}$ **17.** $\frac{4}{5}$ **19.** $\frac{3}{4}$ **21.** $\frac{19}{50}$ **23.** $\frac{101}{1,000}$ **25.** $\frac{243}{500}$

27. $\frac{56}{125}$ **29.** $\frac{1}{125}$ **31.** $10\frac{9}{50}$ **33.** $\frac{1}{3}$ **35.** $\frac{1}{11}$ **37.** =

39. < **41.** 15 pieces **43.** $\frac{15}{18}$ **45.** 0.675

47. 0.125 or $\frac{1}{8}$ **49.** 0.515625 or $\frac{33}{64}$

Pages 166-167 Lesson 5-10

1. $15\frac{5}{8}$ yd **2.** $10\frac{1}{2}$ feet **3.** 264 students **5.** about 14

minutes **7.** about 1,000 miles long **9.** $46\frac{7}{8}$ miles

11. $2\frac{1}{8}$ bushels **13.** 5 monkeys **15.** 66 **16.** 22

17. 378 **18.** 8 **19.** 7 **20.** 15 **21.** 9 **22.** 17

23. Albert

Pages 168-169 Chapter 5 Review

1. i **3.** e **5.** d **7.** 45 **9.** 2 **11.** less than **13.** $\frac{1}{9}$

15. $6\frac{2}{3}$ **17.** $\frac{9}{25}$ **19.** $\frac{4}{9}$ **21.** $\frac{2}{3}$ **23.** $1\frac{1}{20}$ **25.** $27\frac{3}{5}$

27. $16\frac{2}{3}$ **29.** 1 **31.** $\frac{1}{10}$ ton **33.** $\frac{3}{5}$ ton

35. 36 centerpieces **37.** $\frac{4}{3}$ **39.** $\frac{1}{6}$ **41.** $\frac{2}{9}$ **43.** $1\frac{3}{4}$

45. $\frac{3}{4}$ **47.** $\frac{3}{20}$ **49.** $1\frac{2}{9}$ **51.** $\frac{7}{12}$ **53.** $\frac{21}{40}$ **55.** 8

57. $\frac{10}{27}$ **59.** 2 **61.** 0.1 **63.** 0.52 **65.** $0.\overline{01}$ **67.** $\frac{17}{500}$

69. $\frac{1}{4}$ **71.** 3 scoops **73.** $\frac{19}{30}$ of her salary **75.** $6\frac{2}{3}$

Pages 172-173 Cumulative Review/Test

Free Response: 1. $3 \times 3 \times 3 \times 3 = 81$ **3.** $10 \times 10 \times 10 \times 10 \times 10 = 100,000$ **5.** $1 \times 1 \times 1 \times 1 \times 1 \times 1 \times 1 \times 1 \times 1 \times 1 = 1$ **7.** $8 \times 8 = 64$ **9.** 12 **11.** 10 **13.** 156 **15.** 7.8 **17.** 162.7 **19.** 3,888 **21.** 7,680 **23.** 63,000 **25.** 4 **27.** 4 **29.** 400 **31.** $124.96 **33.** 1, 89; prime **35.** 1, 3, 13, 39 **37.** 4; 360 **39.** 1; 225 **41.** $1\frac{3}{4}$ cups **43.** $\frac{1}{21}$

45. $15\frac{5}{9}$ **47.** $1\frac{17}{27}$ **49.** $36\frac{2}{5}$

Multiple Choice: 1. c **3.** b **5.** b **7.** a **9.** b

CHAPTER 6 MEASUREMENT

Pages 178-179 Lesson 6-1

1. thousandths **2.** thousands **3.** hundredths **4.** tens **5.** tenths **6.** hundreds **7.** 10 **8.** 10 **9.** 1,000 **10.** 1,000 **11.** dekagram **12.** centigram **13.** millimeter **14.** decimeter **15.** 1,000 **17.** 0.001 **19.** 0.01 **21.** millimeter **23.** hectogram **25.** dekaliter **27.** kilometer **29.** liter **31.** 1 dekameter **33.** 7 spools **35.** Changes for road signs, record keeping for sports events, pricing per metric unit, packaging, scales, and so on **37.** The current metric system was developed in 1795 by the French Academy of Sciences and expanded in 1960. In the United States, the Metric Conversion Act of 1975 specified a national policy

of *voluntary* use of the metric system and established the U.S. Metric Board to help ease the change. **39.** 463
40. 58.2 **41.** 16,400 **42.** 697 **43.** $5\frac{2}{3}$ **44.** $8\frac{1}{7}$
45. $3\frac{5}{18}$ **46.** $3\frac{1}{10}$ **47.** no; too short

Pages 180-181   Lesson 6-2
1. meter **2.** centimeter **3.** millimeter **4.** meter
5. kilometer **6.** millimeter **7.** 66 cm **9.** 22 cm
11. 120 cm **13.** 1.2 cm, 12 mm **15.** 25 bows
17. 1,149 km **21.** *about* 5 m

Pages 182-183   Lesson 6-3
1. 300 **2.** 8,000 **3.** 20 **4.** 120 **5.** 4 **6.** 960 **7.** 2
8. 4 **9.** 0.0046 **10.** 0.275 **11.** 0.00978 **12.** 567
13. 70 **15.** 3.5 **17.** 400 **19.** 5 **21.** 0.3 **23.** 420
25. 0.0087 **27.** 36 **29.** 1,092 mps **31.** 49.4 km
35. 7.01×10^6 **36.** 8.3×10^5 **37.** 4.8×10^7
38. 5.37×10^4 **39.** 2 **40.** 2, 3 **41.** 3 **42.** 3, 5, 9
43. 2, 3, 5, 9, 10 **44.** pizza

Pages 184-185   Lesson 6-4
1. 1 g **2.** 17 mg **3.** 1,000 kg **4.** 10 g **5.** 80 g
6. 1.4 kg **7.** 0.0049 **8.** 5.862 **9.** 8.754 **10.** 67,000
11. 73,000 **12.** 9,700 **13.** 8,000 **15.** 0.752
17. 9,400 **19.** 0.912 **21.** 24,000 **23.** 0.005
25. 25,800 grams **27.** 2 kg **29.** 12 cans **31.** 129
32. 1,456 **33.** 7,623 **34.** 13,288 **35.** 78,800
36. 30 R3 **37.** 17 R9 **38.** 52 R62 **39.** 51 R51
40. 36 days

Pages 186-187   Lesson 6-5
1. liter **2.** liter **3.** liter **4.** milliliter **5.** 42,000
6. 900 **7.** 36,000 **8.** 0.791 **9.** 0.2356 **10.** 0.004
11. 6,000 **13.** 7,500 **15.** 0.234 **17.** 0.127
19. 28,000 **21.** 0.262 **23.** 17,000 **25.** 0.0968
27. 780,000 mL **29.** 30 mL **31.** 250 cups **33.** 12.3
34. 4.36 **35.** 18.67 **36.** 35 **37.** 36 **38.** 51
39. $37

Pages 188-189   Lesson 6-6
1. 13 pretzels **2.** $5.50 gain **3.** 240 flowers
5. 4 blocks west **7.** 60 **9.** 2 hours longer **10.** 24
11. 63 **12.** 5 **13.** 4 **14.** $2\frac{2}{7}$ **15.** $\frac{1}{2}$ **16.** $\frac{2}{11}$ **17.** $\frac{7}{8}$
18. 20 rungs

Pages 190-191   Lesson 6-7
1. 36 **2.** 6 **3.** 3,520 **4.** 108 **5.** 10,560 **6.** 45 **7.** 2
8. 2 **9.** 5 **10.** 2, 6 **11.** 7, 2 **12.** 6, 4 **13.** 60
15. 72 **17.** 7, 2 **19.** 9 **21.** 5 **23.** 18 **25.** 2, 2

27. $1\frac{1}{2}$ **29.** 6 ft 4 in. **31.** Answers may vary. You may respond to the ease of computing in the metric system or the familiarity of using the customary system.
33. Size 9

Pages 194-195   Lesson 6-8
1. 16 **2.** 8 **3.** 5 **4.** 3 **5.** 5 **6.** $1\frac{1}{2}$ **7.** 4,000 **8.** 3
9. 6 **10.** 80 **11.** 48 **12.** 8 **13.** 32 **15.** 2 **17.** 10
19. 11 **21.** 12 **23.** 8 **25.** 16 **27.** 16 **29.** 96
31. 7 pounds **33.** gallon **35.** cup, ounce
37. 20 weeks **39.** about 59 pounds **41.** 32 **43.** 4
45. Find the number of ounces in one gallon, 128, and multiply by 8.

Pages 196-197   Lesson 6-9
1. 20°F **2.** 30°C **3.** 23°F **4.** 70°F **5.** 140°F
6. 59°F **7.** 122°F **8.** 185°F **9.** 86°F **10.** 5°C
11. 15°C **12.** 65°C **13.** 115°C **14.** 90°C **15.** 212°F
17. 127°F **19.** 252°F **21.** 55°F **23.** 145°F **25.** 35°C
27. 38°C **39.** 100°C **31.** 32°C **33.** 1°C **35.** 35.6°F
37. 20.5° to 23.3°C **39.** 10 hours **41.** 158° **43.** 100°

Pages 198-199   Lesson 6-10
1. 1 h 5 min **2.** 3 days **3.** 6 h **4.** 500 s **5.** 2 days
6. 300 min. **7.** 300 **8.** 480 **9.** 96 **10.** 2 **11.** 6
12. 1 **13.** 3 h 25 min **14.** 2 h 23 min **15.** 240
17. 3 **19.** $\frac{3}{4}$ **21.** 1,440 **23.** 270 **25.** 13 h 30 min
27. 243 sec **31.** 300 times

Pages 200-201   Lesson 6-11
1. 13:35 **2.** 19:20 **3.** 02:10 **4.** 06:40 **5.** 16:50
6. 00:10 **7.** 8 h **8.** 8 h **9.** 8 h **11.** 6 h 30 min
13. 8 h 10 min **15.** 7 h 10 min **17.** no;
20 minutes late **19.** about 278 households **22.** $10\frac{3}{8}$
23. $21\frac{11}{12}$ **24.** $9\frac{13}{45}$ **25.** $17\frac{20}{21}$ **26.** $15\frac{6}{15}$ **27.** 32
28. $7\frac{1}{2}$ **29.** $4\frac{1}{8}$ **30.** 0.36 km

Pages 202-203   Lesson 6-12
1. 7 ft 9 in. **2.** 24 glasses **3.** $3\frac{1}{3}$ h **4.** 8 yd
5. 1 mm too large **7.** 113.5 grams **10.** 900
11. 4,900 **12.** 17.00 **13.** 1,000 **14.** 186
15. 1,716 **16.** 12,600 **17.** 150,381
18. 840 yards

Pages 204-205   Chapter 6 Review
1. 100 **3.** liter **5.** Celsius **7.** time **9.** customary
11. foot **13.** centimeter **15.** milliliter **17.** milligram

19. 4.2 cm; 42 mm   **21.** 5.1 cm; 51 mm   **23.** 50   **25.** 1
27. 0.4   **29.** 34,000   **31.** 4.536   **33.** 8,500   **35.** 36
37. 10   **39.** 12   **41.** 30   **43.** 4,000   **45.** 40
47. 86°F   **49.** 38°C   **51.** 42°C   **53.** 2°C   **55.** 180
57. 168   **59.** 302   **61.** 02:33   **63.** 23:29
65. 6 h 59 min   **67.** 9,600 kg   **69.** 256 min

Pages 208-209   Cumulative Review/Test

Free Response: 1. 6,000   **3.** 690,000   **5.** 5.1, 3.9, 2.7

7. 2, 3, 9   **9.** 2, 3, 5, 9, 10   **11.** >   **13.** <   **15.** $\frac{1}{2}$

17. 0   **19.** $\frac{1}{2}$ **21.** $23\frac{5}{8}$ **23.** $2\frac{1}{4}$ **25.** $\frac{3}{200}$ **27.** 11

29. $\frac{1}{4}$ **31.** 5   **33.** 75,000   **35.** 32,000

37. 4 min 43 sec

Multiple Choice: 1. a   **3.** d   **5.** c   **7.** b   **9.** a

CHAPTER 7   GEOMETRY

Pages 212-213   Lesson 7-1

1. $\overline{CD}$ **2.** $\overrightarrow{BD}$ **3.** $\overleftrightarrow{RS}$ **4.** $\overrightarrow{MN}$ **5.** $\overline{AB}$ **6.** T
7. pin point; thumbtack   **8.** pencil; intersection of the
wall and the floor   **9.** flashlight beam; laser beam
10. wall; floor   **11.** $\overline{GH}$; $\overline{HG}$; line segment GH;
line segment HG **13.** point M; M **23.** 1 line segment
27. infinitely many   **31.** $\frac{3}{4}$ inch   **33.** $\frac{14}{4}$ or $3\frac{1}{2}$ inches
35. 52   **36.** 76   **37.** 3,763   **38.** C   **39.** P
40. P   **41.** C   **42.** 12 gallons

Pages 214-215   Lesson 7-2

1. ∠3; ∠XYZ; ∠ZYX   **2.** ∠1; ∠PQR; ∠RQP
3. ∠2; ∠TUV; ∠VUT   **4.** ∠5; ∠GHI; ∠IHG   **5.** 74°
7. 90°   **9.** 136°   **11.** ∠HIL; ∠HIK; ∠LIJ; ∠LIK;
∠KIJ; There is more than one angle with vertex I.

13. 180°   **17.** 2; 14; 7; 3.5   **18.** 4; 3; $\frac{3}{4}$; $\frac{3}{8}$ **19.** 9

20. 24   **21.** 20   **22.** 4   **23.** 120 pizzas

Pages 216-217   Lesson 7-3

1. 65°; acute   **2.** 90°; right   **3.** 120°; obtuse   **5.** obtuse
7. acute   **9.** acute   **11.** right   **13.** obtuse   **15.** obtuse
19. 30°, 60°   **21.** 6; 10; 15; 21 angles

Page 219   Lesson 7-4

11. acute   **13.** Draw a line. Construct a segment
congruent to $\overline{CD}$. Construct a second segment congruent
to $\overline{CD}$ on the line, extending from the end of the first
segment.

Pages 220-221   Lesson 7-5

1. parallel; $\overleftrightarrow{PQ} \parallel \overleftrightarrow{RS}$ **2.** perpendicular $\overleftrightarrow{ZW} \perp \overleftrightarrow{XY}$
3. skew   **5.** parallel; $\overleftrightarrow{RS} \parallel \overleftrightarrow{TQ}$ **7.** skew   **9.** skew
11. $\overline{TY}$, $\overline{QV}$, $\overline{SX}$ **15.** parallel   **17.** 0.38   **18.** 0.84
19. 4.09   **20.** 9.98   **21.** 13.00   **22.** $2\frac{5}{6}$ **23.** $\frac{19}{25}$

24. $1\frac{17}{40}$ **25.** $5\frac{1}{3}$ **26.** 42°F

Pages 224-225   Lesson 7-6

1. quadrilateral; not regular   **2.** pentagon; not regular
3. octagon; regular   **4.** decagon; not regular
5. quadrilateral; not regular   **7.** triangle; not regular
9. not closed   **11.** not made up of line segments
19. Closed means all line segments must intersect. Plane
means the figures lie in one plane. They have only 2
dimensions, not 3.   **21.** 280.9 feet per second
23. Texas, Oklahoma   **25.** Texas, Alaska
27. $13,776,000,000

Pages 226-227   Lesson 7-7

1. equilateral   **2.** scalene   **3.** isosceles   **4.** scalene
5. right   **6.** acute   **7.** acute   **8.** right   **9.** equilateral;
acute   **11.** isosceles; acute   **13.** isosceles; obtuse
21. true   **23.** false   **25.** true   **27.** △PBD; △APD
29. △APD   **33.** 48°, 90°
35. The sum of the measures of the two right angles is
180°. Since the sum of the measures of the angles in a
triangle is 180°, perpendicular rays will not intersect to
form the third vertex.

Pages 228-229   Lesson 7-8

1. rectangle   **2.** square   **3.** trapezoid   **4.** rhombus
5. square   **7.** rhombus   **13.** false   **15.** false   **17.** false
19. true   **21.** 37 cm   **23.** 11 squares   **24.** $1\frac{1}{5}$

25. $1\frac{4}{5}$ **26.** $2\frac{4}{7}$ **27.** $\frac{3}{5}$ **28.** $\frac{21}{50}$ **29.** $\frac{13}{40}$ **30.** $\frac{319}{500}$

31. 6:50 A.M.

Page 231   Lesson 7-9

1. cone   **2.** cylinder   **3.** rectangular pyramid
4. pentagonal prism   **5.** 4; 4; 6; 7; 10; 15; 5; 6; 9; 5;
5; 8; 6; 6; 10   **7.** $F + V = E + 2$ **9.** 200 vertices
11. Prisms are named by the shape of the figure.

Pyramids are named by the shape of their base.
13. 2; No, a cylinder does not have two line segments that intersect.

1. b, not a ball   **2.** c, not a polygon   **3.** c, not three dimensional   **4.** d, not a prism   **5.** a   **7.** b   **9.** 11
11. 8   **13.** 49, 81; square numbers   **15.** $1\frac{2}{5}$ **16.** $1\frac{5}{12}$
17. $1\frac{6}{35}$ **18.** 1   **19.** 0.042   **20.** 9,500   **21.** 250
22. 20°

1. line segment   **3.** acute   **5.** isosceles   **7.** congruent
9. edges   **19.** 52°; acute   **21.** 96°; obtuse   **25.** skew
27. pentagon; regular   **29.** quadrilateral; not regular
31. isosceles; right   **33.** equilateral; acute   **35.** rectangle
37. trapezoid   **39.** triangular prism   **41.** cylinder
43. c

Free Response: 1. 13   **3.** 2.1   **5.** 500   **7.** 666
9. 217   **11.** \$1,500,000   **13.** $\frac{9}{10}$ **15.** $\frac{3}{7}$ **17.** 25
19. $\frac{1}{8}$ **21.** 2   **23.** 100   **25.** $9\frac{1}{3}$ **27.** 0.85   **29.** 0.30
31. 1.5   **33.** equilateral, acute   **35.** right; scalene
Multiple Choice; 1. d   **3.** c   **5.** c   **7.** d   **9.** c

CHAPTER 8   AREA AND VOLUME

1. 25.1 cm   **2.** 22.0 m   **3.** 81.6 in.   **4.** 1,507.2 mm
5. 3.1 m   **6.** 21.0 ft   **7.** 160.1 cm   **8.** 72.2 in.
9. 182.1 cm   **11.** 113.0 in.   **13.** 18.8 ft or 6.3 yd
15. 17.9   **16.** 130.8   **17.** rhombus   **18.** trapezoid
19. rectangle   **20.** parallelogram   **21.** 3.625 gallons

1. 42 m²   **2.** 60 in²   **3.** 22.5 cm²   **4.** 45 m²
5. 297.5 cm²   **6.** 48 ft²   **7.** 40,920 miles²   **9.** 55.2 cm²
11. 5.6 yd²   **13.** 30.15 m²   **15.** 3,840 in²
17. 2,803.2 m²   **19.** There are thirty-five different arrangements.   **21.** 8 ft

1. 234 cm²   **2.** 36 ft²   **3.** 60 cm²   **4.** 11.25 ft²
5. 11.2 in²   **6.** 1.575 m²   **7.** 110 yd²   **8.** 77 mm²
9. 0.975 cm²   **11.** 22.14 m²   **13.** 51 cm²   **15.** 4.55 in²
17. 336 mm²   **19.** 15.75 m²   **21.** 104 ft²   **25.** 16
26. 103   **27.** 45.5   **28.** 61,300   **29.** 0.4265
30. 2,100   **31.** 33°F

1. 78.5 cm²   **2.** 254.3 m²   **3.** 1,256 yd²   **4.** 2,826 mm²
5. 50.2 cm²   **6.** 78.5 in²   **7.** 706.5 yd²   **8.** 379.9 m²
9. 12.6 cm²   **11.** 132.7 cm²   **13.** 254.3 yd²
15. 283.4 cm²   **17.** 1,017.4 ft²   **19.** 830.5 ft²
21. 5 figures   **23.** 9,852.0 in²

1. \$243   **2.** \$315   **3.** \$456   **4.** \$264   **5.** 127.3 yd²; \$3,048.84   **7.** 30 yd²; \$448.50   **9.** \$4,685.24   **11.** 450
12. 150   **13.** 600   **14.** 1,000   **15.** $\overline{DC}, \overline{EF}, \overline{HG}$
16. 12:00 and 12:30

1. 39 students   **3.** 60 stamps   **5.** 240 doctors
7. \$75.91; calculator   **9.** \$23.48   **10.** 5.31
11. 21.107   **12.** $\overrightarrow{RS}$ **13.** $\overleftrightarrow{XY}$ **14.** $\overline{MN}$ **15.** $\overrightarrow{AB}$
16. 3,140 ft

1. 80 m², 80 m², 48 m², 48 m², 60 m², 60 m², 376 m²
2. 33.6 in², 33.6 in², 12 in², 12 in², 11.2 in², 11.2 in², 113.6 in²   **3.** 1,750 cm²   **5.** 19,518 cm²   **7.** 1,181 in²
9. 5,330 cm²   **11.** 39 ft²   **13.** 2 × 4 × 6 inch-prism
15. 4

1. 153.9 m², 153.9 m²; 219.8 m², 527.6 m²
2. 78.5 cm², 78.5 m², 565.2 cm², 722.2 cm²
3. 979.7 cm²   **5.** 1,132.0 cm²   **7.** 829.0 in²
9. 10 cm radius, 5 cm height   **11.** 706 cans
13. *about* 8 ft diameter

1. 64 m³   **2.** 105,000 cm³   **3.** 297 in³   **4.** 192.5 in³
5. 105 ft³   **6.** 9,000 in³   **7.** 18,900 cm³   **9.** 23,760 ft³
11. 229.5 m³   **13.** 3,375 m³   **15.** 19.5 ft³
17. 10 horsepower   **21.** 12 launches

1. 384 m³   **2.** 300 cm³   **3.** 1,200 in³   **4.** 286 m³
5. 10 cm³   **6.** 367.8 yd³   **7.** 40 m³   **9.** 256 m³
11. 16 cm³   **13.** 21,769 ft³   **15.** The volume is eight times greater.   **17.** 15   **19.** 0.2

1. 2,373.8 cm³   **2.** 5,652 in³   **3.** 769.3 m³   **4.** 62.8 in³
5. 1,356.5 cm³   **6.** 1,271.7 ft³   **7.** 602.9 cm³
9. 39.3 m³   **11.** 11,335.4 in³   **13.** $\approx$ 3,532.5 m³
15. 14 days   **18.** 65°   **19.** 130°   **20.** 28°   **21.** 18,000
22. 9,500   **23.** 0.000028   **24.** 36 ft²

Pages 268-269   Lesson 8-12
1. 50.2 cm³   **2.** 10,048 yd³   **3.** 150.7 m³   **4.** 401.9 m³
5. 100.5 mm³   **6.** 376.8 in³   **7.** 1,004.8 ft³   **9.** 52.$\overline{3}$ cm³
11. 401.9 mm³   **13.** 2,713 ft³   **15.** 10 cm   **17.** Friday

Pages 270-271   Lesson 8-13
1. 11,304 mi²   **2.** 6 ft   **3.** 2.5 in²   **5.** 74.2 ft³
8. 2   **9.** $\frac{1}{2}$ **10.** $\frac{1}{2}$ **11.** $\frac{1}{4}$ **12.** 2 bags

Pages 272-273   Chapter 8 Review
1. b   **3.** j   **5.** c   **7.** f   **9.** k   **11.** 150.7 cm
13. 50.2 in.   **15.** 12 in²   **17.** 113.04 in²
19. 3,215.36 cm²   **21.** 8 chapter 4s   **23.** 1,350 in²
25. 84 cm³   **27.** 197.8 ft³   **29.** 400 ft³
31. *about* 13,541 gallons of milk

Pages 276-277   Chapter 8 Cumulative
Review/Test
Free Response: 1. 6   **3.** 86   **5.** 4,114   **7.** 3,209
9. $9\frac{5}{6}$ **11.** $3\frac{11}{12}$ **13.** $\frac{7}{10}$ **15.** 1   **17.** 36 in.
19. $\overleftrightarrow{AB} \parallel \overleftrightarrow{RS}$ **21.** $\overleftrightarrow{FG} \parallel \overleftrightarrow{HI}$ **23.** parallelogram   **25.** 22
Multiple Choice: 1. c   **3.** b   **5.** c   **7.** b   **9.** c

CHAPTER 9   RATIO, PROPORTION, & PERCENT

Pages 280-281   Lesson 9-1
1. $\frac{30}{11}$ **2.** $\frac{28}{13}$ **3.** $\frac{24}{41}$ **4.** $\frac{1}{30}$ **5.** $\frac{15}{1}$ **6.** $\frac{19}{15}$ **7.** $\frac{4}{3}$ **9.** $\frac{7}{11}$
11. $\frac{2}{15}$ **13.** $\frac{52}{1}$ **15.** $\frac{1}{7}$ **17.** $\frac{3}{4}$ **19.** Tampa Bay
21. $\frac{14}{2}$ or $\frac{7}{1}$ **23.** $\frac{51}{32}$ **25.** Every 10 fingers has two
hands. There are five times as many fingers as there are
hands. Every hand has 5 fingers.   **27.** 12   **28.** 91
29. 36   **30.** 80   **31.** $\frac{9}{5}$ **32.** $\frac{31}{15}$ **33.** $\frac{13}{8}$ **34.** $\frac{3}{7}$
35. 0.006 kg

Pages 282-283   Lesson 9-2
1. $\frac{1}{2}$ **2.** $\frac{1}{8}$ **3.** $\frac{7}{8}$ **4.** $\frac{5}{12}$ **5.** $\frac{3}{12}$ or $\frac{1}{4}$ **6.** $\frac{4}{12}$ or $\frac{1}{3}$
7. $\frac{9}{12}$ or $\frac{3}{4}$ **8.** $\frac{8}{12}$ or $\frac{2}{3}$ **9.** 0   **11.** $\frac{1}{3}$ **13.** $\frac{5}{6}$ **15.** $\frac{6}{6}$
17. $\frac{4}{9}$ **19.** $\frac{7}{9}$ **21.** $\frac{1}{36}$ **23.** $\frac{1}{4}$ **25.** $\frac{1}{4}$ **27.** $\frac{1}{6}$ **29.** $\frac{1}{2}$
31. 7 boats   **33.** A sample answer is 3 *a*'s, 4 *b*'s,
and 5 *c*'s.

Pages 284-285   Lesson 9-3
1. yes   **2.** no   **3.** yes   **4.** no   **5.** yes   **6.** no   **7.** yes
9. yes   **11.** yes   **13.** no   **15.** no   **17.** no
19. They are equivalent. $\frac{8}{20} = \frac{2}{5}$ and $\frac{40}{100} = \frac{2}{5}$ **21.** 8 to 20
23. no   **25.** yes   **27.** no, $\frac{190,000}{1} \neq \frac{316,667}{1}$
29. 88   **31.** 4   **33.** $\frac{1}{4}$

Pages 286-287   Lesson 9-4
1. yes   **2.** no   **3.** no   **4.** yes   **5.** 18   **6.** 42   **7.** 3
8. 9.$\overline{3}$ or $9\frac{1}{3}$ **9.** no   **11.** yes   **13.** 7   **15.** 12   **17.** 72
19. 25.2   **21.** 125   **23.** 66.$\overline{6}$ or $66\frac{2}{3}$
25. No—should divide by 8, not multiply   **27.** 222 perch
29. Sample answers are 2, 4, 8 or 3, 6, 12   **31.** $1\frac{1}{2}$
32. $2\frac{2}{9}$ **33.** $4\frac{11}{24}$ **34.** $500

Page 289   Lesson 9-5
1. 396 ft   **2.** 90 ft   **3.** 117 ft   **4.** 3 in.   **5.** $2\frac{1}{2}$ in.
6. $1\frac{1}{2}$ in.   **7.** $\frac{2}{5}$ in.   **8.** $\frac{1}{8}$ in.   **9.** 20 ft × 15 ft
11. 15 ft × 15 ft   **13.** $1\frac{1}{2}$ in.   **15.** $1\frac{1}{4}$ in.   **17.** $\frac{1}{4}$ in.
19. $\frac{3}{4}$ in.   **21.** 96 pieces

Pages 292-293   Lesson 9-6
1. 10 cm   **2.** 16 in.   **3.** 12.5 cm   **5.** 5 yd   **7.** 13.5 ft
9. 15 m   **11.** 14 cm   **13.** 15.75 m   **15.** 24 m
17. 35 ft   **19.** 2.3 in.   **21.** 67.5 min

Pages 294-295   Lesson 9-7
1. 531 miles   **2.** 13 h 25 min   **3.** Los Angeles & Dallas
5. Denver to Memphis to Jacksonville   **7.** 15.9 h or 15 h
54 min   **9.** 298 kilometers per hour   **11.** 62.5 mi
13. 25 mi   **14.** 19   **15.** 313   **16.** 6,292   **17.** 1,190
18. m   **19.** mm   **20.** Convenient Mart

Page 297 Lesson 9-8

1. $\frac{1}{2}$ **2.** $\frac{1}{10}$ **3.** 2 **4.** $1\frac{1}{2}$ **5.** $\frac{1}{8}$ **6.** $\frac{1}{1,000}$ **7.** 90%

8. 25% **9.** 62.5% **10.** 85% **11.** $33\frac{1}{3}$% **12.** $12\frac{1}{2}$%

13. $\frac{29}{100}$ **15.** $\frac{13}{20}$ **17.** $1\frac{1}{4}$ **19.** 2 **21.** $\frac{1}{4}$ **23.** $\frac{3}{8}$

25. 21% **27.** 25% **29.** 82% **31.** $12\frac{1}{2}$% **33.** $16\frac{2}{3}$%

35. 400% **37.** 100% **39.** $\frac{7}{25}$

Pages 298-299 Lesson 9-9

1. 0.025 **2.** 0.732 **3.** 0.006 **4.** 0.099 **5.** 0.0025
6. 0.75 **7.** 0.07 **8.** 1 **9.** 0.09 **10.** 0.5 **11.** 12%
12. 7% **13.** 1.5% **14.** 10.5% **15.** 90% **16.** 750%
17. 100% **18.** 200% **19.** 615% **20.** 810% **21.** 0.12
23. 0.3 **25.** 0.081 **27.** 8.5 **29.** 0.382 **31.** 0.015
33. 68% **35.** 70% **37.** 1% **39.** 5.6% **41.** 138%
43. 300% **45.** $46\frac{2}{3}$% or $46.\overline{6}$% **47.** 23 free throws
49. $33\frac{1}{3}$% **51.** 0.05 **53.** 1 **55.** 0.005 **56.** 276
57. 11,520 **58.** 8,700 **59.** 38,555 **60.** 102,204
64. 14.72 cm²

Pages 300-301 Lesson 9-10

1. 1-36 roll; 4-24 rolls **2.** 17, 18, and 19
3. 3 quarters, 2 dimes, and 3 nickels **5.** 39
7. 30 and 45 **9.** $7 **11.** left: 0, 1, 2, 6 (also 9), 7, 8;
right: 0, 1, 2, 3, 4, 5 **12.** 48 **13.** 6 **14.** 120
15. obtuse **16.** acute **17.** right **18.** 201 cm²

Pages 302-303 Chapter 9 Review

1. j **3.** f **5.** g **7.** h **9.** $\frac{4}{3}$ **11.** $\frac{21}{1}$ **13.** $\frac{1}{6}$ **15.** $\frac{1}{3}$
17. Y **19.** 15 **21.** 6 **23.** $2.25 **25.** 3 in. **27.** 8 in.
29. 60 cm **31.** 25 m **33.** 236.25 mi **35.** $\frac{4}{25}$ **37.** 9%
39. 0.16 **41.** 89% **43.** $\frac{1}{5}$ **45.** 25% **47.** 53, 46

Pages 306-307 Cumulative Review/Test

Free Response: 1. 50 **3.** 144 **5.** 349.7 **7.** 82.05
9. 4.085 **11.** 108 **13.** 162,081 **15.** 0.0007
17. hexagon, regular **19.** quadrilateral, regular
21. 8 ft **23.** 314 in² **25.** 15 **27.** 0.63 **29.** 125
31. 75 **33.** 125
Multiple Choice: 1. b **3.** b **5.** b **7.** a **9.** d

CHAPTER 10 APPLYING PERCENTS

Pages 312-313 Lesson 10-1

1. 6 **2.** 12.6 **3.** 79.1 **4.** 1.35 **5.** 5 **6.** 2
7. 18 **8.** 6.4 **9.** 3 **11.** 110 **13.** 2 **15.** 0.54
17. 10 **19.** 43 **21.** 732 **23.** 7 **25.** 12 **27.** 60
29. 45 **31.** 30 **33.** false **35.** Using $\frac{1}{3}$ as a multiplier
is like dividing by 3. Since 45 is divisible by 3, $\frac{1}{3}$ of 45 is
15. **37.** 400 light bulbs **39.** $7 \times 77 + 7 \times 77 -$
77 **40.** 53.2 cm **41.** 14.4 in. **42.** $1\frac{3}{10}$ **43.** $3\frac{5}{33}$
44. $3\frac{9}{50}$ **45.** 18 minutes

Pages 314-315 Lesson 10-2

1. 80% **2.** 90% **3.** 15% **4.** 150% **5.** 37.5% or
$37\frac{1}{2}$% **6.** 50% **7.** 160% **8.** 80% **9.** $66.\overline{6}$% or
$66\frac{2}{3}$% **11.** 500% **13.** 2.5% or $2\frac{1}{2}$% **15.** 150%
17. 75% **19.** 75% **21.** 80% **23.** 75% **25.** 40%
27. $1.80 **29.** $2.10

Pages 316-317 Lesson 10-3

1. 20 **2.** 40 **3.** 24 **4.** 54 **5.** 15 **6.** 63 **7.** 40
9. 50 **11.** 68 **13.** 100 **15.** 60 **17.** 17.5 **19.** 50
21. 32 **23.** 75 **25.** 200 votes **27.** 13.67 million
29. 32 **31.** 15 **33.** 6 moves

Pages 318-319 Lesson 10-4

1. 8 families **3.** 18 **5.** $15 **9.** =
10. < **11.** > **12.** > **13.** 40 **14.** 50 **15.** 39
16. 3, 2

Pages 320-321 Lesson 10-5

1. 5 **2.** 30 **3.** 2 **4.** 10 **5.** 240 **6.** $3 **7.** 11
9. 72 **11.** 9 **13.** 24 **15.** 5 **17.** 84 **19.** 1
21. 5 **23.** 18 **25.** 2 billion **27.** *about* $7.50
29. *about* $90 **35.** $0.80; $8.80 **37.** $6.40; $89.40

Pages 324-325 Lesson 10-6

1. 13% **2.** 25% **3.** 6% **4.** 11% **5.** 16%
6. 26% **7.** 8% **8.** 14% **9.** 19% **11.** 23%
13. 29% **15.** 70% **17.** 72% **19.** 25% **21.** 47%
25. 1 m **26.** 1 kL **27.** 4 h 10 m **28.** 11 h 55 m
29. at least 25.12 inches

Pages 326-327 Lesson 10-7

1. $0.85, $7.65 **2.** $13.90, $41.70 **3.** $315, $585
4. 3.25, $3.25 **5.** $11.20, 20% **6.** $1,760, 14%
7. $0.98, $8.82 **9.** $1.44, $27.31 **11.** $49, 25%

13. $4.34, 21.7% **15.** $31.50 **17.** 25% of 60 is larger than 25% of 48, $45 **19.** $75, $50, $80, $70, $60

Pages 328-329 Lesson 10-8
1. $60, $560 **2.** $648, $1,848 **3.** $30.23, $650.23
4. $13.80 **5.** $9.60, $32.81 **7.** $21.60, $111.60
9. $1,200, $3,700 **11.** $18.45, $223.45 **13.** $34.20
15. $544.38 **17.** $18 **19.** $38.50 **23.** $200

Page 331 Lesson 10-9
1. $265.34 **2.** $545 **3.** $265.23 **4.** $546.54
5. $260.10 **7.** $1,453.86 **8.** 60 cm² **9.** 12.96 mm²
10. no **11.** yes **12.** yes **13.** $\frac{2}{9}$

Page 332 Lesson 10-10
1. greater **2.** $74,452 **3.** $71,491 **5.** 10 games
7. 20% **9.** 18% **12.** rhombus **13.** trapezoid
14. rectangle **15.** 24 cm³ **16.** 350 in³ **17.** 24.48 m³
18. 1,160 miles

Pages 334-335 Chapter 10 Review
1. multiply **3.** 20% **5.** discount rate **7.** principal
9. compound **11.** 12.8 **13.** 630 **15.** 7.13
17. 130% **19.** 31.25% **21.** 22.2% **23.** 20
25. 30 **27.** 70 **29.** $0.60 **31.** 13% increase
33. 5% decrease **35.** 8% increase **37.** $93.60
39. $658 **41.** $234 **43.** $13.33 **45.** 5%
47. $3,594.71

Pages 338-339 Cumulative Review/Test
Free Response: 1. 70 **3.** 35 **5.** 20 **7.** 416,845
9. 83,034 **11.** 1,656,207 **13.** 51,752 **15.** 84,348
17. 2,416,444 **19.** 2 **21.** 8 **23.** 5 **25.** 3
27. kilogram **29.** 15 in., 9 in., 9 in. **31.** 6 **33.** 4
35. $\frac{7}{10}$ **37.** $11, $44

Multiple Choice: 1. c **3.** b **5.** d **7.** d **9.** c

CHAPTER 11 STATISTICS AND GRAPHS

Pages 342-343 Lesson 11-1
1. 6, 6, 7 **2.** 14, 11, 6 **3.** 75, 75, 7 **4.** 35.5, 43, 20
5. 53, no mode, 72 **6.** 278.5, no mode, 93 **7.** 2, 2, 8
9. 29, 12, 86 **11.** 68.5, 91, 66 **13.** 41.5, no mode, 3

15. 2.45, 1.8, 1.6 **17.** 8.2, 8.3, 1.4 **19.** 34.5, 48.5, 8.2 **21.** no mode, no mode, 87.4 **23.** $g - 256 = 102$, $g = 358$ **25.** 400 **27.** 18,000

Pages 344-345 Lesson 11-2
1. 10; 13; 17; 10; 9; 16 **3.** 65 in. **5.** 13 in.
7. 3.75 mi **9.** 3 mi **13.** $1.76 **15.** size $10\frac{1}{2}$

Pages 346-347 Lesson 11-3
1. mean, It is the greatest. **2.** mode, It is the most common. **3.** median, Only three earn large salaries.
4. mean **5.** The president, vice-president, managers, and supervisors would probably be excluded. They are probably not union members. **7.** A good answer would refer to example and exercises on page 346. **9.** lower mean—25, median—not affected **11.** median—not affected; lower mean—$74,467 **13.** 22 **14.** 15
15. 3.98 **16.** 17 **17.** 0.0048 **18.** 52.3 **19.** 0.63
20. 3,000 **21.** 15 in. for both

Pages 348-349 Lesson 11-4
7. New York and Tulsa; Chicago and Miami
9. Charlotte **11.** 3 stations **13.** 29 mi

7. 20 cm **9.** 40-50 days **11.** 42 cm **13.** 4.3×10^6
14. 4.21×10^5 **15.** 8.21×10^4 **16.** 9.42×10^2
17. cube **18.** triangular prism **19.** triangular pyramid
20. 3 miles

Pages 354-355 Lesson 11-6
1. 500,000 farms **2.** 1970-1975 **3.** 1975-1980
4. 1,050,000 farms **9.** 10 **11.** 80 **13.** 90 **15.** 9,000
17. 800 **19.** increase **21.** about 500,000 people
24. $\frac{4}{3}$ **25.** 6 **26.** $\frac{21}{26}$ **27.** $6\frac{2}{3}$ **28.** 210 cm²
29. 480 in² **30.** 172 mm² **31.** 6 cars

Pages 356-357 Lesson 11-7
1. 15% **2.** 2-person households **3.** 25% **4.** 58%
11. 12% **13.** 4 times **15.** suits **17.** $2,000

Pages 358-359 Lesson 11-8
1. 15 videos **2.** 48 videos **3.** 21 videos **4.** 3 days
5. 27 **6.** 34 **7.** 30–39 **9.** 1, 2, 3, 5 **11.** 5, 6, 7, 8
13. 98.5; 80 and 99 **17.** 62 **19.** 8 **21.** 10-19 year olds **23.** 301.4 cm² **24.** 395.6 mm² **25.** 150.7 in²
26. $\frac{6}{7}$ **27.** $\frac{1}{25}$ **28.** 60%

Page 361 Lesson 11-9
1. 51.5, 67, 32, 35 **2.** more **3.** 19, 9, 10 **5.** 3, 9, 16

7. the spread of the values **9.** consistent **11.** 3 slices
13. 14 **15.** 13

Pages 362-363 Lesson 11-10

1. 80, 50 **2.** 50, 100 **3.** one fourth of the data **4.** one
half of the data **5.** one half of the data **7.** the mode,
the exact values of some of the data, the mean **9.** one-
fourth of the number of the values, yes, Quartiles divide
the data into four equal parts. **11.** the lower quartile,
interquartile range **15.** 28°C **16.** 16°C **17.** 6°C
18. 43°C **19.** 24 m³ **20.** 196 cm³ **21.** 456 in³
22. 5.25 gallons

Pages 364-365 Lesson 11-11

1. 1,215 plants **2.** 15 high fives **3.** 163 pages
5. 9 zeros and 20 of every other digit **7.** 190 pounds
9. $52 **11.** 50% **12.** 25% **13.** 85% **14.** 20%
15. $33\frac{1}{3}$% **16.** 2.5 **17.** 68 **18.** 243

Pages 366-367 Chapter 11 Review

1. k **3.** c **5.** h **7.** j **9.** b **11.** e **13.** median: 14,
mode: 14, range: 8 **15.** 40 **17.** 75 **19.** median age
25. 10% **27.** 30 participants **29.** 10; 7; 3 **31.** 243
puzzles

Pages 370-371 Cumulative Review/Test

Free Response: 1. 107 **3.** 80 **5.** 48 **7.** 5,737
9. 2,855 **11.** 218 **13.** < **15.** < **17.** < **19.** 12
21. 5,280 **23.** 5 **25.** $\overleftrightarrow{AB}$; line AB **27.** line segment
MN; $\overline{MN}$ **29.** 100.5 cm³ **31.** 1,004.80 in³
33. 17; 18; 15
Multiple Choice: 1. b **3.** d **5.** d **7.** a **9.** b

CHAPTER 12 INTEGERS

Page 377 Lesson 12-1

1. < **2.** > **3.** < **4.** < **5.** > **6.** < **7.** > **8.** <
9. −9, −3, 0, 4, 5 **10.** −8, −6, 0, 2, 6
11. −7, −1, 0, 6, 11 **12.** −5, −3, 0, 3, 5 **13.** <
15. > **17.** > **19.** < **21.** > **23.** < **25.** <
27. > **29.** < **31.** > **33.** −4, 0, 3, 8, 9
35. −4, −2, −1, 0, 1 **37.** −3.5, −3, 0, 3, 3.5
39. −10; 0 **41.** Antarctica **43.** Great Falls, Montana

Page 379 Lesson 12-2

1. −2 + (−3) = −5 **2.** −3 + 6 = 3 **3.** 8 **4.** −7
5. −6 **6.** −2 **7.** 2 **8.** −1 **9.** −1 **11.** −15 **13.** 0
15. −4 **17.** 11 **19.** −33 **21.** 13 **23.** −38 **25.** −0.8
27. −48 **29.** To add integers with different signs, find
the difference of their absolute values. Give the result the
same sign as the integer with the greater absolute
value. **31.** 6°C **33.** −123°C
35. O = 6; D = 5; E = 1; V = 3; N = 0

Pages 380-381 Lesson 12-3

1. 17 **2.** 4 **3.** 20 **4.** 4 **5.** −5 **6.** −1 **7.** −5
8. 1 **9.** −15 **10.** −16 **11.** −12 **12.** −8 **13.** 0
14. −10 **15.** 8 **16.** −1 **17.** −9 **19.** −6 **21.** −6
23. −13 **25.** −4 **27.** 4 **29.** 0 **31.** −12 **33.** −84.3
35. −2.5 **37.** −6 **39.** −7°C **41.** 14,776 feet
45. −2 **47.** 2 **49.** −6 **51.** −7 **53.** 28 **55.** 29

Pages 382-383 Lesson 12-4

1. −39°F **2.** −33°F **3.** 37°F **5.** −48°F **7.** −74°F
9. 20°F **11.** −79°F **13.** 26°F **15.** 39°F **17.** 25 mph
19. 52°F **21.** $221.3 billion **23.** deficit **25.** 66.4
26. 115.06 **27.** 236.66 **28.** 229.3

Page 387 Lesson 12-5

1. 15 **2.** 36 **3.** 9 **4.** 90 **5.** 35 **6.** 84 **7.** 72
8. 30 **9.** −12 **10.** −30 **11.** −32 **12.** −98 **13.** −77
14. −540 **15.** 0 **16.** 0 **17.** 40 **19.** −6 **21.** −42
23. −70 **25.** −54 **27.** 56 **29.** −369 **31.** −210
33. −12.6 **35.** $-\frac{4}{7}$ **37.** −108 **39.** 16.5 feet
41. 84°C **43.** 0 **45.** 48 **47.** $\frac{2}{3}$ **48.** $\frac{3}{7}$ **49.** $\frac{1}{4}$
50. $n = 8m$ **51.** $x = 11\frac{1}{4}$ cm **52.** $y = 3$ in.
53. $247.74

Pages 388-389 Lesson 12-6

1. 7 **2.** 5 **3.** 0.25 **4.** 0.2 **5.** 6 **6.** 6 **7.** 0.5
8. 7.5 **9.** −12 **10.** −9 **11.** −0.5 **12.** 0 **13.** −14
14. −9 **15.** −0.1 **16.** 0 **17.** 4 **19.** −9 **21.** −5
23. −2 **25.** 0 **27.** −6 **29.** 14 **31.** −8 **33.** −3.7
35. −1 **37.** $10\frac{5}{7}$ **39.** −3.5 **41.** −55 **43.** > **45.** <
47. 6 **49.** −3 **51.** −23 **53.** −0.4 or $-\frac{2}{5}$ **55.** Olympic
Games **57.** E **59.** E **61.** E

Pages 390-391 Lesson 12-7

1. 28 apples **3.** 10 microbes **5.** 3 letters; 5 postcards
7. 16 rooms **9.** 3 of the 6-oz boxes **12.** 7,000
13. 0.1481 **14.** 0.0912 **15.** 4 games

Pages 392-393 Chapter 12 Review
1. opposite **3.** whole numbers, integers **5.** negative
7. adding **9.** > **11.** < **13.** = **15.** >
17. −9, −7, 0, 7, 9 **19.** −3 **21.** 0 **23.** 4 **25.** −22
27. −7 **29.** 17 **31.** −9 **33.** −13 **35.** 11 ft rise
37. 44°F **39.** 54 **41.** 99 **43.** −84 **45.** 84 **47.** −9
49. $-1\frac{1}{4}$ **51.** 8 **53.** 0 **55.** 9 **57.** −27°F

Pages 396-397 Cumulative Review/Test
Free Response: 1. 5 **3.** 9 **5.** 527 **7.** 6,172
9. 131,707 **11.** 32 **13.** 3 **15.** $\frac{1}{8}$ **17.** $\frac{1}{12}$ **19.** 4 in.
21. $100.10 **23.** $70.40 **25.** 6.4; 6.5; 9 **27.** 32°
Multiple Choice: 1. c **3.** a **5.** b **7.** b **9.** d

CHAPTER 13 EXTENDING ALGEBRA

Pages 400-401 Lesson 13-1
1. 9 **2.** 6 **3.** −5 **4.** 20 **5.** 0 **6.** 157 **7.** −3 **8.** −9
9. −4 **10.** −8 **11.** −35 **12.** 29 **13.** −2 **15.** 34
17. −31 **19.** 7 **21.** −3 **23.** 15 **25.** 383 **27.** 0
29. −0.8 **31.** −$13 **35.** −0.04; They are the same.
36. 8.1 **37.** 76.61 **38.** 421.2 **39.** 0.3362
40. 25.2 m² **41.** 60.45 cm² **42.** 425.35 mm²
43. 40 free throws

Pages 402-403 Lesson 13-2
1. 2.5 **2.** −5 **3.** 15 **4.** −75 **5.** −50 **6.** 40
7. −192 **8.** 200 **9.** $-\frac{5}{7}$ **11.** −4 **13.** −4 **15.** −144
17. 8.3 **19.** −3 **21.** $-\frac{4}{5}$ **23.** $\frac{1}{4}$ **25.** −42 **27.** $\frac{2}{3}$
29. $\frac{2}{3}$ **31.** $\frac{1}{5}$ **33.** loss of 6 yards **35.** 5, −4 **37.** −8
39. −1 **41.** −243 **43.** 16

Pages 404-405 Lesson 13-3
1. −4 **2.** 6 **3.** −7 **4.** −20 **5.** −5 **6.** −10.5 **7.** 5
9. $-6\frac{1}{4}$ **11.** −4 **13.** −5 **15.** 212 **17.** 22 **19.** 8
21. 16 **23.** $30 **27.** $3\frac{1}{5}$ **28.** $\frac{21}{32}$ **29.** $1\frac{1}{2}$ **30.** 16
31. 1, 2, 3, 4 **32.** 0, 1, 2, 3, 4 **33.** 20, 34

Page 409 Lesson 13−4
1. (2,1) **2.** (−3,2) **3.** (−3,−2) **4.** (3,−3) **5.** (1,3)
6. (1,0) **17.** (−2,3) **19.** (−3,−2) **21.** (4,−4)

35. (2,−4) **37.** It is similar in moving horizontally for
one number and vertically for the other.

Pages 410-411 Lesson 13-5
19. October **21.** 8 pounds **23.** $x - 3 = y$

Pages 412-413 Lesson 13-6
1. Los Encinos St. Historical Park **2.** Highway 110
3. Santa Monica Mun. Airport **4.** B-1 **5.** B-4 **6.** D-4
7. B-3 **8.** E-1 **9.** Griffith Park Zoo **11.** LA
International **13.** A-2 **15.** D-2 **17.** N. on Lincoln
Blvd. to Tijera Blvd.; N. on Tijera Blvd. to Manchester
Ave.; E. on Manchester Ave. to the Forum **19.** Sunset
Blvd. **22.** 75° **23.** 105° **24.** 57° **25.** −2 **26.** −18
27. −6 **28.** 53 pizzas

Pages 414-415 Lesson 13-7
1. $2x + x = 81$; 54 tapes, 27 compact discs **2.** $212 =$
$10.50 + \frac{1}{2}x$; $403 **3.** 39 mums **5.** 10 parakeets, 8
kittens **7.** *about* $45 **9.** 20, 70 **10.** $\frac{1}{2}$ of the data
11. 21 **12.** −5 **13.** −30 **14.** 425 feet

Pages 416-417 Chapter 13 Review
1. same number **3.** ordered pair **5.** horizontal
7. origin **9.** 15 **11.** 19 **13.** 25 **15.** −2 **17.** −13
19. −192 **21.** −4 **23.** −3 **25.** $-1\frac{3}{5}$ **27.** (−4,3)
29. (−2,−3) **43.** River Road **45.** A4 **47.** B6
49. 18 m **51.** $135, $185

Pages 420-421 Cumulative Review/Test
Free Response: 1. 2,100 **3.** 4 **5.** 3,741 cars **7.** $1\frac{1}{7}$
9. $1\frac{5}{8}$ **11.** 64 **13.** 18 ft³ **15.** 1,800 in³ **21.** −7
23. 29 **25.** 28 **27.** −15 **31.** 11 miles, work; 15
miles, school
Multiple Choice: 1. c **3.** b **5.** d **7.** c **9.** c

CHAPTER 14 PROBABILITY

Pages 426-427 Lesson 14-1
3. $6 \times 6 = 36$ **4.** $2 \times 6 = 12$ **5.** 8 outcomes
7. 24 outcomes **9.** $4 \times 4 \times 4 \times 4 = 256$ choices
11. $\frac{1}{100}$ **13.** 5.8 **14.** 11.1 **15.** 8.5 **16.** 137.03

17. 11.969 **18.** $\frac{1}{3}$ **19.** $\frac{1}{2}$ **20.** $\frac{1}{3}$ **21.** 45 singles

Pages 428-429 Lesson 14-2
1. 30% **2.** 20% **3.** 20% **4.** 40% **5.** 20% **6.** 80%
7. 0% **8.** 20% **9.** 60% **11.** 40% **13.** 60%
15. 100% **17.** 0% **19.** 3% **21.** 6% **23.** 17%
25. 11% **27.** $\frac{1}{25}$ or 4%

Pages 430-431 Lesson 14-3
1. $\frac{1}{15}$ **2.** $\frac{1}{10}$ **3.** 0 **4.** $\frac{1}{5}$ **5.** $\frac{1}{4}$ **6.** $\frac{1}{36}$ **7.** $\frac{1}{6}$ **8.** $\frac{1}{18}$
9. $\frac{1}{15}$ **11.** 0 **13.** $\frac{1}{45}$ **15.** $\frac{2}{9}$ **17.** $\frac{3}{50}$ **19.** 0 **21.** $\frac{1}{25}$
23. $\frac{1}{4}$ **25.** $\frac{1}{625}$ **27.** $\frac{1}{2}$, $\frac{1}{2}$

Pages 432-433 Lesson 14-4
1. $\frac{1}{3}$ **2.** $\frac{1}{2}$ **3.** $\frac{2}{3}$ **4.** $\frac{1}{3}$ **5.** $\frac{2}{3}$ **6.** 1 **7.** $\frac{2}{3}$ **8.** $\frac{2}{3}$ **9.** $\frac{3}{4}$
11. $\frac{1}{8}$ **13.** $\frac{3}{8}$ **15.** $\frac{1}{2}$ **17.** $\frac{3}{4}$ **19.** $\frac{5}{8}$ **21.** $\frac{1}{3}$
23. The probability of spinning blue or a number greater than 5 will remain the same if the four adjacent blue wedges are numbered 14, 15, 16 and 1. Any other placement of four adjacent wedges will result in a different probability.

Pages 436-437 Lesson 14-5
1. 125 **2.** 86 **3.** 88 **4.** $\frac{1}{6}$ **5.** a floral design **7.** $\frac{111}{150}$
9. $\frac{23}{75}$ **11.** $\frac{44}{75}$ **13.** $\frac{527}{5,625}$ **15.** $\frac{1,332}{11,250}$ **17.** $\frac{13}{25}$
21. 12 **23.** 240

Pages 438-439 Lesson 14-6
1. 1 to 5 **2.** 1 to 5 **3.** 1 to 5 **4.** 0 to 6 **5.** 3 to 3
6. 3 to 3 **7.** 3 to 3 **8.** 2 to 4 **9.** 2 to 4 **10.** 3 to 3
11. 2 to 4 **12.** 2 to 4 **13.** 4 to 2 **14.** 4 to 2
15. 1 to 5 **16.** 4 to 2 **17.** 3 to 3 **18.** 4 to 2
19. 4 to 2 **21.** 3 to 33 **23.** 31 to 5 **25.** 33 to 3
27. 27 to 9 **29.** 6 to 14; $\frac{3}{10}$ **31.** 14 to 6; $\frac{7}{10}$
33. 11 to 9; $\frac{11}{20}$ **35.** 3 to 1 **37.** $\frac{5}{36}$ **39.** 58 **41.** 67
43. 46

Pages 440-441 Lesson 14-7
1. 504 students **2.** 85 customers **3.** $\frac{1}{4}$ **5.** $\frac{1}{8}$ **7.** $\frac{2}{5}$
9. The sample was not representative. The larger population did not have the same preferences as the sample. Too much time elapsed between taking sample and choices of population. **11.** 153.9 cm² **12.** 201.0 in²
13. 1,808.6 mm² **14.** 38.5 cm² **15.** 41.2 **16.** 38
17. 38

Pages 442-443 Chapter 14 Review
1. ratio **3.** odds **5.** tree **7.** dependent
9. 9 outcomes **11.** 50% **13.** 70% **15.** 80%
17. 100% **19.** $\frac{3}{64}$ **21.** $\frac{1}{15}$ **23.** 1 **25.** $\frac{3}{5}$ **27.** $\frac{9}{30}$
29. $\frac{26}{75}$ **31.** 8 to 8 **33.** 700 owners **35.** $\frac{1}{3}$

Pages 446-447 Cumulative Review/Test
Free Response: 1. 5,010 **3.** 30,100 **5.** 11.4 m
7. 24 yd **9.** 7,010 **11.** 283 **13.** $0.\overline{6}$ **15.** $0.1\overline{6}$
17. 234 **19.** 25, 26 **21.** -9 **23.** -2 **25.** -14
27. $\frac{1}{2}$ **29.** $-\frac{19}{20}$
Multiple Choice: 1. c **3.** d **5.** a **7.** c **9.** d

PHOTO CREDITS

Cover, Aaron Haupt, Image Development/Design, Terry Anderson; **v-xi,** Ken Frick; **A1,** (t) Ken Frick, (b) Aaron Haupt; **A2-A11,** Ken Frick; **2,** (t) Randy Schieber, (o) Ken Frick; **4,** ©Farrell Grehan, FPG; **5,** Paolo Koch, Photo Researchers; **6,** Hickson & Associates Photography; **8,** Ken Frick; **10,** Steven E. Sutton/Duomo; **12,** Ken Frick; **14,** (t) Ken Frick, (b) Janet Adams; **15,** Ken Frick; **16,** Superstock, Inc; **18,** Doug Martin; **19,** Bob Hipp; **20,** Ken Frick; **21,** Doug Martin; **23-24,** Ken Frick; **27,** Studiohio; **28,** Randy Schieber; **30,** (t) Randy Schieber, (b) Ken Frick; **31,** Tim Courlas; **33,** (l) Steve Lissau, (r) Alan Carey; **35,** Bud Fowle; **36,** Studiohio; **38,** CALVIN & HOBBES Copyright 1988 Universal Press Syndicate. Reprinted with permission, all rights reserved; **40,** Kenji Kerins; **42,** Doug Martin; **44,** © Eugene Gebhardt, FPG; **45,** Mary Lou Uttermohlen; **46,** Bud Fowle; **48,** Studiohio; **50,** Doug Martin; **51,** Randy Schieber; **53,** Ken Frick; **54,** Doug Martin; **57,** ©Barry Rosenthal, FPG; **59,** John Griffin/The Image Works; **60,** Doug Martin; **61,** ©Michael Nelson, FPG; **63,** Doug Martin; **64,** ©Dennie Cody, FPG; **65,** Doug Martin; **66,** Ken Frick; **68,** Studiohio; **70,** Doug Martin; **71-72,** Ken Frick; **76,** Bud Fowle; **78, 81,** Ken Frick; **82,** Doug Martin; **83,** ©TRAVELPIX, FPG; **84,** Doug Martin; **85,** Randy Schieber; **89,** Jim Franck; **90,** ©Allsports USA/Brian Drake; **91,** Ken Frick; **94,** Bud Fowle; **95,** Janet Adams; **96,** (t,b) Ken Frick, (c) Jim Franck; **99,** ©Bob Taylor, FPG; **100,** Doug Martin; **101,** ©Lee Balterman, FPG; **104,** FRANK & ERNEST Reprinted by permission of NEA, INC.; **106,** (t) Studiohio, (b) Doug Martin; **109,** Ken Frick; **110,** ©John Terrence Turner, FPG; **111,** Ken Frick; **113,** Bud Fowle; **115,** Ken Frick; **116,** Tim Courlas; **118,** Michael Sargent/The White House; **119,** Ken Frick; **120,** Hank Morgan, Photo Researchers; **122,** Terje Kveen/The Image Bank; **123,** Nancy Durrell McKenna, Photo Researchers; **124,** ©Allsports USA/Rick Stewart; **127,** Ken Frick; **128, 130,** Bud Fowle; **132,** ©Allen B. Seiden, FPG; **133,** Bud Fowle; **134,** (l) Hickson & Associates Photography, (r) Doug Martin; **137,** (l) Steven Lissau, (r) Ken Frick; **139,** ©M. Kozlowski, FPG; **142,** Ken Frick; **144,** (l) Studiohio, (r) Ken Frick; **146,** Michel Tcherevkoff/The Image Bank; **147,** Brian Parker/Tom Stack & Associates; **150,** Doug Martin; **152,** G. Glod/Superstock, Inc.; **154,** Hickson & Associates Photography; **156,** Doug Martin; **157,** ©T.J. Florian/Rainbow; **158,** Ken Frick; **160,** Bud Fowle; **162,** Doug Martin; **165,** ©Allsports USA/Otto Greule; **166,** Tim Courlas; **169,** Lloyd Lemmermann; **170,** Ken Frick; **171,** Doug Martin; **172,** Frank Cezus; **174,** BROOM-HILDA ©Reprinted by permission: Tribune Media Services; **176,** Studiohio; **179,** Doug Martin; **182,** Lloyd Lemmermann; **183,** Ken Frick; **185,** (l) Ken Frick, (r) Superstock, Inc; **187,** Doug Martin; **188-189,** Ken Frick; **191,** Brian Parker/Tom Stack & Associates; **192,** Doug Martin; **193,** Superstock Inc; **194,** Hickson & Associates Photography; **195,** Frank Balthis; **197,** ©Allsports USA/Mike Powell; **199,** Larry Kolvoord/The Image Works; **200,** Ken Frick; **202,** ©Bob Daemmrich; **203,** (t) Ken Frick, (b) Bud Fowle; **205,** ©Allsports USA/Tony Duffy; **206,** Aaron Haupt; **207,** Smithsonian Institution; **208,** Ken Frick; **210,** ©Peter Grigley, FPG; **212, 214,** Ken Frick; **216,** Alan Pitcairn/Grant Heilman Photography; **217,** Ken Frick; **218,** (t) Hickson & Associates Photography, (b) Ken Frick; **222,** Doug Martin; **223,** ©Bob Daemmrich; **224,** ©Michael Tamborrino, FPG; **225,** Cletis Reaves, FPG; **226,** Cincinnati Convention and Visitor's Bureau; **228,** (l) Ken Frick, (r) ©Tom Carroll, FPG; **232,** Bud Fowle; **237,** Doug Martin; **238,** Ken Frick; **240,** CALVIN & HOBBES Copyright 1988 Universal Press Syndicate. Reprinted with permission, all rights reserved; **242,** Lloyd Lemmermann; **242,** Böhm-NBBJ Architects; **244,** Picture Unlimited; **245,** Ken Frick; **246,** Doug Martin; **247,** Janet Adams; **248,** Steve Lissau; **250,** Ted Rice; **251,** Ken Frick; **252,** Doug Martin; **253,** Ken Frick; **254,** Aaron Haupt; **255,** Studiohio; **256,** Hickson & Associates Photography; **257,** Larry Hamill; **258,** Ken Frick; **261,** Lloyd Lemmermann; **262,** Bud Fowle; **264,** ©Paul Degreve, FPG; **265,** ©David Sucsy, FPG; **266,** Ken Frick; **268,** Randy Schieber; **269,** Ken Frick; **270,** (t) Lloyd Lemmermann, (b) Ken Frick; **273,** John Colwell/Grant Heilman Photography; **274,** Studiohio; **276,** Ken Frick; **277,** file photo; **278,** Bud Fowle; **280,** Ken Frick; **281-284,** Ken Frick; **284, 286,** Doug Martin; **287,** David R. Frazier; **291,** Joe Munroe, Photo Researchers; **292,** Pictures Unlimited; **296,** Ted Rice; **298,** Pictures Unlimited; **299,** (t,b) Ken Frick, (c) Roger K. Burnard; **300,** Ken Frick; **303,** Superstock, Inc; **305,** Lloyd Lemmermann; **306,** (l) Bud Fowle, (r) Ken Frick; **308,** FUNKY WINKERBEAN Copyright 1990 North America Syndicate. Reprinted with permission, all rights reserved; **310,** (t) Doug Martin, (b) Jim Franck; **311,** Daniel Erickson; **313,** Grant Heilman/Grant Heilman Photography; **314,** Herbet Eisenberg/Superstock Inc; **315,** (t) Ken Frick, (b) Jim Franck; **316,** (t) Bob Dammerich/The Image Works, (b) Ken Frick; **318,** Superstock Inc; **319,** Hickson & Associates Photography; **320,** Doug Martin; **321,** Hickson & Associates Photography; **322-323,** Doug Martin; **324,** Lloyd Lemmermann; **324, 326,** Ken Frick; **327,** ©Dennis Hallinan, FPG; **329-330,** Doug Martin; **331,** Gerard Photography; **332,** Ken Frick; **335,** Janet Adams; **337,** Ken Frick; **338,** Doug Martin; **340,** (t) Superstock Inc, (c) ©TRAVELPIX, FPG; (b) Aaron Haupt; **342,** Ken Frick; **346,** W. Strode/Superstock Inc; **347,** file photo; **349,** Ken Frick; **350,** Doug Martin; **352,** Studiohio; **353,** David Frazier; **354,** Joseph A. Dichello; **356,** Superstock Inc; **358,** Doug Martin; **360,** Aaron Haupt; **362,** Ken Frick; **364, 366,** Doug Martin; **369,** Mary Lou Uttermohlen; **370,** Ken Frick; **372,** FOR BETTER OR WORSE Copyright 1990 Universal Press Syndicate. Reprinted with permission, all rights reserved; **374,** (t) file photo, (b) Superstock Inc; **376,** Doug Martin; **378,** Runk/Schoenberge/Grant Heilman Photography; **379,** Michael Melford/The Image Bank; **380,** Walter R. Marks/Metro Dade Tourism; **381,** Ken W. Davis/Tom Stack & Associates; **382,** The Image Works; **384,** Bud Fowle; **385,** Owens-Corning Fiberglass Corporation; **386,** ©Dave Bartuff, FPG; **388,** Ken Frick; **389,** ©T. Tracy, FPG; **390,** Studiohio; **392,** file photo; **393,** Superstock Inc; **394,** Pictures Unlimited; **395,** David Frazier; **396,** (l) Ken Frick, (r) Studiohio; **398,** Doug Martin; **399,** Randy Schieber; **400,** Doug Martin; **403,** First Image; **404,** Ken Frick; **405,** Doug Martin; **406,** Lloyd Lemmermann; **407,** Doug Martin; **408, 412,** Ken Frick; **413,** George H. Harrison/Grant Heilman Photography; **415,** ©Dan McCoy/Rainbow; **418,** Brian Parker/Tom Stack & Associates; **419,** Ken Frick; **420,** (l) Ken Frick, (r) Studiohio; **422,** CALVIN & HOBBES Copyright 1990 Universal Press Syndicate. Reprinted with permission, all rights reserved; **424,** (t) Studiohio, (b) Ken Frick; **427,** Cobalt Photography; **428,** Ken Frick; **429,** Brian Parker/Tom Stack & Associates; **430,** Bob Hipp; **431,** Ken Frick; **433,** Kent Dannen, Photo Researchers; **434,** Superstock Inc; **435,** Bud Fowle; **436, 438,** Doug Martin; **439,** Ken Frick; **440,** Ted Rice; **441,** (l) David R. Frazier, Photo Researchers, (cl) First Image, (cr,r) ©Henry Gris, FPG; **442,** Superstock Inc; **443,** ©T.J. Florian/Rainbow; **444,** Ken Frick; **445,** Lloyd Lemmermann; **483,** Ken Frick.

514

TECH PREP

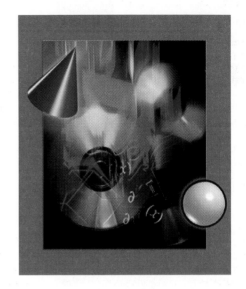

H A N D B O O K

What does a robotics technician do? To find out, you can read pages 525-527. The Tech Prep Handbook provides a career description, a description of technical equipment used in the occupation, information about the type of education required, and a list of resources for each occupation listed. This information can help you determine whether you would be interested in pursuing this occupation.

LASER TECHNICIAN

The narrow beams of light that zip across the stage at a rock concert are produced by lasers. Where do you think the word **laser** comes from? Each letter in the word laser stands for a word.

Light
Amplification by
Stimulated
Emission of
Radiation

Beams of laser light do not spread out because all their waves travel in the same direction.

CAREER DESCRIPTION

Laser technicians operate and adjust all the elements of laser devices to make them work properly in industry, supermarkets, offices, and homes. A laser technician troubleshoots and repairs systems that use lasers, performs tests and measurements using electronic testing equipment, operates laser systems and related equipment, assembles laser devices and systems, and researches and develops laser devices.

TOOLS OF THE TRADE

An oscilloscope, shown below, may be used to troubleshoot a malfunction in a laser system. It does not generate the usual sine wave (wavy line) that you may be familiar with. Instead it measures digital signals like those produced by electronic equipment with a series of ones and zeros.

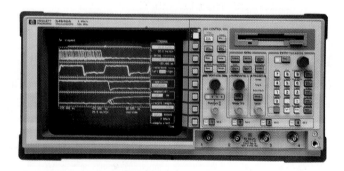

EDUCATION

A good background in mathematics and physics is required. Plan to attend a 2-year laser technician program. Such programs are offered at technical institutes, community colleges, and technical colleges. Also, laser manufacturers may hire graduates with a 2-year associate degree in physics or electronics.

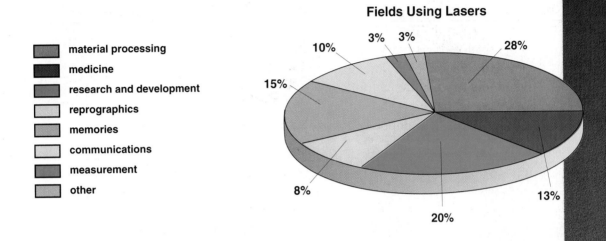

Fields Using Lasers

- material processing
- medicine
- research and development
- reprographics
- memories
- communications
- measurement
- other

28%
3% 3%
10%
15%
8%
20%
13%

EXERCISES

Write in expanded form using 10^0, 10^1, 10^2, and so on.

The following numbers represent the speed at which light travels.

1. 300,000,000 meters per second

2. 186,000 miles per second

Use the circle graph above to answer each question.

3. What field is the largest user of laser applications?

4. What percent of laser applications do medicine and communications make up?

Express in your own words.

5. What kind of skills are required of a laser technician?

6. Do you think you would be interested in being a laser technician? Why or why not?

7. What do you like or not like about the career of laser technician?

PROJECT

UPC Coding and Decoding

Materials: 10 grocery store items each with a UPC bar code and its price

Overview Most grocery products are marked with a bar code system known as a Universal Product Code, or UPC, as shown at the right. Each UPC contains 12 numbers. The first six numbers include one to designate the character set used in the code and five to identify the maker of the product. The second six numbers include five to identify the product followed by one number called the modulo check cipher. For example, omitting the first and last number, the code 51000-00011 identifies a $10\frac{3}{4}$-ounce can of Campbell's tomato soup.

Procedure Make an inventory list like the one below. Remove each UPC code from the grocery items and place it beneath the UPC number on the inventory list.

UPC Number	Maker of Product	Product	Price
51000-00011	Campbell's	$10\frac{3}{4}$-ounce can tomato soup	$0.89

Then make up a shopping list using only the UPC numbers. Use your inventory list to identify, on a separate sheet of paper, the maker of the product, identify the product, and name its price. Exchange shopping lists with other groups. Compare their answers to your shopping list with your answers.

RESOURCES

For further information on a career as a laser technician, write or call the following association.

- Laser Institute of America
  12424 Research Parkway, Suite 130
  Orlando, FL 32826   (407) 380-1553

Check your local library for the following books or magazines.

- Hech, Jeff, The Laser Guidebook

- Laser Focus World
  PennWell Publishing Company
  1 Technology Park Drive
  Westford, MA 01886   (508) 692-0700

AGRIBUSINESS SPECIALIST

CAREER DESCRIPTION

An agribusiness specialist may work in business management. She or he may be employed by a large farm or dairy as a business manager, a credit institution as a lender, or a business as a buyer for farm products.

An agribusiness specialist may work in sales and service. This specialist might work for an aerial crop spraying company, a distribution company for farm products, an insurance company, or any company that offers services to farmers.

A third area is record keeping. Agribusiness specialists may set up and analyze record systems for farmers. He or she can be valuable in helping farmers get the most benefit from the records.

W hat do you think the word **agribusiness** means? Hint: Agribusiness is made up of the words agriculture and business. Any guesses? As you may have guessed, agribusiness is the business end of agriculture. It deals with a wide variety of businesses in agriculture such as farm supplies and services and the processing and marketing of agricultural products.

TOOLS OF THE TRADE

The major tool of the trade for an agribusiness specialist is the computer. The specialist needs programs for record keeping, inventory lists, data bases, spreadsheets, scheduling, payroll, bookkeeping, taxes, graphics, and so on.

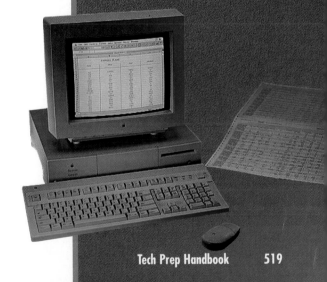

EDUCATION

If you are interested in a career as an agribusiness specialist, you should take courses in mathematics, social studies, agriculture, business, and a laboratory science such as biology, chemistry, or physics. English literature and composition will be helpful since good oral and written communication skills are essential.

After high school, plan to attend a 2-year agricultural or technical college program. The college courses will include business, economics, science, agriculture, and supervised career experience.

EXERCISES

Corn yields in Michigan can be expected to decrease about one bushel per acre for each day corn is planted after May 10 and two bushels per acre for each day corn is planted after May 25.

1. If the Danson family plants corn on May 20, how many bushels per acre less can they expect than if they had planted by May 10?

2. If the Sims family plants corn on May 31, how many bushels per acre less can the Sims expect than if they had planted on May 10?

3. The Danson family plants 700 acres of corn and the Sims family plants 500 acres. How many total bushels of corn less can each family expect to harvest as a result of their late planting? See Exercises 1 and 2 for more information.

4. If the corn price is $3.15 per bushel, how much did the Danson and Sims families lose as a result of their late planting? See Exercise 3 for more information.

Solve. Use the graph at the right.

5. How many sports items can be made from the hide of one cow?

6. What fraction of the cow's hide is used to make baseballs?

7. Name one sports item that makes up $\frac{3}{56}$ of the sports items from the cow's hide.

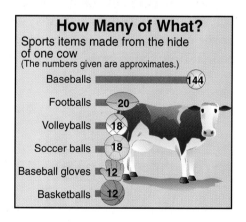

How Many of What?
Sports items made from the hide of one cow
(The numbers given are approximates.)

Baseballs — 144
Footballs — 20
Volleyballs — 18
Soccer balls — 18
Baseball gloves — 12
Basketballs — 12

Express in your own words.

8. Name the skills you have that prepare you to be an agribusiness specialist.

9. Name one area that you would be interested in doing as an agribusiness specialist and tell why.

10. Do you think you would pursue a career as an agribusiness specialist? Why or why not?

PROJECT
Tracking Product Handlers from the Farm to the Grocery Store

Materials: rulers, paper

Overview You will be making a **flowchart** to show the different routes a farm product travels in making its way from the farm to the shelves of a grocery store. A flowchart is a diagram used to show the steps in a procedure.

Procedure Interview a farmer or an agribusiness specialist (at a grain elevator, a feed store, or so on), do research, or write to a food company to determine the handling of a farmer's product from the time the farmer begins working with the product until you take it home from the grocery store.

You may choose various types of products such as grains, animals, nuts, fruits, vegetables, lumber, flowers, trees, and so on.

You may even start backwards by starting with an item from the store and contacting the company on the label.

Once you have collected your data, make a flowchart. Compare your flowcharts with other groups in your class.

RESOURCES

For more information on a career as an agribusiness specialist, write or call the following associations.

- U.S. Department of Agriculture
  Department of Public Information
  Washington, D.C. 20250   (202) 447-3631

- Council for Agricultural Science and
     Technology
  137 Lynn Avenue
  Ames, IA 50010-7197   (515) 292-2125

- Agribusiness Council
  2550 M Street, NW, Suite 275
  Washington, D.C. 20037   (202) 296-4563

- Future Farmers of America
  National FFA Venter
  P.O. Box 15160
  5632 Mt. Vernon Memorial Hwy
  Alexandria, VA 22309   (703) 360-3600

TECH PREP

HANDBOOK

GRAPHIC ARTS PROFESSIONAL

Who helped in the making of the comic book, the color discount store ad, or the magazine you recently read? None other than the graphic arts professional. The graphic arts industry is a part of many industries such as publishing, printing, advertising, photography, manufacturing, packaging, paper-product processing, and so on.

Computers, lasers, and 21st century electronics have dramatically changed the field of graphic arts.

CAREER DESCRIPTION

Since graphic arts is normally done in a small business setting, the graphic arts professional must learn to serve in several different roles. The roles include estimator, process and quality controller, color laboratory technician, darkroom technician, technical illustrator, and so on. The estimator estimates the cost of a project for a customer. The process and quality controller makes sure that each step of the process is done correctly and that the quality of the end product is good. The color laboratory technician works with the color film, separating colors for art and photos, and checks the quality of the color. The darkroom technician works with various kinds of film. The technical illustrator uses the computer to draw technical art such as machinery parts.

TOOLS OF THE TRADE

Two of the many pieces of equipment used in the graphic arts industry are described below.

A technical illustrator uses a computer to draw technical art as seen on the computer screen on the left. The printer uses a color press as shown on the right.

EDUCATION

While in high school, 4 years of English, 2 years of mathematics, 1 year of chemistry or physics, mechanical drawing, basic electronics, and computer classes will prepare you to be a graphic arts professional. Upon completion of high school, enroll in a 2-year graphic arts program at a technical college. Students in technical colleges' graphic arts programs often find employment, or are recruited, before they graduate.

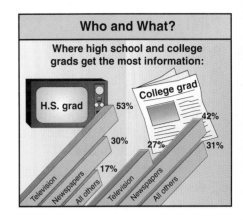

Who and What?

Where high school and college grads get the most information:

H.S. grad 53%

College grad 42%

30% 27% 31%

17%

Television Newspapers All others Television Newspapers All others

Study the graph on the right. Who gets more information from the newspaper? Who watches more TV? Why do you think this is so? What resources do you think might be in the All Others category?

EXERCISES

Measurement is one area that is common to most of the graphic arts industry. The measurement system for graphic arts uses points and picas. Points are used to measure small measures such as type (8-point type, 10-point type, and so on). Picas measure larger measures such as the dimensions of the page or a piece of art. There are 6 picas in 1 inch and 12 points in 1 pica.

Name the number of points in each pica.

1. 2 picas
2. 3 picas
3. 1½ picas

Name the fraction of an inch that represents each pica.

4. 1 pica
5. 3 picas
6. 10 picas

Name the decimal equivalent in inches for each pica. Round to the nearest hundredth.

7. 1 pica
8. 3 picas
9. 10 picas

Express in your own words.

10. Do you think you would be interested in being a graphic arts professional? Why or why not?

11. If you are interested in another part of the graphic arts industry mentioned but not discussed in this section, what could you do?

12. Have you ever discussed your work future with a guidance or vocational counselor at your school? Why or why not?

PROJECT

Producing a Children's Book

Materials: 11-inch by 17-inch white paper, scissors, tape, blue pencils, rulers

Overview You will be writing and producing a 16-page children's book. To begin, you will a make a flat. A flat shows the order of the pages as laid out on a single flat sheet of paper before it is folded.

Procedure Work with a group. Begin by folding the 11×17-inch sheet of paper into two $8\frac{1}{2} \times 11$-inch sheets. Fold the paper again into four $5\frac{1}{2} \times 8\frac{1}{2}$-inch sheets. Then fold the paper again into eight $4\frac{1}{4} \times 5\frac{1}{2}$-inch sheets. After the pages have been folded, label the front page as Front Cover, label the back page as Back Cover, and number the other pages 1 through 14. Remember to number both sides of the pages. Unfold the paper to see the page sequence. This is your flat. Leave the paper unfolded. Use the blue pencil to mark the same-size top, bottom, left, and right margins on every page. Now begin writing the 14 pages for your children's story and the front and back covers. Draw artwork for each page. When the story is written, either type, word process, or print the story onto a piece of paper the size of the pages on your flat. The story and art must fit inside the blue margin lines. Then tape the story and art to the correct pages of the flat.

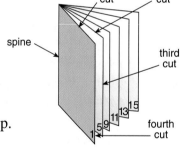

Now fold the paper as you did earlier. Holding the book, vertically cut the paper on the folds as shown in the figure on the right. *Do not cut the fold on the spine.*

The first and second cuts are parallel to the bottom or the top.

The third and fourth cuts are vertical to the spine.

RESOURCES

For further information on a career as a graphic arts professional, write or call the following association.

- Graphic Arts Technical Foundation
  4615 Forbes Avenue
  Pittsburgh, PA 15213   (412) 621-6941

Check your local library for the following magazine.

- Graphic Arts Monthly
  249 W. 17th Street
  New York, NY 10011   (212) 463-6834

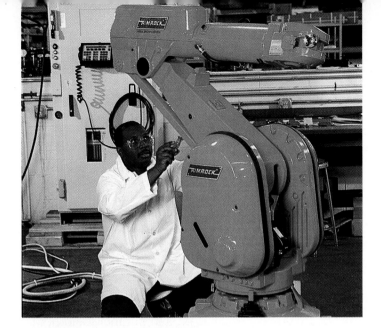

W ould it seem strange for a maid to be a robot? How would you feel if you walked into a warehouse and saw carts going up and down the aisles by themselves? Have you seen robots working on the assembly line in a car factory?

Robotics is a fast-growing technology. It includes the design, maintenance, and use of robots. Robots on an assembly line are machines that work in place of a human hand and arm. Some robots also work as walking machines or teleoperators. Robots usually contain and are operated by tiny computers called **microprocessors**.

CAREER DESCRIPTION

Robotics technicians assist engineers in the design, development, production, testing, and operation of robots. Technicians who are also trained in computer programming sometimes perform programming and reprogramming of robots.

Two computer systems are important for robotics technicians to know. Computer Aided Manufacturing (CAM) is a system that is used to run factories and plants. Computer Aided Design (CAD) is a system used to design parts and products. CAM and CAD are often linked together so engineers and technicians can preview a simulated part or product before it is actually produced.

Robotics technicians often serve as a link between robotics engineers and customers. They may also be responsible for installing robots at manufacturing plants or other sites.

The demand for robotics technicians is high and is expected to remain high.

TOOLS OF THE TRADE

Besides the computer, the robot is also a tool of the trade.

EDUCATION

Most robotics technicians earn a 2-year associate degree in robot technology. Studies usually include hydraulics, pneumatics, electronics, CAD/CAM systems, and microprocessors. Robotics manufacturers generally provide additional on-the-career training.

EXERCISES

Solve. Use the diagram at the right. Use $\frac{22}{7}$ for π.

1. The robot's waist can rotate 315°. The total length of the robot's arm is 406 mm. Find the area the robot's arm will cover as its waist rotates 315°. Use the formula $A = \frac{7}{8} \cdot \pi \cdot r^2$.

2. The shoulder of the robot will rotate vertically 320°. The total length of the robot's arm is 406 mm. Find the area the robot's arm will cover as its shoulder rotates 320°. Use the formula $A = \frac{8}{9} \cdot \pi \cdot r^2$. Round to the nearest whole number.

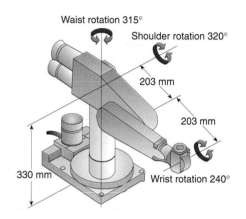

Waist rotation 315°
Shoulder rotation 320°
203 mm
203 mm
330 mm
Wrist rotation 240°

Solve. Use the graph at the right.

3. Is human labor or robot labor more cost effective when a manufacturing plant produces a small volume of products?

4. Is human labor or robot labor more cost effective when a manufacturing plant produces a large volume of products?

5. As a manufacturer, what would you do if your plant produced an intermediate volume of products?

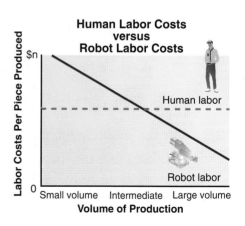

Human Labor Costs versus Robot Labor Costs

Labor Costs Per Piece Produced
$n
Human labor
Robot labor
0 Small volume   Intermediate   Large volume
Volume of Production

Express in your own words.

6. What do you like about this career?

7. Do you think you would want to be a robotics technician? Why or why not?

PROJECT Artificial Intelligence

Materials: three-square by three-square board, one-half inch grid paper, 3 red checkers, 3 black checkers, red pen, black pen, 4 beads each a different color, 4 crayons or 4 colored pencils the same colors as the beads, 1 ketchup cup

Overview Few of today's robots are able to adjust to changing conditions. Providing robots with artificial intelligence may provide robots this ability. Teaching a robot to learn from its mistakes is shown by the following game of Six Piece Checkers.

Playing the Game Work in groups. Choose one member of your group to be the robot. Then choose a member of your group to be the human player.

Line up the checkers on opposite ends of the board as shown. You move first using the red checker. A checker piece can move forward one square or take an opponent's piece on the diagonal. A player wins by getting a checker to the opposite side of the board, taking all the opponent's pieces, or blocking so the opponent cannot move.

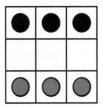

In order to create artificial intelligence for this robot, you need a record of the possible moves the robot can make. To do this, draw three-square by three-square boards on grid paper. Draw a diagram for each checker position after each odd-numbered move (your move) that can be made. Draw an arrow on the diagram to show the robot's legal move. Each arrow on the diagram is the same color as one of the beads. Keep all the (robot's) second-, fourth-, and sixth-move boards together. Do this for several games.

When it's the robot's turn to move, find the board on your paper that matches the gameboard. Put the beads in the ketchup cup and have the robot select a bead without looking. Make the move shown by the arrow that is the same color as the bead. Since each turn can have 1-4 legal moves, place the same number of matching colored beads in the cup as the number of legal moves shown on the paper.

When the game is over, determine who won. If the robot lost, remove the bead from the cup of the last move the robot made. This is how the robot learns from its mistakes. Keep track of this next to the appropriate board on your paper each time the robot loses. The robot soon becomes the perfect player.

RESOURCES

For further information on a career as a robotics technician, write or call the following association.

- Robotics International
  One SME Drive
  P.O. Box 930
  Dearborn, MI 48121   (313) 271-1500

AUTOMOTIVE MECHANIC

Almost every teenager in America dreams of having her or his own car. Did you know that the Arab Oil Embargo of 1973 raised oil prices and forced the U.S. auto industry to redesign cars for greater fuel efficiency? Also, Japan and West Germany became more efficient at manufacturing cars than the U.S. manufacturers. This forced the U.S. into using robots and computer-aided processes to improve their manufacturing and assembly-line operations.

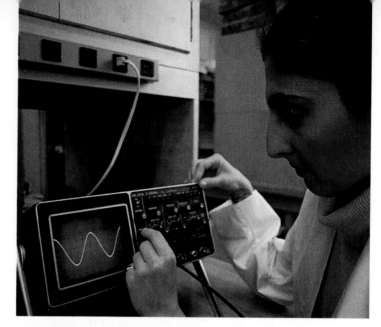

CAREER DESCRIPTION

Automotive mechanics help engineers design, develop, maintain, and repair automotive equipment.

There are five categories of automotive technicians.
1. *Research and Development Mechanics* prepare engines or related equipment for testing.
2. *Service and Sales Representatives* make sure that customers get the maximum performance from engines and advise customers in buying the best engines for their needs.
3. *Mechanics in Related Fields* work for transportation and other companies, oil companies, and insurance companies.
4. *High-Performance Engine Mechanics* are mostly in demand in the car dealerships.
5. *Manufacturing Mechanics* work in manufacturing plants at many different careers.

TOOLS OF THE TRADE

Automotive mechanics use oscilloscopes, as shown on the left, to diagnose engine problems. Oscilloscopes produce various types of sine waves, as shown on the screen. The sine waves represent levels of electrical signals that are received over a certain amount of time.

EDUCATION

To be an automotive mechanic, you must attend a 2-year program at a technical college. There are also several 4-year programs available.

Mathematics and science are important courses for you to take in high school to prepare for the technical college. One year of algebra and one year of geometry are recommended. Also one year of laboratory science such as physics or chemistry should be taken.

EXERCISES

Solve. Use the line graph on the right.

1. What percent of cars that are between two and three years old have a front-end alignment?

2. What percent of cars that are between four and five years old have their shock absorbers replaced?

3. What percent of cars that are between seven and eight years old have a muffler and tailpipe replaced?

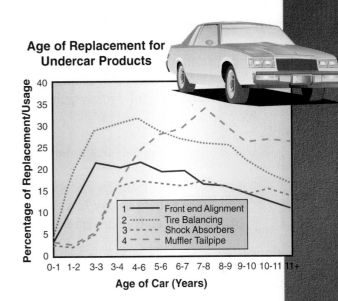

Age of Replacement for Undercar Products

Solve.

4. Suppose a typical sedan weighs 4,000 pounds and 2,240 pounds of that weight rests on the front wheels. Find the wheel weight percentage for the front and back wheels, and the percents for the weights carried by the front and back wheels.

Express in your own words.

5. What do you like about the career of an automotive mechanic?

6. Would you want to be an automotive mechanic? Why or why not?

7. If you did not want to repair cars but wanted to be a part the automotive mechanic career, what else could you do? Explain why you would or wouldn't want to do this.

PROJECT
When More is Less

Materials: poster board, various colored marking pens, straightedge, compass

Overview Suppose you work for a car manufacturer as a automotive mechanic. Your team of mechanics and engineers have just developed an engine that can travel 80 miles on one gallon of gasoline. The research department held focus group meetings and sent out questionnaires, and they have given you the statistical results of their surveys.

Procedure Work in groups. You are giving a presentation before a board of your superiors. You want to convince them that consumers are willing to pay a little more for the price of the car if the car is able to get 80 miles per gallon of gasoline. You also want to convince them of an increase in the price of the car that is agreeable with the majority of consumers. Plan what you are going to say and how you will present your statistics. Use poster board, colored marking pens, a straightedge, and a compass to make a very creative and colorful circle graph, line graph, or bar graph. It could look like a graph you have seen in a newspaper.

Use the following statistics to help make up a presentation.

- 43.2% of the consumers are willing to pay between $1,000 and $2,000 more.
- 42.1% are willing to pay less than $1,000 more.
- 14.7% of the consumers are willing to pay more than $2,000 more.

Give your presentation to the class.

RESOURCES

For further information on a career as an automotive mechanic, write or call the following associations.

- National Automotive Technicians Education Foundation
  13505 Dulles Technology Drive
  Herndon, VA 22071-3415   (703) 713-0100

- Automotive Service Association
  1901 Airport Freeway, Suite 100
  P.O. Box 929
  Bedford, TX 76021-0929   (817) 283-6205

Check your local library for the following magazine.

- Automobile Magazine
  120 E. Liberty
  Ann Arbor, MI 48104   (313) 994-3500

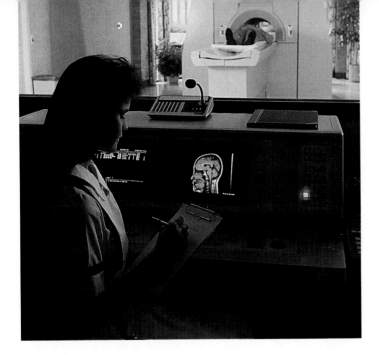

BIOMEDICAL EQUIPMENT SPECIALIST

Are you interested in working in the medical field but don't want to be a doctor or a nurse? There are many careers to choose from in the medical field. One interesting career is a biomedical equipment specialist, sometimes referred to as a medical electronics technician, or a biomedical engineering technician.

CAREER DESCRIPTION

One of the main functions of a biomedical equipment specialist is to repair equipment that is not working properly. He or she determines the problem, determines the extent of the problem, and makes repairs. Another main function is to install equipment. The third area of responsibility is maintenance. He or she tries to find problems before they become serious problems.

In all three areas of responsibility, the specialist consults with physicians, administrators, engineers, and other related professionals.

The specialist must work well under pressure because health matters often demand quick decision-making and prompt repairs. She or he must be extremely precise and accurate in her or his work.

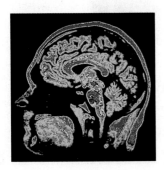

TOOLS OF THE TRADE

The tools of the trade may be as common as the screwdriver, hammer, level, wrench, vacuum sweeper or as precise as a micrometer, monitor, drill, or diagnostic equipment.

EDUCATION

If you are interested in being a biomedical equipment specialist, you should take at least two years of mathematics as well as biology, chemistry, and physics. Courses in English, electronics, and drafting are also helpful. Following high school, you will enroll in a 2-year program to earn an associate degree.

Besides course work, programs include practical experience in repairing and servicing equipment in a laboratory setting. You will learn about electrical components and circuits, the construction of common pieces of machinery, and computer technology.

EXERCISES

Solve. Use the graph on the right.

1. In what city is the cholesterol screening test the most expensive?

2. In what city is the test the least expensive?

3. How much more will it cost to be tested in New York City than in Hastings, Nebraska?

4. Why do think it is more expensive to have the same test using the same kind of equipment in New York City than in Hastings?

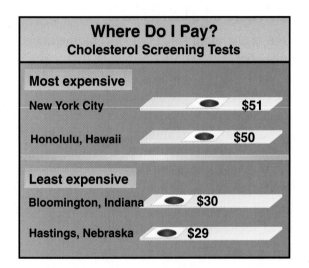

Express in your own words.

5. What aspects of the biomedical equipment specialist's career do you like most?

6. What aspects of this career do you like least?

7. Do you have some skills that you think would make you a good candidate for this career?

8. Do you think you would be interested in being a biomedical equipment specialist? Explain.

PROJECT

Changing Electricity into Sound

Materials: portable radio or portable CD player, earphone plug, 2 meters of thin insulated wire, paper cup, glue, bar magnet

Overview A biomedical equipment specialist might repair an ultrasound machine. Ultrasound is sound above human hearing. Some ultrasound machines change electric energy into ultrasonic (sound) waves. They have a disk that, when charged with electricity, vibrates so rapidly that ultrasonic waves are created. In medicine, this ultrasound machine can be used to destroy brain tumors and kidney stones.

You can change electric energy into sound energy by making a loudspeaker from a radio, earphone plug, insulated wire, paper cup, glue, and a bar magnet. When the magnet is near the coil, it causes the coil to vibrate. The motion of the coil causes the bottom of the cup to vibrate also, creating sound waves.

Procedure **Follow the steps to build a loudspeaker.**

1. Remove 1 centimeter of insulation from each end of a 2-meter length wire.

2. In the middle of the wire, coil the wire about ten times making turns that are 1 centimeter in diameter.

3. Glue the coil securely to the bottom of a paper cup. It may be necessary to allow the glue to dry overnight.

4. Attach one of the stripped ends of the wire to each wire of an earphone plug from a portable radio or CD player.

5. Turn on the radio, and insert the earphone plug into the earphone jack.

6. Hold the magnet very close to the coil.

Choose a team member to demonstrate your group's loudspeaker and to give a simple explanation in your group's own words how electric energy changes into sound energy.

RESOURCES

For further information on a career as a biomedical equipment specialist, write or call the following associations.

- National Society of Biomedical Equipment Technicians
  3330 Washington Boulevard, Suite 400
  Arlington, VA 22201-4598   (703) 525-4890

- American Medical Technologists
  710 Higgins Road
  Park Ridge, IL 60068   (708) 823-5169

CHEMICAL ASSISTANT

D o you ever stop to think where your vitamins, pain relievers, or medicines for a temporary illness come from? They are developed by chemists and doctors.

How do chemists in the field of medicine have time to conduct all the necessary tests and to do all the paperwork that is required for all the new and improved medicines being developed? How do chemists in any of the fields such as petroleum, aerospace, agriculture, and so on get all their work done? The answer is simple. Their help and support comes from chemical assistants.

CAREER DESCRIPTION

There are two types of chemical assistants. One is a chemical laboratory assistant; the other is a chemical engineering assistant.

Chemical laboratory assistants conduct tests to find the chemical content, strength, purity, and so on of a wide range of materials. These assistants test ores, foods, drugs, plastics, paints, petroleum and so on.

Chemical engineering technicians work closely with chemical engineers to develop and improve the equipment and processes used in chemical plants. They prepare tables, charts, diagrams, and flow charts that illustrate the results. These assistants also help build, install, and maintain chemical processing equipment.

TOOLS OF THE TRADE

Some of the tools of the trade are gamma counters that test the amount of protein in cancer cells, laboratory equipment such as Bunsen burners, flasks and tubes, and large industrial machines. The tools are too numerous to mention all of them.